Leslie Copley
Mathematics for the Physical Sciences

Leslie Copley

Mathematics for the Physical Sciences

Managing Editor: Paulina Leśna-Szreter

DE GRUYTER OPEN

Published by De Gruyter Open Ltd, Warsaw/Berlin
Part of Walter de Gruyter GmbH, Berlin/Munich/Boston

ISBN 978-3-11-040945-1
e-ISBN 978-3-11-040947-5

Bibliographic information published by the Deutsche Nationalbibliothek
The Deutsche Nationalbibliothek lists this publication in the Deutsche Nationalbibliografie;
detailed bibliographic data are available in the Internet at http://dnb.dnb.de.

Managing Editor: Paulina Leśna-Szreter

www.degruyteropen.com

Cover illustration: © Istock/VikaSuh

Contents

Definition: Two complex numbers $z_1 \equiv (x_1, y_1)$ and $z_2 \equiv (x_2, y_2)$ are **equal** if and only if their real and imaginary parts are separately equal; that is,

$$z_1 = z_2 \text{ if and only if } x_1 = x_2 \text{ and } y_1 = y_2.$$

Definition: If $z_1 \equiv (x_1, y_1)$ and $z_2 \equiv (x_2, y_2)$, then

$$\begin{aligned}
z_1 + z_2 &\equiv (x_1 + x_2, y_1 + y_2) \\
-z &\equiv (-x, -y) \\
z_1 - z_2 = z_1 + (-z_2) &\equiv (x_1 - x_2, y_1 - y_2) \\
z_1 \cdot z_2 &\equiv (x_1 x_2 - y_1 y_2, x_1 y_2 + x_2 y_1).
\end{aligned} \tag{1.2.1}$$

Armed with these definitions of the fundamental operations, one can readily show that the standard laws of real number arithmetic apply to complex numbers as well:

(i)   the commutative and associative laws of addition,

$$\begin{aligned}
z_1 + z_2 &= z_2 + z_1 \\
z_1 + (z_2 + z_3) &= (z_1 + z_2) + z_3;
\end{aligned} \tag{1.2.2}$$

(ii) the commutative and associative laws of multiplication,

$$\begin{aligned}
z_1 z_2 &= z_2 z_1 \\
z_1 (z_2 z_3) &= (z_1 z_2) z_3 = z_1 z_2 z_3;
\end{aligned} \tag{1.2.3}$$

(iii) the distributive law,

$$(z_1 + z_2) z_3 = z_1 z_3 + z_2 z_3. \tag{1.2.4}$$

Definition: If $z_1 \equiv (x_1, y_1)$ and $z_2 \equiv (x_2, y_2) \neq (0, 0)$, the quotient $z = \dfrac{z_1}{z_2}$ is that complex number (x, y) for which

$$z_1 = z z_2 = (x x_2 - y y_2, x y_2 + x_2 y).$$

From the definition of equality we have

$$\begin{aligned}
x_1 &= x x_2 - y y_2 \\
y_1 &= x y_2 + y x_2
\end{aligned}$$

which is a system of two linear equations in two unknowns, x and y. Solving, we obtain the unique expressions

$$x = \frac{x_1 x_2 + y_1 y_2}{x_2^2 + y_2^2} \text{ and } y = \frac{x_2 y_1 - x_1 y_2}{x_2^2 + y_2^2}. \tag{1.2.5}$$

Notice that division by the complex number $(0, 0)$ is meaningless.

Definition: Complex numbers of the form $(x, 0)$ are called **complex real numbers** or simply **complex reals** while numbers of the form $(0, y)$ are called **pure imaginaries**.

The operations of addition, subtraction, multiplication and division with complex reals only lead to other complex reals. For example,

$$(x, 0) + (y, 0) = (x + y, 0)$$
$$(x, 0) \cdot (y, 0) = (xy, 0).$$

Moreover, complex reals evidently obey exactly the same arithmetic laws as do their real number counterparts. In other words, complex reals can be treated just as though they were real numbers. It is important to recognize that this is a statement of isomorphism or equivalence, **not** of identity. As a consequence of the isomorphism it has become customary to use the same symbol x to denote both the complex real $(x, 0)$ and its real number counterpart. Then, representing the pure imaginary $(0, 1)$ with the symbol i, we obtain the **simplified notation**

$$(x, y) = x + iy, \qquad (1.2.6)$$

since

$$x + iy \equiv (x, 0) + (0, 1) \cdot (y, 0) = (x, 0) + (0, y) = (x, y).$$

Notice that

$$i^2 = (0, 1) \cdot (0, 1) = (-1, 0) = -1$$

so that i may be thought of as the square root of the (complex) real number -1.

When using this notation, we may treat x, y and i as though they are ordinary real numbers provided that we always replace i^2 *by* -1. It is permissible to do so because we understand the logical significance of each symbol that appears in $z = x + iy$. Specifically, we know there is an important distinction between the real number x and the complex number $(x, 0)$ but it is a distinction that admits interchangeable use of their symbols for practical convenience. Moreover, thanks to the pure imaginary i, our ordered pairs of real numbers remain ordered and real number arithmetic becomes applicable.

We shall now introduce two definitions with no counterpart in the real number system. The first one partially makes up for the fact that there is no order on the basis of size among complex numbers.

The phrases "greater than" and "less than" have no meaning when applied to the numbers but do in reference to their **moduli**.

Definition: The **modulus** of $z = x + iy$, written $|z|$, is the real number

$$+\sqrt{x^2 + y^2}.$$

Notice that $|z| = 0$ if and only if $x = 0$ and $y = 0$.

Definition: If $z \equiv x + iy$, then $z^* \equiv x - iy$ is the **complex conjugate** of z.

Definition: The **principal value** of arg z, Arg$z$, is that restricted to the range

$$-\pi < \text{Arg}z \leqslant \pi.$$

Evidently, a complex number $z = x + iy$ can also be represented by a two-dimensional vector of length $|z|$ and components x and y. Thus, since complex numbers obey the same addition rule that applies to vectors in a plane, we can add them graphically by means of the **parallelogram rule.**

From the diagrams in Figure 1.2, we see that the distance between the two points z_1 and z_2 is $|z_1 - z_2|$. Thus, since the sum of the lengths of two sides of a triangle is larger than the length of the third side, we understand why equations (1.2.11) and (1.2.12) are called **triangle inequalities.**

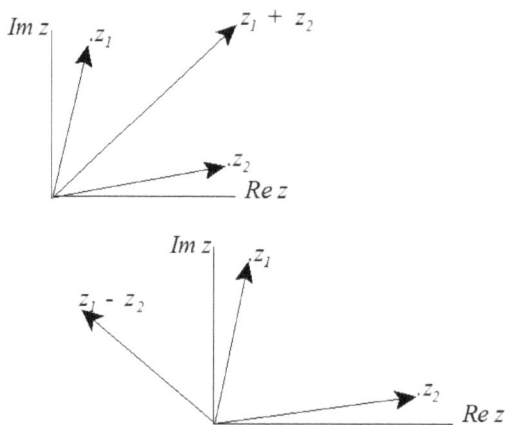

Figure 1.2: The use of the parallelogram law for the addition of vectors to determine graphically the sum and difference of two complex numbers.

While the Cartesian representation is clearly the most useful one for working out sums and differences of complex numbers, it is the polar representation which should be employed when taking products, quotients and powers. For example, if $z_1 = r_1(\cos\theta_1 + i\sin\theta_1)$ and $z_2 = r_2(\cos\theta_2 + i\sin\theta_2)$, then

$$z_1 z_2 = r_1 r_2[(\cos\theta_1\cos\theta_2 - \sin\theta_1\sin\theta_2) + i(\sin\theta_1\cos\theta_2 + \cos\theta_1\sin\theta_2)]$$
$$= r_1 r_2[\cos(\theta_1 + \theta_2) + i\sin(\theta_1 + \theta_2)] \tag{1.2.17}$$

and

$$\frac{z_1}{z_2} = \frac{z_1 z_2^*}{|z_2|^2} = \frac{r_1}{r_2}\left[\cos(\theta_1 - \theta_2) + i\sin(\theta_1 - \theta_2)\right]. \tag{1.2.18}$$

These two equations immediately yield a previously obtained result involving moduli, (see equation (1.2.10)),

$$|z_1 z_2| = |z_1||z_2| \qquad (1.2.19)$$

$$\left|\frac{z_1}{z_2}\right| = \frac{|z_1|}{|z_2|} \qquad (1.2.20)$$

as well as a new one involving arguments,

$$\arg(z_1 z_2) = \arg z_1 + \arg z_2 \qquad (1.2.21)$$

$$\arg\left(\frac{z_1}{z_2}\right) = \arg z_1 - \arg z_2 \, . \qquad (1.2.22)$$

The generalization of (1.2.17) reads

$$z_1 z_2 \ldots z_n = r_1 r_2 \ldots r_n [\cos(\theta_1 + \theta_2 + \ldots + \theta_n) + i \sin(\theta_1 + \theta_2 + \ldots + \theta_n)] \qquad (1.2.23)$$

as can be easily proven by induction. Specializing to the case $z_1 = z_2 = \ldots = z_n = z$, this becomes

$$z^n \equiv [r(\cos\theta + i\sin\theta)]^n = r^n(\cos n\theta + i\sin n\theta). \qquad (1.2.24)$$

This result is known as **de Moivre's theorem** and is valid for both positive and negative integer values of n.

Abraham de Moivre (1667-1754) was born in Vitry in France. A Protestant, he came to England in about 1686 and worked as a teacher. He became known to the leading mathematicians of his time and was elected a Fellow of the Royal Society in 1697. In this capacity he helped to decide the famous controversy between Newton and Leibniz on the origins of the calculus. His principal work is The Doctrine of Chances (1718) on probability theory, but he is best remembered for the fundamental formula given in equation (1.2.23).

At this point it is convenient to note that equations (1.2.19) through (1.2.24) imply that the sum $\cos\theta + i\sin\theta$ possesses all the properties that we would be inclined to associate with an exponential. In other words, if we **define**

$$e^{i\theta} \equiv \cos\theta + i\sin\theta, \qquad (1.2.25)$$

then

$$e^{i\theta_1} \cdot e^{i\theta_2} = e^{i(\theta_1 + \theta_2)}$$

$$\frac{d}{d\theta} e^{i\theta} = i\, e^{i\theta},$$

$$e^{i0} = 1.$$

Equation (1.2.25) is called **Euler's formula**. In due course we shall derive it from the formal definition of the exponential function e^z. We have introduced it now to take advantage of the evident simplification it brings to the polar representation of complex numbers,

$$z = r\, e^{i\theta}.$$

It is useful to remember the following special values of $e^{i\theta}$:

$$e^{\pm i\pi} = -1, \quad e^{\pm i\pi/2} = \pm i.$$

Leonhard Euler (1707-83) was born in Basel where he studied mathematics under Jean Bernoulli. In 1727 he went to St. Petersburg where he became professor of physics (1731) and then professor of mathematics (1733). In 1741 he moved to Berlin at the invitation of Frederick the Great but returned to St. Petersburg in 1766 after a disagreement with the king. He became blind but still continued to publish, remaining in Russia until his death. He was a giant of 18th-century mathematics with over 800 publications, almost all in latin, on every aspect of pure and applied mathematics, physics and astronomy.

Next, consider the algebraic equation $w^n = z$ whose solutions are the nth roots of the complex number z. If we set $w = R\, e^{i\varphi}$ and $z = r\, e^{i\theta}$, then, from de Moivre's theorem, we have

$$w^n = R^n\, e^{in\varphi} = r\, e^{i\theta} = z.$$

Equality between two complex numbers requires separate equality of their moduli and arguments. Thus, we must have

$$R^n = r \quad \text{or} \quad R = r^{1/n},$$

where the root is real and positive and therefore uniquely determined, and

$$n\varphi = \theta + 2k\pi \quad \text{or} \quad \varphi = \theta/n + 2k\pi/n, \quad k = 0, \pm 1, \pm 2, \ldots.$$

Successively substituting the numbers $0, 1, 2, \ldots n-1$ for k, we obtain n distinct values for $z^{1/n}$. Substitution of other values of k only gives rise to repetitions of these values. Thus, we conclude that, for $|z| \neq 0$, $z^{1/n}$ has the n distinct values

$$z^{1/n} = r^{1/n} \exp[i(\theta + 2k\pi)/n], \quad k = 0, 1, \ldots, n-1. \tag{1.2.26}$$

These n values lie on a circle of radius $r^{1/n}$, centre at the origin, and constitute the vertices of a regular $n - sided$ polygon.

Example: $w = \sqrt{z}$ has the two values

$$w_0 = r^{1/2} \exp(i\theta/2)$$
$$w_1 = r^{1/2} \exp(i\theta/2 + i\pi) = -w_0,$$

where θ = arg z and r = $|z|$. These two values are symmetric with respect to the origin. In the specific case of z = i, arg z = $\frac{\pi}{2}$ and $|z|$ = 1, and so the two values of $i^{1/2}$ are

$$w_0 = \exp(i\pi/4) = \frac{1}{\sqrt{2}} + \frac{i}{\sqrt{2}},$$

$$w_1 = \exp(i[\pi/4 + \pi]) = -\frac{1}{\sqrt{2}} - \frac{i}{\sqrt{2}} = -w_0.$$

Notice that we used the principal value of arg i in our calculation of $\sqrt{i}$. This is the conventional choice when evaluating the nth roots of complex numbers but any other choice of value for arg z would yield the same set of numbers for $z^{1/n}$ but in a different order. Had we set arg i = $\pi/2 \pm 2\pi$, for example, we would have obtained

$$w_0 = \exp(i[\pi/4 \pm \pi]) = -\frac{1}{\sqrt{2}} - \frac{i}{\sqrt{2}}$$

$$w_1 = \exp(i[\pi/4 \pm \pi + \pi]) = \frac{1}{\sqrt{2}} + \frac{i}{\sqrt{2}}.$$

Example: If z is a positive (complex) real, so that arg z = 0, then w = $z^{1/3}$ has one (complex) real value

$$w_0 = |z|^{1/3} \exp(0) = |z|^{1/3},$$

and two conjugate complex values

$$w_1 = w_0 \exp(i2\pi/3) = w_0 \left(-\frac{1}{2} + \frac{i\sqrt{3}}{2} \right)$$

$$w_2 = w_0 \exp(i4\pi/3) = w_0 \left(-\frac{1}{2} - \frac{i\sqrt{3}}{2} \right).$$

Notice that this simple example illustrates the important distinction between a complex real and its real number counterpart. The former has exactly three cube roots while the latter has only one, the real number counterpart of w_0.

1.2.3 Curves and Regions in the Complex Plane

A prerequisite for discussing sets of complex numbers or equivalently, sets of points in the complex plane, is to agree upon a basic vocabulary. Thus, the purpose of this section is to introduce some of the terminology which will appear throughout this and subsequent chapters.

Since the distance between two points, z and z_0, is $|z - z_0|$, it follows that a circle C of radius r and with centre at the point z_0 can be represented by the equation

$$|z - z_0| = r. \tag{1.2.27}$$

Consequently, the inequality

$$|z - z_0| < r \tag{1.2.28}$$

holds for any point z **inside C** and thus it represents the interior of C. Such a region is often called an **open (circular) disc,** while the region described by

$$|z - z_0| \leqslant r \tag{1.2.29}$$

is called a **closed (circular) disc**, since it also includes the bounding circle C. Similarly, the inequality

$$|z - z_0| > r \tag{1.2.30}$$

represents the **exterior** of C while

$$r_1 < |z - z_0| < r_2 \tag{1.2.31}$$

is the **annulus** bounded by the concentric circles $|z - z_0| = r_1, |z - z_0| = r_2, r_1 < r_2$.

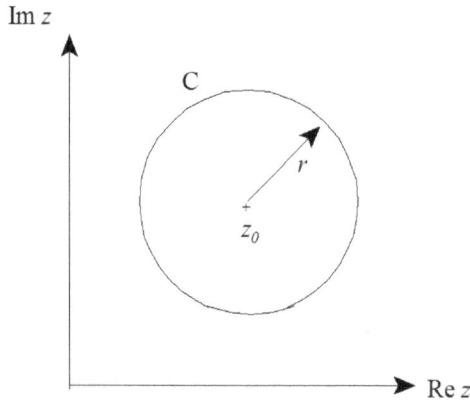

Figure 1.3: The circle C, $|z - z_0| = r$, with centre at z_0 and radius r.

And now we present a few formal definitions:
Definition: A **neighbourhood** of a point z_0 is the set of all points z contained in the open disc $|z - z_0| < \varepsilon$ where ε is a given positive number. A **deleted neighbourhood** of z_0 is one from which the point z_0 itself is omitted: $0 < |z - z_0| < \varepsilon$.
Definition: A point z_0 belonging to a set of points S is called an **isolated point** of S if it has a neighbourhood that does not contain further points of S.
Definition: A point z_0 is called a **limit point** of a set of points S if every deleted neighbourhood of z_0 contains at least one point of S. (This implies that every deleted neighbourhood of a limit point z_0 actually contains an infinite number of points of S for, given $\varepsilon > 0$, the neighbourhood $|z - z_0| < \varepsilon$ contains $z_1 \neq z_0$, the neighbourhood $|z - z_0| < |z_1 - z_0| < \varepsilon$ contains a point $z_2 \neq z_0$ and so on, ad infinitum.)

Examples: The set of points $z = n$, $n = 1, 2, 3, \ldots$, consists entirely of isolated points and has no limit points in the finite plane.

The set of points $z = i/n$, $n = 1, 2, 3, \ldots$, also consists only of isolated points but it does have a limit point. The limit point, $z = 0$, does not belong to the set.

In contrast to the first two examples, the set $|z| < 1$ has no isolated points; every point of the set as well as every point on the unit circle $|z| = 1$ is a limit point.

Evidently, limit points need not belong to the set in question. This observation prompts the next group of definitions.

Definition: A set S is said to be **closed** if every limit point of S belongs to S.

An obvious **example** of a closed set is the closed circular disc $|z| \leqslant 1$.

Definition: A limit point z_0 of S is an **interior point** if there exists a neighbourhood of z_0 which consists entirely of points of S. A limit point which is not an interior point is a **boundary point**.

For **example**, the points lying on the circle $|z| = 1$ are boundary points of the discs $|z| \leqslant 1$ and $|z| < 1$, since no neighbourhood of a point on $|z| = 1$ lies entirely in either set.

Definition: A set which consists entirely of interior points is said to be an **open set**; thus, every point of an open set has a neighbourhood every point of which belongs to the set.

Obvious **examples** are the disc $|z| < 1$ and the points of the right or left half-plane, $\text{Re}\, z > 0$ or $\text{Re}\, z < 0$.

As the set consisting of the disc $|z| < 1$ plus the point $z = 1$ illustrates, a set need not be either open or closed.

Definition: The equation

$$z = z(t) \equiv x(t) + iy(t), \quad t_1 \leqslant t \leqslant t_2, \tag{1.2.32}$$

where $x(t)$ and $y(t)$ are real functions of the real variable t, determines a set of points in the complex plane called, interchangeably, an **arc** or **curve**.

Appropriate adjectives are required to distinguish between curves defined by functions possessing varying degrees of "smoothness". Thus, a **continuous curve** is one for which $x(t)$ and $y(t)$ are continuous in the specified range $t_1 \leqslant t \leqslant t_2$. If, in addition, $x(t)$ and $y(t)$ have continuous first derivatives there then we have a **smooth curve**.

We must also address the possibility of curves intersecting themselves or even closing on themselves. This will occur when more than one value of t in the range $t_1 \leqslant t \leqslant t_2$ yields the same complex number $x(t) + iy(t)$ in which case we say that the curve has a **multiple point**. We shall be interested primarily in continuous curves consisting of a finite number of smooth arcs and possessing no multiple points, except possibly for a double point corresponding to the terminal values of t, t_1 and t_2. We refer to such curves as **simple curves** and, should they possess the double point, as **simple closed curves**. Obvious examples of the latter are a circle, which consists of a single smooth arc, and a polygon, which consists of a finite chain of straight line

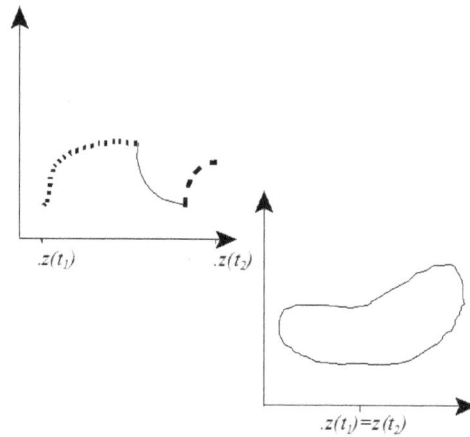

Figure 1.4: The simple curve on the left consists of three smooth arcs and so is piecewise smooth with no self- intersections. The simple closed curve on the right consists of a single, smooth closed arc with no self-intersections.

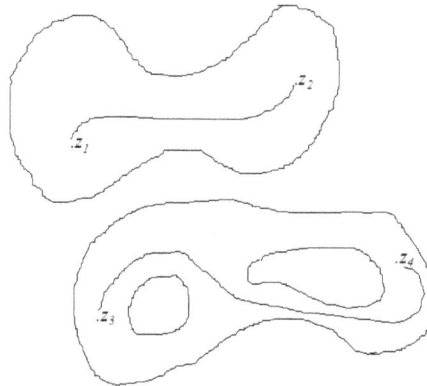

Figure 1.5: The points z_1 and z_2 as well as the points z_3 and z_4 can be connected by a simple curve lying entirely within the shaded areas. When the points z_1 and z_3 are connected, a part of curve necessarily lies outside the shaded areas.

segments. Thus, the adjective **simple** connotes both piecewise smoothness and an absence of self-intersections.

The next group of definitions refers once more to unspecified sets of points and culminates with the important concept of **domain**.

Definition: A set S is said to be **bounded** if there exists a positive real constant M such that $|z| < M$ for every point in S; (that is to say, S is bounded if all points in S lie within a circle of finite radius).

Definition: A set S is said to be **connected** if any two of its points can be joined by a simple curve all of whose points are contained within the set.

Definition: An open connected set of points is called a **domain**. The set obtained by adding to a domain some (all) of its boundary points is a **(closed) region**.

Example: In Figure 1.5, the points belonging to either one of the shaded areas but not lying on one of the boundary curves form a domain. However, the set consisting of points belonging to **both** shaded areas, the union of the two domains, is not itself a domain.

An apparently self-evident theorem states that a simple closed curve divides the complex plane into two domains which have the curve as a common boundary. Of these domains, one is bounded and is called the **interior,** the other is unbounded and called the **exterior**. An immediate **example** is offered by the domains $|z| < r$ and $|z| > r$ which are the interior and exterior, respectively of the circle $|z| = r$.

Definition: A domain is said to be **simply-connected** if its boundary consists of a simple closed curve. Otherwise, it is said to be **multiply connected**. More precisely, a domain is **n-fold connected** if its boundary consists of **n** simple closed curves with no common points. Thus, for **example**, one of the domains in Figure 1.5 is simply connected while the other is three-fold connected.

1.3 Functions of a Complex Variable

1.3.1 Basic Concepts

Definition: Let S be an arbitrary point set in the complex plane. Suppose that to each point z_0 in S there corresponds a complex number (or numbers) $w_0 = f(z_0)$. We then say that w is a **function** of the complex variable z,

$$w \equiv f(z) \text{ for } z \text{ in } S, \tag{1.3.1}$$

and that it defines a **mapping** of S into the complex plane.

Since it is defined over a set of points in a plane, $f(z)$ must be a function of the two real variables x and y. Thus, separating its complex values into their real and imaginary parts, we see that we can always write it in the form

$$w = f(z) = u(x, y) + iv(x, y), \tag{1.3.2}$$

where u and v are real functions of the two real variables x and y. This suggests that the properties of functions of a complex variable should be readily deduced from the theory of functions of two real variables. However, we are quickly disabused of this idea when we recall that the variables x and y are determined by

$$x = \frac{1}{2}(z + z^*) \qquad y = \frac{1}{2i}(z - z^*).$$

Thus, an arbitrary pair of real functions combined according to

$$w = u(x, y) + iv(x, y)$$

will necessarily result in a function that depends explicitly on both z **and** z^*. The latter dependence is unwanted: x and y must always occur in the unique combination $x + iy$. Evidently, this requires the imposition of a restriction that limits the choice of functions $v(x, y)$ that can be paired with a specified $u(x, y)$ and vice versa. But if its real and imaginary parts are so closely interrelated for some range of x and y, $f(z)$ must itself be restricted in the way it varies as a function of z. As we shall see, the required restriction is **differentiability throughout a domain** of the complex plane. That being said, we shall postpone a formal definition of differentiability and introduce instead the concept of a **point at infinity**.

The function $w = \frac{1}{z}$ provides a well-defined, one-to-one mapping of the points of the $z - plane$ onto those of the $w - plane$ with but two exceptions: the point $z = 0$ has no image and the point $w = 0$ has no pre-image. For example, the unit circle $|z| = 1$ is mapped onto itself, $|w| = 1$; its exterior $|z| > 1$ is mapped onto the interior points $|w| < 1$, $w \neq 0$, and its interior $|z| < 1$, $|z| \neq 0$ is mapped onto the exterior points $|w| > 1$. This suggests that the two exceptions are closely related and that they can be eliminated by **defining** the image of $z = 0$ to be the **point at infinity**, denoted by $w = \infty$. Then, since the inverse of our mapping is $z = \frac{1}{w}$, the point $z = \infty$ is the required pre-image of $w = 0$.

The $z - plane$ augmented by $z = \infty$ is referred to as the **extended complex plane**. When we wish to emphasize the exclusion of $z = \infty$, we shall refer to the **finite complex plane**. The function $w = \frac{1}{z}$ maps the **extended** complex plane onto itself without exceptions. In particular, it maps circular discs with centre at the origin onto circular discs with centre at the point at infinity. Thus, a neighbourhood of the point $z = \infty$ is denoted by $|z| > R$, where R is a given positive number. It is useful on occasion to think of the extended complex plane as the surface of an "infinite sphere". In that context the origin and point at infinity are at opposite poles of the sphere and are joined by the rays $\arg z = constant$. All other straight lines in the plane correspond to "circles of infinite radius" intersecting at $z = \infty$.

1.3.2 Continuity, Differentiability and Analyticity

Real variable analysis provides us with a prototypical format to follow in defining what we shall mean by continuity and differentiability. Thus, for a function defined on a domain D we define continuity as follows.

Definition: The function $f(z)$ is continuous at the point z_0 of D if, given any $\varepsilon > 0$, there exists a δ such that

$$|f(z) - f(z_0)| < \varepsilon$$

for all points z in D satisfying

$$|z - z_0| < \delta.$$

An alternative statement, which really defines what we shall mean by a limit, is that $f(z)$ is continuous at z_0 if

$$\lim_{z \to z_0} f(z) = f(z_0) \text{ for } z \text{ in } D. \tag{1.3.3}$$

In geometrical terms this means that we can restrict $f(z)$ to lie within a circle of radius ε about $f(z_0)$ in the $w - plane$ simply by requiring that z lie within a circle of radius δ about z_0 in the $z - plane$. This in turn reveals a critically important feature of limits in complex analysis: the limit in (1.3.3) exists only if the number $f(z_0)$ is obtained regardless of the path followed by z as it approaches z_0.

The same path-independence is required of limits involving **two** or more real variables. Therefore, it should come as no surprise that if $u(x, y)$ and $v(x, y)$ are continuous functions of x and y then

$$f(z) = u(x, y) + iv(x, y)$$

is a continuous function of z. The converse is also true. This means that the imposition of a requirement of continuity does not preclude the possibility of an explicit dependence on z^* as well as on z, the goal we set ourselves at the beginning of the preceding section. Therefore let us move on and introduce differentiation.

Definition: Let $f(z)$ be a single-valued continuous function defined in a domain D. We say that $f(z)$ is **differentiable** at the point z_0 of D if the limit

$$\lim_{z \to z_0} \frac{f(z) - f(z_0)}{z - z_0}, \quad z \text{ in } D \tag{1.3.4}$$

exists as a finite number, independent of how z approaches z_0. The limit, when it exists, is called the **derivative** of $f(z)$ at z_0 and is denoted by $f'(z_0)$.

The path independence of this limit is a much more exacting condition than is its counterpart in the definition of continuity. Consequently, continuity by no means implies differentiability.

Examples: The function

$$f(z) = x + 2iy = \frac{3}{2}z - \frac{1}{2}z^*$$

is continuous everywhere in the finite plane. However, forming the quotient prescribed in (1.3.4), we obtain

$$\frac{f(z) - f(z_0)}{z - z_0} = \frac{3(z - z_0) - (z - z_0)^*}{2(z - z_0)} = \frac{3}{2} - \frac{1}{2}\frac{(z - z_0)^*}{(z - z_0)}.$$

Setting $z - z_0 = |z - z_0|\exp(i\theta)$, $\theta = \arg(z - z_0)$, this becomes

$$\frac{f(z) - f(z_0)}{z - z_0} = \frac{3}{2} - \frac{1}{2}\exp(-2i\theta)$$

which obviously does not tend to a unique value as $z \to z_0$. In particular, if $z \to z_0$ parallel to the real axis then $\theta = 0$ and the quotient tends to $+1$, and if $z \to z_0$ parallel to the imaginary axis, $\theta = \frac{\pi}{2}$ and so the quotient tends to $+2$. Thus, this function is not differentiable anywhere.

Similarly, the continuous function $f(z) = |z|^2$ is differentiable only at $z_0 = 0$. For if $z \neq 0$, we have

$$\frac{f(z) - f(z_0)}{z - z_0} = \frac{z z^* - z_0 z_0^*}{z - z_0} = z^* + z_0 \frac{(z - z_0)^*}{(z - z_0)} = z^* + z_0 \exp(-2i \arg(z - z_0))$$

which again exhibits a dependence on $\arg(z - z_0)$.

By way of contrast, the function $f(z) = z^2$ is differentiable everywhere in the finite plane as can be seen from

$$f'(z_0) = \lim_{z \to z_0} \frac{z^2 - z_0^2}{z - z_0} = \lim_{z \to z_0} z + z_0 = 2 z_0 \,.$$

These examples suggest that our search for a condition which will guarantee that functions have no explicit dependence on z^* may be at an end. To confirm that this is the case we shall determine necessary and sufficient conditions for a function to be differentiable; the question of z^* independence will be resolved as a corollary. However, before we do so, we should note that all the familiar rules of real differential calculus continue to hold. Specifically,

$$\frac{d}{dz}(f \pm g) = \frac{df}{dz} \pm \frac{dg}{dz} \tag{1.3.5}$$

$$\frac{d}{dz}(f \cdot g) = \frac{df}{dz} \cdot g + f \cdot \frac{dg}{dz} \tag{1.3.6}$$

$$\frac{d}{dz}\left(\frac{f}{g}\right) = \frac{1}{g^2}\left(\frac{df}{dz} \cdot g - f \cdot \frac{dg}{dz}\right) \tag{1.3.7}$$

$$\frac{d}{dz}f[g(z)] = \frac{df}{dg} \cdot \frac{dg}{dz} \tag{1.3.8}$$

provided that, in each case, the derivatives on the right hand side exist.

As the next theorem details, the path independence of derivatives implies a relationship between the real and imaginary parts of a differentiable function.

Theorem: Suppose that $f(z) = u(x, y) + iv(x, y)$ is defined and continuous in a neighbourhood of $z = x + iy$. A necessary condition for the existence of $f'(z)$ is that all first partial derivatives of u and v exist and satisfy the **Cauchy-Riemann equations**,

$$\frac{\partial u}{\partial x} = \frac{\partial v}{\partial y} \qquad \frac{\partial u}{\partial y} = -\frac{\partial v}{\partial x}, \tag{1.3.9}$$

at the point (x, y).

Proof: If $f(z)$ is differentiable at z, the limit

$$\lim_{\Delta z \to 0} \frac{f(z + \Delta z) - f(z)}{\Delta z} \equiv f'(z)$$

must exist and be independent of how $\Delta z \to 0$. Since

$$\frac{f(z + \Delta z) - f(z)}{\Delta z} = \frac{[u(x + \Delta x, y + \Delta y) - u(x, y)] + i[v(x + \Delta x, y + \Delta y) - v(x, y)]}{\Delta x + i\Delta y}$$

and since we may take Δz to be real, so that $\Delta y = 0$, it follows that

$$\frac{u(x + \Delta x, y) - u(x, y)}{\Delta x} + i\frac{v(x + \Delta x, y) - v(x, y)}{\Delta x}$$

tends to a definite limit as $\Delta x \to 0$. Therefore, the partial derivatives $\frac{\partial u}{\partial x}$ and $\frac{\partial v}{\partial x}$ must exist at the point (x, y) and the limit is

$$f'(z) = \frac{\partial u}{\partial x} + i\frac{\partial v}{\partial x}. \tag{1.3.10}$$

Similarly, if we take Δz to be a pure imaginary, so that $\Delta x = 0$, we find that $\frac{\partial u}{\partial y}$ and $\frac{\partial v}{\partial y}$ both exist at (x, y) and obtain the limit

$$f'(z) = \frac{\partial v}{\partial y} - i\frac{\partial u}{\partial y}. \tag{1.3.11}$$

Since the two limits must be identical, we can equate real and imaginary parts to obtain

$$\frac{\partial u}{\partial x} = \frac{\partial v}{\partial y} \qquad \frac{\partial u}{\partial y} = -\frac{\partial v}{\partial x}$$

as required.

Baron Augustin Louis Cauchy (1789-1857) was born in Paris. He studied to become an engineer but ill health obliged him to forego engineering and to teach mathematics at the Ecole Polytechnique. Following the 1830 revolution he spent some years in exile in Turin and Prague, returning to Paris in 1838. He did important work on partial differential equations, the wave theory of light, the mathematical theory of elasticity, the theory of determinants and group theory, but is primarily remembered as the founder of the theory of functions of a complex variable.

Georg Friedrich Bernhard Riemann (1826-66) was a German mathematician who studied under Carl Friedrich Gauss. He succeeded Gustav Dirichlet as professor of mathematics at Gottingen in 1859 but was forced to retire by illness in 1862 and subsequently died in Italy of tuberculosis. His first publication (1851) was on the foundations of the theory of functions of a complex variable. In this and a later paper (1857) he introduced the concept of "Riemann surface" to deal with multi-valued functions. Meanwhile, in geometry, he introduced the concept of an n-dimensional curved space which is fundamental to the modern theory of differentiable manifolds and to the general theory of relativity in physics. His name is also associated with the zeta-function which plays an important role in number theory.

Have we stumbled across the relationship between u and v that will ensure that the function of z they comprise does not have an explicit dependence on z^* as well?

To answer, we shall treat z and z^* as independent variables and define the partial derivatives

$$\frac{\partial f}{\partial z} = \frac{\partial f}{\partial x}\frac{\partial x}{\partial z} + \frac{\partial f}{\partial y}\frac{\partial y}{\partial z}$$

$$\frac{\partial f}{\partial z^*} = \frac{\partial f}{\partial x}\frac{\partial x}{\partial z^*} + \frac{\partial f}{\partial y}\frac{\partial y}{\partial z^*}.$$

These lead almost immediately to the equalities

$$\frac{\partial f}{\partial z} = \frac{1}{2}\left[\frac{\partial u}{\partial x} + \frac{\partial v}{\partial y}\right] + \frac{i}{2}\left[\frac{\partial v}{\partial x} - \frac{\partial u}{\partial y}\right]$$

$$\frac{\partial f}{\partial z^*} = \frac{1}{2}\left[\frac{\partial u}{\partial x} - \frac{\partial v}{\partial y}\right] + \frac{i}{2}\left[\frac{\partial v}{\partial x} + \frac{\partial u}{\partial y}\right].$$

Applying the Cauchy-Riemann equations we then obtain

$$\frac{\partial f}{\partial z} = \frac{df}{dz}$$

$$\frac{\partial f}{\partial z^*} = 0. \tag{1.3.12}$$

Thus, a function that is differentiable throughout some domain D cannot have an explicit dependence on z^*.

Our prolonged search for a criterion to limit the range of our study is at an end. It only remains to attach an identifying label to those functions which satisfy it.

Definition: A single-valued continuous function $f(z)$ is said to be an **analytic function** of z (or more simply, to be **analytic**) in a domain D if it is differentiable at every point of D, save possibly for a finite number of exceptional points. The exceptional points are called the **singular points** or **singularities** of $f(z)$ in D. If no point of D is a singularity of $f(z)$ then we say that it is **holomorphic** in D. Further, we say that $f(z)$ is holomorphic at a point $z = z_0$ if it is holomorphic in some neighbourhood of z_0.

The terms *regular* and (with even greater potential for confusion) *analytic* are used by some authors as synonyms of *holomorphic*. Thus, some care is required when reading other texts.

Our next theorem, an extension of the last one, identifies sufficient conditions for a function to be holomorphic. It is presented without proof.

Theorem: The continuous single-valued function $f(z) = u(x,y)+iv(x,y)$ is an analytic function of $z = x+iy$, holomorphic in a domain D, if the four partial derivatives $\frac{\partial u}{\partial x}$, $\frac{\partial v}{\partial x}$, $\frac{\partial u}{\partial y}$ and $\frac{\partial v}{\partial y}$ exist, are continuous , and satisfy the Cauchy-Riemann equations at each point of D.

Definition: An analytic function which is holomorphic in every finite region of the complex plane is said to be **entire**.

Such functions have no singularities in the finite plane but as we shall see, this has implications for their behaviour at infinity.

Definition: An analytic function $f(z)$ is holomorphic at $z = \infty$ if $f(1/w)$ is holomorphic at $w = 0$.

Examples: The function $f(z) = \frac{1}{1-z}$ is holomorphic in any domain that excludes the point $z = 1$. Evidently, the derivative $f'(z) = \frac{1}{(1-z)^2}$ is undefined at $z = 1$. On the other hand, $f(1/w) = \frac{w}{w-1}$ is holomorphic at $w = 0$ and so $f(z)$ is holomorphic at $z = \infty$. Thus, the full domain of holomorphy is the extended plane with the point $z = 1$ removed. In contrast, any polynomial

$$f(z) = c_0 + c_1 z + \ldots + c_n z^n, n \geqslant 1$$

is singular at $z = \infty$. However, this is their only singularity; they are entire functions.

Definition: Two real functions u and v of the two real variables x and y are said to be **conjugate functions** if $f(z) = u(x, y) + iv(x, y)$ is an analytic function of $z = x + iy$.

As a consequence of the Cauchy-Riemann equations, conjugate functions are solutions of Laplace's equation. To see why this is so we need only differentiate the Cauchy-Riemann equations,

$$\frac{\partial^2 u}{\partial x^2} = \frac{\partial^2 v}{\partial x \partial y} \quad \frac{\partial^2 u}{\partial y^2} = -\frac{\partial^2 v}{\partial y \partial x}$$

$$\frac{\partial^2 u}{\partial y \partial x} = \frac{\partial^2 v}{\partial y^2} \quad \frac{\partial^2 u}{\partial x \partial y} = -\frac{\partial^2 v}{\partial x^2},$$

and assume equality of the mixed second derivatives,

$$\frac{\partial^2 u}{\partial x \partial y} = \frac{\partial^2 u}{\partial y \partial x} \quad \frac{\partial^2 v}{\partial x \partial y} = \frac{\partial^2 v}{\partial y \partial x}.$$

We immediately obtain

$$\nabla^2 u \equiv \frac{\partial^2 u}{\partial x^2} + \frac{\partial^2 u}{\partial y^2} = 0,$$

and

$$\nabla^2 v = \frac{\partial^2 v}{\partial x^2} + \frac{\partial^2 v}{\partial y^2} = 0.$$

Since we obtain two solutions of Laplace's equation merely by separating any analytic function of z into its real and imaginary parts, complex analysis has an evident relevance to the solution of potential problems in two dimensions. The converse is true also; in fact, an important insight can be gained at this point from the theory of partial differential equations. As we know, a particular solution of a PDE is completely and uniquely determined by specifying its behaviour at the boundary of the domain in which the partial differential equation obtains. Thus, since $u(x, y)$ and $v(x, y)$ satisfy the equations

$$\nabla^2 u(x, y) = 0 \quad \nabla^2 v(x, y) = 0$$

throughout the domain of the $xy - plane$ in which $f(x + iy) = u(x, y) + iv(x, y)$ is a holomorphic function of $z = x + iy$, it follows that a knowledge of how $f(z)$ behaves at the boundary of its domain of holomorphy is sufficient to determine how it behaves everywhere else in that domain. Or, since the boundary is where the function ceases to be holomorphic, we can assert that **analytic functions are completely determined by their singularities.** This is the basis of the most important applications of complex analysis. Consequently, our exploration of the theory will be directed primarily toward discovering how these determinations can be made in practice.

We conclude this section by noting that the Cauchy-Riemann equations can be used to find the conjugate of any given harmonic function and hence determine an analytic function that has the original harmonic function as either its real or imaginary part.

Example: The function $u(x, y) = x^2 - y^2$ is harmonic and has derivatives $\frac{\partial u}{\partial x} = 2x$ and $\frac{\partial u}{\partial y} = -2y$. Hence, the conjugate of $u(x, y)$ must satisfy

$$\frac{\partial v}{\partial y} = 2x \ \text{ and } \ \frac{\partial v}{\partial x} = 2y.$$

Integrating the first of these with respect to y, we find

$$v(x, y) = 2xy + \varphi(x)$$

where $\varphi(x)$ depends only on x. Substituting this expression into the second of our two equations, we obtain $\varphi'(x) = 0$ and so conclude that $\varphi(x) = c$, constant. Thus, the conjugate of $u(x, y) = x^2 - y^2$ is $v(x, y) = 2xy + c$ and the analytic function which they comprise is

$$f(z) = x^2 - y^2 + i(2xy + c) = z^2 + ic,$$

where c is an arbitrary real constant.

1.4 Power Series

Many of the analytic functions one encounters in mathematical physics are defined by means of power series. This section will provide an introduction to their properties as well as a brief overview of the convergence theorems that apply to complex infinite series.

As usual we begin our discussion with a raft of definitions.

Definition: A series of the form

$$\sum_{m=0}^{\infty} c_m (z - z_0)^m = c_0 + c_1(z - z_0) + c_2(z - z_0)^2 + \ldots, \tag{1.4.1}$$

where z is a variable while $c_0, c_1, c_2, \ldots, c_m, \ldots$ and z_0 are all constants, is called a **power series** about the point $z = z_0$ (or, with **centre** at the point $z = z_0$).

Definition: A series of complex numbers

$$\sum_{m=0}^{\infty} w_m = w_1 + w_2 + w_3 + \ldots + w_m + \ldots$$

is said to converge to a limit S if, for each $\varepsilon > 0$, there exists an integer N such that

$$|S_n - S| < \varepsilon$$

for all $n > N$, where S_n is the series' n^{th} partial sum,

$$S_n = \sum_{m=0}^{n} w_m \,.$$

The number S is called the **value** or **sum** of the series and we write

$$S = \sum_{m=0}^{\infty} w_m \,.$$

A useful test of convergence (or rather, divergence since it involves a necessary but not sufficient condition for convergence) is provided by a theorem which we state without proof.

Theorem: If the series $\sum_{m=0}^{\infty} w_m$ converges, then

$$\lim_{m \to \infty} w_m = 0. \tag{1.4.2}$$

Thus, a series which does not satisfy (1.4.2) necessarily diverges.

Definition: A series $\sum_{m=0}^{\infty} w_m$ of complex numbers is said to be **absolutely convergent** if the (real) series of moduli $\sum_{m=0}^{\infty} |w_m|$ converges. If $\sum_{m=0}^{\infty} w_m$ converges but $\sum_{m=0}^{\infty} |w_m|$ diverges, the series is called **conditionally convergent**.

Theorem: If a series $\sum_{m=0}^{\infty} w_m$ is absolutely convergent, then it converges.

Because of this theorem and the relative ease of working with the moduli of complex numbers rather than the numbers themselves, one generally tries to establish the convergence of a complex series by showing that it is absolutely convergent. The following three tests are the most important means of doing this.

1. **Comparison Test** If one can find a convergent series of positive real terms $\sum_{m=0}^{\infty} u_m$ such that $|w_m| \leqslant u_m$ for each m, then the series $\sum_{m=0}^{\infty} w_m$ is absolutely convergent.

2. **Ratio Test** Assume that $w_m \neq 0$ for all m and that the sequence of ratios $|w_{m+1} / w_m|$ converges to a limit L. If $L < 1$, the series $\sum_{m=0}^{\infty} w_m$ converges absolutely; if $L > 1$, the series diverges; if $L = 1$, the test fails.

Should the sequence of ratios possess more than one limit point, the test can still be used but it requires a more general wording: if the largest value of $\lim_{m \to \infty} |w_{m+1}/w_m| < 1$ then the series $\sum_{m=0}^{\infty} w_m$ converges absolutely ; if the smallest value of $\lim_{m \to \infty} |w_{m+1}/w_m| > 1$ then the series diverges.

3. **Root Test** Assume that the sequence of roots $|w_m|^{\frac{1}{m}}$, $m = 1, 2, 3, \ldots$, converges to a limit L. If $L < 1$, the series $\sum_{m=0}^{\infty} w_m$ converges absolutely; if $L > 1$, the series diverges; if $L = 1$, the test fails. More generally, the series is absolutely convergent or divergent according as the largest or smallest value of $\lim_{m \to \infty} |w_m|^{\frac{1}{m}}$ is respectively less than or greater than one.

Application of these tests to power series leads to a simple characterization of their convergence properties. For example, suppose that it is known that a given power series $\sum_{m=0}^{\infty} c_m(z - z_0)^m$ converges when z is assigned the specific value z_1. Then, since $\lim_{m \to \infty} |c_m(z_1 - z_0)^m| = 0$, there must exist a real positive number M such that

$$|c_m(z_1 - z_0)^m| \leqslant M \quad \text{for all} \quad m.$$

Let us now assign z a second value z_2 which is subject to the single constraint

$$|z_2 - z_0| < |z_1 - z_0|. \tag{1.4.3}$$

We then have

$$|c_m(z_2 - z_0)^m| = |c_m(z_1 - z_0)^m| \cdot \left|\frac{z_2 - z_0}{z_1 - z_0}\right|^m \leq M\left|\frac{z_2 - z_0}{z_1 - z_0}\right|^m$$

Therefore, the series $\sum_{m=0}^{\infty} c_m(z_2 - z_0)^m$ converges absolutely by **comparison** with the geometric series $\sum_{m=0}^{\infty} Mr^m = \frac{M}{1-r}$, $r = \left|\frac{z_2 - z_0}{z_1 - z_0}\right| < 1$.

Since z_2 was arbitrary, subject only to (1.4.3), we have just proven that if a power series converges at a given point z_1 then it converges absolutely for all z for which

$$|z - z_0| < |z_1 - z_0|,$$

that is, at all points z lying closer to z_0 than does z_1. (See Figure 1.6). This result suggests that we attribute to each power series a **radius of convergence** R defined to be the smallest real number such that the distance from the centre z_0 to any point z at which the series converges is at most equal to R. It then follows that the power series converges absolutely for all z for which $|z - z_0| < R$ and diverges for all z for which $|z - z_0| > R$. Nothing definite can be said about the behaviour of a power series on its **circle of convergence**, $|z - z_0| = R$. It may diverge at every point, converge at some points and diverge at others or, converge (absolutely) at every point. If a power series converges for all values of z, then we set $R = \infty$; if it converges only at $z = z_0$, then $R = 0$.

As the next theorem shows the ratio and root tests provide methods for explicitly calculating the radius of convergence for a given power series.

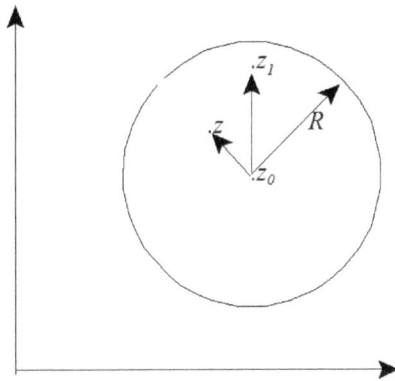

Figure 1.6: Convergence at a point z_1 implies absolute convergence at any point z such that $|z-z_0| < |z_1 - z_0| \leqslant R$.

The Cauchy-Hadamard Theorem: The radius of convergence R of the power series $\sum\limits_{m=0}^{\infty} c_m(z - z_0)^m$ is determined by its coefficients c_m and can be evaluated as follows:

$$R = \lim_{m\to\infty} \left| \frac{c_m}{c_{m+1}} \right| \tag{1.4.4}$$

if the limit exists, (the improper limit $+\infty$ is allowed), or

$$R = \min \lim_{m\to\infty} \frac{1}{|c_m|^{\frac{1}{m}}} \tag{1.4.5}$$

if the limit exists, (the improper limit $+\infty$ is allowed).

Jacques Hadamard (1865-1963) was born in Versailles and educated in Paris. He became a lecturer in Bordeaux (1893-97), the Sorbonne (1897-1909), and then Professor at the College de France and the Ecole Polytechnique until his retirement in 1937. He was a leading figure in French mathematics throughout his career, working in complex analysis, differential geometry and partial differential equations. He was still publishing mathematical work in his eighties.

Power series will be used over and over again as our exposition of complex analysis unfolds. One of the reasons for this is provided by our next theorem.

Theorem: The sum of a power series is a holomorphic function within its circle of convergence.

The standard proof of this theorem shows that the derivative of the power series

$$f(z) = \sum_{m=0}^{\infty} c_m(z - z_0)^m$$

can be obtained by differentiating the series term by term:

$$f'(z) = \sum_{m=0}^{\infty} m \, c_m (z - z_0)^{m-1} \, .$$

It takes little additional effort to show that this series has the same radius of convergence as does $f(z)$.

By repeated application of this theorem we obtain a further important result.

Theorem: Within its circle of convergence, a power series

$$f(z) = \sum_{m=0}^{\infty} c_m (z - z_0)^m$$

has derivatives of all orders, with the kth derivative given by

$$f^{(k)}(z) = \sum_{m=k}^{\infty} m(m-1) \ldots (m-k+1) \, c_m (z - z_0)^{m-k},$$

all of which possess the same radius of convergence as does the original series. Moreover, we have

$$f^{(k)}(z_0) = k! \, c_k$$

and so the original power series is the **Taylor series** of its sum:

$$f(z) = \sum_{m=0}^{\infty} \frac{f^{(m)}(z_0)}{m!} (z - z_0)^m \, .$$

In a subsequent section we shall prove the converse of this theorem. It is the converse that is referred to as Taylor's theorem and it states that a function $f(z)$ may be expanded in a unique power series $\sum_{m=0}^{\infty} c_m (z - z_0)^m$, with non-zero radius of convergence, about any point z_0 at which it is holomorphic. Taken together, these two theorems tell us that a function $f(z)$ may be expanded in a Taylor series about a point z_0 **if and only if** z_0 lies within the function's domain of holomorphy. Moreover, the Taylor series is the only power series expansion about z_0 that is possessed by $f(z)$. This important result is most explicitly exploited in a formulation of complex analysis due to Weierstrass. It begins by defining an analytic function to be one which admits expansion in a power series with non-zero radius of convergence.

Karl Theodor Wilhelm Weierstrass (1815-97) was a German mathematician, educated at the Universities of Bonn and Munster. He became professor at Berlin in 1856. He published relatively little but became famous for his lectures on analysis. In addition to his contributions to complex function theory, he made important advances in the theory of elliptic and abelian functions.

We shall now use power series to define what are known as the "elementary functions".

1.5 The Elementary Functions

1.5.1 Rational Functions

A polynomial in z,

$$w = c_0 + c_1 z + \ldots + c_n z^n = \sum_{m=0}^{n} c_m z^m,$$

is, as we have seen already, an entire function. Therefore, it may be regarded as a power series about $z = 0$ that converges for all values of z. To obtain its power series expansion about any other point z_0, we need only replace z by $z_0 + (z - z_0)$. The result is another polynomial, in powers of $(z - z_0)$ this time, and so the radius of convergence is again $R = \infty$.

The quotient of two polynomials

$$w = \frac{c_0 + c_1 z + \ldots + c_n z^n}{d_0 + d_1 z + \ldots + d_k z^k}$$

is called a **rational function**. It is an analytic function whose only singularities are the zeros of the denominator. One of the simplest but at the same time, most important rational functions is that defined by the geometric series $\sum_{m=0}^{\infty} z^m$. This is a power series about $z = 0$ which, by the Cauchy-Hadamard theorem, has radius of convergence $R = 1$. If we consider its nth partial sum

$$S_n = 1 + z + z^2 + \ldots + z^n$$

and subtract from it

$$z\, S_n = z + z^2 + \ldots + z^{n+1},$$

we find

$$S_n = \frac{1 - z^{n+1}}{1 - z}.$$

Thus, taking the limit as $n \to \infty$, we have

$$\sum_{m=0}^{\infty} z^m = \frac{1}{1 - z}, \quad |z| < 1. \tag{1.5.1}$$

This identity is the basis for a practical prescription for generating the power series expansions possessed by an arbitrary rational function. The prescription will be given in a subsequent and more relevant section.

1.5.2 The Exponential Function

The exponential function $\exp z$ is defined to be the sum function of the series

$$\exp z \equiv e^z \equiv \sum_{m=0}^{\infty} \frac{z^m}{m!} = 1 + z + \frac{z^2}{2!} + \frac{z^3}{3!} + \ldots. \tag{1.5.2}$$

Applying the Cauchy-Hadamard theorem we find that the radius of convergence of this series is infinite. Thus, $\exp z$ is an entire function of z. This is confirmed by term by term differentiation of the series which yields a well-defined derivative

$$\frac{d}{dz} e^z = e^z$$

for all finite z. By multiplication of series one can also prove that the familiar multiplication law

$$e^z \cdot e^w = e^{z+w}$$

holds for complex z and w.

In Section 1.1.2 we arbitrarily introduced $e^{i\theta} \equiv \cos\theta + i\sin\theta$ for reasons of notational convenience. Now that we have defined what is meant by an exponential function of a complex variable, we can derive this expression. Assigning z the pure imaginary value iy, y real, in (1.5.2), we find

$$e^{iy} = \sum_{m=0}^{\infty} \frac{(iy)^m}{m!} = \sum_{n=0}^{\infty}(-1)^n \frac{y^{2n}}{(2n)!} + i\sum_{n=0}^{\infty}(-1)^n \frac{y^{2n+1}}{(2n+1)!}.$$

However, the latter two power series are known to be the Taylor series expansions of the cosine and sine of the real variable y. Thus, as anticipated,

$$e^{iy} = \cos y + i\sin y.$$

Consequently,

$$e^z = e^{x+iy} = e^x\, e^{iy} = e^x(\cos y + i\sin y) \tag{1.5.3}$$

or $|e^z| = e^x$ and $\arg(e^z) = y$.

Since $e^{2\pi i} = 1$, we have $e^{z \pm 2n\pi i} = e^z$, $n = 0, 1, 2, \ldots$. Thus, e^z is a periodic function of period $2\pi i$. This means that every value which e^z can assume is attained in the infinite strip $-\pi < y \leqslant \pi$, or in any strip obtainable from it by a translation parallel to the imaginary axis.

Finally, we note that e^z never vanishes for, if $e^{z_1} = 0$, then e^{-z_1} would be infinite which contradicts the fact that e^z is entire.

1.5.3 The Trigonometric and Hyperbolic Functions

The sine and cosine functions of a complex variable are defined to be the sum-functions of the series

$$\sum_{m=0}^{\infty}(-1)^m \frac{z^{2m+1}}{(2m+1)!} \equiv \sin z,$$

$$\sum_{m=0}^{\infty}(-1)^m \frac{z^{2m}}{(2m)!} \equiv \cos z. \tag{1.5.4}$$

Since each of these power series has an infinite radius of convergence, $\sin z$ and $\cos z$ are entire functions. The other trigonometric functions are defined by

$$\tan z \equiv \frac{\sin z}{\cos z}, \quad \cot z \equiv \frac{1}{\tan z}, \quad \csc z \equiv \frac{1}{\sin z}, \quad \sec z \equiv \frac{1}{\cos z}. \tag{1.5.5}$$

Differentiating the power series (1.5.4) term by term we find that

$$\frac{d}{dz}\sin z = \cos z, \quad \frac{d}{dz}\cos z = -\sin z. \tag{1.5.6}$$

Also, applying to e^{iz} the same manipulations that led us to Euler's formula we obtain

$$e^{\pm iz} = \cos z \pm i \sin z \tag{1.5.7}$$

or, equivalently,

$$\cos z = \frac{1}{2}(e^{iz} + e^{-iz}), \ \sin z = \frac{1}{2i}(e^{iz} - e^{-iz}). \tag{1.5.8}$$

From these formulae, and the multiplication rule for e^z, one can readily deduce that all the trigonometric identities that hold for real variables do so for complex variables as well. In particular,

$$\sin^2 z + \cos^2 z = 1$$
$$\sin(z \pm w) = \sin z \cos w \pm \cos z \sin w$$
$$\cos(z \pm w) = \cos z \cos w \mp \sin z \sin w. \tag{1.5.9}$$

Many of the properties possessed by the trigonometric functions are most easily discerned by expressing the functions in terms of their real and imaginary parts. Setting $z = x + iy$, we have

$$\sin z = \frac{1}{2i}[e^{i(x+iy)} - e^{-i(x+iy)}] = \frac{1}{2i}[e^{-y}(\cos x + i \sin x) - e^{y}(\cos x - i \sin x)],$$

or

$$\sin z = \cosh y \sin x + i \sinh y \cos x, \tag{1.5.10}$$

where we have used the real variable definitions

$$\sinh y \equiv \frac{1}{2}(e^y - e^{-y})$$
$$\cosh y \equiv \frac{1}{2}(e^y + e^{-y}).$$

Similarly,

$$\cos z = \cosh y \cos x - i \sinh y \sin x. \tag{1.5.11}$$

Evidently, $\sin z$ can only vanish if both $\sin x \cosh y = 0$ and $\cos x \sinh y = 0$. Since $\cosh y \geqslant 1$, we require $\sin x = 0$ or $x = n\pi, n = 0, \pm 1, \pm 2, \ldots$. This in turn implies that $\cos x \neq 0$ and hence, that $\sinh y = 0$. The latter condition can only be met if $y = 0$. Thus, we conclude that $\sin z$ vanishes if, and only if $z = n\pi, n = 0, \pm 1, \pm 2, \ldots$. Similarly, $\cos z$ vanishes if, and only if, $z = (n + \frac{1}{2})\pi, n = 0, \pm 1, \pm 2, \ldots$.

Unlike their real variable counterparts, $|\sin z|$ and $|\cos z|$ are not bounded, let alone bounded by unity. This too follows from equations (1.5.10) and (1.5.11) which yield

$$|\sin z|^2 = \sin^2 x + \sinh^2 y \qquad (1.5.12)$$

$$|\cos z|^2 = \cos^2 x + \sinh^2 y \qquad (1.5.13)$$

both of which increase without limit as $y \to \infty$. This behaviour reflects a singularity at the point at infinity which the trigonometric functions have in common with the exponential function.

Other properties that follow immediately from (1.5.10) and (1.5.11) are

$$(\sin z)^* = \sin(z^*) \qquad (\cos z)^* = \cos(z^*)$$
$$\sin(-z) = -\sin z \qquad \cos(-z) = \cos z$$

and, the very important periodicity conditions

$$\sin(z \pm 2n\pi) = \sin z, \quad \cos(z \pm 2n\pi) = \cos z, \quad n = 0, \pm 1, \pm 2, \ldots.$$

The hyperbolic functions of a complex variable are defined by

$$\sinh z = -i \sin iz \qquad \cosh z = \cos iz$$
$$\sin z = -i \sinh iz \qquad \cos z = \cosh iz$$

and so their properties can be deduced from those of the trigonometric functions. In particular, one finds

$$\sinh z = \frac{1}{2}(e^z - e^{-z}), \qquad \cosh z = \frac{1}{2}(e^z + e^{-z})$$

and,

$$\cosh^2 z - \sinh^2 z = 1.$$

1.5.4 The Logarithm

The natural logarithm of the complex variable z, denoted by

$$w = \ln z,$$

is defined for each $|z| \neq 0$ by the equation

$$e^w = z. \tag{1.5.14}$$

We have already seen that the exponential function e^z has period $2\pi i$ and hence, that it maps each of the strips

$$(n-1)\pi < \operatorname{Im} z \leqslant (n+1)\pi, \quad n = 0, \pm 1, \pm 2, \dots$$

onto the entire complex plane (less the origin). Therefore, it should come as no surprise that equation (1.5.14) admits an infinite number of solutions each of which is a logarithm of z. To see how this comes about in detail, set

$$w = u + iv \quad \text{and} \quad z = |z|\, e^{i \arg z}$$

so that (1.5.14) becomes

$$e^w = e^u \cdot e^{iv} = |z|\, e^{i \arg z}.$$

Thus, from the definition of equality,

$$v = \arg z \quad \text{and} \quad e^u = |z|.$$

But, since v and $|z|$ are both real, with $|z| > 0$, the latter equation has the unique solution $u = \ln |z|$. Therefore, we finally obtain

$$\ln z = \ln |z| + i \arg z. \tag{1.5.15}$$

Since $\arg z$ is only determined to within multiples of 2π, $\ln z$ is infinitely many-valued with successive values differing by $2\pi i$.

The **principal value** of $\ln z$, which is obtained by giving $\arg z$ its principal value, will be denoted by $\operatorname{Ln} z$. Thus,

$$\operatorname{Ln} z = \ln |z| + i \operatorname{Arg} z, \quad -\pi < \operatorname{Arg} z \leqslant \pi \tag{1.5.16}$$

and, $\ln z = \operatorname{Ln} z \pm 2n\pi i$, $n = 0, 1, 2, \dots$. The principal value is identical with the real logarithm when z is real and positive.

Since

$$\lim_{\varepsilon \to 0}[\operatorname{Ln}(|z|\, e^{i(\pi - \varepsilon)}) - \operatorname{Ln}(|z|\, e^{i(-\pi + \varepsilon)})] = 2\pi i,$$

$\operatorname{Ln} z$ is discontinuous across the negative real axis. It should be noted however that the exact location of this line of discontinuity was determined by **our** choice for the range of $\operatorname{Arg} z$ in the definition of $\operatorname{Ln} z$. Because of the discontinuity, the single valued function $\operatorname{Ln} z$ can only be holomorphic in a domain that excludes the negative real axis. Certainly, it is differentiable for $|z| \neq 0$, $\operatorname{Arg} z \neq \pm\pi$ as can be seen from

$$\frac{d}{dz}\operatorname{Ln} z = \frac{\partial}{\partial x} \ln \sqrt{x^2 + y^2} + i \frac{\partial}{\partial x} \tan^{-1} \frac{y}{x} = \frac{x}{x^2 + y^2} - i \frac{y}{x^2 + y^2} = \frac{1}{z}. \tag{1.5.17}$$

Thus, its domain of holomorphy must be the domain $0 < |z| < \infty$, $-\pi < \operatorname{Arg} z < \pi$ which is the **cut plane** obtained by removing the origin and negative real axis.

1.5.5 The General Power z^α

We define the power z^α, where z and α denote any complex numbers, to be

$$z^\alpha \equiv e^{\alpha \ln z}, \quad |z| \neq 0 \tag{1.5.18}$$

and its principal value to be $e^{\alpha \, \mathrm{Ln}\, z}$.

Since $\ln z$ is infinitely many-valued, z^α might reasonably be expected to be so too. In general, this is indeed the case; the only exceptions occur, as we have seen already, when α assumes integer or rational values. This is most easily discerned by writing (1.5.18) in the form

$$z^\alpha = e^{\alpha[\mathrm{Ln}\, z + 2k\pi i]}, \quad k = 0, \pm 1, \pm 2, \dots. \tag{1.5.19}$$

If $\alpha = n, n = \pm 1, \pm 2, \dots$ then, since $e^{\pm 2kn\pi i} = 1$, equation (1.5.19) yields the single value

$$z^n = |z|^n \, e^{in\mathrm{Arg}\, z},$$

which is precisely what one obtains from the de Moivre theorem. If $\alpha = m/n$, with the integer n not a divisor of the integer m then, as expected, equation (1.5.19) yields the n values

$$z^{m/n} = |z|^{m/n} \exp\left[i\frac{m}{n}(\mathrm{Arg}\, z + k2\pi)\right], \quad k = 0, 1, \dots, n-1.$$

Finally, if α is a complex or real irrational number then $e^{2k\alpha\pi i} \neq 1$ for any value of $k = \pm 1, \pm 2, \dots$ and so each value of k yields a distinct value of z^α.

The principal value of z^α (for non-integer α) is discontinuous across the negative real axis. Hence, like $\mathrm{Ln}\, z$, it is holomorphic only in the cut plane $0 < |z| < \infty$, $-\pi < \mathrm{Arg}\, z < \pi$.

Example: We conclude this Section with an evaluation of what might appear to be the epitome of a complex number, i^i. From (1.5.19) we have

$$i^i = e^{i \ln i} = e^{i[\mathrm{Ln}\, i + 2k\pi i]}, \quad k = 0, \pm 1, \pm 2, \dots.$$

But, $|i| = 1$ and $\mathrm{Arg}\, i = \pi/2$ so that $\mathrm{Ln}\, i = \frac{\pi}{2}i$. Thus, i^i is in fact the infinity of **real** numbers

$$i^i = \exp\left[-\frac{\pi}{2} - 2k\pi\right], \quad k = 0, \pm 1, \pm 2, \dots$$

and has the principal value $e^{-\frac{\pi}{2}}$.

1.6 Multivalued Functions and Riemann Surfaces

Our definition of analyticity requires functions to be single-valued. However, as we have just seen, some important elementary functions, $\ln z$, z^α, and any of the **inverse** trigonometric and hyperbolic functions, are multivalued. Thus, one might suppose that we require two distinct theories of functions of a complex variable: one for analytic functions and the other for a suitably defined class of multivalued functions. Fortunately, this is not the case. As we will now show, a geometrical construction due to Riemann can be used to ensure that the theory of analytic functions can be applied to both single and multivalued functions.

The definition of the various **principal value** functions in the preceding Section demonstrates that one can dissociate a multivalued function into a series of single valued ones that are individually holomorphic in a cut plane. This provides us with an important first clue as to how to proceed. A second and decisive clue is presented by the way in which the multivalued character of a function manifests itself in the behaviour of its single valued constituents.

For clarity, we shall consider the specific function

$$w = z^{1/2} = |z|^{1/2} \exp\left[\frac{i}{2} \arg z\right] = |z|^{1/2} \exp\left[\frac{i}{2}(\mathrm{Arg} z + 2k\pi)\right], \quad k = 0, 1. \quad (1.6.1)$$

We shall denote the values corresponding to $k = 0$ by w_1 and those corresponding to $k = 1$ by w_2.

Let us see what happens to w as z describes a closed path encircling the origin. Starting from a point $z_0 = (|z_0|, \theta_0)$, we perform a complete anticlockwise cycle around the origin, returning again to z_0. The function $w = z^{1/2}$ changes continuously as we follow the closed curve but, after completion of a full cycle, $z_0^{1/2}$ differs from its initial value by a factor of (-1):

$$[z_0^{1/2}]_i = |z_0|^{1/2} e^{i\theta_0/2}$$
$$[z_0^{1/2}]_{f_1} = |z_0|^{1/2} e^{i(\theta_0 + 2\pi)/2} = -|z_0|^{1/2} e^{i\theta_0/2} = -[z_0^{1/2}]_i.$$

Moreover, to regain the initial value of $z_0^{1/2}$ we see that we must perform yet another complete cycle:

$$[z_0^{1/2}]_{f_2} = |z_0|^{1/2} e^{i(\theta_0 + 4\pi)/2} = |z_0|^{1/2} e^{i\theta_0/2} = [z_0^{1/2}]_i.$$

This requirement of two complete cycles to regain the function's initial value is clearly a manifestation of the function's double-valuedness.

Definition: A point of the complex plane having the property that after the completion of any cycle around it a given function is not restored to its initial value is called a **branch point** of the function. A branch point is of $n^{\mathbf{th}}$ **order** if after making not less than $(n + 1)$ complete cycles around it we restore the function to its initial value. Otherwise, a branch point is said to be of infinite order.

Example: The point $z = 0$ is a branch point of order one of the function $z^{1/2}$. It follows that $\zeta = 0$ is similarly a branch point of order one of $\zeta^{-1/2}$ and so $z = \infty$ is a second branch point of order one of $z^{1/2}$. These are the only two branch points that this function possesses.

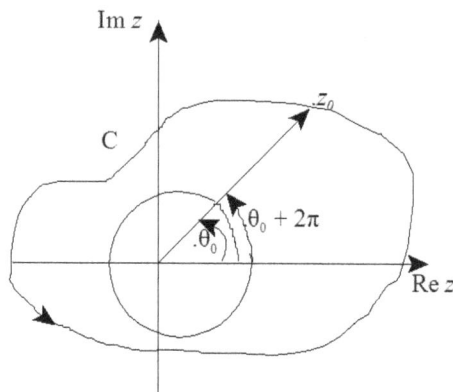

Figure 1.7: After one complete cycle about the origin the argument of z_0 is increased by 2π.

We shall now follow the progress of $z^{1/2}$ as z cycles around the branch point $z = 0$ in a little more detail. If θ_0 is assigned a specific value of $-\pi < \text{Arg}z \leqslant \pi$, $[z_0^{1/2}]_i$ assumes one of the values possessed by the single valued function $w_1(k = 0)$. As we proceed around the curve C in Figure 1.7 , $z^{1/2}$ varies continuously through values that correspond to w_1 until we reach the negative real axis. On crossing the axis $z^{1/2}$ still varies continuously but its values are now those that correspond to $w_2(k = 1)$. A further cycle around the curve sees $z^{1/2}$ vary continuously through the values of w_2 until the negative real axis is again encountered, at which point it re-assumes values corresponding to w_1. This is shown in Figure 1.8 in which we have used a unit circle on each of two superposed planes to picture the two complete cycles.

In summary, $z^{1/2}$ varies continuously through the values of w_1 and w_2 as z goes from $(|z_0|, \theta_0)$ to $(|z_0|, \theta_0 + 4\pi)$ because, even though w_1 and w_2 are separately discontinuous across the negative real axis,

$$\lim_{\varepsilon \to 0}[w_k(|z|, \pi - \varepsilon) - w_k(|z|, -\pi + \varepsilon)] = \pm 2i|z|^{1/2}, \quad k = 1, 2, \qquad (1.6.2)$$

the value of $w_1(w_2)$ just above the axis is the same as the value of $w_2(w_1)$ just below it:

$$\lim_{\varepsilon \to 0^+} w_1(|z|, \pi - \varepsilon) = \lim_{\varepsilon \to 0^+} w_2(|z|, -\pi + \varepsilon) = i|z|^{1/2}$$
$$\lim_{\varepsilon \to 0^+} w_2(|z|, \pi - \varepsilon) = \lim_{\varepsilon \to 0^+} w_1(|z|, -\pi + \varepsilon) = -i|z|^{1/2}. \qquad (1.6.3)$$

This fact, plus the knowledge that w_1 and w_2 are holomorphic in the domain $0 < |z| < \infty$, $-\pi < \text{Arg}z < \pi$ is all we require in order to construct a domain of definition on which $z^{1/2}$ will enjoy the benefits of analyticity.

The first step in the construction is to choose a curve joining the two branch points of $z^{1/2}$; a line starting from $z = 0$ and extending to $z = \infty$ is an obvious example.

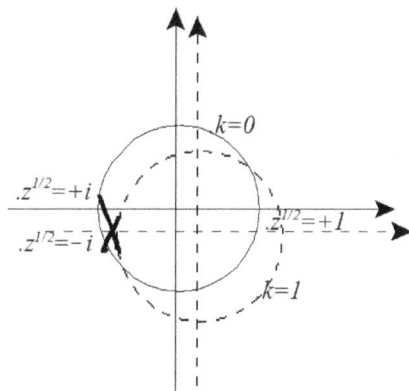

Figure 1.8: The value of $w_1(w_2)$ just above the real axis is the same as the value of $w_2(w_1)$ just below it.

Next, we cut the complex plane by removing all the points that lie on this curve; the curve is then referred to as a **branch cut**. This provides a domain of definition for two single valued functions or **branches** which together reproduce all the values that $z^{1/2}$ can assume, except of course those corresponding to values of z lying on the cut. To conform with our earlier conventions, we shall take the cut to lie along the negative real axis. Our two branches are thus the familiar functions

$$w_1 = |z|^{1/2} e^{i\theta/2}, \quad -\pi < \theta < \pi, 0 < |z| < \infty$$
$$w_2 = -|z|^{1/2} e^{i\theta/2}, \quad -\pi < \theta < \pi, 0 < |z| < \infty \tag{1.6.4}$$

which are analytic throughout the cut plane. (Notice that we have opted to use θ in place of Arg$z$.)

To complete our construction we now superpose two cut planes joined edge to edge at the cut, the upper edge of each being joined continuously with the lower edge of the other. The two planes are then referred to as **Riemann sheets** and the surface resulting from their superposition is called a **Riemann surface**. As the following argument shows, this surface provides a domain of holomorphy for $z^{1/2}$.

Let arg z vary from $-\pi$ to $+\pi - \varepsilon$ for fixed $|z|$. Then, z describes a circle about the origin on the top sheet and $z^{1/2}$ assumes the values of w_1. At the end of the circle, (arg $z = \pi$), we cross to the second sheet where, as arg z increases from $\pi+\varepsilon$ to $3\pi-\varepsilon$, z describes another circle about the origin and $z^{1/2}$ assumes the values of w_2. At the end of this circle, (arg $z = 3\pi$), we cross back to our starting point on the first sheet. Because of equation 1.6.3, $z^{1/2}$ varies **continuously** along the whole of this path, tracing out a complete circle of radius $|z|^{1/2}$ in the w-plane. Thus, cuts are no longer necessary and the function $z^{1/2}$ is holomorphic over the whole Riemann surface, except at the two branch points $z = 0$ and $z = \infty$.

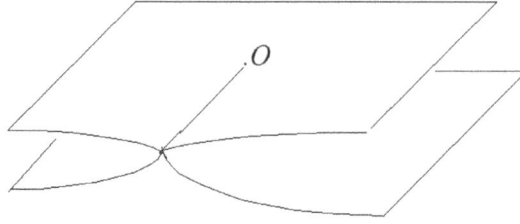

Figure 1.9: The Riemann surface for the function $w = z^{1/2}$.

Since the mapping aspect of functions of a complex variable can enhance our intuitive appreciation, we note that we have just shown that the function $z = w^2$ maps the w-plane onto a two- sheeted Riemann surface for z.

Recall that the branch cut joining $z = 0$ and $z = \infty$ is a line of "man-made" singularities for the two branches w_1 and w_2 and as such, has an entirely arbitrary location. Any simple curve connecting these two points could serve as an acceptable cut along which the two Riemann sheets of $z^{1/2}$ can be joined. For example, suppose that we choose the ray $\arg z = \theta_0$; (straight lines are always the easiest curves to deal with). The two branches of $z^{1/2}$ must then be defined as

$$w_1 = |z|^{1/2}\, e^{i\theta/2}, \quad \theta_0 < \theta < \theta_0 + 2\pi$$
$$w_2 = -|z|^{1/2}\, e^{i\theta/2}, \quad \theta_0 < \theta < \theta_0 + 2\pi;$$

that is, we merely change the range of variation of θ. A definition of $z^{1/2}$ that makes it holomorphic everywhere, except at its branch points $z = 0$ and $z = \infty$, is now obtained by constructing a two-sheeted Riemann surface, cut and joined along the line $\arg z = \theta_0$. Some of the values of $z^{1/2}$ which occurred for z on the **first** Riemann sheet when the cut was made along the negative real axis now correspond to the **second** Riemann sheet and vice versa. This illustrates that the Riemann construction is merely a way of classifying the values of a function in a single valued manner and the details of the classification are matters of convention.

For the more general case of the n-valued functions $z^{1/n}$ and $z^{m/n}$, it can be readily shown that $z = 0$ and $z = \infty$ are branch points of order $(n-1)$ and that the appropriate Riemann surface is a closed n-sheeted structure with the nth sheet reconnected to the first. Such a surface is difficult to visualize and impossible to sketch, except in the most schematic way. Rather than try to do so the reader should keep in mind that the surface is a mathematical construction whose sole purpose is to classify the n values. On the kth sheet, for example, the function $z^{1/n}$ assumes the values of its kth branch

$$w_k = |z|^{1/n} \exp\left\{ \frac{i}{n}[\theta + 2(k-1)\pi] \right\}, \quad -\pi < \theta < \pi, \quad 0 < |z| < \infty$$

where we have again chosen the branch cut to lie along the negative real axis. Thus, as $\arg z$ increases through π, z crosses from the first to the second sheet and $z^{1/n}$ varies continuously from the values of w_1 to those of w_2. As $\arg z$ increases through 3π, z again encounters the cut and so drops down from the second to the third sheet. This process continues until we reach the nth sheet, which corresponds to $(2n-3)\pi < \arg z < (2n-1)\pi$. As $\arg z$ increases through $(2n-1)\pi$, z moves through the cut back up to the first sheet and $z^{1/n}$ varies in a continuous fashion from the values of w_n to those of w_1.

Paradoxically, the Riemann surface for the infinitely many valued logarithm function

$$\ln z = \ln|z| + i\arg z = \text{Ln}\,z + 2k\pi i, \quad k = 0, \pm 1, \pm 2, \ldots \qquad (1.6.5)$$

is somewhat easier to visualize. The function's branch points are readily seen to be $z = 0$ and $z = \infty$ again. For example, if one completes a counterclockwise cycle about the origin one leaves $\ln|z|$ unchanged but $\arg z$ is increased by 2π and so, $\ln z$ changes value by an amount $2\pi i$. Moreover, no matter how many times one encircles the origin, the logarithm **never** regains its original value. Thus, in this case, $z = 0$ and $z = \infty$ are branch points of **infinite** order. This means that a single valued definition of the logarithm requires a Riemann surface consisting of an infinite number of sheets, each one joined to the one above and below it by a cut running from $z = 0$ to $z = \infty$.

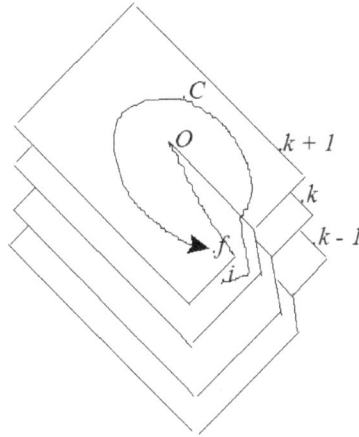

Figure 1.10: A part of the Riemann surface for the logarithm. The path C encircles the branch point at the origin and hence, necessarily crosses from one sheet to the next.

Choosing the cut to again lie along the negative real axis, the single valued branches of $\ln z$ that correspond to this construction are

$$w_k = \ln|z| + i\theta + 2\pi ki, \quad -\pi < \theta < \pi, 0 < |z| < \infty, k = 0, \pm 1, \pm 2, \ldots. \qquad (1.6.6)$$

These functions are holomorphic everywhere in the cut plane and the nth member of the set, w_n, takes on the same values that $\ln z$ assumes when $(2n - 1)\pi < \arg z <$

$(2n + 1)\pi$. Although each w_k has a jump discontinuity of

$$\lim_{\varepsilon \to 0}[w_k(|z|, \pi - \varepsilon) - w_k(|z|, -\pi + \varepsilon)] = 2\pi i$$

across the cut, the value of w_k just above the cut is the same as the value of w_{k+1} just below it:

$$\lim_{\varepsilon \to 0^+} w_k(|z|, \pi - \varepsilon) = \lim_{\varepsilon \to 0^+} w_{k+1}(|z|, -\pi + \varepsilon).$$

Thus, with the upper lip of the cut in the kth sheet connected to the lower lip of the cut in the $(k + 1)^{\text{th}}$ sheet, the Riemann surface provides a domain of definition on which $\ln z$ varies **continuously** through all the values it can assume and hence, on which $\ln z$ is holomorphic everywhere, except for the two branch points $z = 0$ and $z = \infty$.

The preceding discussion provide a good basis from which to tackle more complicated functions.

For **example,** $w = (z - a)^{1/2}$ has a branch point at $z = a$ rather than $z = 0$ but is otherwise identical in behaviour to $z^{1/2}$. Thus, its Riemann surface is a closed, two-sheeted structure with the sheets cut and joined along an arbitrary line extending from $z = a$ to $z = \infty$. Figure 1.11 shows a more or less random choice for this line. To utilize our experience with $z^{1/2}$ we introduce the variable $\theta_a = \arg(z - a)$. Then, the single valued branches corresponding to our particular choice of branch cut are $w_1 = |z - a|^{1/2} e^{i\theta_a/2}$ and $w_2 = -|z - a|^{1/2} e^{i\theta_a/2}$ with $|z| < \infty$, $|z - a| > 0$ and $\theta_0 - 2\pi < \theta_a < \theta_0$, which completes the definition of a Riemann surface for this function.

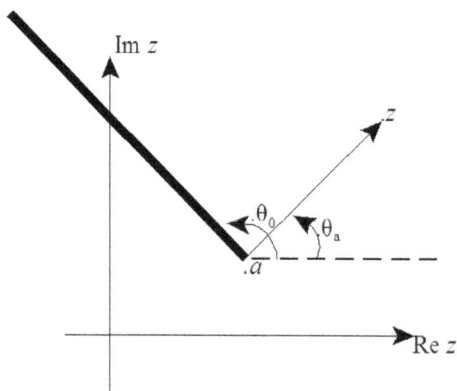

Figure 1.11: A possible choice of branch cut for the function $w = (z - a)^{1/2}$.

A more interesting as well as more challenging **example** is posed by the function $w = [(z - a)(z - b)]^{1/2}$. To identify its branch points we introduce the variables $r_a = |z - a|$, $r_b = |z - b|$, $\theta_a = \arg(z - a)$ and $\theta_b = \arg(z - b)$ as shown in Figure 1.12. Only two of these four variables are needed to specify z but, by using all four, we can write w in the suggestive form

$$w = (r_a\, r_b\,)^{1/2} \exp\left[\frac{i}{2}(\theta_a + \theta_b)\right] = [r_a^{1/2}\, e^{i\,\theta_a\,/2}][r_b^{1/2}\, e^{i\,\theta_b\,/2}]. \qquad (1.6.7)$$

From our experience with $z^{1/2}$ we can assert that $z = a$ and $z = b$ are both branch points of order one. This is readily confirmed by performing a cycle around either, but **not** both, of the two points. If we encircle $z = a$ on a curve that does not enclose $z = b$, θ_a increases by 2π while θ_b increases and then decreases and finally returns to its initial value. Thus, after one full cycle, w changes its sign and after two full cycles it returns to its initial value.

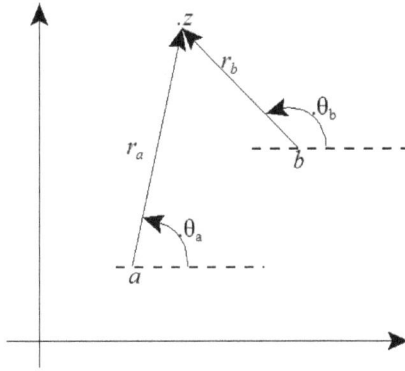

Figure 1.12: The definition of the variables $r_a, r_b, \theta_a, \theta_b$.

It is both interesting and important to notice that a cycle enclosing both $z = a$ and $z = b$ causes no change in w. This is because both θ_a and θ_b increase by 2π causing a change in the argument of w of

$$\frac{1}{2}(\theta_a + \theta_b) \rightarrow \frac{1}{2}(\theta_a + \theta_b + 4\pi) = \frac{1}{2}(\theta_a + \theta_b) + 2\pi. \qquad (1.6.8)$$

This implies that $z = \infty$ is not a branch point although going on past experience alone, we might mistakenly have assumed that it is. To confirm this, we set $z = \frac{1}{\zeta}$ and note that $w = [\frac{1}{\zeta^2} - 1]^{1/2} \rightarrow \pm\frac{1}{\zeta}$ as $\zeta \rightarrow 0$. Thus, although w is singular at $z = \infty(\zeta = 0)$, and double valued in any neighbourhood of that point, it is unchanged by a cycle about it and so $z = \infty$ is not a branch point.

The next step in defining a Riemann surface for this function is to choose a branch cut joining $z = a$ and $z = b$. As Figure 1.13 illustrates, this may be done in two distinct ways:

(a) a curve of finite length terminating at $z = a$ and $z = b$;

(b) a curve which terminates at $z = a$ and $z = b$ but passes through $z = \infty$ in between.

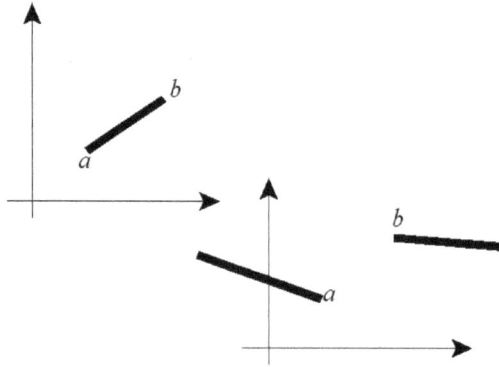

Figure 1.13: In the diagram on the left the z plane is cut along the straight line segment that joins $z = a$ and $z = b$. In the one on the right the plane is cut along two straight lines that meet at the point at infinity.

To discuss these two options as clearly as possible, we shall specialize to the function

$$w = (z^2 - 1)^{1/2} = [(z - 1)(z + 1)]^{1/2} = (r_+ r_-)^{1/2} e^{i(\theta_+ + \theta_-)/2}$$

whose branch points are at $z = \pm 1$. Given past conventions, the two obvious choices for a branch cut for this function are the real axis segments

(a) $-1 \leqslant x \leqslant 1$, and

(b) $x \leqslant -1, x \geqslant 1$

as shown in Figure 1.14. (This cut structure is also relevant to the function $w = \ln \frac{z+1}{z-1}$ which we will encounter in subsequent sections of the book.) Since each of these is a straight line, the corresponding branches of $(z^2 - 1)^{1/2}$ are given by

$$w_{1,2} = \pm(r_+ r_-)^{1/2} e^{i(\theta_+ + \theta_-)/2},$$

with θ_+ and θ_- restricted to prevent us crossing whichever cut we have decided to work with. The cut along $-1 \leqslant x \leqslant 1$ is avoided if **either** θ_+ is restricted to $-\pi < \theta_+ < \pi$ **or** θ_- is restricted to $0 < \theta_- < 2\pi$. However, although w_1 and w_2 are discontinuous across the cut, they must be continuous everywhere else. In particular, they must be continuous across the line segments $x < -1$ and $x > 1$. This latter restriction can only be satisfied if θ_+ and θ_- have the same range of variation; (cf. equation 1.6.8 and the discussion that led up to it). To conform with convention, we therefore require $-\pi < \theta_{\pm} < \pi$. The closed, two-sheeted Riemann surface that corresponds to this choice of cut is now completely specified by assigning to $(z^2 - 1)^{1/2}$ the values $w_1 = |z - 1|^{1/2} |z + 1|^{1/2} e^{i(\theta_+ + \theta_-)/2}$, $-\pi < \theta_\pm < \pi$, $|z \pm 1| > 0$ for z on the first sheet, and the values $w_2 = -w_1$ for z on the second sheet of the surface.

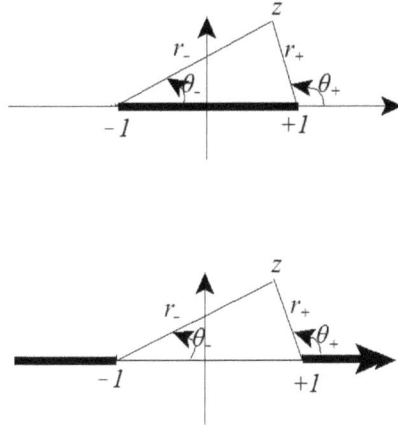

Figure 1.14: Two obvious choices for the branch cut for the function $(z^2 - 1)^{1/2}$.

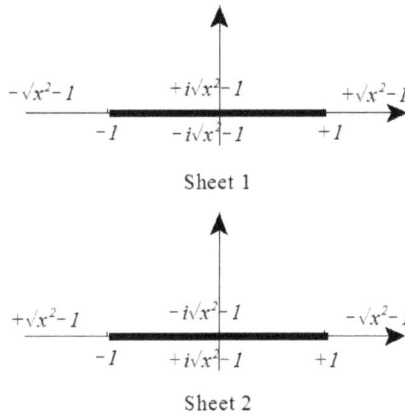

Figure 1.15: The values assumed by $(z^2 - 1)^{1/2}$ along the real axes of its two Riemann sheets when the branch cut is chosen to lie along $-1 \leqslant x \leqslant 1$.

It is instructive to calculate the values assumed by $(z^2 - 1)^{1/2}$ for values of z close to the real axis on the two Riemann sheets. This is done below and the results displayed in Figure 1.15. For $-1 \leqslant x \leqslant 1, y = 0^{\pm}$, we have $\theta_- = 0^{\pm}, \theta_+ = \pi^{\mp}$ and so

$$w_1 = \sqrt{1 - x^2} \, e^{\pm i\pi/2} = \pm i\sqrt{1 - x^2}.$$

For $x < -1, y = 0^{\pm}$, we have $\theta_- = \theta_+ = \pm \pi^{\mp}$ and

$$w_1 = \sqrt{x^2 - 1} \, e^{\pm \pi i} = -\sqrt{x^2 - 1}.$$

For $x > 1, y = 0^{\pm}$, we have $\theta_- = \theta_+ = 0^{\pm}$ and

$$w_1 = +\sqrt{x^2 - 1}.$$

If the cut is chosen to lie along the line segment $x \leqslant -1, x \geqslant 1$, there is only one way to avoid crossing it: θ_+ and θ_- must be restricted to the ranges $0 < \theta_+ < 2\pi$ and $-\pi < \theta_- < \pi$. This conclusion follows from noting that, with this choice of cut, $r_+^{1/2} e^{i\theta_+/2}(r_-^{1/2} e^{i\theta_-/2})$ behaves like $z^{1/2}$ when its cut is taken to lie along the positive (negative) real axis. Thus, in this case, the Riemann surface is completely specified by defining the branches of $(z^2 - 1)^{1/2}$ to be $w_1 = |z - 1|^{1/2} |z + 1|^{1/2} e^{i(\theta_+ + \theta_-)/2}$, and $w_2 = -w_1, 0 < \theta_+ < 2\pi, -\pi < \theta_- < \pi$ and $|z - 1| > 0$. The values assumed by $(z^2 - 1)^{1/2}$ near the real axes of this surface are shown in Figure 1.16.

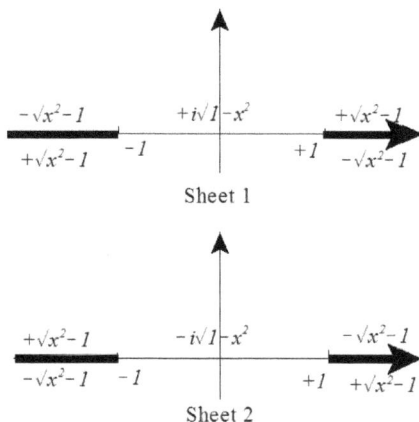

Figure 1.16: The values assumed by $(z^2 - 1)^{1/2}$ along the real axes of its two Riemann sheets when the branch cut is chosen to lie along $x \leqslant -1, x \geqslant +1$.

As Figures 1.15 and 1.16 affirm, both Riemann surfaces are closed. Moreover, they are constructed in such a way that one changes sheets after any cycle that encloses only **one** of the branch points but is returned to one's starting point after a cycle that encloses **both** branch points.

It is now easy to generalize to the case $w = \sqrt{(z - a_1)(z - a_2) \ldots (z - a_n)}$. If n is even, there are n branch points located at $z = a_1, a_2, \ldots, a_n$ while, if n is odd, there are $n + 1$ branch points including the point at infinity. The branch points must be joined pair-wise by branch cuts and then a closed two-sheeted Riemann surface can be constructed by means of two interconnections along each cut. Assuming straight line branch cuts, the surface is completely specified simply by defining the branches of the function to be

$$w_1 = (r_{a_1} \ldots r_{a_n})^{1/2} \, e^{i(\theta_{a_1} + \ldots + \theta_{a_n})/2}, \; w_2 = -w_1, r_{a_i} = |z - a_i|, \; \theta_{a_i} = \arg(z - a_i),$$

with each of $\theta_{a_1}, \ldots, \theta_{a_n}$ restricted to ranges of length 2π as determined by the particular choice of cuts that has been made.

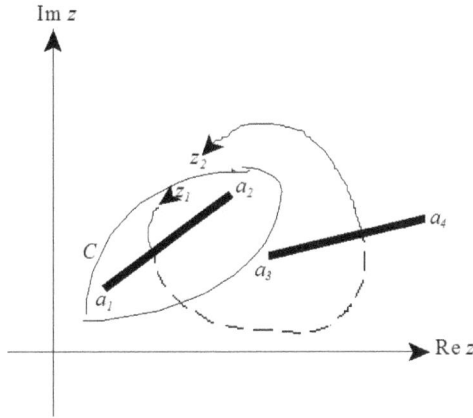

Figure 1.17: If z moves from the point z_1, **inside** a closed curve C on the top sheet of a surface for $\sqrt{(z - a_1)(z - a_2)(z - a_3)(z - a_4)}$ so that it crosses to the bottom sheet via the cut enclosed by C, it can cross back to the top sheet via the second cut and thus end up at the point z_2 , **outside** C , without intersecting C itself.

We conclude with a couple of **examples** of the kind of curious phenomena that can occur with Riemann surfaces.

On a surface for a function with four or more branch points in the finite plane one can move continuously from the inside to the outside of a closed curve without actually crossing the curve itself. How this comes about is shown in Figure 1.17 using a surface for the function $\sqrt{(z - a_1) \ldots (z - a_4)}$.

Of more practical interest is the behaviour exhibited by the function

$$w = \frac{z^{1/2} + ia}{z^{1/2} - ia}, \quad a > 0$$

which is singular when $z^{1/2} = ia$. The Riemann surface defined for $z^{1/2}$ can be used for this function as well. Thus, we see that w has a singularity at $z = -a^2$ on the **first** sheet of the surface but on the **second** sheet it has a zero there. This illustrates that while a function may have a singularity (which is not a branch point) on some sheets of a Riemann surface, it may be perfectly well-behaved at the same point(s) on the other sheets.

We have devoted what might seem to be an inordinate amount of space to multi-valued functions. This has not been done because the subject is difficult but because, on the contrary, it is relatively simple once it is recognized that to attempt to visualize Riemann surfaces as three dimensional objects will only impede understanding; the surfaces merely provide a classification scheme, a means of separating and ordering the many values. Thus, what almost amounts to an overexposure at this juncture is intended to make the reader feel if not at home, then at least comfortable with the concepts of branch cuts, sheets and surfaces when they reappear in subsequent sections.

1.7 Conformal Mapping

The mapping defined by an analytic function $w = f(z)$ has the property that it is *conformal* (angle-preserving) except at points where $f'(z) = 0$. As we shall see, this has an important application in the solution of two dimensional boundary value problems by transforming a given complicated region into a simpler one.

To see how conformality comes about, consider a smooth curve C passing through the point z_0 in the $z - plane$. A function $w = f(z)$ that is holomorphic at $z = z_0$ will map z_0 onto a point $w = w_0$ and C onto a curve C_w passing through w_0. Next, consider a near-by point z_1 on C and its image $w_1 = f(z_1)$ on C_w. Denoting $z - z_0$ by Δz and $w - w_0$ by Δw, we take the limit

$$\frac{\Delta w}{\Delta z}$$

which by definition is the derivative $f'(z_0)$. Examining first the modulus and then the argument of this limit, we obtain

$$\frac{|\Delta w|}{|\Delta z|} = |f'(z_0)| \tag{1.7.1}$$

and

$$\varphi_0 - \theta_0 \equiv [\arg \Delta w - \arg \Delta z] = (\arg \frac{\Delta w}{\Delta z}) = \arg f'(z_0) \equiv \propto \tag{1.7.2}$$

where θ_0 and φ_0 are the angles that the tangents to the curves C and C_w make with the real axis in their respective planes. This means that C is rotated through an angle $\alpha = \arg f'(z_0)$ when it is mapped onto C_w but this angle of rotation is the *same* for *all* curves passing through z_0. Thus, the angle formed by two intersecting curves at z_0 remains *unchanged* both in magnitude and direction. At the same time, (1.7.1) tells us that the *magnification* of an infinitesimal arc of C that occurs when it is mapped onto C_w is $|f'(z_0)|$ and so is the *same* for *all* curves passing through z_0. This means that an infinitesimal circle about z_0 is mapped onto an infinitesimal circle about w_0 and the ratio of the radii is $|f'(z_0)|$.These two geometrical properties, invariance of angles and of the shape of infinitesimal circles, are what define a *conformal* transformation. Thus, we have established the

Theorem: If $f(z)$ is holomorphic at z_0 and $f'(z_0) \neq 0$, then the mapping $z \to w = f(z)$ is conformal at z_0.

An immediate application follows from the orthogonality of the straight lines $u = $ constant and $v = $ constant in the $w - plane$. The conjugate harmonic functions $u(x, y) = $ constant and $v(x, y) = $ constant must form a system of orthogonal curves (called *level curves*) in the $z - plane$. But they are also solutions of Laplace's equation in two dimensions, $\nabla^2 u = 0$ and $\nabla^2 v = 0$. Thus, if $u(x, y)$ is an electrostatic potential then $u(x, y) = $ constant represents an *equipotential surface* and the curve $v(x, y) = $ constant represents a *line of force*.

The practical significance of conformal mapping stems from the following

Theorem: A harmonic function $\phi(x, y)$ remains harmonic under a change of variables resulting from a one to one conformal mapping defined by an analytic function

$$w = f(z).$$

The proof is straightforward. If $\phi(x, y)$ is harmonic ($\nabla^2 \emptyset = 0$) within the domain of holomorphy D of $w = f(z)$ then it has a conjugate $\psi(x, y)$ such that $\Phi(x, y) = \phi(x, y) + i\psi(x, y)$ is an analytic function of $z = x + iy$ in D. Since $w = f(z)$ is holomorphic with a non-vanishing derivative in D, it maps D onto a domain D_w in the $w - plane$ where there exists a unique inverse function $z = F(w)$ which has the derivative

$$\frac{dF}{dw} = \frac{1}{df/dw}$$

and maps D_w onto D conformally. Hence, $\Phi(F(w))$ is an analytic function of w in D_w. Its real part is

$$\phi(x(u, v), y(u, v))$$

and is a harmonic function of u and v in D_w. The theorem is used as follows. Suppose that it is required to solve Laplace's equation in a given domain D subject to the imposition of specific values on the boundary of D. It may be possible to identify a conformal mapping which transforms D into a simpler domain such as a circular disc or a half-plane. Then we can solve Laplace's equation in the $w - plane$ subject to the transformed boundary conditions. The resulting solution transformed back to D will be the solution of the original problem. The catch is one needs to have a detailed knowledge of the mapping properties of a great many analytic functions. Catalogues of mappings have been compiled and can be consulted for exactly this purpose.

Just to illustrate this technique we shall consider one particular example of what are known as *linear fractional transformations*,

$$w = \frac{az + b}{cz + d}, \quad (ad - bc \neq 0). \tag{1.7.3}$$

The reason for the condition $ad - bc \neq 0$ is to ensure that $w' \neq 0$. We state without proof the

Theorem: Every linear fractional transformation maps the totality of circles and straight lines in the $z - plane$ onto the totality of circles and straight lines in the $w - plane$.

Our example is the function

$$w = \frac{b}{z} = \frac{b}{x^2 + y^2}(x - iy) = u + iv.$$

A standard approach to exploring the properties of a mapping is to find the *level curves* in the $z - plane$ that map onto lines of constant u or constant v in the $w - plane$. In this case the straight lines $u = c$ = constant have the *pre-image*

$$x^2 + y^2 = \frac{b}{c}x \quad \text{or}$$

$$\left(x - \frac{b}{2c}\right)^2 + y^2 = \frac{b^2}{4c^2}.$$

This is the equation of circles with centres at $(\frac{b}{2c}, 0)$ and radii $\frac{|b|}{2|c|}$ which confirms the theorem above. Notice that the centre of the circle maps onto the point $w = 2c$ and more generally, the interior of the circle is mapped onto the half-plane $u > c$ while the exterior is mapped onto $u < c$ if $c > 0$.

In a similar vein, the lines $v = k$ = constant correspond to the level curves

$$x^2 + y^2 = -\frac{b}{k}y \quad \text{or}$$

$$x^2 + (y + \frac{b}{2k})^2 = \frac{b^2}{4k^2}.$$

These are circles with centres at $(0, -\frac{b}{2k})$ and radii $\frac{|b|}{2|k|}$. Notice that they are orthogonal to the previous set of circles. Notice also that the axes $u = 0$ and $v = 0$ are the images of $x = 0$ and $y = 0$, respectively but with the origin mapped onto the point at infinity and vice versa. Once again, the interior of the circles maps onto $v > k$ while the exterior maps onto $v < k$ if $k > 0$.

An **example** that makes use of this analysis is one consisting of an infinite metal cylinder of radius R maintained at electric potential V and resting on top of but separated by a line of insulation from a grounded metal sheet lying in the $y = 0$ plane. Since the circular cross-section of the cylinder can be located in the upper half plane with centre at $(0, R)$, it will be mapped by $w = \frac{b}{z}$ onto the line $v = -\frac{b}{2R}$. Thus, if we take $b = 2R^2$, the line will be $v = -R$. As we have seen, the real axis $y = 0$ maps onto the real axis $v = 0$ and the exterior of the cylinder above the sheet maps onto the area between $v = 0$ and $v = -R$. Thus, the original problem is mapped onto a parallel plate capacitor with one plate at potential V and located at $v = -R$ and the other grounded and located at $v = 0$. We can write down the solution there immediately:

$$\phi = -\frac{V}{R}v.$$

But, under our mapping, $v = -\frac{2R^2}{x^2+y^2}y = -\frac{2R^2}{r}\sin\theta$ where we have switched to polar coordinates. Therefore, the solution to the original problem is

$$\phi(r,\theta) = \frac{2RV}{r}\sin\theta.$$

This is the real part of the complex potential $\Phi(z) = i\frac{2RV}{z}$. Thus, the equipotential surfaces are given by

$$\frac{2RV}{r}\sin\theta = \text{constant}$$

while the lines of force are described by

$$\text{Im}\,\Phi(z) = \frac{2RV}{r}\cos\theta = \text{constant}.$$

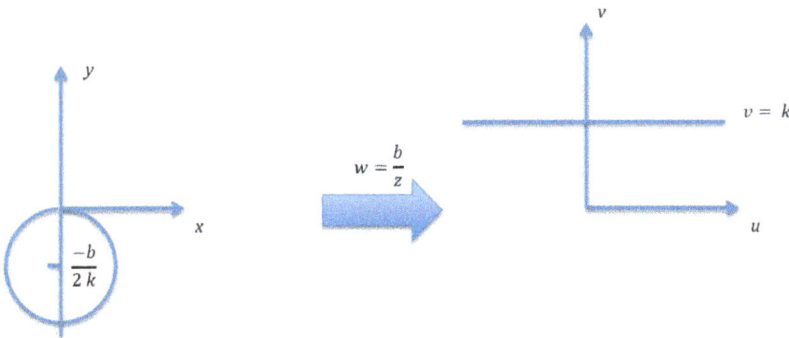

Figure 1.18: The function $w = b/z$ maps the circle centered at $y = \frac{-b}{2k}$ and passing through the origin onto the straight line $v = k$.

2 Cauchy's Theorem

2.1 Complex Integration

We shall now confront the problem of defining an integral over a single complex variable z when the z-plane is, in fact, a **two-dimensional** continuum. There are infinitely many ways of integrating between two values of z and, at this juncture, we have no way of knowing whether they will or should yield the same number. The solution, suggested by the correspondence between complex numbers and two-dimensional vectors, is to define an integral along a particular path or **contour** joining the two points in question.

We shall require that all contours be simple curves. Thus, a contour joining the points a and b, and consisting of k smooth arcs, can be specified by two piecewise smooth real functions $x(t)$, $y(t)$ of the real variable t such that

$$z = z(t) \equiv x(t) + iy(t), \quad t_0 \leqslant t \leqslant t_k \tag{2.1.1}$$

with $z(t_0) = a$ and $z(t_k) = b$. So defined, a contour is rectifiable with a length

$$L = \sum_{i=1}^{k} \int_{t_{i-1}}^{t_i} \left[\left(\frac{dx}{dt} \right)^2 + \left(\frac{dy}{dt} \right)^2 \right]^{1/2} dt \tag{2.1.2}$$

where the sum is over the contour's constituent arcs.

Definition: Let C denote a contour with end-points a and b as shown in Figure 2.1 Sub-divide C into n segments by introducing the $n+1$ points $a = z_0, z_1, z_2, \ldots, z_n = b$. Then, introduce an additional set of points $\zeta_1, \zeta_2, \ldots, \zeta_n$ taken along C in such a way that ζ_j lies between z_{j-1} and z_j. We now form the sum

$$I_n = \sum_{j=1}^{n} f(\zeta_j)(z_j - z_{j-1}) \tag{2.1.3}$$

where $f(z)$ is the function to be integrated. If this sum approaches a limit I as $n \to \infty$ in such a way that

$$|z_j - z_{j-1}| \to 0 \tag{2.1.4}$$

for all j, and if this limit is independent of the manner in which we have chosen the points z_j and ζ_j, then I is said to be the **contour integral** of $f(z)$ along C and is written as

$$I \equiv \int_C f(z)dz. \tag{2.1.5}$$

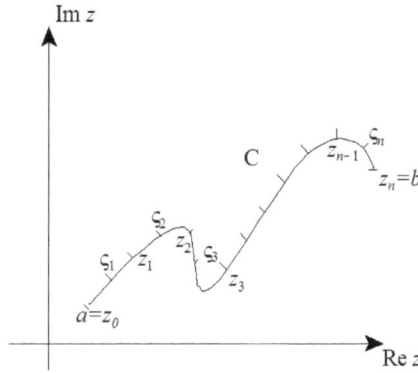

Figure 2.1: A contour of integration C joining the points a and b.

Separating $f(z)$ and z into their real and imaginary parts, we can rewrite the sum in (2.1.3) as

$$I_n = \sum_{j=1}^{n} [u(\xi_j, \eta_j)(x_j - x_{j-1}) - v(\xi_j, \eta_j)(y_j - y_{j-1}) + i[v(\xi_j, \eta_j)(x_j - x_{j-1}) + u(\xi_j, \eta_j)(y_j - y_{j-1})]$$

$$(2.1.6)$$

where we have set $\zeta_j = \xi_j + i\eta_j$. The limiting procedure defined by (2.1.4) implies that $|x_j - x_{j-1}| \to 0$ and $|y_j - y_{j-1}| \to 0$ for all j. Thus, (2.1.6) tells us that the integral I can be expressed in terms of real line integrals:

$$I = \int_C (u\,dx - v\,dy) + i \int_C (v\,dx + u\,dy). \qquad (2.1.7)$$

This, in turn, can be transformed into a real definite (Riemann) integral with respect to the parameter t. Using the parameterization (2.1.1) we can write (2.1.7) as

$$I = \int_{t_0}^{t_k} \left(u\frac{dx}{dt} - v\frac{dy}{dt} \right) dt + i \int_{t_0}^{t_k} \left(v\frac{dx}{dt} + u\frac{dy}{dt} \right) dt. \qquad (2.1.8)$$

Since $\frac{dx}{dt} + i\frac{dy}{dt} = \frac{dz}{dt}$, we may also write this as

$$I = \int_{t_0}^{t_k} (u + iv)\frac{dz}{dt}\,dt = \int_{t_0}^{t_k} f(z(t))\frac{dz(t)}{dt}\,dt. \qquad (2.1.9)$$

This allows us to use the theory of real integral calculus to determine the conditions under which I will exist. Since C is a simple curve it is sufficient to demand that $f(z)$ be continuous on C. This ensures that the points at which the constituent arcs of

C join, which are points of discontinuity for the derivatives $\frac{dx}{dt}$ and $\frac{dy}{dt}$, will cause no trouble.

There are a number of properties of the contour integral that follow from (2.1.9) by virtue of the corresponding properties of real definite integrals. For example,
(i) the integral is linear with respect to the integrand,

$$\int_C [\alpha f_1(z) + \beta f_2(z)]dz = \alpha \int_C f_1(z)dz + \beta \int_C f_2(z)dz, \qquad (2.1.10)$$

(ii) the integral is additive with respect to the contour,

$$\int_{C_1+C_2} f(z)dz = \int_{C_1} f(z)dz + \int_{C_2} f(z)dz, \qquad (2.1.11)$$

where $C_1 + C_2$ denotes the simple curve consisting of C_1 followed by C_2,
(iii) reversing the orientation of the path replaces the integral by its negative,

$$\int_{C(a\to b)} f(z)dz = - \int_{C(b\to a)} f(z)dz, \qquad (2.1.12)$$

(iv) the following inequalities hold,

$$\left| \int_C f(z) \right| \le \int_C |f(z)| \, |dz| \le \max |f(z)| \cdot L(C), \qquad (2.1.13)$$

where $L(C)$ is the length of C.

This last result is known as the **Darboux Inequality** and will prove to be very useful. To derive it one invokes the generalized triangle inequality,

$$|I_n| \le \sum_{j=1}^{n} |f(\zeta_j)| \cdot |z_j - z_{j\,1}| \le \max |f(z)| \cdot \sum_{j=1}^{n} |z_j - z_{j-1}|$$

plus the definition of $L(C)$ given in equation (2.1.2). Then, on taking the limit $n \to \infty$, one immediately obtains (2.1.13).

Examples: Before presenting general theorems on the integration of functions of a complex variable, we shall work out a few examples using the rather limited set of tools currently at our disposal. We start with the integral

$$I = \int_C \cos z \, dz$$

which we wish to evaluate for two contours possessing the same end-points, $z = 0$ and $z = 1 + i$. The particular contours we shall use are shown in Figure 2.2.

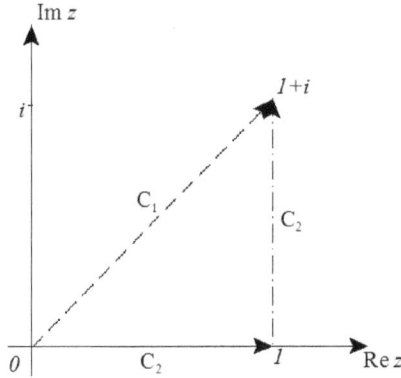

Figure 2.2: Two contours of integration.

Since $\cos z = \cosh y \cos x - i \sinh y \sin x$, equation (2.1.7) yields

$$I = \int_C [\cosh y \cos x dx + \sinh y \sin x dy] + i \int_C [-\sinh y \sin x dx + \cosh y \cos x dy].$$

Along the contour C_1 $x = y$. Therefore,

$$I_1 = \int_{C_1} \cos z dz = (1 + i) \int_0^1 \cosh x \cos x dx + (1 - i) \int_0^1 \sinh x \sin x dx.$$

Integrating by parts, we have

$$\int_0^1 \cosh x \cos x dx = \cosh x \sin x \Big|_0^1 - \int_0^1 \sinh x \sin x dx$$

$$\int_0^1 \sinh x \sin x dx = -\sinh x \sin x \Big|_0^1 + \int_0^1 \cosh x \cos x dx.$$

Thus,

$$I_1 = \frac{1}{2}(1 + i)[\cosh x \sin x + \sinh x \cos x] \Big|_0^1 - \frac{1}{2}(1 - i)[\sinh x \cos x + \cosh x \sin x] \Big|_0^1$$

$$= \cosh 1 \sin 1 + i \sinh 1 \cos 1 = \sin(1 + i).$$

The contour C_2 consists of two smooth arcs. Along the first of them $y = 0$ and $dy = 0$ while along the second, $x = 1$ and $dx = 0$. Thus,

$$I_2 = \int_{C_2} \cos z dz = \int_0^1 \cos x dx + \int_0^1 \sinh y \sin 1 dy + i \int_0^1 \cosh y \cos 1 dy.$$

Integrating, we have

$$I_2 = \sin x \Big|_0^1 + \cosh y \Big|_0^1 \sin 1 + i \sinh y \Big|_0^1 \cos 1$$

$$= \cosh 1 \sin 1 + i \sinh 1 \cos 1 = \sin(1 + i).$$

We see that I has the same value for the two paths followed. In fact, we have obtained the same value that results from using indefinite integration as would be suggested by the rules of real variable calculus:

$$I_1 = I_2 = \int_0^{1+i} \cos z \, dz = \sin z \Big|_0^{1+i} = \sin(1 + i).$$

Both of these observations, the path-independence of the integral and the applicability of the "fundamental theorem of calculus", are explained by **Cauchy's Theorem**. As we shall see, they must obtain in any simply connected domain in which the integrand is holomorphic and, in the present case, that means everywhere in the finite plane.

Notice that the path-independence of the integral of $\cos z$ implies that if we had integrated around a closed contour, $C_1 + (-C_2)$ for example, the result would have been zero. This should be contrasted with our next example which involves the function z^*. Choosing the unit circle taken in the counterclockwise direction as the contour, we can set $z = e^{i\theta}$, $-\pi \leqslant \theta < \pi$. This means the contour integral is

$$I = \int_C z^* \, dz = \int_{-\pi}^{\pi} e^{-i\theta} i e^{i\theta} \, d\theta = 2\pi i.$$

Since $z^* = \frac{1}{z}$ on this particular contour, we have also evaluated

$$\int_C \frac{1}{z} dz = 2\pi i.$$

As we shall see, Cauchy's Theorem provides an explanation of this result too. Because z^* is not an analytic function, its integrals are path dependent everywhere in the complex plane. Consequently, its integral around a closed contour is non-zero and, in fact, each closed contour yields a different non-zero value. On the other hand, $\frac{1}{z}$ is an analytic function with a single singularity at $z = 0$. Its closed contour integrals only admit two values $2\pi i$ if the contour encloses $z = 0$; 0 if it does not.

As a final example we shall evaluate

$$I = \int_C (z - z_0)^n \, dz, \quad n = 0, \pm 1, \pm 2, \ldots$$

with C being a circle taken in the counterclockwise direction about $z = z_0$ with an arbitrary radius r. On the contour we can set $z = z_0 + re^{i\theta}$, $-\pi \leqslant \theta < \pi$. Thus,

$$I = \int_{-\pi}^{\pi} r^n e^{in\theta} ire^{i\theta} d\theta = ir^{n+1} \int_{-\pi}^{\pi} e^{i(n+1)\theta} d\theta,$$

or,

$$I = \begin{cases} 2\pi i, & n = -1 \\ 0, & n = 0, 1, \pm 2, \pm 3, \ldots \end{cases} \tag{2.1.14}$$

If it is feasible to use indefinite integration to evaluate contour integrals of analytic functions, we should be able to reproduce this result for I by applying the formula

$$\int_C f(z)dz = F(b) - F(a), \tag{2.1.15}$$

where $F(z)$ is an antiderivative (indefinite integral) of $f(z)$, $\frac{dF(z)}{dz} = f(z)$, and a and b are the end-points of the contour C. An antiderivative of $f(z) = (z - z_0)^n$ is

$$F(z) = \begin{cases} \ln(z - z_0), & n = -1 \\ \frac{1}{n+1}(z - z_0)^{n+1}, & n \neq -1 \end{cases} \tag{2.1.16}$$

Therefore, integrating $f(z)$ around a closed contour for the case $n \neq -1$, we obtain the value zero because the antiderivative is single valued and the end-points coincide. This value arises without regard to whether z_0 is contained within the contour C. Thus, for **any** closed contour C we have

$$\int_C (z - z_0)^n dz = 0, \quad n = 0, 1, \pm 2, \pm 3, \cdots$$

For $n = -1$, the antiderivative is multivalued with a branch point at z_0. Therefore, we must distinguish between contours that enclose z_0 and those that do not. A closed contour that encircles z_0 must cross from one sheet of the Riemann surface for $\ln(z - z_0)$ to an adjacent one. Since we are integrating in a counterclockwise direction, this means that the argument of z will increase by 2π as we proceed from the end-point $z = a$ on the initial sheet to $z = b$ on the sheet above. Therefore, rather than having $a = b$, we now have $|a| = |b|$ and $\arg(b) - \arg(a) = 2\pi$ and so, $F(b) - F(a) = 2\pi i$ in (2.1.15). Notice that this result is quite independent of the detailed nature of the contour; it need only be closed and contain z_0. If, on the other hand, z_0 is not contained within C, the contour must return to its starting point on the initial sheet, no matter how the branch cut from $z = z_0$ to $z = \infty$ is chosen. Since the end-points once more coincide, the value of the integral is zero. Thus, in summary, we have for **any** closed contour C

$$\int_C \frac{1}{z - z_0} dz = \begin{cases} 2\pi i, & z_0 \text{ inside C} \\ 0, & z_0 \text{ outside C} \end{cases}$$

The ease with which we have generalized the results in (2.1.14) by means of indefinite integration underlines the importance of finding out precisely when we can or cannot make use of it. The answer is provided by Cauchy's Theorem which, in a very real sense, is the basis of the entire theory of analytic functions.

2.2 Cauchy's Theorem

2.2.1 Statement and Proof

Cauchy's Theorem: If $f(z)$ is holomorphic in a simply connected, bounded domain D, then

$$\int_C f(z)dz = 0 \tag{2.2.1}$$

for every simple closed path C in D.

We shall run through an over-simplified proof that makes the theorem plausible as well as renews the correspondence between complex numbers and vectors in a plane. It is implicitly based on Green's Theorem in two dimensions,

$$\int_C (Pdx + Qdy) = \int\int_S \left(\frac{\partial Q}{\partial x} - \frac{\partial P}{\partial y} \right) dxdy,$$

or equivalently, on Stokes' Theorem,

$$\int_C A \cdot dl = \int\int_S (\nabla \times A) \cdot dS,$$

where S is the surface bounded by C.

From equation (2.1.7) we have

$$\int_C f(z)dz = \int_C (udx - vdy) + i \int_C (udy + vdx). \tag{2.2.2}$$

Let us consider the first term on the right hand side and divide the area S into strips parallel to the imaginary axis as shown in Figure 2.3. It then follows that

$$\int_C u[x, y(x)]dx = \int_a^b u[x, y_1(x)]dx + \int_a^b u[x, y_2(x)]dx = - \int_a^b (u[x, y_2(x)] - u[x, y_1(x)])dx.$$

Thus,

$$\int_C u[x, y(x)]dx = - \int_a^b \int_{y_1(x)}^{y_2(x)} \frac{\partial u}{\partial y} dydx = - \int\int_S \frac{\partial u}{\partial y} dxdy. \tag{2.2.3}$$

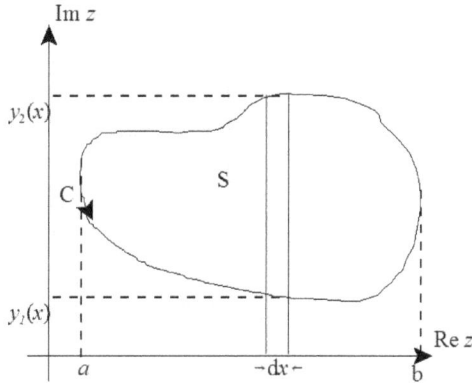

Figure 2.3: A closed contour C containing surface area S which is divided into infinitesimal strips that parallel the imaginary axis.

The other terms on the right hand side of (2.2.2) may be similarly transformed to give

$$\int_C f(z)dz = \int\int_S \left[\left(-\frac{\partial u}{\partial y} - \frac{\partial v}{\partial x}\right) + i\left(\frac{\partial u}{\partial x} - \frac{\partial v}{\partial y}\right)\right] dxdy. \qquad (2.2.4)$$

However, the integrand of the surface integral vanishes by virtue of the Cauchy-Riemann equations. Therefore,

$$\int_C f(z)dz = 0.$$

The steps taken in the derivation of (2.2.3) are valid only if $\frac{\partial u}{\partial y}$ is a continuous function of x and y. Hence, this particular proof requires an assumption that is not given in the statement of the theorem; namely, that $f'(z)$ is continuous throughout the domain D. However, Cauchy's Theorem will enable us to establish that a holomorphic function possesses continuous derivatives of all orders which suggests that such an assumption must be superfluous. This was confirmed first by E. Goursat whose proof of Cauchy's Theorem can be found in, for example, Copson's Theory of Functions of a Complex Variable.

To continue with the story of Augustin Louis Cauchy, he was appointed to the French Academy in 1816 after it had been purged following the restoration of the French monarchy. At the same time he was made professor at the Polytechnic and his lectures there on algebraic analysis, calculus and the theory of curves were published as text books. However, the revolution of 1830 meant that he had to go into exile. He returned to France in 1837 but because he refused to take a loyalty oath he was denied a teaching appointment. It was not until 1851 and by special dispensation from the Emperor that he was permitted to occupy a chair in mathematics without taking the oath of allegiance. During this period his productivity was extraordinary; from 1830 to his death in 1857, he published over 600 original papers and about 150 reports. Included in this prodigious body of work was the foundation of complex analysis.

The most immediate consequences of Cauchy's Theorem are sufficient to explain the observations we made while evaluating the example integrals in the last Section. We shall conclude the present Section with their description.

2.2.2 Path Independence

Let C_1 and C_2 be any two curves having the same end-points and lying in a simply connected domain in which the function $f(z)$ is holomorphic. We denote by $(-C_2)$ the contour obtained from C_2 by reversing the direction of integration. Then, since $C_1 + (-C_2)$ is a closed contour, we may apply Cauchy's Theorem together with (2.1.11) and (2.1.12) to obtain

$$0 = \int_{C_1+(-C_2)} f(z)dz = \int_{C_1} f(z)dz + \int_{(-C_2)} f(z)dz = \int_{C_1} f(z)dz - \int_{C_2} f(z)dz.$$

Hence,

$$\int_{C_1} f(z)dz = \int_{C_2} f(z)dz. \tag{2.2.5}$$

Since C_1 and C_2 are arbitrary, we have established that the integral of $f(z)$ between two points a and b is path independent in any simply connected domain that includes a and b and excludes the singularities of $f(z)$. Under such conditions we may denote the integral by

$$\int_a^b f(z)dz$$

since the specification of a contour is irrelevant. Thus, in our example involving $\cos z$ we need only have written $\int_0^{1+i} \cos z\, dz$ for the integrals in question since any integral of this function is path independent everywhere in the finite plane.

Path independence can also be established for integrals around **closed** contours in **multiply**-connected domains. Suppose that we have a function $f(z)$ that is holomorphic in a doubly connected domain like that shown in Figure 2.4. Nothing is assumed about the behaviour of $f(z)$ in the area interior to the inner boundary of the domain but presumably it is singular at one or more points of this region. We introduce two contours C_1 and C_2 connected by a narrow tube consisting of the straight lines L_1 and L_2. The exact location of these curves is arbitrary so long as they all lie fully within the outer boundary and C_1 and C_2 both encircle the inner boundary of the domain. By construction, $f(z)$ is holomorphic at all points **within and on** the simple closed curve

$C = C_1 + L_1 + C_2 + L_2$. Thus, C is contained in a **simply** connected domain in which $f(z)$ is everywhere holomorphic and we can apply Cauchy's Theorem to obtain

$$\int_{C_1} + \int_{L_1} + \int_{C_2} + \int_{L_2} f(z)dz = 0. \tag{2.2.6}$$

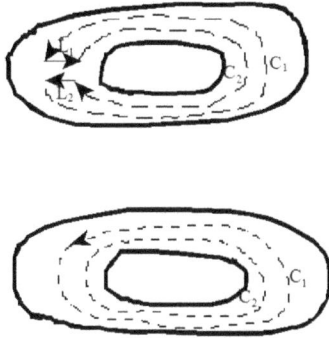

Figure 2.4: (upper panel) C_1 and C_2 are connected by L_1 and L_2 to form a single closed contour; (bottom panel) L_1 and L_2 are removed leaving C_1 and C_2 as closed contours encircling the inner boundary of the domain

However, the contributions from L_1 and L_2 cancel if we now let the separation between them tend to zero. Therefore, in this limit, (2.2.6) becomes

$$\int_{C_1} + \int_{C_2} f(z)dz = 0,$$

with C_1 and C_2 being closed contours traversed in opposite directions. Reversing the direction of one of them we finally obtain

$$\int_{C_1} f(z)dz = \int_{C_2} f(z)dz. \tag{2.2.7}$$

This is an important result and one that was anticipated in our example involving the function $f(z) = (z - z_0)^{-1}$. It shows that an integral has the **same** value for **all** closed contours that contain the **same** singularities of its integrand. This means that any contour can be arbitrarily deformed so long as we do not cross a singularity of the integrand without changing the value of the integral. Hence, we need not specify a contour very precisely, only its relationship to the singularities of the function being integrated.

A generalization of the argument leading to equation (2.2.7) shows that if our function $f(z)$ has an n-fold connected domain of holomorphy then we can write

$$\int_C f(z)dz = \int_{C_1} f(z)dz + \int_{C_2} f(z)dz + \ldots + \int_{C_{n-1}} f(z)dz \qquad (2.2.8)$$

where each of the $n-1$ "holes" in the domain is enclosed by one of the contours C_j, $j = 1, 2, \ldots n-1$, and all are enclosed by the contour C. The integrals in (2.2.8) are all taken in the same (counterclockwise) direction. Figure 2.5 shows an appropriate set of contours for a four-fold connected domain.

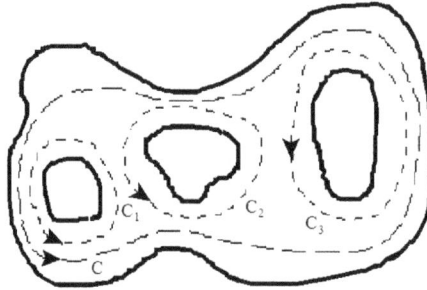

Figure 2.5: A four-fold connected domain of holomorphy for the function $f(z)$; $\int_C f(z)dz = \int_C f(z)dz + \int_{C2} f(z)dz + \int_{C3} f(z)dz$.

2.2.3 The Fundamental Theorem of Calculus

Let $f(z)$ be a function which is holomorphic in some simply connected domain D. It then follows that the integral

$$F(z) = \int_{z_0}^{z} f(\zeta)d\zeta, \text{ for fixed } z_0 \text{ in } D, \qquad (2.2.9)$$

defines for all z in D a unique function which is independent of the path of integration from z_0 to z. Since D is a domain, there must exist a neighbourhood N of z that lies entirely within it. Let $z + \Delta z$ be a point in N and let us form the quotient

$$\frac{F(z+\Delta z) - F(z)}{\Delta z} = \frac{1}{\Delta z}\int_{z_0}^{z} f(\zeta)d\zeta - \frac{1}{\Delta z}\int_{z_0}^{z} f(\zeta)d\zeta = \frac{1}{\Delta z}\int_{z}^{z+\Delta z} f(\zeta)d\zeta$$

$$= \frac{f(z)}{\Delta z}\int_{z}^{z+\Delta z} d\zeta + \frac{1}{\Delta z}\int_{z}^{z+\Delta z} [f(\zeta) - f(z)]d\zeta. \qquad (2.2.10)$$

By construction, the straight line joining z to $z + \Delta z$ must lie within D and so can be used as a contour for the integrals in (2.2.10). The first integral on the right hand side of the final equality is then easily evaluated by parameterizing this straight line and yields

$$\int_{z}^{z+\Delta z} d\zeta = \Delta z.$$

Thus, (2.2.10) can be rewritten as

$$\frac{F(z + \Delta z) - F(z)}{\Delta z} = f(z) + \frac{1}{\Delta z} \int_{z}^{z+\Delta z} [f(\zeta) - f(z)]d\zeta. \tag{2.2.11}$$

However, using the same straight line contour and the Darboux inequality we have

$$\left| \frac{1}{\Delta z} \int_{z}^{z+\Delta z} [f(\zeta) - f(z)]d\zeta \right| \leq \max |f(\zeta) - f(z)|$$

which, because $f(z)$ is continuous in D, tends to zero as Δz approaches zero in any direction. Therefore,

$$\frac{dF}{dz} \equiv \lim_{\Delta z \to 0} \frac{F(z + \Delta z) - F(z)}{\Delta z} = f(z). \tag{2.2.12}$$

In other words,

$$F(z) = \int_{z_0}^{z} f(\zeta)d\zeta, \quad z_0 \text{ and } z \text{ in } D,$$

is an **antiderivative** of $f(z)$ and is itself holomorphic in D. Hence, for **all** paths in D joining any two points a and b of D, we can write

$$\int_{a}^{b} f(z)dz = \int_{z_0}^{b} f(z)dz - \int_{z_0}^{a} f(z)dz = F(b) - F(a). \tag{2.2.13}$$

Thus, we have finally identified the circumstances under which we can employ **indefinite integration** to evaluate integrals. This is a practical boon of some importance since recognizing functions as being the derivatives of other functions is the only generally applicable integration technique one ever learns.

2.3 Further Consequences of Cauchy's Theorem

2.3.1 Cauchy's Integral

The most important of all the consequences of Cauchy's Theorem is an integral representation that is basic to the further development of the theory of analytic functions. It also has a number of physical applications.

Theorem: Let $f(z)$ be holomorphic in a simply connected domain D and let C be any simple closed curve within D. If z is a point within C, then

$$f(z) = \frac{1}{2\pi i} \int_C \frac{f(\zeta)}{\zeta - z} d\zeta, \tag{2.3.1}$$

the integration being taken in the counterclockwise direction.

Proof: Using one of the manipulations that helped us prove the fundamental theorem of calculus we rewrite the integral in (2.3.1) in the form

$$\int_C \frac{f(\zeta)}{\zeta - z} d\zeta = f(z) \int_C \frac{d\zeta}{\zeta - z} + \int_C \frac{f(\zeta) - f(z)}{\zeta - z} d\zeta. \tag{2.3.2}$$

We have already evaluated the first integral on the right hand side of (2.3.2). For any closed contour encircling the point $\zeta = z$,

$$\int_C \frac{d\zeta}{\zeta - z} = 2\pi i. \tag{2.3.3}$$

Since the integrand of the second integral is holomorphic everywhere within C, except possibly at $\zeta = z$, we can use equation (2.2.7) to replace C by a small circle γ about z with radius r sufficiently small that it lies entirely within C. Thus,

$$\int_C \frac{f(\zeta) - f(z)}{\zeta - z} d\zeta = \int_\gamma \frac{f(\zeta) - f(z)}{\zeta - z} d\zeta = i \int_{-\pi}^{\pi} [f(z + re^{i\theta}) - f(z)] d\theta \tag{2.3.4}$$

where we have used the fact that on γ, $\zeta = z + re^{i\theta}$, $-\pi \leqslant \theta < \pi$.

Since $f(z)$ is continuous, we have $|f(\zeta) - f(z)| < \varepsilon$ whenever $|\zeta - z| < \delta(\varepsilon)$. If we choose the radius r to be less than $\delta(\varepsilon)$, then Darboux's inequality applied to (2.3.4) yields

$$\left| \int_\gamma \frac{f(\zeta) - f(z)}{\zeta - z} d\zeta \right| \leq 2\pi\varepsilon.$$

Hence, by taking r small enough, this modulus can be made smaller than any preassigned number. On the other hand, the value of the integral must be independent of r.

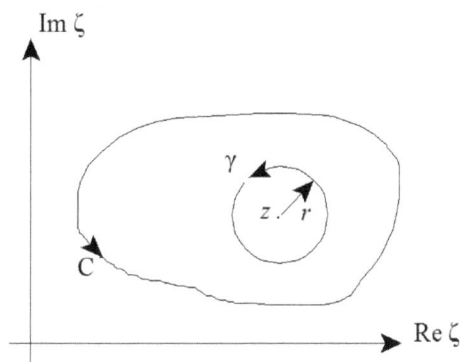

Figure 2.6: The contours used in equation (2.3.4).

Therefore,

$$\int_\gamma \frac{f(\zeta) - f(z)}{\zeta - z} d\zeta = 0$$

which, together with equations (2.3.3) and (2.3.4), dictates that (2.3.2) read

$$\int_C \frac{f(\zeta)}{\zeta - z} d\zeta = 2\pi i f(z)$$

as required.

As well as being one of the most useful results in mathematical physics, Cauchy's Integral is one of the most remarkable. If a function is holomorphic within and on a simple closed curve C, its value at **every** point within C is determined solely by its values on that curve. By means of a simple extension, we see that this implies that a function must be completely determined by its values at the boundary of its domain of holomorphy and hence, by its singularities. We first encountered this idea in Section 1.6 and will return to it often throughout the remainder of this Chapter.

A purely practical use of Cauchy's Integral is in the evaluation of closed contour integrals. **Examples:** Consider the function $f(z) = \frac{z^2+1}{z^2-1}$ and suppose that we wish to integrate it around the circles $|z-1| = 3/2$, $|z| = 3/2$, $|z+1| = 3/2$. As a first step, we use partial fractions to write $f(z)$ as

$$f(z) = \frac{z^2 + 1}{2} \left[\frac{1}{z-1} - \frac{1}{z+1} \right].$$

Next, we note that $z^2 + 1$ is holomorphic on and within all three of the proposed contours. Thus, we may use (2.3.1) to evaluate the integrals

$$I_+ = \int_C \frac{z^2 + 1}{z - 1} dz \quad I_- = \int_C \frac{z^2 + 1}{z + 1} dz$$

for each C.

In the case of $|z - 1| = 3/2$, $z = 1$ is included, and $z = -1$ excluded by the contour. Thus, $I_+ = 2\pi i[z^2 + 1]_{z=1} = 4\pi i$ by Cauchy's Integral, $I_- = 0$ by Cauchy's Theorem, and so

$$\int_{|z-1|=3/2} \frac{z^2 + 1}{z^2 - 1} dz = \frac{1}{2} 4\pi i = 2\pi i.$$

In the case of $|z| = 3/2$, $z = \pm 1$ are both enclosed by the contour. Thus, $I_\pm = 2\pi i[z^2 + 1]_{z=\pm 1} = 4\pi i$, and so

$$\int_{|z|=3/2} \frac{z^2 + 1}{z^2 - 1} dz = 0.$$

With $|z + 1| = 3/2$, $z = 1$ is now excluded and $z = -1$ included. Thus, $I_+ = 0$, $I_- = 4\pi i$ and

$$\int_{|z+1|=3/2} \frac{z^2 + 1}{z^2 - 1} dz = -2\pi i.$$

Since Cauchy's Integral plays such a central role in complex analysis, we shall adopt the convention that every integration along a closed contour will be taken in the counterclockwise direction or, such that the interior of the contour is always on the left hand side.

2.3.2 Cauchy's Derivative Formula

Our next theorem is almost as remarkable as the last and every bit as foreign to our experience with functions of a real variable. It asserts that the existence of the derivative of a function in some simply connected domain necessarily implies the existence of derivatives of **all** orders in that domain.

Theorem: If $f(z)$ is holomorphic in a simply connected domain D, then all of its derivatives exist and are themselves holomorphic in D. Moreover, if z is any point in D, then

$$f'(z) = \frac{1}{2\pi i} \int_C \frac{f(\zeta)}{(\zeta - z)^2} d\zeta$$

$$f''(z) = \frac{2!}{2\pi i} \int_C \frac{f(\zeta)}{(\zeta - z)^3} d\zeta$$

$$\vdots$$

$$f^{(n)}(z) = \frac{n!}{2\pi i} \int_C \frac{f(\zeta)}{(\zeta - z)^{n+1}} d\zeta$$

$$\vdots$$

$$(2.3.5)$$

where C is any simple closed curve in D which encloses z.

Proof: By definition,

$$f'(z) = \lim_{\Delta z \to 0} \frac{f(z + \Delta z) - f(z)}{\Delta z}$$

and so, using Cauchy's Integral (2.3.1), we have

$$f'(z) = \lim_{\Delta z \to 0} \frac{1}{2\pi i} \int_C \left[\frac{f(\zeta)}{\zeta - z - \Delta z} - \frac{f(\zeta)}{\zeta - z} \right] \frac{d\zeta}{\Delta z}$$

$$= \lim_{\Delta z \to 0} \frac{1}{2\pi i} \int_C \frac{f(\zeta)}{(\zeta - z - \Delta z)(\zeta - z)} d\zeta.$$

Hence,

$$f'(z) - \frac{1}{2\pi i} \int_C \frac{f(\zeta)}{(\zeta - z)^2} d\zeta = \lim_{\Delta z \to 0} \frac{1}{2\pi i} \int_C f(\zeta) \left[\frac{1}{(\zeta - z - \Delta z)(\zeta - z)} \right] d\zeta$$

$$= \lim_{\Delta z \to 0} \frac{1}{2\pi i} \Delta z \int_C \frac{f(\zeta)}{(\zeta - z - \Delta z)(\zeta - z)^2} d\zeta.$$

The modulus of the integrand on the left hand side of this expression is

$$\left| \frac{f(\zeta)}{(\zeta - z - \Delta z)(\zeta - z)^2} \right| = \frac{|f(\zeta)|}{|\zeta - z - \Delta z| \, |\zeta - z|^2} \leq \frac{|f(\zeta)|}{(|\zeta - z| - |\Delta z|) \, |\zeta - z|^2}.$$

Replacing $|\zeta - z|$ by its minimum value m, and $|f(\zeta)|$ by its maximum value M for ζ on C, we can apply the Darboux inequality to obtain

$$\left| f'(z) - \frac{1}{2\pi i} \int_C \frac{f(\zeta)}{(\zeta - z)^2} d\zeta \right| \leq \frac{1}{2\pi} \lim_{\Delta z \to 0} |\Delta z| \frac{ML}{(m - |\Delta z|)m^2} = 0$$

where L is the length of the contour. Thus, we have proved

$$f'(z) = \frac{1}{2\pi i} \int_C \frac{f(\zeta)}{(\zeta - z)^2} d\zeta$$

and, by using induction, one can readily establish that, in general,

$$f^{(n)}(z) = \frac{n!}{2\pi i} \int_C \frac{f(\zeta)}{(\zeta - z)^{n+1}} d\zeta.$$

The fact that the derivative of a holomorphic function is itself holomorphic is all that one needs to prove a converse of Cauchy's Theorem. It is called **Morera's Theorem**.

Theorem: If $f(z)$ is continuous in some simply connected domain D and if the integral

$$\int_C f(z)dz$$

vanishes for any closed contour C in D, then $f(z)$ is holomorphic in D.

Proof: The vanishing of an integral along any closed path within a simply connected domain is sufficient to establish the path independence of an integral between any two points in the domain. Thus, as in the Fundamental Theorem of Calculus,

$$F(z) = \int_{z_0}^{z} f(\zeta)d\zeta, \qquad \text{for fixed } z_0 \text{ in } D,$$

defines a unique function for all z in D. Moreover, since $f(z)$ is continuous throughout D, the proof of the Fundamental Theorem applies here as well since, of the various properties possessed by holomorphic functions, only those of continuity and path independent integration were used in the proof of the Fundamental Theorem. Thus, $F(z)$ is an anti-derivative of $f(z)$, $\frac{dF(z)}{dz} = f(z)$, for all z in D. This means that $F(z)$ is holomorphic throughout D and, by the preceding theorem, so is its first derivative $f(z)$.

2.3.3 The Maximum Modulus Principle

Our next theorem further illustrates the surprising properties possessed by analytic functions. Although not as dramatic in appearance as are its predecessors, this theorem has a number of important applications a particular example of which can be found in the derivation of the Method of Steepest Descents in Section 6.3.

Theorem: The modulus of an analytic function $f(z)$ cannot have a local maximum within the domain of holomorphy of the function.

Proof: Let z_0 be an arbitrary point in the function's domain of holomorphy. By definition, there must exist a neighbourhood of z_0 which also lies entirely within that domain. Let C denote a circle, of radius r and centre at $z = z_0$, lying within that neighbourhood. Then, invoking Cauchy's Integral, we have

$$f(z_0) = \frac{1}{2\pi i}\int_C \frac{f(z)}{z - z_0}dz.$$

Using the Darboux Inequality, this implies that

$$|f(z_0)| \leq \frac{1}{2\pi}max\left|\frac{f(z)}{z - z_0}\right| \cdot 2\pi r, \qquad \text{for } z \text{ on } C.$$

Since $|z - z_0| = r$, this simplifies to

$$|f(z_0)| \leqslant \max|f(z)|, \qquad \text{for } z \text{ on } C.$$

Thus, there exists at least one point on C such that $|f(z)| \geqslant |f(z_0)|$. Since we may choose r to be arbitrarily small, this means that in any neighbourhood of z_0, no matter how small, there always exists at least one point at which

$$|f(z)| \geqslant |f(z_0)|. \tag{2.3.6}$$

Hence, $|f(z)|$ cannot have a local maximum at z_0.

A number of corollaries follow almost immediately. By applying the theorem to $\exp[f(z)]$ and $\exp[if(z)]$, we find that neither

$$\left| e^{f(z)} \right| = e^{\operatorname{Re} f(z)} \quad \text{nor} \quad \left| e^{if(z)} \right| = e^{\operatorname{Im} f(z)}$$

can have a local maximum in the domain of holomorphy of $f(z)$. Since the real exponential function is monotonic, this implies that the real and imaginary parts of an analytic function (which is to say, any harmonic function of two real variables) cannot have local maxima in the function's domain of holomorphy (harmonicity).

Similarly, applying the theorem to $\frac{1}{f(z)}$, we find that $|f(z)|$, as well as $\operatorname{Re} f(z)$ and $\operatorname{Im} f(z)$, cannot have a local **minimum** in the domain of holomorphy of $f(z)$, except at points $z = z_0$ where $f(z_0) = 0$.

2.3.4 The Cauchy-Liouville Theorem

We conclude this Section with a theorem that places a limit on how well-behaved an analytic function can be.

The theorem is named after Joseph Liouville as well as the ubiquitous Augustin-Louis Cauchy. Liouville (1809-82) was born in St. Omer, France and studied engineering at the Ecole Polytechnique and the Ecole des Ponts et Chausees. However, it was as a mathematician and mathematical physicist that he made his mark. An individual of exceptional intellectual breadth, he made significant contributions in theoretical dynamics and celestial mechanics, in the theory of differential equations, in algebra and algebraic function theory and in number theory where he introduced new methods of investigating transcendental numbers. Not surprisingly for someone with such extensive knowledge, he founded the Journal de Mathematiques and edited it for almost 40 years. It continues to be one of the leading French mathematical journals.

Theorem: A bounded entire function must be a constant.

Proof: Using the Derivative Formula (2.3.5), we set

$$f'(z) = \frac{1}{2\pi i} \int_C \frac{f(\zeta)}{(\zeta - z)^2} d\zeta$$

where, because $f(z)$ is entire, C may be chosen to be a very large circle of radius R, centred at z. Darboux's inequality then yields

$$|f'(z)| \leqslant \frac{1}{2\pi} \frac{\max |f(\zeta)|}{R^2} 2\pi R < \frac{K}{R}, \quad \text{for } \zeta \text{ on } C,$$

where K is the assumed upper bound on $|f(z)|$, $|f(z)| < K$ for all values of z. Thus, given any value of z, we can make $|f'(z)|$ smaller than any preassigned number ε simply by taking R sufficiently large. Therefore,

$$f'(z) = 0$$

for all values of z. This in turn implies that $f(z)$ is a constant for, given any two points z_1 and z_2, the Fundamental Theorem of Calculus yields

$$0 = \int_{z_1}^{z_2} \frac{df(\zeta)}{d\zeta} d\zeta = f(z_2) - f(z_1)$$

or, $\quad f(z_2) = f(z_1)$.

Two obvious corollaries of this theorem are that "not-constant" functions that are bounded at infinity must have at least one singularity in the finite plane while those that are entire must be singular at the point at infinity. Thus, the sine and cosine functions are not unbounded by coincidence but rather, by virtue of their holomorphy.

2.4 Power Series Representations of Analytic Functions

2.4.1 Uniform Convergence

Absolute convergence is not sufficient to guarantee the legitimacy of some of the operations that we shall want to perform on power series. Stated in the simplest terms, this failure is due to the rate of convergence of an absolutely convergent series being itself a function of z and, quite conceivably, a very sensitive one. Thus, for example, the series resulting from term by term integration of an absolutely convergent series may not converge at all, let alone to the integral of the sum function.

The problem is best illustrated by considering the geometric series

$$\sum_{m=0}^{\infty} z^m = \frac{1}{1-z} = S(z), \quad |z| < 1,$$

whose circle of (absolute) convergence is $|z| = 1$. The error committed in approximating its sum function by the nth partial sum of the series is

$$R_n(z) = |S(z) - S_n(z)| = \left| \sum_{m=n+1}^{\infty} z^m \right| = \left| \frac{z^{n+1}}{1-z} \right|.$$

This becomes arbitrarily large for z sufficiently close to one. Consequently, the closer we are to $z = 1$, the more terms must be included in the partial sum and hence, the larger n must be, in order that the error $R_n(z)$ be less than some preassigned number.

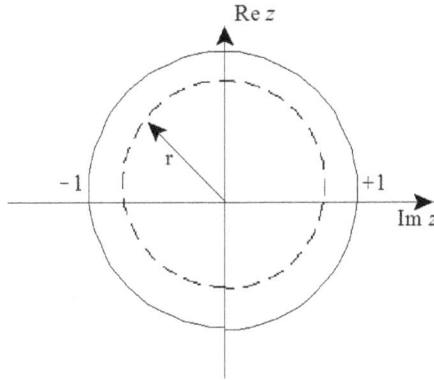

Figure 2.7: The geometric series $\sum_{m=0}^{\infty} z^m$ converges absolutely in the open disc $|z| < 1$ and uniformly in any closed disc $|z| \leqslant r < 1$.

In more mathematical language, if the only restriction on z is that it be confined to $|z| < 1$ then, given any $\varepsilon > 0$, one can**not** find an N that **depends only on** ε such that $R_n(z) < \varepsilon$ for all $n > N(\varepsilon)$. On the other hand, if we impose the further restriction that z be confined to the **closed** disc $|z| \leqslant r < 1$, such an N **can** be found because we have now specified exactly how close one can get to the dangerous point $z = 1$. Indeed, choosing N so that $R_n(r) < \varepsilon$ for all $n > N(\varepsilon)$ we also ensure that $R_n(z) < \varepsilon$ for all $n > N(\varepsilon)$ and **all** z in $|z| \leqslant r$. Thus, the geometric series converges in a z-independent or **uniform** manner in any closed disc with radius less than one.

Definition: Consider the series

$$\sum_{m=0}^{\infty} f_m(z) = f_0(z) + f_1(z) + \ldots$$

Let $F(z)$ be the sum and $S_n(z)$ the nth partial sum of the series. If, given any $\varepsilon > 0$, there exists a number $N(\varepsilon)$, independent of z, such that

$$|F(z) - S_n(z)| < \varepsilon$$

for all $n > N(\varepsilon)$ and all z in some region $\mathbb{R}$ of the complex plane, then the series is **uniformly convergent** in $\mathbb{R}$.

The following theorem generalizes our experience with the geometric series. It is stated without proof.

Theorem: A power series $\sum_{m=0}^{\infty} c_m(z - z_0)^m$, with non-zero radius of convergence R, is uniformly convergent in every closed disc $|z - z_0| \leqslant r$ with $r < R$.

The importance of uniform convergence and hence of the preceding theorem is made manifest throughout this Section. To begin with, consider the generic series of functions

$$\sum_{m=0}^{\infty} f_m(z) = F(z)$$

which we will assume to be uniformly convergent in some region $\mathbb{R}$. Let C be a simple curve of length L that lies entirely in $\mathbb{R}$. Then, provided that all integrals exist, we can show that

$$\int_C F(z)dz \equiv \int_C \sum_{m=0}^{\infty} f_m(z)dz = \sum_{m=0}^{\infty} \int_C f_m(z)dz. \qquad (2.4.1)$$

Setting $S_n(z) = \sum_{m=0}^{n} f_m(z)$, we can rewrite (2.4.1) as

$$\int_C F(z)dz \equiv \int_C \lim_{n\to\infty} S_n(z)dz = \lim_{n\to\infty} \int_C S_n(z)dz. \qquad (2.4.2)$$

In other words, we must prove that the order of the limit and the integration can be reversed. Darboux's Inequality makes the proof almost immediate since it yields

$$\left| \int_C [F(z) - S_n(z)]dz \right| \le \max |F(z) - S_n| \cdot L, \quad \text{for } z \text{ on } C. \qquad (2.4.3)$$

Because, and only because the curve C lies in the region where the series that sums to $F(z)$ is uniformly convergent, we can make the right hand side of (2.4.3) smaller than any preassigned number ε by taking n sufficiently large. Thus, the series on the right and side of (2.4.1) converges to the integral on the left.

Uniform convergence also makes term by term differentiation permissable. Indeed, the next **theorem**, due to **Weierstrass**, goes a step further and, in doing so, supercedes the theorem in Section 1.4 which established that every power series defines a holomorphic function within its circle of convergence.

Theorem: Let $F(z) = \sum_{m=0}^{\infty} f_m(z)$. If each term $f_m(z)$ of the series is holomorphic within a domain D and if the series is uniformly convergent throughout every region $\mathbb{R}$ interior to D, then $F(z)$ is holomorphic within D and

$$\frac{dF(z)}{dz} = \sum_{m=0}^{\infty} \frac{df_m(z)}{dz} \qquad (2.4.4)$$

where the series on the right hand side of (2.4.4) is also uniformly convergent.

Proof: Since the functions $f_m(z)$ are all holomorphic, we have

$$\int_C F(z)dz = \sum_{m=0}^{\infty} \int_C f_m(z)dz = 0$$

for any closed contour C in D and so, by Morera's Theorem, (the proof that $F(z)$ is continuous is left to the reader), $F(z)$ is holomophic in D.

Let z be an arbitrary point in D and C be a closed contour lying entirely in D and encircling z. Integrating along C, we have

$$\frac{1}{2\pi i}\int_C \frac{F(\zeta)}{(\zeta-z)^2}d\zeta \equiv \frac{1}{2\pi i}\int_C \sum_{m=0}^{\infty}\frac{f_m(\zeta)}{(\zeta-z)^2}d\zeta = \sum_{m=0}^{\infty}\frac{1}{2\pi i}\int_C \frac{f_m(\zeta)}{(\zeta-z)^2}d\zeta.$$

Thus, using the Derivative Formula (2.3.5), we find

$$\frac{dF(z)}{dz}=\sum_{m=0}^{\infty}\frac{df_m(z)}{dz}$$

as required. It now only remains to establish that the convergence of this series is uniform.

Let $S_n(z) = \sum_{m=0}^{n} f_m(z)$ so that (2.4.4) can be rewritten in the form

$$\frac{dF(z)}{dz}\equiv\frac{d}{dz}\left[\lim_{n\to\infty}S_n(z)\right]=\lim_{n\to\infty}\frac{dS_n(z)}{dz}.$$

Using the Derivative Formula again, we have

$$\left|\frac{dF(z)}{dz}-\frac{dS_n(z)}{dz}\right|=\left|\frac{1}{2\pi i}\int_C\frac{[F(\zeta)-S_n(\zeta)]}{(\zeta-z)^2}d\zeta\right|$$

where we are free to choose C to be a circle with centre at the point z and radius R, provided that R is sufficiently small that the circle lies entirely within D. Applying Darboux's Inequality, we have

$$\left|\frac{dF(z)}{dz}-\frac{dS_n(z)}{dz}\right|\leq \max\frac{|F(\zeta)-S_n(z)|}{R}, \quad \text{for } \zeta \text{ on } C. \tag{2.4.5}$$

Since $\lim_{n\to\infty}S_n(z)=F(z)$ uniformly in any region interior to D and since R is finite, we can make the right hand side of (2.4.5) smaller than any preassigned number $\varepsilon > 0$ by taking n larger than some number $N(\varepsilon)$ that depends only on ε. Therefore, the series in (2.4.4) converges uniformly to $\frac{dF(z)}{dz}$.

2.4.2 Taylor's Theorem

The Weierstrass Theorem tells us that every power series with non-zero radius of convergence defines a homomorphic function within its circle of convergence. The converse is one of the key theorems of complex analysis. It is called **Taylor's Theorem**.

Brook Taylor (1685-1731) was born in Edmonton, England and read mathematics at St. John's College, Cambridge. In 1715 he published a monograph entitled Methodus Incrementorum *which contained his theorem on power series expansions.*

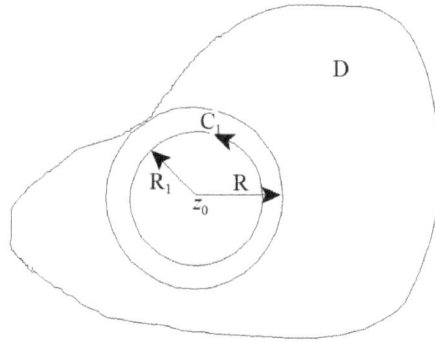

Figure 2.8: The radius of convergence R of the Taylor Series for $f(z)$ about $z = z_0$ is equal to the distance from z_0 to the nearest singularity of $f(z)$.

Theorem: Let $f(z)$ be holomorphic in a simply connected domain D and $z = z_0$ be any point in D. Let R be the radius of the largest circle with centre at z_0 and having its interior in D. Then, there is a power series

$$\sum_{m=0}^{\infty} c_m(z - z_0)^m \qquad (2.4.6)$$

which converges uniformly to $f(z)$ in every closed disc $|z - z_0| \leqslant r < R$. Furthermore, the coefficients of the series are given by

$$c_m = \frac{f^{(m)}(z_0)}{m!} = \frac{1}{2\pi i} \int_C \frac{f(z)}{(z - z_0)^{m+1}} dz \qquad (2.4.7)$$

where C is a simple closed curve in D enclosing $z = z_0$. This series, the **Taylor series** for $f(z)$ about $z = z_0$, is unique ; (it is the **only** power series representation of $f(z)$ with centre $z = z_0$).

Proof: Let C_1 be the circle $|z - z_0| = R_1$ where $R_1 < R$. Since $f(z)$ is holomorphic within and on C_1, we may invoke Cauchy's Integral to write

$$f(z) = \frac{1}{2\pi i} \int_{C_1} \frac{f(\zeta)}{\zeta - z} d\zeta \qquad (2.4.8)$$

where z is any point enclosed by C_1. Recalling the properties of the geometric series, we expand the denominator in (2.4.8) as

$$\frac{1}{\zeta - z} = \frac{1}{(\zeta - z_0) - (z - z_0)} = \frac{1}{\zeta - z} \cdot \frac{1}{1 - \frac{z-z_0}{\zeta-z_0}} = \frac{1}{\zeta - z_0} \sum_{m=0}^{\infty} \frac{(z - z_0)^m}{(\zeta - z_0)^m}. \qquad (2.4.9)$$

Since ζ is on C_1 and z within C_1, we are guaranteed that $\left|\frac{z-z_0}{\zeta-z_0}\right| \le r < 1$ and hence, that the series in (2.4.9) is uniformly convergent. Substituting (2.4.9) into (2.4.8) and integrating term by term, we find

$$f(z) = \frac{1}{2\pi i}\sum_{m=0}^{\infty}(z-z_0)^m \int_{C_1} \frac{f(\zeta)}{(\zeta-z_0)^{m+1}}\,d\zeta. \tag{2.4.10}$$

However, we also know that

$$\frac{1}{2\pi i}\int_{C_1}\frac{f(\zeta)}{(\zeta-z_0)^{m+1}}\,d\zeta = \frac{1}{2\pi i}\int_{C}\frac{f(\zeta)}{(\zeta-z_0)^{m+1}}\,d\zeta = \frac{1}{m!}f^{(m)}(z_0)$$

where we have invoked path independence to replace C_1 by any closed contour C in D that encloses $z = z_0$. Thus, (2.4.10) can be rewritten in the required form

$$f(z) = \sum_{m=0}^{\infty}\frac{f^{(m)}(z_0)}{m!}(z-z_0)^m. \tag{2.4.11}$$

The radius of convergence of this series is determined by how large we can make the radius of C_1. Thus, it is equal to the radius R of the largest circle with centre at $z = z_0$ that has its interior entirely in D or, in other words, it is equal to the distance from $z = z_0$ to the nearest singularity of $f(z)$. For, if the radius $R_1 \ge R$, C_1 would either pass through or encircle a singularity of $f(z)$ and the Cauchy Integral would no longer provide a representation of $f(z)$.

To prove that the Taylor series is unique, suppose that $f(z)$ can be represented by some other power series, $f(z) = \sum_{m=0}^{\infty} c_m(z-z_0)^m$ with $c_m \ne \frac{1}{m!}f^{(m)}(z_0)$ for at least one value of m, in some neighbourhood of $z = z_0$. Since it is uniformly convergent in a closed neighbourhood of $z = z_0$, we can perform term by term differentiation any number of times. Doing so m times and setting $z = z_0$, we obtain $c_m = \frac{1}{m!}f^{(m)}(z_0)$ for arbitrary m, which completes the proof of the theorem.

Examples: While we have already encountered examples of Taylor series in Section 1.5, a few more are appropriate at this point if only to punctuate a steady stream of theorems.

Consider the function $f(z) = z^{-1}$ whose derivatives are $f^{(m)}(z) = (-1)^m m! z^{-m-1}$, $m = 1, 2, \ldots$ The only singularity of $f(z)$ is at $z = 0$. Thus, it possesses a Taylor series about any other point $z = z_0$, with circle of convergence $|z - z_0| = |z_0|$. In particular, if we choose $z = 1$ as the centre, the Taylor series is

$$\frac{1}{z} = \sum_{m=0}^{\infty}(-1)^m(z-1)^m$$

which is valid for all z in $|z - 1| < 1$.

Next, consider the function $f(z) = \text{Ln}(1 + z)$ which has branch points at $z = -1$ and $z = \infty$. If we choose the branch cut to lie along the negative real axis, (actually, the only restriction we need to place on the choice of cut is that it not intersect the unit circle except at the branch point $z = -1$), $f(z)$ will be holomorphic within the unit circle and so will possess a Taylor series representation about $z = 0$. The mth derivative is easily found to be

$$f^{(m)}(z) = \frac{(-1)^{m-1}(m-1)!}{(1+z)^m}.$$

Thus, the Taylor series with centre $z = 0$ is

$$Ln(1+z) = \sum_{m=1}^{\infty} \frac{(-1)^m z^m}{m} = z - \frac{z^2}{2} + \frac{z^3}{3} - + \ldots \tag{2.4.12}$$

Either by applying the ratio test or simply by noting that the nearest singularity to $z = 0$ is $z = -1$, we see that this representation of $Ln(1+z)$ is valid for all $|z| < 1$.

If we replace z by $-z$ and multiply both sides of (2.4.12) by -1, we obtain

$$-Ln(1-z) = \sum_{m=1}^{\infty} \frac{z^m}{m} = z + \frac{z^2}{2} + \frac{z^3}{3} + \ldots \tag{2.4.13}$$

This too is valid for all $|z| < 1$ and therefore corresponds to a cut joining the two branch points of $Ln(1-z)$, $z = 1$ and $z = \infty$, along the positive real axis.

Adding (2.4.12) and (2.4.13) yields a third series,

$$Ln\left(\frac{1+z}{1-z}\right) = 2 \sum_{m=1}^{\infty} \frac{z^{2m-1}}{2m-1} = 2\left(z + \frac{z^3}{3} + \frac{z^5}{5} + \ldots\right),$$

for all $|z| < 1$. By construction, this series defines the principal branch of $\ln\left(\frac{1+z}{1-z}\right)$, for $|z| < 1$, when the two branch points $z = \pm 1$ are joined by a cut that passes through the point at infinity. If the branch points were joined along the real axis segment $-1 \leqslant x \leqslant 1$, $Ln\left(\frac{1+z}{1-z}\right)$ would not possess a Taylor series about $z = 0$. However, as we shall soon see, it would still be possible to provide it with another type of power series representation, valid for $|z| > 1$ and with centre $z = 0$. This latter type of series, expanded about a singular point, is called a **Laurent series**; we shall construct one in the next example.

The rational function

$$f(z) = \frac{2z+1}{z^3+z^2} = \frac{1}{z^2} \cdot \left(\frac{2z+1}{z+1}\right)$$

is singular at $z = 0$ and $z = -1$ and therefore cannot be represented by a Taylor series about either point. However, it factors into the product of an inverse power of z and a function

$$g(z) = \frac{2z+1}{z+1}$$

which is holomorphic within $|z| < 1$ and hence, can be expanded in a Taylor series there. Thus, we can determine a power series expansion about $z = 0$ for $f(z)$ which will be valid for $0 < |z| < 1$ but, unlike a Taylor series, it will contain both direct and inverse powers of z.

By a simple partial fraction decomposition, we can write

$$g(z) = 2 - \frac{1}{1+z}.$$

Replacing z by $-z$ in the geometric series, we then have

$$g(z) = 2 - \sum_{m=0}^{\infty}(-1)^m z^m, \quad |z| < 1$$

$$= 1 + z - z^2 + z^3 - + \dots$$

Hence, $f(z)$ has the representation

$$f(z) = \frac{1}{z^2} + \frac{1}{z} - 1 + z - z^2 + - \dots$$

$$= \sum_{m=1}^{2} \frac{1}{z^m} + \sum_{m=0}^{\infty}(-1)^{m-1} z^m, \quad 0 < |z| < 1.$$

Since this expression contains only powers of z, it is a power series about $z = 0$; it is the **Laurent series** for $f(z)$ in the annulus $0 < |z| < 1$.

2.4.3 Laurent's Theorem

As the last two examples intimate, it is often desirable and sometimes necessary to expand a function about one of its singularities. Moreover, the second of these examples suggests that it should be possible to find an extension of Taylor's Theorem which will cover such an eventuality. The required theorem is due to **Laurent** and, as its statement will make clear, it handles the problem by assuming nothing about the behaviour of the function at the point that is chosen as the centre of the power series. In fact, it replaces the very limiting condition that this point lie within a simply connected domain of holomorphy of the function with one that requires merely that it be the centre of an **annulus** that is contained within the domain of holomorphy, the domain remaining unrestricted with respect to multiple connectedness.

Theorem: If $f(z)$ is holomorphic in the annulus $0 \leqslant R_1 < |z - z_0| < R_2 \leqslant \infty$, then

$$f(z) = \sum_{m=0}^{\infty} a_m(z - z_0)^m + \sum_{m=1}^{\infty} \frac{b_m}{(z - z_0)^m} \equiv \sum_{m=-\infty}^{\infty} c_m(z - z_0)^m, \qquad (2.4.14)$$

with

$$a_m = \frac{1}{2\pi i} \int_C \frac{f(\zeta)}{(\zeta - z)^{m+1}} d\zeta, \quad b_m = \frac{1}{2\pi i} \int_C (\zeta - z)^{m-1} f(\zeta) d\zeta, \quad m = 0, 1, 2, \dots$$

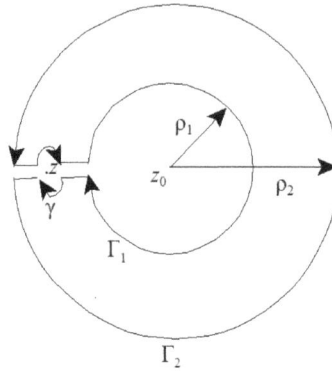

Figure 2.9: The contour Γ consists of the circles Γ_1, Γ_2 and γ connected by parallel straight line segments.

or, equivalently,

$$c_m = \frac{1}{2\pi i} \int_C \frac{f(\zeta)}{(\zeta - z)^{m+1}} d\zeta, \quad m = 0, \pm 1, \pm 2, \ldots, \tag{2.4.15}$$

where C is any contour lying within the annulus and enclosing the point $z = z_0$. The series is uniformly convergent in any closed annulus $R_1 < R_3 \leqslant |z - z_0| \leqslant R_4 < R_2$.

Proof: Working within the ζ-plane, let us define circles Γ_1 and Γ_2 centred at $\zeta = z_0$ with radii ρ_1 and ρ_2 such that $R_1 < \rho_1 < |z - z_0| < \rho_2 < R_2$. Next, we define a small circle γ centred at $\zeta = z$ and contained entirely within the annulus bounded by Γ_1 and Γ_2. We now consider the closed contour Γ constructed from Γ_1, Γ_2 and γ as shown in Figure 2.9

The quotient $\frac{f(\zeta)}{\zeta - z}$ is holomorphic everywhere within and on Γ and so, by Cauchy's Theorem, we have

$$\int_\Gamma \frac{f(\zeta)}{\zeta - z} d\zeta = 0. \tag{2.4.16}$$

The parallel straight line segments of Γ can lie arbitrarily close to each other and thus give contributions which cancel. Therefore, (2.4.16) can be rewritten as

$$\int_{\Gamma_2} \frac{f(\zeta)}{\zeta - z} d\zeta - \int_{\Gamma_1} \frac{f(\zeta)}{\zeta - z} d\zeta - \int_\gamma \frac{f(\zeta)}{\zeta - z} d\zeta = 0 \tag{2.4.17}$$

where, in compliance with our convention, all three integrals are taken in the counterclockwise direction.

From Cauchy's Integral, we know that the integration around γ yields $2\pi i f(z)$. Therefore, (2.4.17) becomes

$$f(z) = \frac{1}{2\pi i} \int_{\Gamma_2} \frac{f(\zeta)}{\zeta - z} d\zeta - \frac{1}{2\pi i} \int_{\Gamma_1} \frac{f(\zeta)}{\zeta - z} d\zeta. \tag{2.4.18}$$

The first integral in (2.4.18) can be expanded in positive powers of $(z - z_0)$ exactly as was done in the proof of Taylor's Theorem. The result is

$$\frac{1}{2\pi i} \int_{\Gamma_2} \frac{f(\zeta)}{\zeta - z} d\zeta = \sum_{m=0}^{\infty} a_m (z - z_0)^m \tag{2.4.19}$$

where

$$a_m = \frac{1}{2\pi i} \int_{\Gamma_2} \frac{f(\zeta)}{(\zeta - z_0)^{m+1}} d\zeta. \tag{2.4.20}$$

An expansion of the second integral can be obtained as follows. We set

$$\frac{1}{\zeta - z} = \frac{-1}{(z - z_0) - (\zeta - z_0)} = \frac{-1}{(z - z_0)} \cdot \frac{1}{\left(1 - \frac{\zeta - z_0}{z - z_0}\right)} = \frac{-1}{(z - z_0)} \cdot \sum_{m=1}^{\infty} \left(\frac{\zeta - z_0}{z - z_0}\right)^{m-1}.$$

This sum is uniformly convergent for $\left|\frac{\zeta - z_0}{z - z_0}\right| \leq r < 1$. Therefore, since $|\zeta - z_0| = \rho_1$ and $\rho_1 < |z - z_0| < \rho_2$, we can integrate term by term to obtain

$$\frac{1}{2\pi i} \int_{\Gamma_1} \frac{f(\zeta)}{\zeta - z} d\zeta = -\sum_{m=1}^{\infty} \frac{b_m}{(z - z_0)^m} \tag{2.4.21}$$

where

$$b_m = \frac{1}{2\pi i} \int_{\Gamma_1} (\zeta - z_0)^{m-1} f(\zeta) d\zeta. \tag{2.4.22}$$

We now observe that the integrals in (2.4.20) and (2.4.22) are independent of the path of integration provided that it lies in the annulus $R_1 < |\zeta - z_0| < R_2$ and encloses $\zeta = z_0$. Therefore, using equations (2.4.19) through to (2.4.22), we can rewrite equation (2.4.18) as

$$f(z) = \sum_{m=-\infty}^{\infty} c_m (z - z_0)^m \tag{2.4.23}$$

where, for all $m = 0, \pm 1, \pm 2, \ldots$,

$$c_m = \frac{1}{2\pi i} \int_C \frac{f(\zeta)}{(\zeta - z_0)^{m+1}} d\zeta \tag{2.4.24}$$

C being any contour that encloses z_0 and lies in the annular region of holomorphy of $f(z)$.

The expansions in equations (2.4.19) to (2.4.21) all converge absolutely for all z within the annulus $\rho_1 < |z - z_0| < \rho_2$, and uniformly for all z within the closed annulus $\rho_1 < R_3 \leqslant |z - z_0| \leqslant R_4 < \rho_2$. However, ρ_1 and ρ_2 are arbitrary, subject to the condition $R_1 < \rho_1 < \rho_2 < R_2$. Therefore, it follows that the Laurent series (2.4.23) converges absolutely to $f(z)$ in $R_1 < |z - z_0| < R_2$ and uniformly to $f(z)$ in any closed annulus interior to $R_1 < |z - z_0| < R_2$.

The Laurent series representation of a given function associated with a given annulus is unique: there is exactly one series (2.4.14) that converges to $f(z)$ in $R_1 < |z - z_0| < R_2$. If $f(z)$ has more than one distinct annulus of holomorphy with centre at $z = z_0$, it has a correspondingly distinct Laurent representation in each one of them. This is best brought home by means of an example.

Examples: The function $f(z) = \operatorname{cosec} z$ is singular at the points $z = n\pi$, $n = 0, 1, 2, \ldots$. Therefore, there is an infinity of distinct annuli with centre at $z = 0$ in which $\operatorname{cosec} z$ is holomorphic, namely

$$n\pi < |z| < (n+1)\pi, \quad n = 0, 1, 2, \ldots.$$

Each annulus admits a unique and distinct Laurent series about $z = 0$, thus providing $\operatorname{cosec} z$ with infinitely many such representations.

To illustrate the use of equations (2.4.14) and (2.4.15) in the determination of a Laurent series, consider the function $(z^2 - 1)^{-1/2}$. As we have seen, this has branch points at $z = \pm 1$. We can choose the cut between these two points to run along the real axis segment $-1 \leqslant x \leqslant 1$ so that the function's principal branch is

$$f(z) = |z + 1|^{-1/2} |z - 1|^{-1/2} e^{-i/2(\theta_+ + \theta_-)}, \quad -\pi < \theta_\pm < \pi.$$

This is holomorphic everywhere except on the cut. Therefore, it must possess a Laurent representation in the annulus $0 < |z| < \infty$ given by

$$(z^2 - 1)^{-1/2} = \sum_{m=0}^{\infty} a_m z^m + \sum_{m=1}^{\infty} b_m z^{-m}$$

with

$$a_m = \frac{1}{2\pi i} \int_C \frac{(\zeta^2 - 1)^{-1/2}}{\zeta^{m+1}} d\zeta, \quad b_m = \frac{1}{2\pi i} \int_C \zeta^{m-1} (\zeta^2 - 1)^{-1/2} d\zeta$$

where C is any contour in $0 < |\zeta| < \infty$ that encloses $\zeta = 0$.

To evaluate the coefficients a_m we choose C to be a circle of arbitrarily large radius. It is then easily shown by means of an application of the Darboux Inequality that the integral vanishes; thus, $a_m = 0$ for all $m = 0, 1, 2, \ldots$.

For the evaluation of b_m, we choose the "dog bone" contour shown in Figure 2.10. It consists of infinitesimally small circles centred at $z = \pm 1$ and two parallel straight

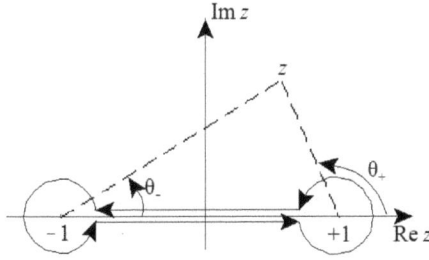

Figure 2.10: The "dogbone" contour used to evaluate the Laurent series coefficients for $(z^2 - 1)^{1/2}$.

lines running just above and below the cut on the real axis. The contribution to b_m from the circles is vanishingly small as one can again show with the help of Darboux's Inequality. Thus, using the fact that the principal branch of $(z^2 - 1)^{-1/2}$ assumes the values $(1 - x^2)^{-1/2} e^{-i\pi/2}$ just above the cut, and $(1 - x^2)^{-1/2} e^{+i\pi/2}$ just below it, we have

$$b_m = \frac{1}{2\pi i} \left[\int_{-1}^{+1} \frac{x^{m-1} e^{i\pi/2}}{\sqrt{1-x^2}} dx + \int_{+1}^{-1} \frac{x^{m-1} e^{-i\pi/2}}{\sqrt{1-x^2}} dx \right] = \frac{1}{\pi} \int_{-1}^{+1} \frac{x^{m-1}}{\sqrt{1-x^2}} dx.$$

This vanishes for even values of m, while for odd values of m, we can write

$$b_m = \frac{2}{\pi} \int_0^1 \frac{x^{m-1}}{\sqrt{1-x^2}} dx = \frac{2}{\pi} \int_0^{\pi/2} \sin^{m-1}\theta \, d\theta = \frac{\pi}{2} \frac{(m-1)!}{2^{m-1} \left(\frac{m-1}{2}! \right)^2}.$$

Thus, setting $m = 2k + 1$, we have

$$(z^2 - 1)^{-1/2} = \sum_{k=0}^{\infty} \frac{(2k)!}{2^{2k}(k!)^2} \frac{1}{z^{2k+1}} = \frac{1}{z} + \frac{1}{2}\frac{1}{z^3} + \frac{3}{8}\frac{1}{z^5} + \frac{5}{16}\frac{1}{z^7} + \dots, \quad 1 < |z| < \infty.$$

Had we chosen the branch cut to lie along the real axis segments $-\infty < x \leqslant -1$ and $1 \leqslant x < \infty$, $(z^2 - 1)^{-1/2}$ would not possess a Laurent series representation about $z = 0$ but it would have a Taylor series that converges for $|z| < 1$. For completeness we shall evaluate its coefficients as well.

With this choice of cut the principal branch is defined by

$$f(z) = |z + 1|^{-1/2} |z - 1|^{-1/2} e^{-i/2(\theta_+ + \theta_-)}, \quad 0 < \theta_+ < 2\pi, -\pi < \theta_- < \pi.$$

It is holomorphic everywhere in the cut plane and so possesses the Taylor series

$$(z^2 - 1)^{-1/2} = \sum_{m=0}^{\infty} c_m z^m, \quad |z| < 1$$

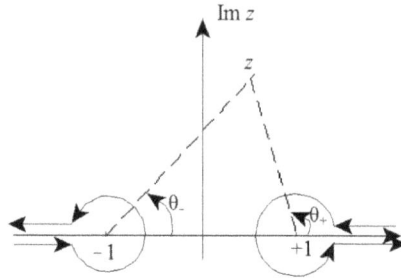

Figure 2.11: The "dogbone" contour used to evaluate the Taylor series coefficients for $(z^2 - 1)^{1/2}$.

with

$$c_m = \frac{1}{2\pi i} \int_C \frac{(\zeta^2 - 1)^{-1/2}}{\zeta^{m+1}} d\zeta$$

where C is any contour in the cut plane that encloses $\zeta = 0$. We shall choose another dog bone contour as shown in Figure 2.11.

Once again the contributions from the infinitesimally small circles vanish. Thus, since $(z^2 - 1)^{-1/2}$ assumes the values $\pm(x^2 - 1)^{-1/2}$ for $1 \leqslant x < \infty, y = 0^{\pm}$, and $\mp(x^2 - 1)^{-1/2}$ for $-\infty < x \leqslant -1, y = 0^{\pm}$, we have

$$c_m = \frac{1}{2\pi i} \int_{-\infty}^{-1} \frac{(x^2 - 1)^{-1/2}}{x^{m+1}} dx - \frac{1}{2\pi i} \int_{-1}^{-\infty} \frac{(x^2 - 1)^{-1/2}}{x^{m+1}} dx$$

$$+ \frac{1}{2\pi i} \int_{+\infty}^{+1} \frac{(x^2 - 1)^{-1/2}}{x^{m+1}} dx - \frac{1}{2\pi i} \int_{+1}^{\infty} \frac{(x^2 - 1)^{-1/2}}{x^{m+1}} dx.$$

This vanishes if m is odd while, if $m = 2k$, it yields

$$c_{2k} = \frac{2i}{\pi} \int_1^{\infty} \frac{(x^2 - 1)^{-1/2}}{x^{2k+1}} dx = \frac{2i}{\pi} \int_0^{\pi/2} \sin^{2k}\theta d\theta = i\frac{(2k)!}{2^{2k}(k!)^2}.$$

Thus, our final result is

$$(z^2 - 1)^{-1/2} = i\sum_{k=0}^{\infty} \frac{(2k)!}{2^{2k}(k!)^2} z^{2k} = i\left[z + \frac{1}{2}z^2 + \frac{3}{8}z^4 + \frac{5}{16}z^6 + \ldots\right], \quad |z| < 1.$$

2.4.4 Practical Methods for Generating Power Series

If one had to go through such a lengthy procedure as that used in the preceding examples, the task of determining power series representations would be rather daunting. Fortunately, there are a number of practical methods available that obviate direct application of Taylor's or Laurent's Theorem. These are best introduced by means of examples which we will label according to the type of function that each addresses.

Examples:

1. **Rational Functions:** Every rational function $f(z) = \frac{P(z)}{Q(z)}$, where $P(z)$ and $Q(z)$ are polynomials, can be decomposed into a sum of a polynomial (if the degree of P is greater than or equal to the degree of Q) and a finite number of partial fractions of the form $\frac{a}{(bz-c)^n}$, where a, b and c are complex constants and n is an integer. The number $z = \frac{c}{b}$ is a root of the polynomial Q and thus is a singularity of the function $f(z)$. Our starting point then is to learn how to expand the fraction

$$g(z) = \frac{1}{bz - c}, \quad b \neq 0$$

about an arbitrary point $z = z_0$.

If $bz_0 - c = 0$, $g(z)$ has only one expansion about $z = z_0$: the Laurent series

$$g(z) = \frac{1/b}{z - c/b}, \quad |z - c/b| > 0.$$

If $bz_0 - c \neq 0$, we can employ a simple trick that was used in the proof of both Taylor's and Laurent's Theorem. We write

$$\frac{1}{bz - c} = \frac{1}{b(z - z_0) + bz_0 - c} = \frac{1}{(c - bz_0)[\frac{b(z-z_0)}{c-bz_0} - 1]}$$

$$= \frac{1}{b(z - z_0)[1 - \frac{c-bz_0}{b(z-z_0)}]}. \tag{2.4.25}$$

The last two expressions in (2.4.25) are completely equivalent and allow us to use the geometric series to obtain alternative power series representations about $z = z_0$ with mutually exclusive domains of convergence. Specifically, we have

$$g(z) = \frac{-1}{c - bz_0} \sum_{m=0}^{\infty} \left(\frac{b(z - z_0)}{c - bz_0} \right)^m$$

$$= -\sum_{m=0}^{\infty} \frac{b^m}{(c - bz_0)^{m+1}} (z - z_0)^m, \quad |z - z_0| < |z_0 - c/b| \tag{2.4.26}$$

and

$$g(z) = \frac{1}{b(z - z_0)} \sum_{m=0}^{\infty} \left(\frac{c - bz_0}{b(z - z_0)} \right)^m$$

$$= \sum_{m=0}^{\infty} \frac{(c - bz_0)^m}{b^{m+1}} \frac{1}{(z - z_0)^{m+1}}, \quad |z - z_0| > |z_0 - c/b|. \tag{2.4.27}$$

Thus, inside the circle $|z - z_0| = |z_0 - c/b|$ there is a Taylor series representation (2.4.26) for $g(z)$ while outside the circle there is the Laurent series representation (2.4.27). The

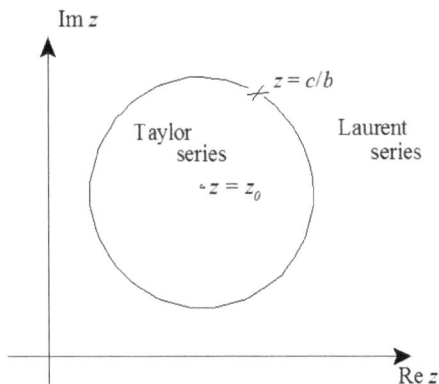

Figure 2.12: The function $g(z) = (bz - c)^{-1}$ is holomorphic within the disc $|z - z_0| < |z_0 - c/b|$ and the annulus $|z - z_0| > |z_0 - c/b|$ and so possesses respectively, a Taylor and a Laurent expansion about $z = z_0$ in these two domains.

common boundary of these two domains, the circle itself, is dictated by the location of the singularity of $g(z)$, $z = c/b$. (See Figure 2.12).

So much for partial fractions corresponding to simple roots. Higher order fractions of the form

$$g(z) = \frac{1}{(bz - c)^n}, \quad n \geqslant 2$$

require little additional effort. We simply note that

$$\frac{1}{(bz - c)^n} = \frac{1}{(n - 1)!} \frac{(-1)^{n-1}}{b^{n-1}} \frac{d^{n-1}}{dz^{n-1}} \frac{1}{(bz - c)}. \tag{2.4.28}$$

Thus, once we have found the uniformly convergent power series for $\frac{1}{bz-c}$ term by term differentiation will do the rest.

As a concrete example, consider the function

$$f(z) = \frac{2z^2 + 9z + 15}{z^3 + z^2 - 8z - 12}$$

and suppose that we wish to find all of its power series representations about the point $z = 0$. The denominator has a double root, $z = -2$, and a simple root, $z = 3$. Therefore, drawing circles with center at the origin which pass through these two singularities, we see that $f(z)$ possesses three expansions about $z = 0$:

1. a Taylor series for $|z| < 2$,
2. a Laurent series for $2 < |z| < 3$, and
3. a second Laurent series for $|z| > 3$.

To determine these three series we perform a partial fraction decomposition and write

$$f(z) = \frac{1}{(z + 2)^2} + \frac{2}{z - 3}.$$

Then, applying (2.4.26) and (2.4.27), we have

$$\frac{1}{z-3} = -\sum_{m=0}^{\infty} \frac{z^m}{3^{m+1}}, \quad |z| < 3$$

$$\frac{1}{z-3} = \sum_{m=0}^{\infty} \frac{3^m}{z^{m+1}}, \quad |z| > 3$$

and,

$$\frac{1}{z+2} = \sum_{m=0}^{\infty} \frac{(-1)^m}{2^{m+1}} z^m, \quad |z| < 2$$

$$\frac{1}{z+2} = \sum_{m=0}^{\infty} \frac{(-2)^m}{z^{m+1}}, \quad |z| > 2.$$

Invoking (2.4.28), we find

$$\frac{1}{(z+2)^2} = \sum_{m=1}^{\infty} \frac{(-1)^{m+1}}{2^{m+1}} m z^{m-1}, \quad |z| < 2$$

$$\frac{1}{(z+2)^2} = \sum_{m=0}^{\infty} \frac{(-2)^m(m+1)}{z^{m+2}}, \quad |z| > 2.$$

Thus, the three series representations are

i. $f(z) = \sum_{m=0}^{\infty} \left[\frac{(-1)^m(m+2)}{2^{m+2}} - \frac{2}{3^{m+1}} \right] z^m, \quad |z| < 2$

ii. $f(z) = \sum_{m=0}^{\infty} \frac{(-2)^m(m+1)}{z^{m+2}} - 2 \sum_{m=0}^{\infty} \frac{z^m}{3^{m+1}}, \quad 2 < |z| < 3$

iii. $f(z) = 2 \sum_{m=0}^{\infty} [3^m - (-2)^{m-2} m] \frac{1}{z^{m+1}}, \quad |z| > 3.$

Notice that term by term differentiation provides a useful method whenever one seeks a series for a function $f(z)$ and one already knows the corresponding expansion for $F(z)$ where $f(z) = c \frac{d^n}{dz^n} F(z)$, c = a constant, for some n.

2. **Exponential, Trigonometric and Hyperbolic Functions:** Because of the simplicity of the differentiation rules that apply to these functions, determining Taylor series for them is seldom a problem. However, their Laurent series are a very different matter. One useful technique is to make use of the **Cauchy product** of two series,

$$\left(\sum_{m=0}^{\infty} a_m \right) \cdot \left(\sum_{m=0}^{\infty} b_m \right) = \sum_{m=0}^{\infty} \sum_{k=0}^{m} a_k b_{m-k} = \sum_{m=0}^{\infty} \sum_{k=0}^{m} a_{m-k} b_k. \tag{2.4.29}$$

As a concrete example we shall determine the Laurent series representation of cosec z in the annulus $0 < |z| < \pi$.

The function $\frac{\sin z}{z}$ has the Taylor series

$$\frac{\sin z}{z} = 1 - \frac{z^2}{3!} + \frac{z^4}{5!} - + \ldots = \sum_{m=0}^{\infty} \frac{(-1)^m z^{2m}}{(2m+1)!}, \quad |z| < \infty$$

and so is entire. Therefore, its inverse $\frac{z}{\sin z}$ is holomorphic within the disc $|z| < \pi$ and must possess a Taylor series there. Thus, we set

$$\frac{z}{\sin z} = \frac{1}{\displaystyle\sum_{m=0}^{\infty} \frac{(-1)^m z^{2m}}{(2m+1)!}} = \sum_{m=0}^{\infty} c_m z^m.$$

Cross-multiplying, we then have

$$1 = \left(\sum_{m=0}^{\infty} \frac{(-1)^m z^{2m}}{(2m+1)!} \right) \cdot \left(\sum_{m=0}^{\infty} c_m z^m \right) \tag{2.4.30}$$

$$= \sum_{m=0}^{\infty} \sum_{k=0}^{m} \frac{(-1)^k}{(2k+1)!} c_{m-k} z^{m+k}. \tag{2.4.31}$$

Since power series representations are unique, we can equate coefficients of like powers of z appearing on either side of equation (2.4.31). This yields the infinite set of equations

$$c_0 = 1$$
$$c_1 = 0$$
$$\frac{-c_0}{3!} + c_2 = 0$$
$$\frac{-c_1}{3!} + c_3 = 0$$
$$\frac{c_0}{5!} - \frac{c_2}{3!} + c_4 = 0$$
$$\frac{c_1}{5!} - \frac{c_3}{3!} + c_5 = 0$$
$$\frac{-c_0}{7!} + \frac{c_2}{5!} - \frac{c_4}{3!} + c_6 = 0 \ldots.$$

whose solutions are

$$c_0 = 1, \quad c_1 = 0, c_2 = \frac{1}{6}, \quad c_3 = 0,$$

$$c_4 = \frac{7}{360}, \quad c_5 = 0, \quad c_6 = \frac{31}{15120}, \quad c_7 = 0, \ldots.$$

Thus, we find that

$$\frac{z}{\sin z} = 1 + \frac{z^2}{6} + \frac{7}{360} z^4 + \frac{31}{15120} z^6 + \ldots, \quad |z| < \pi$$

and hence, that

$$\operatorname{cosec} z = \frac{1}{z} + \frac{1}{6}z + \frac{7}{360}z^3 + \frac{31}{15120}z^5 + \ldots, \quad 0 < |z| < \pi. \qquad (2.4.32)$$

A second useful technique and in fact, one which is useful in dealing with any type of function, is **substitution**. The next two examples will illustrate what is involved.

From the Laurent expansion (2.4.32) about $z = 0$ the substitution $z \to z + k\pi$, $k = \pm 1, \pm 2, \ldots$ immediately yields the local expansion of cosec z about any of its other singularities:

$$\operatorname{cosec}(z + k\pi) = (-1)^k \operatorname{cosec} z = \frac{1}{(z + k\pi)} + \frac{1}{6}(z + k\pi) + \frac{7}{360}(z + k\pi)^3 + \ldots,$$

$$0 < |z + k\pi| < \pi.$$

Just as simple is the conversion of the Taylor series

$$\sinh z = z + \frac{z^3}{3!} + \frac{z^5}{5!} + \ldots, |z| < \infty$$

into the Laurent series

$$\sinh \frac{1}{z} = \frac{1}{z} + \frac{1}{3!z^3} + \frac{1}{5!z^5} + \ldots, \quad |z| > 0$$

under the substitution $z \to \frac{1}{z}$. The simplicity of these examples is characteristic of all applications of the substitution technique.

3. **Logarithms:** Because the derivatives of many logarithms are rational functions, term by term integration is an obvious technique to use in determining their power series expansions. We have already generated the Taylor series representation

$$\operatorname{Ln}\left(\frac{1+z}{1-z}\right) = 2\sum_{m=1}^{\infty} \frac{z^{2m-1}}{2m-1} = 2\left(z + \frac{z^3}{3} + \frac{z^5}{5} + \ldots\right), \quad |z| < 1 \qquad (2.4.33)$$

which corresponds to a choice of branch cut running along the real axis segments $-\infty < x \leqslant -1$ and $1 \leqslant x < \infty$. This is the only power series about $z = 0$ possessed by $\operatorname{Ln}\left(\frac{1+z}{1-z}\right)$ when its branch points $z = \pm 1$ are joined in this way. However, if one makes the other obvious choice, a cut along the segment $-1 \leqslant x \leqslant 1, y = 0$, then $\operatorname{Ln}\left(\frac{1+z}{1-z}\right)$ again possesses only one expansion about $z = 0$ but this time it is a Laurent series in the annulus $|z| > 1$. To determine this series we note that

$$\frac{d}{dz} \operatorname{Ln}\left(\frac{1+z}{1-z}\right) = \frac{1}{1+z} + \frac{1}{1-z} = \frac{2}{1-z^2}.$$

The power series expansions of $\frac{1}{1 \pm z}$ in the annulus $|z| > 1$ are readily found from equation (2.4.27). We have

$$\frac{1}{1+z} = \sum_{m=0}^{\infty} \frac{(-1)^m}{z^{m+1}} \quad \text{and} \quad \frac{1}{1-z} = -\sum_{m=0}^{\infty} \frac{1}{z^{m+1}}, \quad |z| > 1$$

and hence,

$$\frac{d}{dz} \operatorname{Ln}\left(\frac{1+z}{1-z}\right) = -2\sum_{m=0}^{\infty} \frac{1}{z^{2m+2}} = -2\left[\frac{1}{z^2} + \frac{1}{z^4} + \frac{1}{z^6} + \ldots\right], \quad |z| > 1.$$

Integrating term by term we then obtain

$$\operatorname{Ln}\left(\frac{1+z}{1-z}\right) = 2\left[\frac{1}{z} + \frac{1}{3z^3} + \frac{1}{5z^5} + \ldots\right] + c = 2\sum_{m=1}^{\infty} \frac{1}{2m-1}\frac{1}{z^{2m-1}} + c, \quad |z| > 1$$

where c is a constant of integration. Since

$$\operatorname{Ln}\left(\frac{1+z}{1-z}\right) = \ln\frac{|1+z|}{|1-z|} + i[\arg(z+1) - \arg(z-1) + \pi], \quad -\pi < \arg(z \pm 1) < \pi$$

we see that c must be set equal to $+i\pi$ in order to obtain the correct value as $|z| \to \infty$. Thus, absorbing the $+i\pi$ into the argument of the logarithm, we finally obtain

$$\operatorname{Ln}\left(\frac{z+1}{z-1}\right) = 2\sum_{m=1}^{\infty} \frac{1}{2m-1}\frac{1}{z^{2m-1}} = 2\left(\frac{1}{z} + \frac{1}{3z^3} + \frac{1}{5z^5} + \ldots\right), \quad |z| > 1. \quad (2.4.34)$$

2.5 Zeros and Singularities

As we have seen, if a function $f(z)$ is holomorphic within a simply connected domain D then it can be expanded in a Taylor series about any point $z = z_0$ of D and

$$f(z) = \sum_{m=0}^{\infty} c_m(z - z_0)^m, \quad c_m = \frac{1}{m!}\frac{d^m}{dz^m}f(z)\bigg|_{z=z_0}. \quad (2.5.1)$$

Definition: If $f(z)$ vanishes at $z = z_0$, this point is called a **zero** of $f(z)$. Moreover, $f(z)$ is said to have a **zero of ordern** at $z = z_0$ if

$$f(z_0) = \frac{df(z)}{dz}\bigg|_{z=z_0} = \ldots = \frac{d^{n-1}}{dz^{n-1}}f(z)\bigg|_{z=z_0} = 0, \quad \frac{d^n}{dz^n}f(z)|_{z=z_0} \neq 0, \quad (2.5.2)$$

that is, if the first n coefficients of the Taylor series (2.5.1) are all zero but the $(n+1)^{th}$ coefficient is non-zero.

Thus, if there is an nth order zero at $z = z_0$, the Taylor series (2.5.1) assumes the form

$$f(z) = c_n(z - z_0)^n + c_{n+1}(z - z_0)^{n+1} + \ldots$$

$$= (z - z_0)^n \sum_{k=0}^{\infty} c_{n+k}(z - z_0)^k$$

$$= (z - z_0)^n g(z), \quad (2.5.3)$$

where $g(z)$ is holomorphic and non-zero at $z = z_0$. Its holomorphy implies that $g(z)$ is continuous and hence, that it is non-vanishing in some neighbourhood $|z - z_0| < R$ as well as at $z = z_0$. Specifically, if $g(z_0) = \kappa$, then there must exist an R such that $|g(z) - g(z_0)| < \frac{\kappa}{2}$ for $|z - z_0| < R$. Invoking a triangle inequality, we then have

$$|g(z)| \geqslant |\{|g(z_0)| - |g(z) - g(z_0)|\}| > \frac{\kappa}{2}$$

for the same range $|z - z_0| < R$ and so $g(z)$ certainly does not vanish there. Therefore, $f(z)$ itself must be non-vanishing in the deleted neighbourhood $0 < |z - z_0| < R$. This establishes that **the zeros of an analytic function are isolated.** Thus, if the zeros of an analytic function $f(z)$ have a limit point $z = z_l$, then either $f(z) \equiv 0$ or $f(z)$ is singular (discontinuous) at $z = z_l$.

Notice that the order of a zero can be determined by calculating

$$\lim_{z \to z_0} \frac{f(z)}{(z - z_0)^n}$$

for successive values of $n = 1, 2, 3, \ldots$; the lowest n for which this is non-vanishing is the order of the zero.

Definition: If a function $f(z)$ has an **isolated singularity** at a point $z = z_0$ then there must exist a finite deleted neighbourhood of that point, $0 < |z - z_0| < R$ for some R, in which $f(z)$ is holomorphic.

Such a neighbourhood constitutes an annulus on which Laurent's Theorem is applicable and so, for z in this neighbourhood, $f(z)$ can be represented by a Laurent expansion about $z = z_0$:

$$f(z) = \sum_{m=0}^{\infty} a_m (z - z_0)^m + \sum_{m=1}^{\infty} \frac{b_m}{(z - z_0)^m}, \qquad 0 < |z - z_0| < R. \tag{2.5.4}$$

The negative powers of $(z - z_0)$ in this expansion determine the character of the singularity at $z = z_0$. There are two cases corresponding to two new **definitions**.

1. A Finite Number of Negative Powers: Suppose that $b_m = 0$, $m > n$ but $b_n \neq 0$ for some n. The series in (2.5.4) then becomes

$$f(z) = \frac{b_n}{(z - z_0)^n} + \frac{b_{n-1}}{(z - z_0)^{n-1}} + \ldots + \frac{b_1}{(z - z_0)} + \sum_{m=0}^{\infty} a_m (z - z_0)^m. \tag{2.5.5}$$

The function $f(z)$ is said to have a **pole of order n** at $z = z_0$. The sum of negative powers

$$\frac{b_n}{(z - z_0)^n} + \frac{b_{n-1}}{(z - z_0)^{n-1}} + \ldots + \frac{b_1}{(z - z_0)}$$

is called the **principal part** of $f(z)$ at $z = z_0$; it becomes infinite as $z \to z_0$.

The representation of $f(z)$ in (2.5.5) can be rewritten as

$$f(z) = (z - z_0)^{-n} h(z) \tag{2.5.6}$$

where $h(z)$ is holomorphic and non-zero at $z = z_0$,

$$h(z) = b_n + b_{n-1}(z - z_0) + \ldots + b_1(z - z_0)^{n-1} + \sum_{m=0}^{\infty} a_m(z - z_0)^{m+n}.$$

Thus, the order of the pole can be determined by evaluating

$$\lim_{z \to z_0} (z - z_0)^n f(z)$$

for successive values of $n = 1, 2, \ldots$; the lowest n for which this limit exists is the order of the pole.

2. Infinitely Many Negative Powers: When there is an infinite number of coefficients b_m that are non-zero, $f(z)$ is said to have an **isolated essential singularity** at $z = z_0$.
Definition: A function $f(z)$ is said to be **meromorphic** in a domain D if it has no singularities other than poles in D.

The difference between a pole and an essential singularity is reflected most dramatically in the behaviour exhibited by a function in the neighbourhood of each type of singularity. From (2.5.6) it is clear that if $f(z)$ has a pole at $z = z_0$ then $|f(z)| \to \infty$ as $z \to z_0$ in any manner. Thus, in a small neighbourhood of a pole, a function must be **uniformly** large. To see what may happen at an essential singularity, consider the function

$$e^{1/z} = \sum_{m=0}^{\infty} \frac{1}{m! z^m}, \quad |z| > 0,$$

which has an isolated essential singularity at $z = 0$. If we let $z \to 0$ along
1. the negative real axis, $|e^{1/z}| \to 0$,
2. the positive real axis, $|e^{1/z}| \to \infty$,
3. the imaginary axis, $|e^{1/z}|$ remains constant but $\arg(e^{1/z}) \to \infty$ so that $e^{1/z}$ oscillates wildly.

Thus, we cannot assign a specific value to $e^{1/z}$ at $z = 0$ since it evidently takes on every possible non-zero value in any neighbourhood of that point. As the next theorem shows, this behaviour is characteristic of essential singularities.

Picard's Theorem: If a function $f(z)$ has an isolated essential singularity at $z = z_0$ then, in an arbitrarily small neighbourhood of $z = z_0$, $f(z)$ assumes infinitely many times **every** complex value, with at most one exceptional value.

The one exceptional value in the case of $e^{1/z}$ is zero.

This last theorem is named for the French mathematician Charles Emile Picard (1856-1941) who was especially noted for his work in complex analysis and integral and differential equations.

As we have seen already, a function which has an infinite sequence of zeros with a limit point $z = z_l$ must either be singular at that point or be identically zero. We now recognize that if the function is not identically zero then, since it also does not tend uniformly to infinity as $z \to z_l$, the limit point is an isolated essential singularity.

So far we have restricted our discussion to singularities in the finite plane. However, we can also use Laurent expansions to investigate the behaviour of functions at infinity. The usual first step in any investigation of what happens at the point at infinity is to make the substitution $z = \frac{1}{\zeta}$ followed by an examination of how the function in question behaves as $\zeta \to 0$. Therefore, we say that $f(z)$ is holomorphic, or has a pole of order n, or has an essential singularity at $z = \infty$ according as $f(1/\zeta)$ has the corresponding property at $\zeta = 0$. Thus, if $f(z)$ has a pole of order n at $z = \infty$, then $f(1/\zeta)$ must have a Laurent expansion about $\zeta = 0$ of the form

$$f(1/\zeta) = \sum_{m=0}^{\infty} a_m \zeta^m + \frac{b_1}{\zeta} + \frac{b_2}{\zeta^2} + \ldots + \frac{b_n}{\zeta^n}, \quad 0 < |\zeta| < \frac{1}{R}$$

for some R. Hence, $f(z)$ admits the Laurent expansion

$$f(z) = \sum_{m=0}^{\infty} \frac{a_m}{z^m} + b_1 z + b_2 z^2 + \ldots + b_n z^n, \quad R < |z| < \infty \qquad (2.5.7)$$

about $z = 0$. Evidently, the principal part of $f(z)$ at infinity is the polynomial

$$f_P(z) = b_1 z + b_2 z^2 + \ldots + b_n z^n.$$

Similarly, if $f(z)$ has an isolated essential singularity at $z = \infty$, it must have a Laurent expansion about $z = 0$ of the form

$$f(z) = \sum_{m=0}^{\infty} \frac{a_m}{z^m} + \sum_{m=1}^{\infty} b_m z^m, \quad R < |z| < \infty \qquad (2.5.8)$$

for some R, and the principal part at $z = \infty$ is the entire function

$$f_P(z) = \sum_{m=0}^{\infty} b_m z^m.$$

Finally, if $f(z)$ is holomorphic at $z = \infty$ its principal part there will be zero and it will possess a Laurent expansion about $z = 0$ of the form

$$f(z) = \sum_{m=0}^{\infty} \frac{a_m}{z^m}, \quad R < |z| < \infty. \qquad (2.5.9)$$

In each of the three cases (2.5.7) to (2.5.9), R is the distance from the origin to the furthest finite singularity of the function.

Suppose now that we expand $f(z)$ about a point $z = z_0$ located such that $R < |z_0| < \infty$. Since, for each m,

$$\frac{1}{z^m} = \frac{(-1)^m}{(m-1)!} \frac{1}{(z-z_0)^m} \sum_{k=0}^{\infty} \frac{z_0^k}{(z-z_0)^k} \frac{(k+m-1)!}{k!}$$

which has no positive powers of $(z - z_0)$, and since the principal part of $f(z)$ at $z = \infty$, $f_P(z)$, is always entire, the series must assume the form

$$f(z) = \sum_{m=0}^{\infty} \frac{\alpha_m}{(z - z_0)^m} + \sum_{m=1}^{\infty} \beta_m (z - z_0)^m, \quad |z_0| - R < |z - z_0| < \infty \qquad (2.5.10)$$

with $\sum_{m=1}^{\infty} \beta_m (z - z_0)^m = f_P(z)$ and $\beta_m = 0$ for all $m \geq 1$ if $f(z)$ is holomorphic at $z = \infty$, $\beta_m = 0$ for all $m > n$, $\beta_n \neq 0$ if $f(z)$ has a pole of order n at $z = \infty$, $\beta_m \neq 0$ for an infinite number of values of m if $f(z)$ has an essential singularity at $z = \infty$. In other words, **any** Laurent expansion whose **outer** radius of convergence is infinite yields both the nature of the singularity, if any, and the function's principal part at $z = \infty$.

Before we consider some examples it remains to point out that there are two principal types of **non-isolated** singularities. The most obvious of these is a **branch point**, every neighbourhood of which contains a segment of a cut where all branches of the function are discontinuous. The second type is a **non-isolated essential singularity** which is simply the limit point of an infinite sequence of isolated singularities (usually poles).

Examples: It is fairly obvious from (2.5.10) that any polynomial of nth degree has a pole of order n at $z = \infty$. On the other hand, since the degree of its numerator is less than the degree of its denominator, the rational function

$$f(z) = \frac{1}{z(z - 2)^5} + \frac{3}{(z - 2)^3}$$

is holomorphic at $z = \infty$. Its only singularities are a simple pole at $z = 0$ and a fifth order pole at $z = 2$.

The location and nature of the singularities of the last function were easily determined at a single glance. However, a function's appearance can sometimes be deceiving. For example, knowing that $\frac{\sin z}{z}$ is holomorphic at the origin, one might be tempted to claim that

$$f(z) = \left(\frac{3}{z^2} - \frac{1}{z} \right) \sin z - \frac{3}{z^2} \cos z$$

has a second order pole at $z = 0$. Such a claim would be erroneous. Using

$$\sin z = z - \frac{z^3}{3!} + \frac{z^5}{5!} - + \ldots$$

$$\cos z = 1 - \frac{z^2}{2!} + \frac{z^4}{4!} - + \ldots,$$

we see that $f(z)$ possesses the Taylor series

$$f(z) = \frac{8}{5!} z^2 - \frac{24}{7!} z^4 + \frac{48}{9!} z^6 - + \ldots = 4 \sum_{m=1}^{\infty} \frac{(-1)^{m+1} m(m + 1)}{(2m + 3)!} z^{2m}, \quad |z| < \infty$$

about $z = 0$. Thus, $f(z)$ is an entire function with a second order **zero** at $z = 0$ and, of course, an essential singularity at $z = \infty$.

Next, consider the function $\sin \frac{1}{z}$ whose zeros $z = \frac{1}{n\pi}$, $n = \pm 1, \pm 2, \ldots$, have a limit point at $z = 0$. Our earlier analysis suggests that $z = 0$ is an isolated essential singularity of $\sin \frac{1}{z}$ and we easily confirm this by noting that the function has the Laurent expansion

$$\sin \frac{1}{z} = \sum_{m=0}^{\infty} \frac{(-1)^m}{(2m+1)!} \frac{1}{z^{2m+1}}, \quad |z| > 0.$$

The zeros of $\tan \frac{1}{z}$ are the same as those of $\sin \frac{1}{z}$. However, in this case $z = 0$ is not only the limit point of a sequence of zeros, it is also the limit point of a sequence of poles located at $z = \frac{2}{n\pi}$, $n = \pm 1, \pm 3, \ldots$ Therefore, for $\tan \frac{1}{z}$, $z = 0$ is a **non-isolated** essential singularity.

The assertion that "analytic functions are determined by their singularities" should be becoming a familiar refrain. The next theorem provides an example of how the "determination" occurs for a relatively simple class of functions.

Theorem: A function $f(z)$ which is meromorphic throughout the extended plane is necessarily a rational function.

Proof: Since $f(z)$ has no singularities other than poles, and since an infinite number of poles implies the existence of a non-isolated essential singularity, it follows that the poles must be finite in number. Suppose that these are located at the points $z_1, z_2, \ldots, z_m$ and ∞.

The principal part of $f(z)$ at $z = z_k$ may be written

$$\frac{b_{k_1}}{z - z_k} + \frac{b_{k_2}}{(z - z_k)^2} + \ldots + \frac{b_{k_{n_k}}}{(z - z_k)^{n_k}},$$

where n_k is the order of the pole at $z = z_k$, while the principal part at $z = \infty$ is of the form

$$b_1 z + b_2 z^2 + \ldots + b_n z^n.$$

Therefore, let us consider the function

$$D(z) = f(z) - \sum_{k=1}^{m} \left[\frac{b_{k_1}}{z - z_k} + \ldots + \frac{b_{k_{n_k}}}{(z - z_k)^{n_k}} \right] - [b_1 z + \ldots + b_n z^n].$$

Since $D(z)$ has a Taylor expansion about every point in the extended plane, including the points $z_1, z_2, \ldots, z_m$ and $z = \infty$, it is a bounded, entire function. Thus, by Liouville's Theorem, $D(z) = b_0$, a constant. This means that $f(z)$ has the partial fraction representation

$$f(z) = \sum_{k=1}^{m} \left[\frac{b_{k_1}}{z - z_k} + \ldots + \frac{b_{k_{n_k}}}{(z - z_k)^{n_k}} \right] + b_0 + b_1 z + \ldots + b_n z^n$$

and so, is a rational function.

To complete this Section we now visit a second theorem on meromorphic functions, one that involves their zeros as well as their poles.

Theorem: Let $f(z)$ be meromorphic in a simply connected domain D and $g(z)$ be any function which is holomorphic in D. Further, let C be a closed contour in D which does not pass through any of the poles or zeros of $f(z)$. If $f(z)$ has, within C, Z zeros at $z = a_j$ of order m_j, $j = 1, 2, \ldots, Z$, and P poles at $z = b_k$ of order n_k, $k = 1, 2, \ldots, P$, then

$$\frac{1}{2\pi i} \int_C g(z) \frac{f'(z)}{f(z)} dz = \sum_{j=1}^{Z} m_j g(a_j) - \sum_{k=1}^{P} n_k g(b_k). \qquad (2.5.11)$$

Proof: Let z_k be either a zero or a pole of $f(z)$ of order m_k or n_k, respectively. Then, in the neighbourhood of this point, we can set

$$f(z) = (z - z_k)^{l_k} \varphi(z)$$

where $\varphi(z)$ is holomorphic and non-zero at z_k, and $l_k = m_k$ if z_k is a zero, while $l_k = -n_k$ if z_k is a pole. Differentiating, we have

$$f'(z) = l_k (z - z_k)^{l_k - 1} \varphi(z) + (z - z_k)^{l_k} \varphi'(z)$$

and therefore,

$$\frac{f'(z)}{f(z)} = \frac{l_k}{z - z_k} + \frac{\varphi'(z)}{\varphi(z)} = \frac{l_k}{z - z_k} + \text{(a holomorphic function)}.$$

Since $g(z)$ is holomorphic throughout D, Cauchy's Integral and Cauchy's Theorem then yield

$$\frac{1}{2\pi i} \int_{C_k} g(z) \frac{f'(z)}{f(z)} dz = \frac{l_k}{2\pi i} \int_{C_k} \frac{g(z)}{z - z_k} dz = g(z_k) l_k \qquad (2.5.12)$$

for any closed contour C_k which encircles z_k alone of all the zeros and poles of $f(z)$.

Let us now consider a curve C in D which encloses all such contours C_k. Since $g(z) \frac{f'(z)}{f(z)}$ is singular only at the points $z_1, z_2, \ldots, z_n$, we can use the generalized Cauchy Theorem, equation (2.2.8), to write

$$\int_C g(z) \frac{f'(z)}{f(z)} dz = \sum_{k=1}^{Z+P} \int_{C_k} g(z) \frac{f'(z)}{f(z)} dz. \qquad (2.5.13)$$

Thus, combining (2.5.12) and (2.5.13), we find

$$\frac{1}{2\pi i} \int_C g(z) \frac{f'(z)}{f(z)} dz = \sum_{j=1}^{Z} m_j g(a_j) - \sum_{k=1}^{P} n_k g(b_k)$$

as required.

An interesting special case obtains when $g(z) \equiv 1$. Equation (2.5.11) then reads

$$\frac{1}{2\pi i} \int_C \frac{f'(z)}{f(z)} dz = M - N \qquad (2.5.14)$$

where $M = \sum_{j=1}^{Z} m_j$ is the number of zeros and $N = \sum_{k=1}^{P} n_k$ is the number of poles of $f(z)$ inside C, zeros and poles being counted with their proper multiplicities. Since

$$\frac{d}{dz} \ln f(z) = \frac{f'(z)}{f(z)},$$

we can write this result in the form

$$\int_C \frac{f'(z)}{f(z)} dz = \Delta_C[\ln f(z)] = 2\pi i (M - N)$$

where Δ_C denotes the variation of $\ln f(z)$ around the contour C. Writing $\ln z = \ln |f(z)| + i \arg(f(z))$ and noting that $\ln |f(z)|$ is single-valued, we see that equation (2.5.14) in fact states that

$$\Delta_C[\arg(f(z))] = 2\pi (M - N) \qquad (2.5.15)$$

which means that as z describes the simple closed path C, the argument of $f(z)$ changes by an integer multiple of 2π according to the number of zeros and poles of $f(z)$ contained within C.

This result is known as the **principle of the argument**.

3 The Calculus of Residues

3.1 The Residue Theorem

A very important application of the theory of analytic functions involves the evaluation of **real** definite integrals. The key ingredients in the evaluation procedure are the concept of a **residue** and an associated theorem. Thus, our first task is to familiarize ourselves with both.

Definition: Let f(z) be holomorphic in some deleted neighbourhood of $z = z_0$, $0 < |z - z_0| < R$ say, and let C be any closed contour within this neighbourhood and enclosing $z = z_0$. Then, the integral

$$\frac{1}{2\pi i} \int_C f(z)dz = \text{Res}[f(z_0)] \tag{3.1.1}$$

is independent of the choice of C and is called the **residue** of $f(z)$ at $z = z_0$.

Since $f(z)$ is holomorphic in $0 < |z - z_0| < R$, it must possess the Laurent series representation

$$f(z) = \sum_{m=-\infty}^{\infty} c_m(z - z_0)^m, \quad 0 < |z - z_0| < R,$$

$$c_m = \frac{1}{2\pi i} \int_C \frac{f(\zeta)}{(\zeta - z_0)^{m+1}} d\zeta \tag{3.1.2}$$

where C is any closed contour in the annulus $0 < |\zeta - z_0| < R$ that encloses $\zeta = z_0$. Comparing (3.1.1) and (3.1.2) we see that an equivalent definition of the residue of $f(z)$ at $z = z_0$ is

$$\text{Res}[f(z_0)] = c_{-1}. \tag{3.1.3}$$

Equation (3.1.3) applies only to points in the finite plane. However, our first definition (3.1.1) can be applied **at infinity** as well, provided one does so with care. If $f(z)$ is holomorphic or has an isolated singularity at $z = \infty$, it must be possible to define a large circle C that encloses **all** the finite singularities of $f(z)$. The circle C lies in an annulus $R < |z| < \infty$ in which $f(z)$ is holomorphic and it encloses the point at infinity. Thus, (3.1.1) may be used with this curve to define

$$\text{Res}[f(\infty)] = \frac{-1}{2\pi i} \int_C f(z)dz \tag{3.1.4}$$

where the minus sign is due to an anticlockwise circuit with respect to $z = \infty$ being a clockwise circuit with respect to the origin. Let us now apply the transformation $z = \frac{1}{\zeta}$ to the integral in (3.1.4). Since this transformation again reverses the sense of the integration, and $dz = \frac{-d\zeta}{\zeta^2}$, we obtain

$$\text{Res}[f(\infty)] = \frac{-1}{2\pi i} \int_{C'} f(\zeta^{-1}) \frac{d\zeta}{\zeta^2} \tag{3.1.5}$$

where C' is a small circle about $\zeta = 0$. From Cauchy's Integral we then have

$$\text{Res}[f(\infty)] = \lim_{\zeta \to 0} \left[\frac{-f(\zeta^{-1})}{\zeta} \right] = \lim_{z \to \infty} [-zf(z)], \tag{3.1.6}$$

provided this limit exists.

This formula brings out an interesting distinction between residues at infinity and residues at points in the finite plane. If $f(z)$ is holomorphic at $z = z_0$, then $\text{Res}[f(z_0)] = 0$ by Cauchy's Theorem. But if $f(z)$ is holomorphic at infinity, then in general $\text{Res}[f(\infty)] \neq 0$. For example, the rational function $f(z) = \frac{c}{z}$ has $\text{Res}[f(0)] = c$ and $\text{Res}[f(\infty)] = -c$ even though it is clearly holomorphic at the latter point. **Cauchy's Residue Theorem** implies a relationship between a function's residue at infinity and its residues in the finite plane which in turn, determines whether the former will vanish.

Theorem: If $f(z)$ is holomorphic on and within a closed contour C except for a finite number of isolated singularities at $z = z_1, z_2, \ldots, z_n$ inside C, then

$$\int_C f(z)dz = 2\pi i \sum_{k=1}^{n} \text{Res}[f(z_k)]. \tag{3.1.7}$$

Proof: The proof of this theorem involves little more than an application of the generalized Cauchy Theorem.

As we did for our last theorem, we individually enclose each singularity $z = z_k$ with a small circle C_k contained within C. Then, since $f(z)$ is holomorphic within and on the boundary of the $(n + 1)$-fold connected domain bounded by $C, C_1, \ldots, C_n$, we can apply equation (2.2.8) and write

$$\int_C f(z)dz = \sum_{k=1}^{n} \int_{C_k} f(z)dz.$$

Invoking the definition of a residue, equation (3.1.1), this immediately yields

$$\int_C f(z)dz = 2\pi i \sum_{k=1}^{n} \text{Res}[f(z_k)]$$

as required.

If a function's singularities are all isolated, then it follows from (3.1.4) and (3.1.7) that

$$\text{Res}[f(\infty)] = -\sum_k \text{Res}[f(z_k)]$$

where the sum is taken over all singular points in the finite plane. Thus, it is the function's behaviour throughout the finite plane that determines whether its residue at infinity will vanish.

An immediate application of the residue theorem is the evaluation of closed contour integrals. Provided that the integrand possesses only isolated singularities within (or alternatively, without) the contour, the evaluation is reduced to the considerably less arduous task of calculating residues.

3.2 Calculating Residues

The basis of all residue calculus techniques is equation (3.1.3) which identifies the residue of a function $f(z)$ at the point $z = z_0$ with the coefficient of the first inverse power of $(z - z_0)$ in the Laurent expansion of $f(z)$ about $z = z_0$. The most direct approach and the only one available when $z = z_0$ is an essential singularity is to use one of the **practical methods** (see Section 2.4.4) available for generating power series and determine the **one** coefficient we need. However, if $z = z_0$ is a pole of order n, there is an alternative approach which is frequently but by no means invariably more convenient. It is based on the fact that within the annulus of convergence, $0 < |z - z_0| < R$, of the Laurent expansion of $f(z)$ about $z = z_0$, we may set

$$f(z) = (z - z_0)^{-n} g(z) \tag{3.2.1}$$

where $g(z)$ is holomorphic within $|z - z_0| < R$ and is non-zero at $z = z_0$. Putting (3.2.1) into the defining equation (3.1.1), we have

$$\text{Res}[f(z_0)] = \frac{1}{2\pi i} \int_C \frac{g(\zeta)}{(\zeta - z_0)^n} d\zeta = \frac{1}{(n-1)!} \frac{d^{n-1}}{dz^{n-1}} g(z) \Big|_{z=z_0}$$

where we have used Cauchy's Differentiation Formula (2.3.5) in the last step. Thus, substituting for $g(z)$ we obtain

$$\text{Res}[f(z_0)] = \lim_{z \to z_0} \frac{1}{(n-1)!} \frac{d^{n-1}}{dz^{n-1}} [(z - z_0)^n f(z)]. \tag{3.2.2}$$

In the case of a **simple pole** ($n = 1$) this expression reduces to

$$\text{Res}[f(z_0)] = \lim_{z \to z_0} (z - z_0) f(z) \tag{3.2.3}$$

and, in the case of a function of the form $f(z) = \frac{g(z)}{h(z)}$ with $h(z_0) = 0$, $\frac{dh(z_0)}{dz} \neq 0$ and $g(z_0) \neq 0$, (3.2.3) in turn reduces to

$$\text{Res}[f(z_0)] = \frac{g(z_0)}{h'(z_0)}. \tag{3.2.4}$$

Simple poles arising from simple zeros are sufficiently common that this last special case will become memorized through use.

Examples: Consider the rational function

$$f(z) = \frac{z^2 - 3z + 5}{(z-2)^2(z+1)}.$$

By performing a partial fraction decomposition,

$$f(z) = \frac{1}{(z-2)^2} + \frac{1}{z+1},$$

we can read off the values of the residues at $z = 2$ and $z = -1$. Since the second term is holomorphic at $z = 2$, there can be no term involving the power $(z-2)^{-1}$ in the Laurent expansion of $f(z)$ about $z = 2$. Therefore, Res$[f(2)] = 0$. Similarly, since the first term is holomorphic at $z = -1$, we have Res$[f(-1)] = 1$. These results can be verified by using equation (3.2.2) . For example,

$$\text{Res}[f(2)] = \lim_{z \to 2} \frac{d}{dz} \left[\frac{z^2 - 3z + 5}{z + 1} \right] = \lim_{z \to 2} \frac{z^2 + 2z - 8}{(z + 1)^2} = 0.$$

Next, suppose that we wish to calculate the residue of

$$f(z) = \frac{e^{3z}}{(z + 2)(z - 1)^4}$$

at $z = 1$ where it has a fourth-order pole. Use of (3.2.2) would involve the calculation of the third derivative of $\frac{e^{3z}}{(z+2)}$ which, while not overly difficult, is sufficiently lengthy to pose some risk of error. On the other hand, the Laurent series approach in this case is relatively straightforward and hence, less likely to give rise to lost minus signs or factors of two.

We need to expand both e^{3z} and $(z + 2)^{-1}$ about $z = 1$. The exponential is entire and has the mth derivative

$$\frac{d^m}{dz^m} e^{3z} \Big|_{z=1} = 3^m e^{3x} \big|_{x=1} = 3^m e^3.$$

Therefore,

$$e^{3z} = e^3 \sum_{m=0}^{\infty} \frac{3^m}{m!} (z - 1)^m, \quad |z - 1| < \infty.$$

The function $(z + 2)^{-1}$ is holomorphic in $|z - 1| < 3$ and so admits the Taylor series

$$\frac{1}{z + 2} = \frac{1}{3} \sum_{m=0}^{\infty} \frac{(-1)^m}{3^m} (z - 1)^m, |z - 1| < 3.$$

Now, to obtain the coefficient of $(z - 1)^{-1}$ in the Laurent expansion of $f(z) = \frac{e^{3z}}{(z-1)^4(z+2)}$ about $z = 1$, we need only determine the coefficient of $(z - 1)^3$ in the product series

$$\frac{e^{3z}}{z + 2} = \left[e^3 \sum_{m=0}^{\infty} \frac{3^m}{m!} (z - 1)^m \right] \left[\frac{1}{3} \sum_{m=0}^{\infty} \frac{(-1)^m}{3^m} (z - 1)^m \right] = \frac{e^3}{3} \sum_{m=0}^{\infty} \sum_{k=0}^{m} \frac{3^k}{k!} \frac{(-1)^{m-k}}{3^{m-k}} (z - 1)^m.$$

Thus,

$$\mathrm{Res}[f(1)] = \frac{e^3}{3}\sum_{k=0}^{3}\frac{3^{2k-3}}{k!}(-1)^{3-k} = \frac{e^3}{3}\left[\frac{3^3}{3!} - \frac{3}{2!} + \frac{1}{3} - \frac{1}{3^3}\right] = \frac{89}{81}e^3.$$

Suppose now that we wish to integrate

$$f(z) = \frac{\tan z}{z}$$

around the circle $|z| = 2$. Writing $f(z)$ in the form

$$f(z) = \frac{(\sin z)/z}{\cos z}$$

we see that it is holomorphic at $z = 0$ but possesses first order poles at $z = \pm\pi/2$ arising from the first order zeros of $\cos z$ at these two points. We can use equation (3.2.4) to calculate the residues at the poles; we find

$$\mathrm{Res}[f(\pi/2)] = \frac{\sin(\pi/2)/(\pi/2)}{\frac{d}{dz}\cos z\Big|_{z=\pi/2}} = -\frac{2}{\pi}$$

$$\mathrm{Res}[f(-\pi/2)] = \frac{-\sin(-\pi/2)/(\pi/2)}{\frac{d}{dz}\cos z\Big|_{z=-\pi/2}} = \frac{2}{\pi}.$$

Without further effort, the Residue Theorem gives us the following value for the integral in question:

$$\int_{|z|=2}\frac{\tan z}{z}dz = 2\pi i\left[\frac{2}{\pi} - \frac{2}{\pi}\right] = 0.$$

A somewhat more difficult and more interesting problem is posed by the integral

$$I = \int_C \frac{1}{z}\sin\frac{1}{z}\,\mathrm{cosec}\,z\,dz$$

where C is the unit circle, $|z| = 1$. Both $\frac{1}{z}$ and $\mathrm{cosec}\,z$ have first order poles at $z = 0$, but $\sin\frac{1}{z}$ has an essential singularity there. Therefore, we have no choice in this case: the Laurent series method of calculating residues is the only one that is applicable.

We require the expansions of $\mathrm{cosec}\,z$ and $\sin\frac{1}{z}$ about $z = 0$. These were found in Section 2.2.4 and are

$$\sin\frac{1}{z} = \sum_{m=0}^{\infty}\frac{(-1)^m}{(2m+1)!}\frac{1}{z^{2m+1}}, \quad |z| > 0$$

$$\mathrm{cosec}\,z = \frac{1}{z} + \frac{z}{6} + \frac{7}{360}z^3 + \frac{31}{15120}z^5 + \dots, \quad 0 < |z| < \pi.$$

Because of the presence of the $\frac{1}{z}$ factor in the integrand, we now must find the coefficient of $(z)^0$ in the Cauchy product of these two series. This is easily accomplished and yields the convergent series

$$\frac{1}{6}\frac{(-1)^0}{1!} + \frac{7}{360}\frac{(-1)^1}{3!} + \frac{31}{15120}\frac{(-1)^2}{5!} + \dots.$$

Therefore, our integral has the value

$$\int\limits_{|z|=1} \frac{1}{z}\sin\frac{1}{z}\cosec z\, dz = 2\pi i\left[\frac{1}{6} - \frac{7}{3!360} + \frac{31}{5!15120} - + \dots\right].$$

The most important application of the Residue Theorem is in the evaluation of certain types of **real definite integrals.**

3.3 Evaluating Definite Integrals

3.3.1 Angular Integrals

Integrals of the type

$$\int\limits_0^{2\pi} f(\cos\theta, \sin\theta)d\theta$$

can often be addressed successfully by making the substitution $e^{i\theta} = z$. One then has

$$\cos\theta = \frac{1}{2}\left(z + \frac{1}{z}\right), \quad \sin\theta = \frac{1}{2i}\left(z - \frac{1}{z}\right), \quad d\theta = \frac{dz}{iz}$$

and hence,

$$\int\limits_0^{2\pi} f(\cos\theta, \sin\theta)d\theta = \int\limits_C F(z)dz$$

where C is the unit circle $|z| = 1$. Of particular interest are integrands $f(\cos\theta, \sin\theta)$ that are rational functions of $\cos\theta$ and $\sin\theta$. The corresponding functions $F(z)$ will themselves be rational functions of z and so the Residue Theorem can be applied to give

$$\int\limits_0^{2\pi} f(\cos\theta, \sin\theta)d\theta = \int\limits_C F(z)dz = 2\pi i\sum_C \mathrm{Res}[F(z)] \qquad (3.3.1)$$

where the sum is taken over those poles of $F(z)$ that lie inside the unit circle.
Example: Let us evaluate the integral

$$I \equiv \int\limits_0^{2\pi} \frac{\sin^2\theta}{a + b\cos\theta}d\theta, \quad a > |b| > 0$$

where the restriction on a and b is needed to prevent the integrand from becoming undefined at any point on the interval of integration.

On making the recommended change of variable, the integral becomes

$$I = \int_C \frac{(z^2 - 1)^2}{(2iz)^2} \frac{1}{a + \frac{b}{2z}(z^2 + 1)} \frac{dz}{iz} = \frac{i}{2b} \int_C \frac{(z^2 - 1)^2}{z^2 \left(z^2 + \frac{2a}{b}z + 1\right)} dz$$

where C is the unit circle, $|z| = 1$. The integrand

$$F(z) = \frac{(z^2 - 1)^2}{z^2 \left(z^2 + \frac{2a}{b}z + 1\right)}$$

has a double pole at $z = 0$ and simple poles at $z = \zeta_1$ and $z = \zeta_2$ where $\zeta_1 = \frac{-a}{b} + \sqrt{\frac{a^2}{b^2} - 1}$ and $\zeta_2 = \frac{-a}{b} - \sqrt{\frac{a^2}{b^2} - 1}$ are the roots of the quadratic $z^2 + \frac{2a}{b}z + 1 = 0$. Since the product $\zeta_1 \zeta_2$ must equal unity, only the root with the smallest absolute value can lie within the unit circle. With $a > |b| > 0$, this is evidently ζ_1. Thus, we need only find the residues of $F(z)$ at $z = \zeta_1$ and $z = 0$.

From equation (3.2.3) we have

$$\text{Res}[F(\zeta_1)] = \lim_{z \to \zeta_1} (z - \zeta_1)F(z) = \lim_{z \to \zeta} \frac{(z^2 - 1)^2}{z^2(z - \zeta_2)} = \frac{(\zeta_1 - \zeta_1^{-1})^2}{(\zeta_1 - \zeta_2)} = (\zeta_1 - \zeta_2) = 2\sqrt{\frac{a^2}{b^2} - 1},$$

where we have made use of $\zeta_1 \zeta_2 = 1$ and hence, $\zeta_1^{-1} = \zeta_2$.

Then, using equation (3.2.2), we find

$$\text{Res}[F(0)] = \lim_{z \to 0} \frac{d}{dz}[z^2 F(z)] = \lim_{z \to 0} \frac{d}{dz} \left[\frac{(z^2 - 1)^2}{z^2 + \frac{2a}{b}z + 1} \right]$$

or,

$$\text{Res}[F(0)] = \lim_{z \to 0} \frac{\left(z^2 + \frac{2a}{b}z + 1\right) 4z(z^2 - 1) - (z^2 - 1)^2 \left(2z + \frac{2a}{b}\right)}{\left(z^2 + \frac{2a}{b}z + 1\right)^2} = -\frac{2a}{b}.$$

Thus, we finally obtain

$$I = \frac{i}{2b} 2\pi i \sum_C \text{Res}[F(z)] = -\frac{\pi}{b} \left[-\frac{2a}{b} + 2\sqrt{\frac{a^2}{b^2} - 1} \right] = \frac{2\pi}{b^2}[a - \sqrt{a^2 - b^2}].$$

3.3.2 Improper Integrals of Rational Functions

Real integrals with an infinite interval of integration are called **improper** and are defined by the limiting procedure

$$\int_{-\infty}^{\infty} f(x)dx = \lim_{a \to -\infty} \int_a^0 f(x)dx + \lim_{b \to \infty} \int_0^b f(x)dx. \tag{3.3.2}$$

If **both** of these limits exist, we can set

$$\int_{-\infty}^{\infty} f(x)dx = \lim_{R\to\infty} \int_{-R}^{R} f(x)dx. \tag{3.3.3}$$

This is not an alternative definition of the improper integral; the limit in (3.3.3) may exist even if the integral does not. For example,

$$\lim_{R\to\infty} \int_{-R}^{R} xdx = \lim_{R\to\infty} \frac{x^2}{2}\Big|_{-R}^{R} = 0$$

but $\lim_{b\to\infty} \int_{0}^{b} xdx$ is unbounded and therefore, so is $\int_{-\infty}^{\infty} xdx$. However, since our interest is in the evaluation of integrals that are presumed to exist, we shall always be able to invoke (3.3.3).

Theorem: Let $f(z)$ satisfy the following conditions:

i $f(z)$ is meromorphic in the upper (lower) half-plane,

ii $f(z)$ has no poles on the real axis,

iii $zf(z) \to 0$ uniformly as $|z| \to \infty$, for $0 \le \arg z \le \pi(-\pi \le \arg z \le 0)$,

iv $\int_{-\infty}^{\infty} f(x)dx$ exists, where $f(x) = \lim_{y\to 0} f(x + iy)$.

Then,

$$\int_{-\infty}^{\infty} f(x)dx = 2\pi i \sum_{+} \text{Res}[f(z)] \tag{3.3.4}$$

$$\left(\int_{-\infty}^{\infty} f(x)dx = -2\pi i \sum_{-} \text{Res}[f(z)] \right) \tag{3.3.5}$$

where $\sum_{+} \left(\sum_{-} \right)$ denotes the sum over all the poles of $f(z)$ in the upper (lower) half-plane.

Proof: As shown in Figure 3.1, we define a semicircular contour C_R, centre the origin and radius R, with R being sufficiently large that C_R and the real axis segment $-R \le x \le R$ together enclose all the upper half-plane poles of $f(z)$. Then, by the Residue Theorem

$$\int_{-R}^{R} f(x)dx + \int_{C_R} f(z)dz = 2\pi i \sum_{+} \text{Res}[f(z)]. \tag{3.3.6}$$

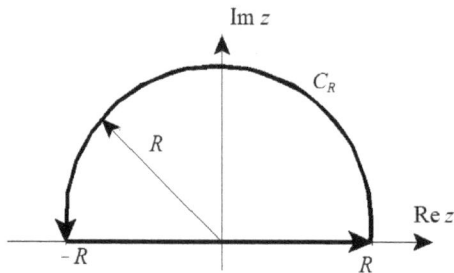

Figure 3.1: The contour used in the integration performed in equation (3.3.6).

From (iii) , we must have $|zf(z)| < \varepsilon(R)$ for all points on C_R, where $\varepsilon(R)$ is a positive number that depends only on R and tends to zero as $R \to \infty$. Thus,

$$\left| \int_{C_R} f(z)dz \right| = \left| \int_0^\pi f(Re^{e\theta})Re^{i\theta}id\theta] \right| < \varepsilon(R) \int_0^\pi d\theta$$

and, as $R \to \infty$, the integral around C_R tends to zero. Therefore, if (iv) is satisfied it follows that

$$\int_{-\infty}^\infty f(x)dx = \lim_{R\to\infty} \int_{-R}^R f(x)dx = 2\pi i \sum_+ \text{Res}[f(z)].$$

If the conditions in the statement of the theorem hold in the lower half-plane rather than the upper, then the semi-circular contour must be chosen to lie in that half-plane and equation (3.3.6) becomes

$$-\int_{-R}^R f(x)dx + \int_{C_R} f(z)dz = 2\pi i \sum_- \text{Res}[f(z)]. \tag{3.3.7}$$

The minus sign in front of the integral along the real axis is necessary to preserve a counter-clockwise direction of integration. Thus, in the limit as $R \to \infty$, we obtain

$$\int_{-\infty}^\infty f(x)dx = -2\pi i \sum_- \text{Res}[f(z)].$$

The condition $xf(x) \to 0$ as $|x| \to \infty$ is not in itself sufficient to guarantee the existence of $\int_{-\infty}^\infty f(x)dx$. Thus, conditions (iii) and (iv) are **both** required in the statement of the theorem.

A particular class of functions which satisfy all four conditions are the rational functions $f(z) = P(z)/Q(z)$ where

i   the degree of Q exceeds that of P by at least 2 , and

ii Q has no real zeros.

We shall illustrate the use of the theorem by evaluating two improper integrals whose integrands are rational functions of this type.

Consider the integral

$$I = \int_0^\infty \frac{x^2}{(x^2 + 4)^2(x^2 + 9)} dx$$

which, because its integrand is an even function of x, can be rewritten as

$$I = \frac{1}{2} \int_{-\infty}^\infty \frac{x^2}{(x^2 + 4)^2(x^2 + 9)} dx.$$

The numerator of

$$f(z) = \frac{z^2}{(z^2 + 9)(z^2 + 4)^2}$$

has degree 2, while its denominator has degree 6. In addition, the denominator has simple zeros at $z = \pm 3i$ and second order zeros at $z = \pm 2i$. Thus, the conditions of the theorem are satisfied and so we need only calculate the residues of $f(z)$ at its two poles in the upper half- plane to evaluate I.

Using equation (3.2.3) , we have

$$\text{Res}[f(3i)] = \lim_{z \to 3i} \frac{z^2}{(z + 3i)(z^2 + 4)^2} = \frac{3i}{50}$$

while, from (3.2.2) ,

$$\text{Res}[f(2i)] = \lim_{z \to 2i} \frac{d}{dz} \left[\frac{z^2}{(z^2 + 9)(z + 2i)^2} \right]$$

$$= \lim_{z \to 2i} \frac{(z^2 + 9)(z + 2i)^2 2z - z^2[2z(z + 2i)^2 + 2(z^2 + 9)(z + 2i)]}{(z^2 + 9)^2(z + 2i)^4} = -\frac{13i}{200}.$$

Equation (3.3.4) then yields

$$I = \frac{1}{2} \int_{-\infty}^\infty f(x)dx = \pi i \left[\frac{3i}{50} - \frac{13i}{200} \right] = \frac{\pi}{200}.$$

Next, consider the integral

$$I = \int_0^\infty \frac{1}{x^4 + a^4} dx, \quad a > 0.$$

We again take advantage of the evenness of the integrand to write

$$I = \frac{1}{2} \int_{-\infty}^\infty \frac{1}{x^4 + a^4} dx$$

and then consider the analytic properties of the function $f(z) = (z^4 + a^4)^{-1}$. As in the preceding example the degree of the denominator exceeds that of the numerator by four and none of the zeros of the denominator are real. Indeed, the zeros, which are poles of $f(z)$, are all simple and are located at the points $z = ae^{i\pi/4}, ae^{i3\pi/4}, ae^{i5\pi/4}, ae^{i7\pi/4}$. Of these, only the first two are in the upper half-plane.

The simple form of $f(z)$ makes equation (3.2.4) the most convenient method for calculating its residues. We find

$$\text{Res}[f(\alpha)] = \lim_{z \to \alpha} \frac{1}{\frac{d}{dz}(z^4 + a^4)} = \frac{1}{4\alpha^3}$$

where α denotes either $ae^{i\pi/4}$ or $ae^{i3\pi/4}$. Thus, $\text{Res}[f(ae^{i\pi/4})] = \frac{e^{-i3\pi/4}}{4a^3} = -\frac{e^{i\pi/4}}{4a^3}$ and $\text{Res}[f(ae^{i3\pi/4})] = \frac{e^{-i9\pi/4}}{4a^3} = \frac{e^{-i\pi/4}}{4a^3}$ which yields, from equation (3.3.4),

$$I = \frac{1}{2}\int_{-\infty}^{\infty} f(x)dx = \frac{\pi i}{4a^3}\left[e^{-\pi/4} - e^{i\pi/4}\right] = \frac{\pi}{2a^3}\sin\frac{\pi}{4} = \frac{\pi}{2\sqrt{2}a^3}.$$

3.3.3 Improper Integrals Involving Trigonometric Functions

Fourier transforms play an important role in the description of wave motion, from acoustics through optics to quantum mechanics. Concomitant with this importance is a need for proficiency in the evaluation of real improper integrals of the form

$$\int_{-\infty}^{\infty} f(x)e^{ikx}dx, \quad \int_{0}^{\infty} f(x)\cos kxdx, \quad \int_{0}^{\infty} f(x)\sin kxdx.$$

Thus, a signal application of complex analysis follows from the recognition that if the complex analogue of $f(x), f(z)$, satisfies the conditions stated in the theorem proved in the preceding subsection, the first of these integrals can be evaluated by means of the formula

$$\int_{-\infty}^{\infty} f(x)e^{ikx}dx = \begin{cases} 2\pi i \sum_{+} \text{Res}[f(z)e^{ikz}], & k > 0 \\ -2\pi i \sum_{-} \text{Res}[f(z)e^{ikz}], & k < 0. \end{cases} \tag{3.3.8}$$

Separating both sides of this equation into their real and imaginary parts, we obtain statements that address the other two integrals:

$$\int_{-\infty}^{\infty} f(x)\cos kxdx = \begin{cases} -2\pi \sum_{+} \text{Im}\{\text{Res}[f(z)e^{ikz}]\}, & k > 0 \\ 2\pi \sum_{-} \text{Im}\{\text{Res}[f(z)e^{ikz}]\}, & k < 0. \end{cases} \tag{3.3.9}$$

and,

$$\int\limits_{-\infty}^{\infty} f(x) \sin kx\, dx = \begin{cases} 2\pi \sum\limits_{+} \text{Re}\{\text{Res}[f(z)e^{ikz}]\}, & k > 0 \\ -2\pi \sum\limits_{-} \text{Re}\{\text{Res}[f(z)e^{ikz}]\}, & k < 0. \end{cases} \tag{3.3.10}$$

The recognition referred to above is that if $k > 0$ and if z lies on a semi-circle C_R in the upper half-plane, then $|e^{ikz}| = |e^{ikx}|e^{-ky} = e^{-ky} \le 1$ and, hence $|f(z)e^{ikz}| \le |f(z)|$ for all z on C_R. Similarly, if $k < 0$ and C_R is in the **lower** half-plane, then we again have $|f(z)e^{ikz}| = |f(z)|e^{-|k||y|} \le |f(z)|$ for all z on C_R. Thus, using one or the other of the contours of the preceding theorem according to the sign of k, the contribution from integration along C_R goes to zero even faster than before as $R \to \infty$ and equation (3.3.8) results. Indeed, since e^{ikz} evidently acts as a "convergence factor", one might surmise that equation (3.3.8) applies to a wider class of functions than that defined by the conditions of Section 3.3.2. A theorem known as **Jordan's Lemma** confirms this expectation.

Theorem: If C_R is a semi-circle in the upper (lower) half-plane, centre the origin and radius R, and if $f(z)$ satisfies the conditions

i $f(z)$ is meromorphic in the upper (lower) half-plane,

ii $f(z) \to 0$ uniformly as $|z| \to \infty$ for $0 \le \arg z \le \pi (-\pi \le \arg z \le 0)$,

then, $\int_{C_R} e^{ikz} f(z) dz \to 0$ as $R \to \infty$ where k is any real positive (negative) number.

Proof: By (ii) we have, for all points on C_R, $|f(z)| < \varepsilon(R)$ where $\varepsilon(R)$ is a positive number that depends only on R and tends to zero as $R \to \infty$. Now, for z on C_R,

$$|e^{ikz}| = |e^{ikR(\cos\theta + i\sin\theta)}| = e^{-kR\sin\theta}.$$

Hence,

$$\left| \int\limits_{C_R} f(z)e^{ikz}\, dz \right| = \left| \int\limits_0^{\pi} f(Re^{i\theta})e^{ikRe^{i\theta}} iRe^{i\theta}\, d\theta \right| < \varepsilon(R) \int\limits_0^{R} e^{-kR\sin\theta} R\, d\theta$$

$$= 2R\varepsilon(R) \int\limits_0^{\pi/2} e^{-kR\sin\theta}\, d\theta$$

where the last step follows from the symmetry of $\sin\theta$ about $\theta = \pi/2$. But, $(\sin\theta)/\theta$ decreases steadily from 1 to $\frac{2}{\pi}$ as θ increases from 0 to $\frac{\pi}{2}$. Thus, if $0 \le \theta \le \pi/2$, $\sin\theta \ge \frac{2\theta}{\pi}$. Hence,

$$\left| \int\limits_{C_R} f(z)e^{ikz}\, dz \right| \le 2R\varepsilon(R) \int\limits_0^{\pi/2} e^{-2kR\theta/pi}\, d\theta = \frac{\pi\varepsilon(R)}{k}[1 - e^{-kR}] < \frac{\pi\varepsilon(R)}{k},$$

from which the theorem follows.

We owe this theorem to Camille Jordan (1838-1922) whose Cours d'analyse *was an influential textbook for the French school of analysts. His fame rested much more on his contributions to group theory than on those to analysis however. He was the leading group theorist of his day, applying group theory to geometry and linear differential equations. Camille Jordan was born in Lyons and was rewarded for his accomplishments by appointments as Professor at the Ecole Polytechnique and the College de France.*

Jordan's Lemma makes evident the proof of the following theorem.

Theorem: Let $f(z)$ be subject to the following conditions:

i $f(z)$ is meromorphic in the upper half-plane,

ii $f(z)$ has no poles on the real axis,

iii $f(z) \to 0$ uniformly as $|z| \to \infty$, for $0 \le \arg z \le \pi$.

Then, for $k > 0$,

$$\int_{-\infty}^{\infty} f(x)e^{ikx}\,dx = 2\pi i \sum_{+} \text{Res}[f(z)e^{ikz}]. \tag{3.3.11}$$

If $k < 0$, and if $f(z)$ is meromorphic in the lower half-plane and tends to zero uniformly as $|z| \to \infty$ for $-\pi \le \arg z \le 0$, then

$$\int_{-\infty}^{\infty} f(x)e^{ikx}\,dx = -2\pi i \sum_{-} \text{Res}[f(z)e^{ikz}]. \tag{3.3.12}$$

A particular set of functions that evidently satisfy the conditions of this theorem is the rational functions $f(z) = P(z)/Q(z)$, where

i   the degree of $Q(z)$ exceeds that of $P(z)$ by at least one, and

ii $Q(z)$ has no real zeros.

Example: Consider the integral

$$I = \int_{0}^{\infty} \frac{\cos kx}{x^2 + a^2}\,dx, \quad a > 0.$$

Invoking the evenness of its integrand, we can set

$$I = \frac{1}{2} \int_{-\infty}^{\infty} \frac{\cos kx}{x^2 + a^2}\,dx.$$

The function $f(z) = (z^2 + a^2)^{-1}$ has simple poles at $z = \pm ai$. Using equation (3.2.4) we find the residues at these poles

$$\text{Res}[f(z)e^{ikz}]_{z=\pm ai} = \lim_{z\to\pm ai} \frac{e^{ikz}}{2z} = \mp i\frac{e^{\mp ka}}{2a}.$$

Therefore, applying our new theorem, we have

$$I = \frac{1}{2} \int\limits_{-\infty}^{\infty} f(x) \cos kx \, dx = \begin{cases} -\pi \, \text{Im}[-ie^{-ka}/2a], & k > 0 \\ +\pi \, \text{Im}[+ie^{+ka}/2a], & k < 0 \end{cases}$$

or, $I = \frac{\pi}{2a} e^{-|k|a}$.

Often, some care is needed when replacing $\cos kx$ or $\sin kx$ by the real or imaginary part of e^{ikx} in an application of (3.3.11) or (3.3.12). For **example**, although the improper integral

$$\int\limits_{0}^{\infty} \frac{\sin kx}{x} \, dx = \frac{1}{2} \int\limits_{-\infty}^{\infty} \frac{\sin kx}{x} \, dx$$

is convergent, it cannot be equated to the imaginary part of

$$\int\limits_{-\infty}^{\infty} \frac{e^{ikx}}{x} \, dx$$

because the pole at $x = 0$ renders the latter undefined. Fortunately, the analytic properties of $\frac{\sin kz}{z}$ offer an alternative route to the successful evaluation of the integral. Since $\frac{\sin kz}{z}$ is entire, its integrals are independent of path everywhere in the finite plane. In particular, we are free to **deform** the contour along the real axis to obtain one that avoids the troublesome point $z = 0$. Any finite deformation will do and, as shown in Figure 3.2, a semicircle with centre at $z = 0$ makes a convenient choice. Denoting the resulting contour by C, we then have

$$I \equiv \frac{1}{2} \int\limits_{-\infty}^{\infty} \frac{\sin kx}{x} \, dx = \lim_{R \to \infty} \frac{1}{2} \int\limits_{-R}^{R} \frac{\sin kx}{x} \, dx = \lim_{R \to \infty} \frac{1}{2} \int\limits_{C} \frac{\sin kz}{z} \, dz.$$

With $z = 0$ thus avoided, we can now make use of $\sin kz = \frac{1}{2i}(e^{ikz} - e^{-ikz})$ to give us

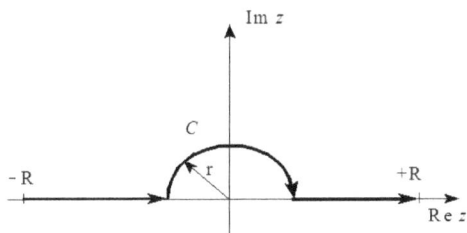

Figure 3.2: A contour along the real axis is deformed by insertion of a semicircle with centre at the origin.

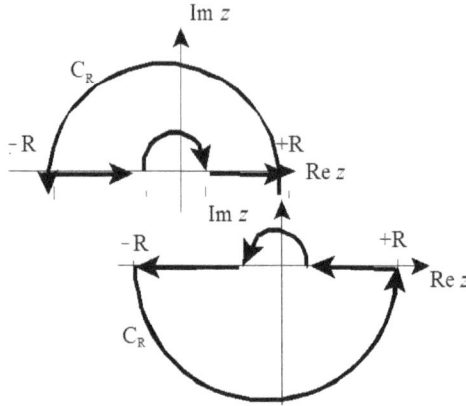

Figure 3.3: The closed contours used to evaluate I_1 and I_2, respectively.

$I = \frac{1}{4i}[I_1 - I_2]$, where $I_1 = \lim\limits_{R\to\infty} \int_C \frac{e^{ikz}}{z}\,dz$ and $I_2 = \lim\limits_{R\to\infty} \int_C \frac{e^{-ikz}}{z}\,dz$. Taking k to be positive, we can evaluate I_1 by closing the contour C with a semi-circle C_R, provided that the latter lies in the upper half-plane. Similarly, I_2 can be evaluated by closing with a semi- circle in the lower half-plane. These choices are dictated by Jordan's Lemma which assures us that the contributions from both semi-circles go to zero as $R \to \infty$. The closed contours are shown in Figure 3.3 which reveals that the simple pole at $z = 0$ lies outside the one chosen for I_1 but inside that used for I_2. Thus, successively using Cauchy's Theorem and the Residue Theorem, we conclude that $I_1 = 0$ and $I_2 = -2\pi i\,\mathrm{Res}\left[\frac{e^{-ikz}}{z}\right]_{z=0} = -2\pi i$, where the minus sign results from preserving a counterclockwise direction of integration. Therefore,

$$I \equiv \int\limits_0^\infty \frac{\sin kx}{x}\,dx = \frac{1}{4i}(2\pi i) = \frac{\pi}{2}, \quad k > 0.$$

Since $\sin kx$ is an odd function of k, our final result is

$$\int\limits_0^\infty \frac{\sin kx}{x}\,dx = \begin{cases} \pi/2, & k > 0 \\ -\pi/2, & k < 0. \end{cases}$$

Evidently, deforming the integration contour can be another valuable aid in the evaluation of real integrals. As we will see in Section 3.3.6 , it can be useful even when the residue theorem is not required.

3.3.4 Improper Integrals Involving Exponential Functions

With exponential or hyperbolic functions present in the integrand, life gets somewhat more complicated due to an unsatisfactory behaviour of the integrands at infinity.

Since these functions have an essential singularity at $z = \infty$, the integrands have no uniform limit as $|z| \to \infty$. Consequently, one cannot make use of a semi-circular arc to close the initial contour. Instead, one defines a rectangular contour that exploits the periodicity of these functions. The following example demonstrates what is involved.

Example: The improper integral

$$I \equiv \int_{-\infty}^{\infty} \frac{e^{ax}}{\cosh \pi x} \, dx, \, -\pi < a < \pi$$

can be shown to converge provided that a is confined to the range indicated. The function $f(z) = e^{az} / \cosh \pi z$ does not vanish uniformly as $|z| \to \infty$ and, in fact, if $z \to \infty$ along the imaginary axis, it does not vanish at all. Therefore, to evaluate I we shall take advantage of $\cosh \pi(z \pm 2i) = \cosh \pi z$. Thus,

$$\frac{e^{a(x+2i)}}{\cosh \pi(x + 2i)} = e^{2ai} \left[\frac{e^{ax}}{\cosh \pi x} \right]$$

and so,

$$\int_{-R}^{R} \frac{e^{a(x+2i)}}{\cosh \pi(x + 2i)} \, dx \equiv \int_{-R+2i}^{R+2i} \frac{e^{az}}{\cosh \pi z} \, dz, \, \text{Im } z = 2$$

$$= e^{2ai} \int_{-R}^{R} \frac{e^{ax}}{\cosh \pi x} \, dx$$

for any R. This suggests that we use the contour shown in Figure 3.4, since we have just established that the integrals along its two horizontal segments are related by a multiplicative constant. Therefore, if we can prove that the integrals along the vertical segments give vanishing contributions in the limit $R \to \infty$, (which is not an unreasonable expectation given the convergence of I) , then we can evaluate I by means of a simple application of the residue theorem.

The integrals along the vertical segments are given by

$$I_V = \pm \int_{0}^{2} \frac{e^{a(\pm R+iy)}}{\cosh \pi(\pm R + iy)} i \, dy.$$

Now,

$$| \cosh \pi(\pm R + iy)|^2 = \cosh^2 \pi R - \sin^2 \pi y \geq \cosh^2 \pi R - 1 = \sinh^2 \pi R.$$

Thus,

$$| \cosh \pi(\pm R + iy)| \geq \sinh \pi R = \frac{1}{2} e^{\pi R}(1 - e^{-2\pi R})$$

and so, by Darboux's Inequality,

$$|I_V| \leq \frac{4e^{-R(\pi \mp a)}}{1 - e^{-2\pi R}} \to 0$$

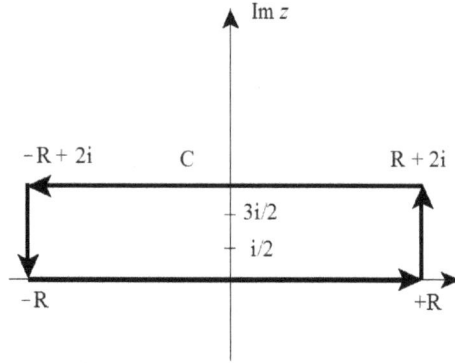

Figure 3.4: The contour used in the evaluation of $\int_{-\infty}^{\infty} \frac{e^{ax}}{\cosh \pi x} dx$, $-\pi < a < \pi$.

as $R \to \infty$, provided that $-\pi < a < \pi$.

Hence, applying the residue theorem to the integral around the **closed** contour shown in Figure 3.4 and taking the limit as $R \to \infty$, we obtain

$$I - e^{2ai}I = 2\pi i \sum_{\text{poles within } C} \text{Res}\left[\frac{e^{az}}{\cosh \pi z}\right],$$

where the first term on the left hand side comes from $\lim_{R\to\infty} \int_{-R}^{R} \frac{e^{ax}}{\cosh \pi x} dx$ and the second from $\lim_{R\to\infty} \int_{R+2i}^{-R+2i} \frac{e^{az}}{\cosh \pi z} dz$, $\text{Im } z = 2$. The singularities of $\frac{e^{az}}{\cosh \pi z}$ in the finite plane are simple poles located at the points $z = \frac{2n+1}{2}i$, $n = 0, \pm 1, \pm 2, \ldots$, of which only two, $z = i/2$ and $z = 3i/2$, lie inside the contour. The residues at these two poles are

$$\text{Res}\left[\frac{e^{az}}{\cosh \pi z}\right]_{z=i/2} = \frac{e^{az}}{\frac{d}{dz}\cosh \pi z}\bigg|_{z=i/2} = \frac{e^{ai/2}}{\pi i},$$

$$\text{Res}\left[\frac{e^{az}}{\cosh \pi z}\right]_{z=3i/2} = \frac{-e^{3ai/2}}{\pi i}.$$

Thus,

$$I(1 - e^{2ai}) = 2\pi i\left[\frac{e^{ai/2}}{\pi i} - \frac{e^{3ai/2}}{\pi i}\right]$$

or,

$$I = \frac{2e^{ai/2}(1 - e^{ai})}{1 - e^{2ai}} = \frac{2e^{ai/2}}{1 - e^{2ai}}$$

and hence, our final result is

$$\int_{-\infty}^{\infty} \frac{e^{ax}}{\cosh \pi x} dx = \sec\frac{a}{2}, \quad -\pi < a < \pi.$$

3.3.5 Integrals Involving Many-Valued Functions

If the integrand of an integral from 0 to ∞ contains a factor like x^α, where α is not an integer, then we are apparently faced with the added complication that its complex analogue z^α is a multi-valued function. However, the supposed complication can often be turned to one's advantage and the means of doing so is to choose the function's branch cut so that it too runs along the real axis from 0 to ∞. The reason for this paradoxical choice is that any single valued branch of the function has a known discontinuity across the cut and so integrals whose contours parallel the real axis immediately above and below the cut are related by a simple multiplicative constant. The next example illustrates how this can be very effectively exploited.

Example: Suppose that we wish to evaluate an integral of the form

$$\int_0^\infty x^{\alpha-1} f(x)\,dx, \quad \alpha \neq \text{ an integer},$$

where $f(z)$ is a rational function which

i   has no poles on the positive real axis, and
ii   satisfies the condition $z^\alpha f(z) \to 0$ uniformly as $|z| \to 0$ and as $|z| \to \infty$.

We start by considering the complex integral

$$I \equiv \int_C z^{\alpha-1} f(z)\,dz \tag{3.3.13}$$

where $z^{\alpha-1}$ denotes a specific branch corresponding to a choice of branch cut running along the positive real axis and where C is a closed contour consisting of

i   a large circle C_R, centre the origin and radius R,
ii   a small circle C_r, centre the origin and radius r, and
iii two parallel straight lines, L_1 and L_2, which join C_R and C_r on either side of the cut.

Although C looks rather dissimilar to the contours in Figures 2.10 and 2.11, it shares the distinction of being a "dogbone" contour. However, its most important feature at this point is that it does not cross the cut thus permitting a single valued definition of the integrand in (3.3.13).

The final result will be independent of our choice of branch for $z^{\alpha-1}$ and so we choose the simplest to work with and set $z^{\alpha-1} = |z|^{\alpha-1} e^{i(\alpha-1)\theta}$, $0 < \theta < 2\pi$. Thus, just above the real axis, on L_1, $z^{\alpha-1} = x^{\alpha-1}$ while, just below the real axis, on L_2, $z^{\alpha-1} = x^{\alpha-1} e^{2\pi i(\alpha-1)}$.

Application of Darboux's Inequality to the integrals around C_R and C_r establishes that they vanish as $R \to \infty$ and $r \to 0$ because we have required that $|z^\alpha f(z)| \to 0$

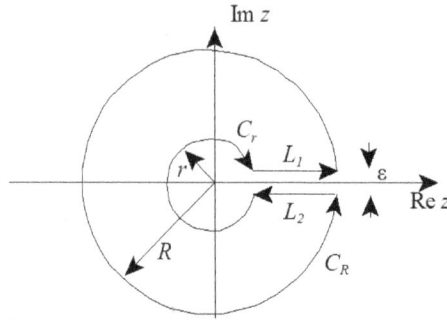

Figure 3.5: The closed contour of integration for the integral in equation (3.3.13).

when both $|z| \to \infty$ and $|z| \to 0$. Therefore,

$$\lim_{R\to\infty} \lim_{r\to 0} \int_C z^{\alpha-1} f(z) dz = -\int_0^\infty e^{2\pi i(\alpha-1)} x^{\alpha-1} f(x) dx + \int_0^\infty x^{\alpha-1} f(x) dx$$

where the first integral on the right is the contribution from L_2 and the second is the contribution from L_1. Together these integrals give

$$\left[1 - e^{2\pi i(\alpha-1)}\right] \int_0^\infty x^{\alpha-1} f(x) dx = \frac{-2i \sin \pi\alpha}{e^{-\alpha \pi i}} \int_0^\infty x^{\alpha-1} f(x) dx.$$

At the same time, the Residue Theorem gives

$$\lim_{R\to\infty} \lim_{r\to 0} \int_C z^{\alpha-1} f(z) dz = 2\pi i \sum_{\text{all poles}} \text{Res}[z^{\alpha-1} f(z)].$$

Thus, we finally obtain

$$\int_0^\infty x^{\alpha-1} f(x) dx = \frac{-\pi e^{-i\pi\alpha}}{\sin \pi\alpha} \sum_{\text{all poles}} \text{Res}[z^{\alpha-1} f(z)]. \qquad (3.3.14)$$

For a specific application of this result, consider the integral $\int_0^\infty \frac{x^{\alpha-1}}{x^2+1} dx$. In order that $z^\alpha/(z^2 + 1) \to 0$ as $|z| \to 0$ and as $|z| \to \infty$, we must restrict α so that it is greater than zero but less than two. Then, since $(z^2 + 1)^{-1}$ has simple poles at $z = \pm i$, we need to calculate

$$\text{Res}\left[\frac{z^{\alpha-1}}{z^2+1}\right]_{z=\pm i} = \frac{z^{\alpha-1}}{2z}\bigg|_{z=\pm i} = -\frac{1}{2}(\pm i)^\alpha.$$

But the principal values of $(\pm i)^\alpha$ are $e^{i\alpha\pi/2}$ and $e^{i3\alpha\pi/2}$, respectively. Thus, we find

$$\int_0^\infty \frac{x^{\alpha-1}}{x^2+1} dx = \frac{\pi e^{-i\pi\alpha}}{2 \sin \pi\alpha} \left[e^{i\alpha\pi/2} + e^{i3\alpha\pi/2}\right] = \frac{\pi}{2} \frac{2\cos \alpha\pi/2}{\sin \alpha\pi} = \frac{\pi}{2} \text{cosec } \alpha\pi/2, \quad 0 < \alpha < 2.$$

Example: Let us now consider the integral

$$I \equiv \int_C f(z) \ln z \, dz$$

where C is the same closed contour used in the preceding example (see Figure 3.5) , $\ln z$ denotes the single valued branch $\ln z = \ln |z| + i\theta$, $0 < \theta < 2\pi$, and $f(z)$ is a rational function with

i   no poles on the real axis, and

ii the degree of its denominator polynomial exceeds that of its numerator by at least two.

The first restriction on $f(z)$ ensures that $z = 0$ is not a zero of its denominator and hence, that $zf(z) \to 0$ uniformly as $z \to 0$. This is sufficient to ensure in turn that the contribution from C_r again vanishes in the limit $r \to 0$. Similarly, the second restriction ensures that the contribution from C_R vanishes in the limit $R \to \infty$. Thus, we have

$$\lim_{R \to \infty} \lim_{r \to 0} \int_C f(z) \ln z \, dz = \int_{\infty L_2}^0 f(z) \ln z \, dz + \int_{0 L_1}^\infty f(z) \ln z \, dz \qquad (3.3.15)$$

while, from the Residue Theorem, we know that

$$\lim_{R \to \infty} \lim_{r \to 0} \int_C f(z) \ln z \, dz = 2\pi i \sum_{\text{all poles}} \text{Res}[f(z) \ln z]. \qquad (3.3.16)$$

It may seem that we are on the threshold of deriving a formula for the evaluation of improper integrals of the type $\int_0^\infty f(x) \ln x \, dx$. However, since $\ln z = \ln x$ on L_1 and $\ln z = \ln x + 2\pi i$ on L_2, equation (3.3.15) becomes

$$\lim_{R \to \infty} \lim_{r \to 0} \int_C f(z) \ln z \, dz = -2\pi i \int_0^\infty f(x) \, dx$$

which, together with (3.3.16) , yields

$$\int_0^\infty f(x) \, dx = - \sum_{\text{all poles}} \text{Res}[f(z) \ln z]. \qquad (3.3.17)$$

In other words, we have found a new method of evaluating integrals over the **half** range $(0, \infty)$ and, unlike our previous methods, it does **not** require the integrand to be an even function of x.

We shall illustrate the use of (3.3.17) with the integral

$$I \equiv \int_0^\infty \frac{1}{x^3 + a^3} \, dx, \quad a > 0.$$

The function $f(z) = (z^3 + a^3)^{-1}$ has simple poles at $z = z_k$ where $z_k = ae^{i(\pi/3 + 2k\pi/3)}$, $k = 0, 1, 2$. The residues at these poles are

$$\text{Res}[f(z_k)] = \frac{1}{3z^2}\bigg|_{z=z_k} = \frac{1}{3a^2}e^{-i(2\pi/3 + 4k\pi/3)}, \quad k = 0, 1, 2.$$

Thus,

$$\sum_{k=0}^{2} \text{Res}[f(z_k) \ln z_k] = \frac{1}{3a^2}e^{-i2\pi/3}(\ln a + i\pi/3) + \frac{1}{3a^2}e^{-i2\pi}(\ln a + i\pi)$$

$$+ \frac{1}{3a^2}e^{-i10\pi/3}(\ln a + i5\pi/3) = \frac{i\pi}{9a^2}[e^{-i2\pi/3} + 3 + 5e^{+i2\pi/3}]$$

$$= \frac{-2\sqrt{3}\pi}{9a^2}$$

and we obtain

$$\int_0^\infty \frac{1}{x^3 + a^3}\,dx = \frac{2\sqrt{3}\pi}{9a^2}, \quad a > 0.$$

While (3.3.17) is to some extent an unexpected bonus, it leaves us with the problem of evaluating $\int_0^\infty f(x)\ln x\,dx$ unresolved. If we impose the additional constraints that $f(x)$ be an even function and possess no poles on the negative as well as the positive real axis, this can be addressed by using as a contour a large semi-circle in the upper half-plane indented at the origin. The branch cut associated with $\ln z$ is then chosen to lie in the lower half-plane. We leave as an exercise to use this construction to derive

$$\int_0^\infty f(x)\ln x\,dx = \pi i \sum_+ \text{Res}[f(z)\ln z] + \frac{\pi^2}{2}\sum_+ \text{Res}[f(z)]. \tag{3.3.18}$$

3.3.6 Deducing Integrals from Others

It is often possible to evaluate an integral merely by deforming its contour of integration until it coincides with a path on which the value of the integral has already been determined, picking up $2\pi i$ times the residue of the integrand at any isolated singularities encountered along the way. Our next two examples illustrate what is involved.

Examples: The improper integral $I \equiv \int_{-\infty}^{\infty} e^{ikx - ax^2}\,dx$, k, a real and $a > 0$ is named after *Carl Friedrich Gauss* on whom we will elaborate later. It may be recognizable as the *Fourier transform* of the *Gaussian function* e^{-ax^2}. By setting

$$-ax^2 + ikx = -a\left(x - \frac{ik}{2a}\right)^2 - \frac{k^2}{4a},$$

Gauss's integral becomes the complex line integral,

$$I = e^{-k^2/4a} \int_{-\infty-ik/2a}^{\infty-ik/2a} e^{-az^2}\, dz, \quad z = x - ik/2a,$$

whose integrand is e^{-az^2} and whose contour of integration runs parallel to the real axis and lies a distance $k/2a$ below it. The value of the integral of e^{-az^2} is well known when the integration contour is the real axis; specifically,

$$\int_{-\infty}^{\infty} e^{-ax^2}\, dx = \sqrt{\frac{\pi}{a}}.$$

Therefore, we shall attempt to evaluate I by integrating e^{-az^2} around a rectangle with vertices $-R - i\frac{k}{2a}, +R - i\frac{k}{2a}, +R, -R$ and then taking the limit as $R \to \infty$.

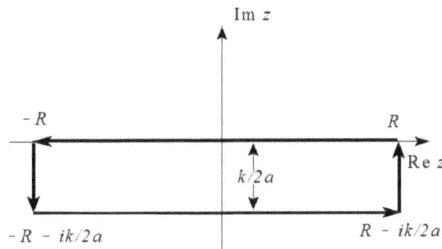

Figure 3.6: The contour used to evaluate Gauss's Integral.

Since e^{-az^2} is entire, we obtain from Cauchy's Theorem

$$e^{k^2/4a} I = \int_{-\infty}^{\infty} e^{-ax^2}\, dx + \lim_{R\to\infty} \int_{-R-ik/2a}^{-R} e^{-az^2}\, dz + \lim_{R\to\infty} \int_{R}^{R-ik/2a} e^{-az^2}\, dz. \qquad (3.3.19)$$

But on the vertical segments of the contour,

$$\left| \int e^{az^2}\, dz \right| \leq \max \left| e^{-ax^2} \right| \frac{k}{2a} = e^{-a(R^2-k^2/4a^2)} \frac{k}{2a}$$

which vanishes in the limit $R \to \infty$. Thus, (3.3.19) yields

$$I = e^{-k^2/4a} \int_{-\infty}^{\infty} e^{-ax^2}\, dx = \sqrt{\frac{\pi}{a}} e^{-k^2/4a}. \qquad (3.3.20)$$

For our second example we shall consider the class of integrals

$$I \equiv \int_{0}^{\infty} x^{\alpha-1} \left\{ \begin{array}{c} \cos x \\ \sin x \end{array} \right. dx, \quad 0 < \alpha < 1. \qquad (3.3.21)$$

These may be viewed as integrals of $z^{\alpha-1} \operatorname{Re}(e^{iz})$ and $z^{\alpha-1} \operatorname{Im}(e^{iz})$ along the positive real axis. Thus, we require a contour on which $z^{\alpha-1} e^{iz}$ assumes a simpler and more familiar form. If we set $z = iy$, then $z^{\alpha-1} e^{iz}$ becomes

$$e^{i(\alpha-1)\pi/2} y^{\alpha-1} e^{-y} = -i \left(\cos \frac{\pi\alpha}{2} + i \sin \frac{\pi\alpha}{2} \right) y^{\alpha-1} e^{-y}.$$

Thus, the integral of $z^{\alpha-1} e^{iz}$ along the positive imaginary axis is

$$\int_0^\infty -i \left(\cos \frac{\pi\alpha}{2} + i \sin \frac{\pi\alpha}{2} \right) e^{-y} y^{\alpha-1} i \, dy = \left(\cos \frac{\pi\alpha}{2} + i \sin \frac{\pi\alpha}{2} \right) \int_0^\infty e^{-y} y^{\alpha-1} dy.$$

The integral on the right hand side is so well known it has been assigned a name, **Euler's Integral of the second kind.** It provides a definition of the **gamma function** which is the continuous variable generalization of the factorial function,

$$\Gamma(z) = \int_0^\infty e^{-y} y^{z-1} dy, \operatorname{Re} z > 0$$

with

$$\Gamma(n) = (n-1)!, n = 1, 2, \ldots.$$

Evaluation of this integral provides values of $\Gamma(\alpha)$ for non-integer α. These can be found in tabulated form in any standard compendium.

To evaluate I in (3.3.22), we shall integrate $z^{\alpha-1} e^{iz}$ around the first quadrant of a circle of radius R and then let $R \to \infty$. However, since $z = 0$ is a branch point of $z^{\alpha-1}$, we must subtract from the interior of the contour the first quadrant of a small circle of radius r and take the additional limit $r \to 0$. The complete contour is shown in Figure 3.7.

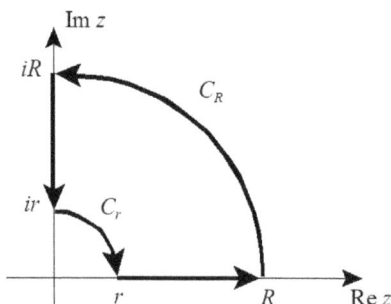

Figure 3.7: The contour used to evaluate the integral equation (3.3.22).

On C_r, $z = re^{i\theta}$ and we have

$$\left| \int_{C_r} e^{iz} z^{a-1} dz \right| = \left| \int_{\pi/2}^{0} e^{-r\sin\theta} r^{a-1} e^{(a-1)i\theta} re^{i\theta} d\theta \right| \leq r^a \left| \int_{\pi/2}^{0} d\theta \right| = \frac{\pi}{2} r^a,$$

since $|e^{-r\sin\theta}| \leq 1$ for all $r > 0$ and $0 \leq \theta \leq \pi/2$. It follows immediately that $\int_{C_r} e^{iz} z^{a-1} dz \to 0$ as $r \to 0$, if $a > 0$.

If $a < 1$, $z^{a-1} \to 0$ uniformly as $|z| \to \infty$ and so, by the same argument as was used to prove Jordan's Lemma, we have $\int_{C_R} e^{iz} z^{a-1} dz \to 0$ as $R \to \infty$.

Therefore, if $0 < a < 1$, an application of Cauchy's Theorem to the integral around the closed contour in Figure 3.7 yields, in the limit as $r \to 0$ and $R \to \infty$,

$$\int_0^\infty e^{ix} x^{a-1} dx + \int_\infty^0 e^{-y} y^{a-1} e^{i(a-1)\pi/2} i \, dy = 0,$$

or,

$$\int_0^\infty x^{a-1}(\cos x + i\sin x) dx = \int_0^\infty e^{-y} y^{a-1} \left(\cos\frac{\pi a}{2} + i\sin\frac{\pi a}{2} \right) dy.$$

Thus, equating real and imaginary parts, we conclude that

$$\int_0^\infty x^{a-1} \cos x \, dx = \Gamma(a) \cos\frac{\pi a}{2}$$

$$\int_0^\infty x^{a-1} \sin x \, dx = \Gamma(a) \sin\frac{\pi a}{2}. \tag{3.3.22}$$

3.3.7 Singularities on the Contour and Principal Value Integrals

An additional and rather serious type of integration impropriety is to have the integration path pass through a singularity of the integrand. In real variable analysis this situation is addressed as follows.

Definition: If a function of a real variable $f(x)$ increases without limit as $x \to c$, $a < c < b$, we define the improper integral of $f(x)$ from a to b, to be

$$\int_a^b f(x) dx = \lim_{\varepsilon \to 0} \int_a^{c-\varepsilon} f(x) dx + \lim_{\delta \to 0} \int_{c+\delta}^b f(x) dx, \tag{3.3.23}$$

provided that **both** limits exist.

Notice the analogy between this definition and that given by (3.3.2) for an improper integral with an improper range of integration. The analogy continues with the **definition** of the **Cauchy principal value** integral,

$$\wp \int_a^b f(x)dx = \lim_{\varepsilon \to 0} \left[\int_a^{c-\varepsilon} f(x)dx + \int_{c+\varepsilon}^b f(x)dx \right]. \tag{3.3.24}$$

The principal value may exist even if the improper integral does not. However, if the latter does exist then the two are equal.

As soon as we avail ourselves of the added flexibility of complex variables, the simplicity of this picture becomes muddied by the introduction of alternative ways of defining the improper integral. These correspond to the various ways available of deforming the contour of integration to avoid the singular point. Two obvious choices are to stop the integration just in front of the singularity and then pass by it on a semi-circle of vanishingly small radius in either the clockwise (upper half-plane) or the counter-clockwise direction (lower half-plane).

Even with only these two choices of how to avoid a singular point, one faces a multiplicity of possible definitions of an improper integral if its integrand has several singular points on the contour. For example, in Section 3.3.3 we found that

$$\int_{-\infty}^\infty \frac{e^{ikx}}{x^2 + a^2}dx = \frac{\pi}{a}e^{-|k|a}, \quad a > 0.$$

Suppose that we change the relative sign in the denominator so that this integral becomes

$$I \equiv \int_{-\infty}^\infty \frac{e^{ikx}}{x^2 - a^2}dx. \tag{3.3.25}$$

Because the integrand now has simple poles at $z = \pm a$, there are four possible definitions of this integral corresponding to the four indented contours shown in Figure 3.8.

Each of the four integrals may be evaluated by an application of the Residue Theorem if we close the contour and make use of Jordan's Lemma. However, to do the latter we must close each contour in the upper half-plane if $k > 0$, and in the lower half-plane if $k < 0$. Thus, these integrals may receive contributions from none, one or both of the poles, depending on the sign of k.

The residues at the poles are

$$\text{Res}\left[\frac{e^{ikz}}{z^2 - a^2} \right]_{z=\pm a} = \pm \frac{e^{\pm ika}}{2a}.$$

Hence, the values of the integrals are

$$I_{(i)} = \begin{cases} 0, & k > 0 \\ \frac{2\pi}{a}\sin ka, & k > 0 \end{cases} \tag{3.3.26}$$

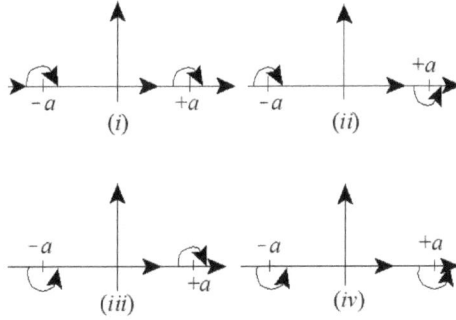

Figure 3.8: Four ways of avoiding the poles of $(z^2 - a^2)^{-1}$.

$$I_{(ii)} = \frac{\pi i}{a} e^{i|k|a} \tag{3.3.27}$$

$$I_{(iii)} = \frac{-\pi i}{a} e^{-i|k|a} \tag{3.3.28}$$

$$I_{(iv)} = \begin{cases} \frac{-2\pi}{a} \sin ka, & k > 0 \\ 0, & k < 0. \end{cases} \tag{3.3.29}$$

In addition to these four definitions of the improper integral I, we still have the possibility of using the principal value $\wp \int\limits_{-\infty}^{\infty} \frac{e^{ikx}}{x^2 - a^2} dx$, whose value we shall calculate momentarily.

In the midst of so much ambiguity how can one attach a unique meaning to such an integral when it arises in a physical problem? Fortunately, the physics of the problem will always contain information that dictates how the singularities are to be avoided and hence, lifts the ambiguity that is inherent to the mathematics. A classic example that we will encounter when we discuss Green's Functions involves an integral almost identical to the I in (3.3.25) . We shall discover that causality, the principle that cause must precede effect, is all that is required to make a unique choice of definition for the integral.

When an improper integral has only simple poles on its contour of integration, its principle value can be related in a very straight forward way to the values obtained by indenting the contour to avoid the poles. To be specific, let us consider the integral

$$I \equiv \int\limits_{-\infty}^{\infty} \frac{f(x)}{x - x_0} dx$$

where there is no loss of generality in having chosen the real axis as contour but having done so we now require that $f(z)$ have no singularities there and that it be suitably behaved at infinity. If we indent the contour to avoid $z = x_0$ by means of a semi-circle in the upper half-plane, (a clockwise semi-circle), I assumes the value

$$I_\cap = \int_{-\infty}^{x_0-\varepsilon} \frac{f(x)}{x - x_0} dx + \int_{x_0+\varepsilon}^{\infty} \frac{f(x)}{x - x_0} dx + \int_C \frac{f(z)}{z - x_0} dz \tag{3.3.30}$$

where C denotes a semi-circle with radius ε and centre at $z = x_0$.

In the limit $\varepsilon \to 0$, the first two terms in (3.3.30) yield the principal value of I while the third term can be evaluated as follows. Since $f(z)$ is holomorphic at $z = x_0$, it can be expanded in a Taylor series about this point. Thus,

$$\lim_{\varepsilon\to 0}\int_C \frac{f(z)}{z-x_0}dz = \lim_{\varepsilon\to 0}\int_C \frac{1}{z-x_0}\sum_{m=0}^{\infty}\frac{f^{(m)}(x_0)}{m!}(z-x_0)^m dz$$

$$= -\int_0^{\pi} if(x_0)d\theta - \lim_{\varepsilon\to 0}\sum_{m=1}^{\infty}\frac{f^{(m)}(x_0)}{m!}\varepsilon^m\int_0^{\pi} ie^{im\theta}d\theta = -i\pi f(x_0) \tag{3.3.31}$$

where we have used the fact that on C $z - x_0 = \varepsilon e^{i\theta}$, $dz = i\varepsilon e^{i\theta}d\theta$, $\pi \geq \theta \geq 0$. Hence, (3.3.30) becomes

$$I_\cap = \wp\int_{-\infty}^{\infty}\frac{f(x)}{x-x_0}dx - i\pi f(x_0). \tag{3.3.32}$$

Similarly, if we indent the contour **below** the pole by means of a counter-clockwise semi-circle,

$$I_\cap = \wp\int_{-\infty}^{\infty}\frac{f(x)}{x-x_0}dx + i\pi f(x_0). \tag{3.3.33}$$

Adding (3.3.32) and (3.3.33), we see that

$$\wp\int_{-\infty}^{\infty}\frac{f(x)}{x-x_0}dx = \frac{1}{2}(I_\cap + I_\cup) \tag{3.3.34}$$

which, after appropriate generalization, provides a means of defining principal value integrals for contours other than the real axis. And returning to our last **example**, it gives us

$$\wp\int_{-\infty}^{\infty}\frac{e^{ikx}}{x^2-a^2}dx = -\frac{\pi}{a}\sin|k|a$$

merely by averaging either (3.3.26) and (3.3.29) or (3.3.27) and (3.3.28).

To obtain a further perspective on these issues, note that because $f(z)$ is holomorphic in a neighbourhood of the real axis and well behaved at infinity, the indented

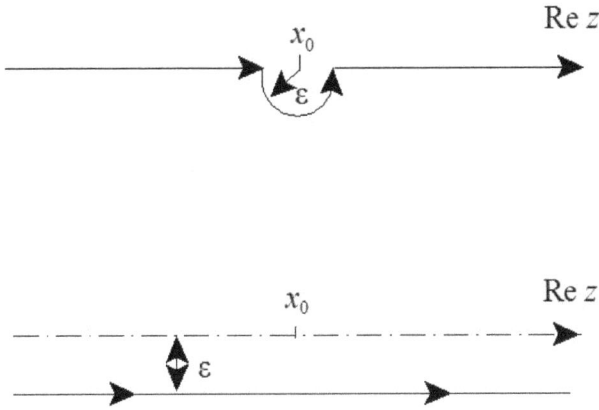

Figure 3.9: The contour that defines $I_\cup$ and an equivalent contour obtained by stretching it out below the axis.

contour used for $I_\cup$ is completely equivalent to one which parallels the real axis a distance ε below it. These two alternatives are pictured in Figure 3.9. Hence, we may write

$$I_\cup = \lim_{\varepsilon \to 0} \int_{-\infty-i\varepsilon}^{\infty-i\varepsilon} \frac{f(z)}{z - x_0} dz, \, z = x_0 - i\varepsilon$$

$$= \lim_{\varepsilon \to 0} \int_{-\infty}^{\infty} \frac{f(x)}{x - x_0 - i\varepsilon} dx. \tag{3.3.35}$$

The second equality shows that lowering the contour to avoid the singularity is equivalent to raising the singularity (to $z = x_0 + i\varepsilon$) to avoid the contour.

Using this result together with the corresponding equation for $I_\cap$, , we can transform (3.3.32) and (3.3.34) to read

$$\lim_{\varepsilon \to 0} \int_{-\infty}^{\infty} \frac{f(x)}{x - x_0 \pm i\varepsilon} dx = \wp \int_{-\infty}^{\infty} \frac{f(x)}{x - x_0} dx \mp i\pi f(x_0). \tag{3.3.36}$$

On the purely practical side, the concept of a principal value integral can be used to simplify the evaluation of integrals like $\int_0^\infty \frac{\sin x}{x} dx$. The method involves a simple extension of the theorems of Sections 3.3.2 and 3.3.3.

Suppose that $f(z)$ has a simple pole at $z = x_0$. In some neighbourhood of that point we can set $f(z) = (z-x_0)^{-1} g(z)$ where $g(z)$ is holomorphic and non-zero at $z = x_0$. Thus, if this is the only singularity on the real axis, we have

$$\int_{-\infty}^{\infty} f(x)dx \Bigg|_{\cap} = \wp \int_{-\infty}^{\infty} f(x)dx + \lim_{\varepsilon \to 0} \int_C \frac{g(z)}{z - x_0} dz = \wp \int_{-\infty}^{\infty} f(x)dx - i\pi g(x_0)$$

$$= \wp \int_{-\infty}^{\infty} f(x)dx - i\pi \text{Res}[f(x_0)],$$

where the $\cap$ signifies that we are using a contour along the real axis indented into the upper half-plane to avoid $z = x_0$. But, if $zf(z) \to 0$ uniformly as $|z| \to \infty$ and $f(z)$ is meromorphic in the upper half-plane, we can use the Residue Theorem to find

$$\int_{-\infty}^{\infty} f(x)dx \Bigg|_{\cap} = 2\pi i \sum_+ \text{Res}[f(z)].$$

Thus,

$$\wp \int_{-\infty}^{\infty} f(x)dx = \pi i \text{Res}[f(x_0)] + 2\pi i \sum_+ \text{Res}[f(z)]. \tag{3.3.37a}$$

If meromorphy occurs in the lower rather than upper half-plane, this is replaced by

$$\wp \int_{-\infty}^{\infty} f(x)dx = \pi i \text{Res}[f(x_0)] - 2\pi i \sum_- \text{Res}[f(z)]. \tag{3.3.37b}$$

Generalizing, we see that if a function $f(z)$ has the requisite behaviour at infinity, is meromorphic in the appropriate half-plane, and has only simple poles on the real axis, then

$$\wp \int_{-\infty}^{\infty} f(x)e^{ikx} dx = \pi i \sum_0 \text{Res}[f(x)e^{ikx}] + \begin{cases} 2\pi i \sum_+ \text{Res}[f(z)e^{ikz}], & k \geq 0 \\ 2\pi i \sum_- \text{Res}[f(z)e^{ikz}], & k \leq 0. \end{cases} \tag{3.3.38}$$

where $\sum_0$ denotes a sum over the poles on the real axis.

Example: As stated earlier, this result simplifies the evaluation of integrals like $\int_0^{\infty} \frac{\sin x}{x} dx$. We start by considering $\frac{e^{iz}}{z}$ whose only singularity in the finite plane is a simple pole at $z = 0$. The residue at this pole is unity and so, from (3.3.38), we have

$$\wp \int_{-\infty}^{\infty} \frac{e^{ix}}{x} dx = \pi i.$$

By equating real and imaginary parts, this yields

$$\wp \int_{-\infty}^{\infty} \frac{\cos x}{x} dx = 0$$

$$\wp \int_{-\infty}^{\infty} \frac{\sin x}{x} dx = \pi.$$

Thus, since $\frac{\sin x}{x}$ is entire, we must have

$$\wp \int_{-\infty}^{\infty} \frac{\sin x}{x} dx = \int_{-\infty}^{\infty} \frac{\sin x}{x} dx$$

and so, with much less effort than before, we find

$$\int_{0}^{\infty} \frac{\sin x}{x} dx = \frac{\pi}{2}.$$

3.4 Summation of Series

The Residue Theorem can also be used to calculate the sums of certain infinite series. The principle is simple. The function $\sin \pi z$ has an infinite sequence of zeros at the points $z = 0, \pm 1, \pm 2, \ldots$ and so, cosec πz and cot πz each have an infinite sequence of simple poles at the same points. Their residues are

$$\text{Res}[\text{cosec } \pi z]_{z=\pm n} = \left. \frac{1}{\pi \cos \pi z} \right|_{z=\pm n} = \frac{(-1)^n}{\pi}, \quad n = 0, 1, 2, \ldots.$$

$$\text{Res}[\cot \pi z]_{z=\pm n} = \left. \frac{\cos \pi z}{\pi \cos \pi z} \right|_{z=\pm n} = \frac{1}{\pi}, \quad n = 0, 1, 2, \ldots.$$

Thus, if $f(z)$ is a meromorphic function whose poles, $z = z_j, j = 1, 2, \ldots, k$, are distinct from the zeros of $\sin \pi z$, and if C is a contour that encloses the points $z = l, l+1, \ldots, m$, then

$$\sum_{n=l}^{m} f(n) = \frac{1}{2\pi i} \int_{C} \pi \cot \pi z f(z) dz - \pi \sum_{z_j \text{ inside } C} \text{Res}[\cot \pi z f(z)] \qquad (3.4.1)$$

and,

$$\sum_{n=l}^{m} (-1)^n f(n) = \frac{1}{2\pi i} \int_{C} \pi \text{ cosec } \pi z f(z) dz - \pi \sum_{z_j \text{ inside } C} \text{Res}[\text{cosec } \pi z f(z)]. \qquad (3.4.2)$$

If we can ensure that the contour integrals go to zero as $n \to \infty$, these two equations can be used to sum the infinite series $\sum_n f(n)$ and $\sum_n (-1)^n f(n)$, respectively. With this goal in mind, let us take $f(z)$ to be a rational function, none of whose zeros or poles are integers and such that $|f(z)| \to 0$ as $|z| \to \infty$. We then choose C to be the square with vertices at $z = (N + 1/2)(\pm 1 \pm i)$. This choice is based on the observation that

$$|\cot \pi z| = \left| \left(\frac{\cos^2 \pi x + \sinh^2 \pi y}{\sin^2 \pi x + \sinh^2 \pi y} \right)^{\frac{1}{2}} \right|$$

so that on the vertical sides of the contour,

$$|\cot \pi z| = \left|\left(\frac{\sinh^2 \pi y}{1 + \sinh^2 \pi y}\right)^{\frac{1}{2}}\right| = |\tanh \pi y| \le 1,$$

while on the horizontal sides of the contour,

$$|\cot \pi z| = \left|\left(\frac{\sinh^2 \pi(N + 1/2) + 1}{\sinh^2 \pi(N + 1/2)}\right)^{\frac{1}{2}}\right| = |\coth \pi(N + 1/2)|$$

which tends to one as $N \to \infty$. Thus, $|\cot \pi z|$ is bounded on C for all N. We shall denote its upper bound by M.

Next, we note that for N sufficiently large and z on C, $f(z)$ admits an expansion of the form

$$f(z) = \sum_{m=1}^{\infty} \frac{a_m}{z^m} = \frac{a_1}{z} + \frac{a_2}{z^2} + \dots.$$

In addition, since

$$\text{Res}\left[\frac{\cot \pi z}{z}\right]_{z=\pm n} = \pm\frac{1}{\pi n}, \quad n = 1, 2, \dots$$

and $\frac{\cot \pi z}{z} = \frac{1}{z^2} - \frac{\pi}{3} - \frac{\pi^3}{45}z^2 - \dots$, $|z| < 1$ so that

$$\text{Res}\left[\frac{\cot \pi z}{z}\right]_{z=0} = 0,$$

the integral

$$\int_C \frac{\cot \pi z}{z} dz = 2\pi i \sum_{n=-N}^{N} \text{Res}\left[\frac{\cot \pi z}{z}\right]_{z=n} = 0.$$

Thus,

$$\int_C \cot \pi z f(z) dz = \int_C \cot \pi z \left(f(z) - \frac{a_1}{z}\right) dz.$$

But, $\left|z\left(f(z) - \frac{a_1}{z}\right)\right| < \epsilon(N)$ where $\varepsilon(N) \to 0$ as $N \to \infty$. Therefore, invoking Darboux's Inequality, we have

$$\left|\int_C z f(z) \cot \pi z \frac{dz}{z}\right| \le \epsilon(N) M \frac{1}{N + 1/2} 8(N + 1/2)$$

where we have used the fact that the minimum value of $|z|$ on C is $(N + 1/2)$ and that the length of C is $8(N + 1/2)$. Thus, in the limit as $N \to \infty$, the integral vanishes and (3.4.1) becomes

$$\sum_{n=-\infty}^{\infty} f(n) = -\pi \sum_{j=1}^{k} \text{Res}[\cot \pi z f(z)]_{z=z_j}. \tag{3.4.3}$$

Evidently, the "background" integral continues to vanish when $\cot \pi z$ is replaced by $\operatorname{cosec} \pi z$ and so, for the same set of rational functions $f(z)$ we have

$$\sum_{n=-\infty}^{\infty} (-1)^n f(n) = -\pi \sum_{j=1}^{k} \operatorname{Res}[\operatorname{cosec} \pi z f(z)]_{z=z_j}. \tag{3.4.4}$$

This rather clever technique for summing infinite series will resurface when we use it to sum a partial wave expansion of a quantum mechanical scattering amplitude. Such expansions are infinite series involving Legendre polynomials $P_l(x)$, with the index l being an angular momentum quantum number. Thus, this particular application will introduce the idea of treating angular momentum as a complex variable, an innovation that had a major impact on subatomic physics in the 1960's.

We shall now consider a few applications that specifically involve equations (3.4.3) and (3.4.4).

Examples: To sum $\sum_{n=-\infty}^{\infty} \frac{1}{(n+\zeta)^2}$ we need the function $f(z) = (z+\zeta)^{-2}$ which has a double pole at $z = -\zeta$. The residue of $\frac{\cot \pi z}{(z+\zeta)^2}$ at $z = -\zeta$ is

$$\operatorname{Res}\left[\frac{\cot \pi z}{(z+\zeta)^2}\right]_{z=-\zeta} = \lim_{z \to -\zeta} \frac{d}{dz} \cot \pi z = -\pi \operatorname{cosec}^2 \pi \zeta.$$

Thus, (3.4.3) immediately gives us

$$\sum_{n=-\infty}^{\infty} \frac{1}{(n+\zeta)^2} = -\pi \operatorname{cosec}^2 \pi \zeta. \tag{3.4.5}$$

Now consider the series $\sum_{n=1}^{\infty} \frac{1}{n^2+\zeta^2}$ and $\sum_{n=0}^{\infty} \frac{(-1)^n}{n^2+\zeta^2}$. In both cases the relevant function is $f(z) = (z^2 + \zeta^2)^{-1}$. It vanishes uniformly as $|z| \to \infty$ and has simple poles at $z = \pm i\zeta$ with residues $\operatorname{Res}[f(\pm i\zeta)] = \pm \frac{1}{2\zeta i}$. Thus, from (3.4.3), we have

$$\sum_{n=-\infty}^{\infty} \frac{1}{n^2+\zeta^2} = \frac{1}{\zeta^2} + 2\sum_{n=1}^{\infty} \frac{1}{n^2+\zeta^2} = -\pi\left[\frac{\cot i\pi\zeta}{2\zeta i} - \frac{\cot(-i\pi\zeta)}{2\zeta i}\right] = \frac{\pi}{\zeta}\coth \pi\zeta$$

and so,

$$\sum_{n=1}^{\infty} \frac{1}{n^2+\zeta^2} = \frac{\pi}{2\zeta}\coth \pi\zeta - \frac{1}{2\zeta^2}. \tag{3.4.6}$$

Notice that if we set $\zeta = \frac{x}{\pi}$, this becomes a series expansion of the Langevin function

$$\coth x - \frac{1}{x} = \sum_{n=1}^{\infty} \frac{2x}{x^2 + n^2\pi^2}. \tag{3.4.7}$$

Similarly, starting from (3.4.4) we find

$$\sum_{n=-\infty}^{\infty} \frac{(-1)^n}{n^2+\zeta^2} = 2\sum_{n=0}^{\infty} \frac{(-1)^n}{n^2+\zeta^2} - \frac{1}{\zeta^2} = -\pi\left[\frac{\operatorname{cosec} i\pi\zeta}{2\zeta i} - \frac{\operatorname{cosec}(-i\pi\zeta)}{2\zeta i}\right] = \frac{\pi}{\zeta}\operatorname{cosech} \pi\zeta$$

and hence,

$$\sum_{n=0}^{\infty} \frac{(-1)^n}{n^2 + \zeta^2} = \frac{1}{2\zeta^2} + \frac{\pi}{2\zeta} \operatorname{cosech} \pi\zeta. \tag{3.4.8}$$

As a final application we shall determine the sum of the series that corresponds to the simplest rational function satisfying our asymptotic constraint, $f(z) = (z - \zeta)^{-1}$. The series is $\sum_{n=-\infty}^{\infty} \frac{1}{n-\zeta}$ and (3.4.3) immediately determines its sum to be

$$\sum_{n=-\infty}^{\infty} \frac{1}{\zeta - n} = \pi \cot \pi\zeta. \tag{3.4.9}$$

Unlike the series in our previous examples, this one is not absolutely convergent. However, we can make it so by adding $\frac{1}{n}$ to the nth term for each $n \neq 0$. The added quantities cancel pair wise and so we obtain

$$\pi \cot \pi\zeta = \frac{1}{\zeta} + \sum_{n=-\infty, n\neq 0}^{\infty} \left[\frac{1}{\zeta - n} + \frac{1}{n} \right]. \tag{3.4.10}$$

If we now treat ζ as a variable, we can integrate this series term by term to obtain

$$\ln \frac{\sin \pi z}{\pi z} = \int_0^z \left[\pi \cot \pi\zeta - \frac{1}{\zeta} \right] d\zeta = \sum_{n=-\infty, n\neq 0}^{\infty} \int_0^z \left[\frac{1}{\zeta - n} + \frac{1}{n} \right] d\zeta$$

$$= \sum_{n=-\infty, n\neq 0}^{\infty} \ln \left[\left(1 - \frac{z}{n} \right) e^{z/n} \right]$$

or,

$$\sin \pi z = \pi z \prod_{n=-\infty, n\neq 0}^{\infty} \left(1 - \frac{z}{n} \right) e^{\frac{z}{n}}.$$

Combining factors which are symmetric with respect to $n = 0$, this becomes the simple product

$$\sin \pi z = \pi z \prod_{n=1}^{\infty} \left(1 - \frac{z^2}{n^2} \right) \tag{3.4.11}$$

which is a complete factorization of the sine function in terms of its zeros.

3.5 Representation of Meromorphic Functions

In much of the preceding Section we represented constants with the symbol ζ even though we have normally reserved it to represent a variable. We have done so because if we now take it to be a variable , each of equations (3.4.5), (3.4.6), (3.4.8) and (3.4.10) becomes a partial fraction decomposition of a meromorphic function and thus, provides an explicit representation of that function in terms of its singularities. As such, they are special cases of a result due to **Mittag-Leffler** who showed that any meromorphic function can be expressed as a sum of an entire function and a series of rational functions. We shall prove a restricted version of that theorem.

Let $f(z)$ be a function whose only singularities in the finite plane are simple poles at $z = z_k (k = 1, 2, \ldots)$ where $0 < |z_1| \le |z_2| \le \ldots \le |z_n| \le \ldots$. Let $C_1, C_2, C_3, \ldots$ be a sequence of circles with centre at the origin such that

i C_n encloses only the poles $z_1, z_2, \ldots, z_n$ and does not pass through any other poles of $f(z)$, and

ii the radius R_n of C_n tends to infinity with n.

Then, denoting the residue of $f(z)$ at $z = z_k$ by r_k, we have

$$\frac{1}{2\pi i} \int_{C_n} \frac{f(\zeta)}{\zeta - z} d\zeta = f(z) + \sum_{k=1}^{n} \frac{r_k}{z_k - z}, \tag{3.5.1}$$

where z is any point other than $\zeta = z_k, k = 1, 2, \ldots, n$, within C_n. Writing this equation for $z = 0$ and subtracting the result from (3.5.1), we obtain

$$f(z) = f(0) + \sum_{k=1}^{n} r_k \left[\frac{1}{z - z_k} + \frac{1}{z_k} \right] + \frac{z}{2\pi i} \int_{C_n} \frac{f(\zeta)}{\zeta(\zeta - z)} d\zeta. \tag{3.5.2}$$

Suppose now that the upper bound of $|f(z)|$ on C_n is itself bounded by M as $n \to \infty$. Then, applying the Darboux Inequality to the integral in (3.5.2) we find

$$\left| \int_{C_n} \frac{f(\zeta)}{\zeta(\zeta - z)} d\zeta \right| \le \frac{2\pi R_n M}{R_n(R_n - |z|)}$$

which goes to zero as $R_n \to \infty$. Hence, since $R_n \to \infty$ as $n \to \infty$, (3.5.2) yields the **Mittag- Leffler representation**

$$f(z) = f(0) + \sum_{k=1}^{\infty} r_k \left[\frac{1}{z - z_k} + \frac{1}{z_k} \right]. \tag{3.5.3}$$

One can show that this series is uniformly convergent in any finite region which does not contain any of the poles of $f(z)$.

Applying this theorem to $f(z) = \pi \cot \pi z - \frac{1}{z}$, we readily recover

$$\pi \cot \pi z = \frac{1}{z} + \sum_{n=-\infty, n\neq 0}^{\infty} \left[\frac{1}{z - n} + \frac{1}{n} \right]$$

which we obtained earlier by treating z as a constant and summing the series. As a more general application, let $f(z)$ be an entire function which has only simple zeros, none of them being at the origin. Its logarithmic derivative $\frac{d}{dz} \operatorname{Ln} f(z) = \frac{f'(z)}{f(z)}$ is meromorphic, possesses simple poles at the simple zeros of $f(z)$, and is easily shown to be suitably bounded away from the poles. Thus, we can apply (3.5.3) to obtain

$$\frac{d}{dz} \operatorname{Ln} f(z) = \frac{f'(0)}{f(0)} + \sum_{k=1}^{\infty} \left[\frac{1}{z - z_k} + \frac{1}{z_k} \right]$$

where we have used the fact that $\operatorname{Res} \left[\frac{f'(z)}{f(z)} \right]_{z=z_k} = \left. \frac{f'(z)}{f'(z)} \right|_{z=z_k} = 1$ for all $k = 1, 2, \ldots$. Integrating term by term and then exponentiating, this yields

$$f(z) = f(0)e^{cz} \prod_{k=1}^{\infty} \left[1 - \frac{z}{z_k} \right] e^{z/z_k} \tag{3.5.4}$$

where $c = \frac{f'(0)}{f(0)}$ and so is a constant.

The complete factorization of the sine function contained in equation (3.4.11) is just a special case of this result. The corresponding factorization for the cosine is

$$\cos z = \prod_{k=-\infty}^{\infty} \left[1 - \frac{2z}{(2k + 1)\pi} \right] e^{2z/(2k+1)\pi}. \tag{3.5.5}$$

4 Dispersion Representations

4.1 From Cauchy Integral Representation to Hilbert Transforms

A recurrent theme thus far is that an analytic function is completely determined by its singularities. We have seen this principle manifested successively for constants (Liouville's Theorem), rational functions and most recently, meromorphic functions with simple poles. In this Section we shall take another step by discovering how functions with branch points can be represented in terms of their behaviour along the associated branch cuts. In so doing, we shall encounter a mathematical tool that has been found to be very useful in several fields of modern physics.

Suppose that we have a function $f(z)$ which is holomorphic throughout the complex plane except for a cut extending from x_0 to ∞ along the positive real axis. Suppose further that $f(z)$ satisfies the conditions

$$f(z^*) = f^*(z), \ |f(z)| \to 0 \text{ as } |z| \to \infty, \text{ and } |(z - x_0)f(z)| \to 0 \text{ as } |z - x_0| \to 0.$$

The first of these conditions is sometimes referred to as a reality condition since, as we shall see in Section 4.2, it implies that $f(x)$ is real for $x < x_0$.

The Cauchy Integral representation of $f(z)$ at a point z not on the cut is

$$f(z) = \frac{1}{2\pi i} \int_C \frac{f(\zeta)}{\zeta - z} d\zeta \tag{4.1.1}$$

where C is a contour like that shown in Figure 4.1. The contribution to the integral from the large circle in C goes to zero as the circle's radius $R \to \infty$. Moreover, the contribution from the small circle about $z = x_0$ vanishes as $\varepsilon \to 0$. Thus, in this double limit we have

$$f(z) = \lim_{\varepsilon \to 0} \frac{1}{2\pi i} \left\{ \int_{x_0 + i\varepsilon}^{\infty + i\varepsilon} \frac{f(\zeta)}{\zeta - z} d\zeta - \int_{x_0 - i\varepsilon}^{\infty - i\varepsilon} \frac{f(\zeta)}{\zeta - z} d\zeta \right\}$$

$$= \lim_{\varepsilon \to 0} \frac{1}{2\pi i} \left\{ \int_{x_0}^{\infty} \frac{f(\xi + i\varepsilon)}{\xi - z + i\varepsilon} d\xi - \int_{x_0}^{\infty} \frac{f(\xi - i\varepsilon)}{\xi - z - i\varepsilon} d\xi \right\} .$$

Since z is not on the cut, we can neglect the $\pm i\varepsilon$ in the denominators of the last two integrals and write

$$f(z) = \frac{1}{2\pi i} \int_{x_0}^{\infty} \lim_{\varepsilon \to 0} \frac{[f(\xi + i\varepsilon) - f(\xi - i\varepsilon)]}{\xi - z} d\xi. \tag{4.1.2}$$

The numerator of the integrand in (4.1.2) is the discontinuity of $f(z)$ across the cut. Thus, all we need in order to compute $f(z)$ at any point where it is holomorphic

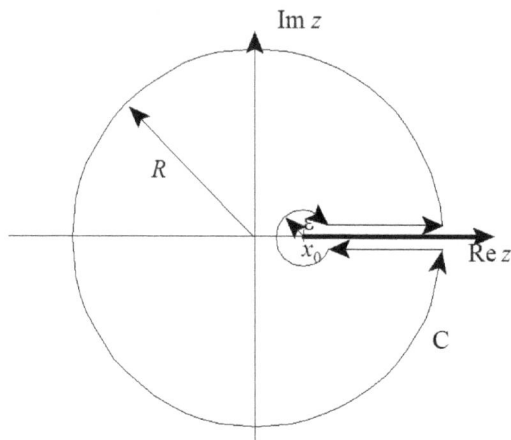

Figure 4.1: The contour used to derive a dispersion representation for a function with a cut along the real; axis segment $x_0 \le x < \infty$.

is a knowledge of its behaviour along its branch cut singularity. One of our initial assumptions allows us to cast this relationship in a somewhat simpler form. Since $f^*(z) = f(z^*)$, we have

$$\lim_{\varepsilon \to 0}[f(x + i\varepsilon) - f(x - i\varepsilon)] = \lim_{\varepsilon \to 0}[f(x + i\varepsilon) - f^*(x + i\varepsilon)]$$

$$= \lim_{\varepsilon \to 0} 2i \operatorname{Im} f(x + i\varepsilon) = 2i \operatorname{Im} f_+(x) \qquad (4.1.3)$$

where we have defined

$$f_+(x) = \lim_{\delta \to 0^+} f(x + i\delta). \qquad (4.1.4)$$

Inserting (4.1.3) into (4.1.2) gives us

$$f(z) = \frac{1}{\pi} \int_{x_0}^{\infty} \frac{\operatorname{Im} f_+(\xi)}{\xi - z} d\xi. \qquad (4.1.5)$$

This is an example of what physicists call a **dispersion relation**, a title which has its origin in the theory of optical and X-ray dispersion at the turn of the (19th) century. In optics a dispersion relation is an integral relationship between the refractive (real) and absorptive (imaginary) parts of the refractive index, the variable of integration being the frequency of the incident radiation. Whether mindful of the historic and scientific significance of the first such relationship, derived by Kramers and Kronig, or appreciative of the virtues of a spare vocabulary, physicists apply the term "dispersion relation" to any integral equation that links the real and imaginary parts of an analytic function.

For reasons which we will explore at a later date the classical dispersion relations were derived for functions for which one can only assume holomorphy in the upper

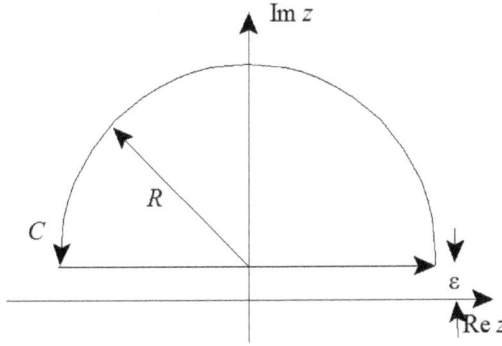

Figure 4.2: The contour used to derive the dispersion relation for a function $f(z)$ that is holomorphic only for $\operatorname{Im} z > 0$.

half-plane, $\operatorname{Im} z > 0$. Thus, in place of the contour of Figure 4.1, we must use a semicircle like that shown in Figure 4.2. Again using Cauchy's Integral Representation as our starting point and taking the double limit as $R \to \infty$ and $\varepsilon \to 0$, we obtain in place of (4.1.2)

$$f(z) = \lim_{\varepsilon \to 0} \frac{1}{2\pi i} \int_{-\infty}^{\infty} \frac{f(\xi + i\varepsilon)}{\xi - z + i\varepsilon} d\xi = \frac{1}{2\pi i} \int_{-\infty}^{\infty} \frac{f_+(\xi)}{\xi - z} d\xi. \qquad (4.1.6)$$

We now let z approach the real axis from above so that (4.1.6) becomes

$$f_+(x) = \lim_{\delta \to 0^+} \frac{1}{2\pi i} \int_{-\infty}^{\infty} \frac{f_+(\xi)}{\xi - x + i\delta} d\xi.$$

But, using equation (3.3.36), this can be rewritten in terms of a principal value integral:

$$f_+(x) = \frac{1}{2\pi i} \wp \int_{-\infty}^{\infty} \frac{f_+(\xi)}{\xi - x} d\xi + \frac{1}{2} f_+(x)$$

or,

$$f_+(x) = \frac{1}{\pi i} \wp \int_{-\infty}^{\infty} \frac{f_+(\xi)}{\xi - x} d\xi. \qquad (4.1.7)$$

Finally, we take the real and imaginary parts of this equation to obtain the classical dispersion relations

$$\operatorname{Re} f_+(x) = \frac{\wp}{\pi} \int_{-\infty}^{\infty} \frac{\operatorname{Im} f_+(\xi)}{\xi - x} d\xi \qquad (4.1.8)$$

$$\operatorname{Im} f_+(x) = \frac{-\wp}{\pi} \int_{-\infty}^{\infty} \frac{\operatorname{Re} f_+(\xi)}{\xi - x} d\xi. \tag{4.1.9}$$

From the point of view of a mathematician these relations establish that the real and imaginary parts of $f_+(x)$ are reciprocal **Hilbert transforms**.

David Hilbert (1862-1943) was a German mathematician who laid the foundations for much of modern mathematics including algebraic geometry and algebraic number theory. Over his lifetime, he made important contributions to the axiomatic foundations of geometry, integral equations, the calculus of variations, theoretical physics (which he thought was too difficult to be the domain of physicists alone), and mathematical logic.

In most applications $\operatorname{Im} f_+(x)$ is known from experiment and $\operatorname{Re} f_+(x)$ is then determined by application of (4.1.8). However, since x is typically a frequency or energy variable, measurements can only be made for values of $x \geq 0$. In that event a **crossing symmetry** such as $f_+(-x) = f_+^*(x)$ has to be invoked to allow (4.1.8) and (4.1.9) to be written

$$\operatorname{Re} f_+(x) = \frac{2}{\pi} \wp \int_0^{\infty} \frac{\xi \operatorname{Im} f_+(\xi)}{\xi^2 - x^2} d\xi \tag{4.1.10}$$

and

$$\operatorname{Im} f_+(x) = \frac{-2x}{\pi} \wp \int_0^{\infty} \frac{\operatorname{Re} f_+(\xi)}{\xi^2 - x^2} d\xi \tag{4.1.11}$$

respectively, which should look familiar to readers who have been introduced to the Kramers-Kronig relation for indices of refraction.

4.2 Adding Poles and Subtractions

Returning to functions whose analytic properties are known throughout the complex plane, we shall generalize the dispersion representation of equation (4.1.5) to accommodate functions possessing poles as well as cuts. Suppose that we have a function $f(z)$ which, in addition to the properties that resulted in (4.1.5), has simple poles at the points $z = z_k$, $k = 1, 2, \ldots, n$, none of which lie on the branch cut. Denoting the residue of $f(z)$ at $z = z_k$ by r_k, equation (4.1.2) now becomes

$$f(z) + \sum_{k=1}^{n} \frac{r_k}{z_k - z} = \lim_{\varepsilon \to 0} \frac{1}{2\pi i} \left\{ \int_{x_0+i\varepsilon}^{\infty+i\varepsilon} \frac{f(\zeta)}{\zeta - z} d\zeta - \int_{x_0-i\varepsilon}^{\infty-i\varepsilon} \frac{f(\zeta)}{\zeta - z} d\zeta \right\}.$$

Thus, (4.1.5) is replaced by

$$f(z) = \sum_{k=1}^{n} \frac{r_k}{z - z_k} + \frac{1}{\pi} \int_{x_0}^{\infty} \frac{\operatorname{Im} f_+(\xi)}{\xi - z} d\xi. \tag{4.2.1}$$

Notice that the sum in this equation is that of the principal parts of $f(z)$ at the poles $z = z_k$. One can show that this continues to be the case should any of the poles be of higher order than one.

A further generalization of (4.1.5) we should consider obtains when $f(z)$ does not vanish but is polynomial bounded as $|z| \to \infty$. If the only constraint at infinity is that $|f(z)/z^n| \to 0$ for some $n > 0$, there is a variety of possible dispersion representations for $f(z)$ depending on what other information we have about the function. For example, if we know the value of $f(z)$ and its first $(n-1)$ derivatives at some point $z = z_0$, we can determine a dispersion representation for $f(z)/(z - z_0)^n$ and from it deduce a representation for $f(z)$ itself. Should we lack such detailed information about $f(z)$ at a specific point but at least possess knowledge of its value at each of a set of n points, $z_1, \ldots, z_n$, we can invoke (4.2.1) to obtain a dispersion representation for $\frac{f(z)}{(z-z_1)\ldots(z-z_n)}$ and hence, one for $f(z)$ as well.

To illustrate what is involved, let us assume that $f(z)$ possesses all the properties that led us to the representation in (4.1.5) save one; instead of $|f(z)| \to 0$ as $|z| \to \infty$, we have $|f(z)| \to$ a non-zero constant. To compensate, we add one further piece of information about $f(z)$: its value at the point $z = 0$. We now have sufficient information to apply (4.2.1) to the function $f(z)/z$ which has a single simple pole at $z = 0$ with residue $f(0)$. Thus, we obtain

$$\frac{f(z)}{z} = \frac{f(0)}{z} + \frac{1}{\pi} \int_{x_0}^{\infty} \frac{\operatorname{Im} f_+(\xi)}{\xi(\xi - z)} d\xi$$

or,

$$f(z) = f(0) + \frac{z}{\pi} \int_{x_0}^{\infty} \frac{\operatorname{Im} f_+(\xi)}{\xi(\xi - z)} d\xi. \tag{4.2.2}$$

This is called a **subtracted** dispersion relation because it can be obtained (notionally) by assuming the unsubtracted dispersion relation (4.1.5),

$$f(z) = \frac{1}{\pi} \int_{x_0}^{\infty} \frac{\operatorname{Im} f_+(\xi)}{\xi - z} d\xi,$$

subtracting from it its value at $z = 0$,

$$f(0) = \frac{1}{\pi} \int_{x_0}^{\infty} \frac{\operatorname{Im} f_+(\xi)}{\xi} d\xi,$$

and then combining the two integrals by means of

$$\frac{1}{\xi - z} - \frac{1}{\xi} = \frac{z}{\xi(\xi - z)}.$$

Subtracted dispersion relations are often used even when the asymptotic behaviour of the function of interest does not make them mandatory. This is because the integral in (4.2.2) is less sensitive to high values of ξ than is its counterpart in (4.1.5). Thus, the error committed in omitting experimentally inaccessible values of $\mathrm{Im}\, f_+(\xi)$ at high ξ is decreased.

4.3 Mathematical Applications

The first application we shall consider is in the construction of functions once their singularities are known.

Example: Suppose that we wish to find an explicit expression, preferably in closed form, for the function $f(z)$ that possesses the following properties:

1. it is holomorphic everywhere except for a branch cut along the real axis segments $-\infty < x \le -1$ and $1 \le x < \infty$ and a simple pole of residue -1 at $z = 0$;
2. it goes uniformly to zero as $|z| \to \infty$;
3. it satisfies $f^*(z^*) = f(z)$ and so is real on the real axis segment $-1 < x < 1$;
4. its discontinuity across the cut is given by $\mathrm{Im}\, f_+(x) = +(x^2 + 1)^{-1}$ for $x > 1$ and $\mathrm{Im}\, f_+(x) = -(x^2 + 1)^{-1}$ for $x < -1$.

By means of an obvious generalization of (4.2.1) to deal with a cut running along two real axis segments we can ascribe to this function the representation

$$
f(z) = \frac{\mathrm{Res}[f(0)]}{z} + \frac{1}{\pi} \int_{-\infty}^{-1} \frac{\mathrm{Im}\, f_+(\xi)}{\xi - z} d\xi + \frac{1}{\pi} \int_{1}^{\infty} \frac{\mathrm{Im}\, f_+(\xi)}{\xi - z} d\xi
$$

$$
= -\frac{1}{z} - \frac{1}{\pi} \int_{-\infty}^{-1} \frac{1}{(\xi^2 + 1)(\xi - z)} d\xi + \frac{1}{\pi} \int_{1}^{\infty} \frac{1}{(\xi^2 + 1)(\xi - z)} d\xi
$$

$$
= -\frac{1}{z} + \frac{1}{\pi} \int_{1}^{\infty} \frac{1}{(\xi^2 + 1)} \left[\frac{1}{\xi + z} + \frac{1}{\xi - z} \right] d\xi.
$$

Using partial fractions, we have

$$
\int^{x} \frac{1}{(\xi^2 + 1)(\xi + z)} d\xi = \frac{1}{z^2 + 1} \ln \frac{x + z}{\sqrt{x^2 + 1}} + \frac{z}{z^2 + 1} \arctan x
$$

and

$$
\int^{x} \frac{1}{(\xi^2 + 1)(\xi - z)} d\xi = \frac{1}{z^2 + 1} \ln \frac{x - z}{\sqrt{x^2 + 1}} - \frac{z}{z^2 + 1} \arctan x.
$$

Thus,

$$
f(z) = -\frac{1}{z} + \frac{1}{\pi} \frac{1}{z^2 + 1} \mathrm{Ln} \frac{x^2 - z^2}{x^2 + 1} \Big|_{1}^{\infty} = -\frac{1}{z} - \frac{1}{\pi} \frac{1}{z^2 + 1} \mathrm{Ln} \frac{1 - z^2}{2}.
$$

The only limit that is placed on the diversity of applications of this sort is our capacity to conjure up unique combinations of singular behaviour.

A further application of dispersion relations involves the evaluation of principal value integrals. If we let z approach the cut on the real axis from above, (4.1.5) becomes

$$f_+(x) = \lim_{\delta \to 0^+} \frac{1}{\pi} \int_{x_0}^{\infty} \frac{\operatorname{Im} f_+(\xi)}{\xi - x - i\delta} d\xi = \frac{1}{\pi} \wp \int_{x_0}^{\infty} \frac{\operatorname{Im} f_+(\xi)}{\xi - x} d\xi + i \operatorname{Im} f_+(x)$$

or,

$$\operatorname{Re} f_+(x) = \frac{1}{\pi} \wp \int_{x_0}^{\infty} \frac{\operatorname{Im} f_+(\xi)}{\xi - x} d\xi, \quad x > x_0. \tag{4.3.1}$$

A similar result obtains when this limiting procedure is applied to (4.1.2) :

$$\operatorname{Re} f_+(x) = \sum_{k=1}^{n} \frac{r_k}{x - z_k} + \frac{1}{\pi} \wp \int_{x_0}^{\infty} \frac{\operatorname{Im} f_+(\xi)}{\xi - x} d\xi, \quad x > x_0. \tag{4.3.2}$$

Thus, integrals of the type $\wp \int_c^{\infty} \frac{g(\xi)}{\xi - x} d\xi, x > c, g(x)$ continuous on $c \le x < \infty$, can be evaluated by identifying the harmonic conjugate of $g(x, y)$, where $g(x, 0) \equiv g(x)$. The process of identification involves intuition, guesswork, trial and error, or a combination of all three which means that the evaluation of an arbitrary integral of this type is by no means a straightforward exercise. Fortunately, it is neither difficult nor particularly time-consuming to generate a table of standard integrals to serve as an intuitive guide. The next two examples will demonstrate just how easy is this latter task.

Examples: Consider the function $f(z) = \frac{(z-1)^{\alpha-1}}{(z+1)^{\alpha}}, 0 < \alpha < 1$ whose behaviour at the branch points $z = \pm 1$ and at infinity permits an unsubtracted dispersion representation and hence, use of (4.3.1). If we choose the branch cut to lie along the real axis segment $-1 \le x \le 1$, the principal branch of this function (the branch that is real for real z not on the cut) will be $f(z) = \frac{r_+^{\alpha-1}}{r_-^{\alpha}} e^{i[(\alpha-1)\theta_+ - \alpha\theta_-]}, -\pi < \theta_\pm < \pi$ where $r_\pm$ and $\theta_\pm$ are defined as shown below.

Just above the cut $r_+ = \lim_{y \to 0} \sqrt{(1-x)^2 + y^2} = (1-x), r_- = 1+x, \theta_+ = \pi^-$ and $\theta_- = 0^+$.

Therefore,

$$f_+(x) = \frac{(1-x)^{\alpha-1}}{(1+x)^{\alpha}} e^{i\pi(\alpha-1)} = -\frac{(1-x)^{\alpha-1}}{(1+x)^{\alpha}} e^{i\pi\alpha}.$$

Taking its real and imaginary parts and substituting them into (4.3.1), modified for the finite length of the cut, we obtain the simple result

$$\wp \int_{-1}^{1} \frac{(1-\xi)^{\alpha-1}}{(1+\xi)^{\alpha}} \frac{1}{\xi - x} d\xi = \pi \cot \pi\alpha \frac{(1-x)^{\alpha-1}}{(1+x)^{\alpha}}.$$

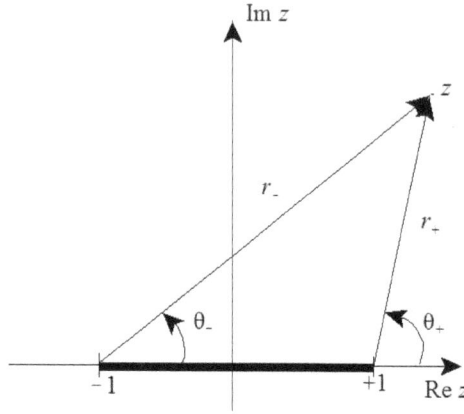

Figure 4.3: The variables used to specify the branches of $f(z) = (z-1)^{\alpha-1}/(z+1)^{\alpha}$.

As a final application, we consider the function $f(z) = \frac{\text{Ln}(a-z)}{z}$, $a > 0$ which has branch points at $z = a$ and $z = \infty$ and a simple pole at $z = 0$ with residue Ln a. In addition, with its cut chosen to lie along the positive real axis, it has the requisite behaviour to admit a dispersion representation and hence, application of (4.3.2).

With this choice of cut,

$$\text{Ln}(a-z) = \ln|a-z| + i \arg(a-z), \quad -\pi < \arg(a-z) < \pi.$$

and so, just above the cut

$$f_+(x) = \frac{1}{x} \ln|a-x| - \frac{i\pi}{x}.$$

Therefore, (4.3.2) immediately yields

$$\frac{1}{x} \ln|a-x| = \frac{\ln a}{x} + \frac{1}{\pi} \wp \int_a^\infty \frac{-\pi}{\xi} \frac{1}{\xi-x} d\xi, \quad x > a$$

or,

$$\wp \int_a^\infty \frac{1}{\xi(\xi-x)} d\xi = -\frac{1}{x} \ln \frac{x-a}{a}.$$

5 Analytic Continuation

5.1 Analytic Continuation

Our focus in the last chapter was on the construction of an analytic function from a knowledge of its singularities. More often than not, however, we are confronted with the inverse problem: given some knowledge of a function in a restricted region of its domain of holomorphy, determine its singularities. This will be the focus of the present chapter.

We have seen repeatedly that one need not know all that much about an analytic function in order to determine its value everywhere in the complex plane or on its Riemann surface. Cauchy's Integral Representation can be viewed as the embodiment of this property and thus far it has provided the key to exploiting it. We are now going to find out what constitutes a **minimal** set of information for the determination of an analytic function. The answer is one that is best exploited not by Cauchy's Integral but by one of its consequences, the Taylor series. In so doing, we shall also find out how to use a representation of a function that is valid in one domain of the complex plane to determine its values at points outside the domain or indeed, at any points where it is holomorphic.

Our starting point is the following theorem which, despite its innocuous appearance, is one of the most remarkable results of complex analysis.

Theorem: Let $f_1(z)$ and $f_2(z)$ be holomorphic in a domain D of the complex plane. If the two functions coincide in any neighbourhood, however small, of a point z in D, or even on a point set with an accumulation point in D, then they coincide throughout D.

Proof: The function $f_1(z) - f_2(z)$ is holomorphic throughout D and has a set of zeros consisting of the points where $f_1(z)$ and $f_2(z)$ coincide, with an accumulation point in D. We know that in any domain where it is holomorphic a function either has isolated zeros or it is identically zero. Thus, $f_1(z) - f_2(z) \equiv 0$ or, $f_1(z) \equiv f_2(z)$, for all z in D.

What this theorem establishes is that a holomorphic function is uniquely determined **everywhere** within its domain of holomorphy by its behaviour in the neighbourhood of an arbitrary point of that domain. But how can one exploit this remarkable property? Obviously not by means of a Cauchy Integral or dispersion representation or anything else of that ilk as we lack the necessary input information. However, what we do have is precisely the information needed to determine a Taylor series representation.

Suppose that we know the value of the function $f(z)$ throughout a neighbourhood of the point $z = z_0$ which is a point lying within the function's domain of holomorphy, D. This is sufficient to permit calculation of the coefficients

$$c_0 = f(z_0), c_1 = f'(z_0), \ldots, c_m = \frac{1}{m!} f^{(m)}(z_0), \ldots$$

of the Taylor expansion of $f(z)$ about $z = z_0$. Hence, we can set

$$f(z) = \sum_{m=0}^{\infty} c_m (z - z_0)^m, \qquad (5.1.1)$$

for all values of z within a circle C_0, $|z - z_0| = R_0$. The quantity R_0, the radius of convergence of the series, is equal to the distance from z_0 to the nearest singularity of $f(z)$ and so will normally be larger than the radius of the neighbourhood from which we started. Thus, equation (5.1.1) provides us with an **analytic continuation** of $f(z)$ which means that it determines the behaviour of $f(z)$ outside its initial domain of definition. Moreover, our theorem guarantees that this analytic continuation is unique.

The disc in which (5.1.1) converges likely represents a small fraction of the domain of holomorphy D and so may not include a particular point $z = z_0'$ that is of interest to us. To analytically continue $f(z)$ to such a distant point requires either a different representation, with a much larger domain of validity, or a process involving the repeated generation of Taylor series with overlapping circles of convergence. We shall consider the second option first.

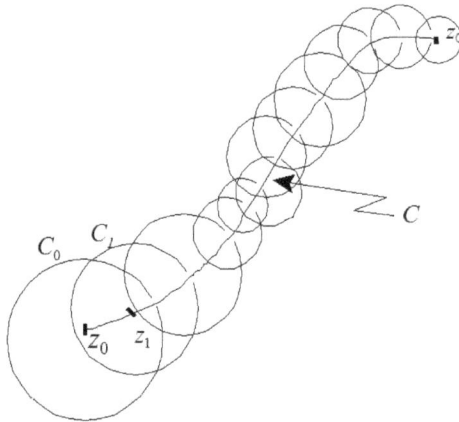

Figure 5.1: A function can be analytically continued along a curve C by means of repeated Taylor expansions about appropriately chosen points of C.

From the definition of a domain we know that the points z_0 and z_0' can be connected by a simple curve C that lies entirely within D. As shown in Figure 5.1, let us take a point z_1 on C such that $|z_1 - z_0| < R_0$, that is, such that z_1 lies within the circle of convergence C_0 of the Taylor series in (5.1.1).

Since the series converges uniformly in every closed disc $|z - z_0| \le r < R_0$, it can be differentiated term by term to determine all derivatives of $f(z)$ at all points of such a disc. In particular, since we may choose an r for which $|z_1 - z_0| < r < R_0$, we can calculate

$$f(z_1) = \sum_{m=0}^{\infty} c_m (z_1 - z_0)^m, f'(z_1) = \sum_{m=1}^{\infty} c_m\, m (z_1 - z_0)^{m-1}, \dots,$$

$$\frac{1}{n!} f^{(n)}(z_1) = \sum_{m=n}^{\infty} c_m \frac{m!}{n!(m-n)!} (z_1 - z_0)^{m-n}, \dots. \qquad (5.1.2)$$

But these are just the coefficients $c_m^{(1)}$, $m = 0, 1, \dots, n, \dots$ of the Taylor series expansion of $f(z)$ about the point $z = z_1$. Hence, we can now set

$$f(z) = \sum_{m=0}^{\infty} c_m^{(1)} (z - z_1)^m \qquad (5.1.3)$$

which is valid for all z within a circle C_1 with centre at $z = z_1$ and radius R_1.

Since $f(z)$ is holomorphic at all points on C, R_1 must be non-zero and hence, it must be possible to choose the point $z = z_1$ so that C_1 lies partly outside C_0. Therefore, (5.1.3) provides a unique analytic continuation of $f(z)$ from C_0 to the somewhat larger domain formed by the union of C_0 and C_1. On the segment of the curve C that lies outside C_0 but within C_1 we now choose a new point $z = z_2$. Then, using the uniformly convergent series in (5.1.3) to calculate the coefficients of the Taylor expansion of $f(z)$ about $z = z_2$, we obtain a further analytic continuation of $f(z)$. Repeating this argument over and over, we proceed along C with overlapping circles $C_0, C_1, C_2, \dots$ until one of the circles finally covers the point $z = z_0'$ thus enabling us to find the Taylor expansion of $f(z)$ about it as well. The behaviour of $f(z)$ in the neighbourhood of $z = z_0'$ is then determined and our analytic continuation procedure is completed.

It is not possible to analytically continue through a singular point of $f(z)$ since the radii of the circles C_i tend to zero as we approach it. However, it is possible to analytically continue around the singularity and in the process, determine its location; (see Figure 5.2). Thus, in principle at least, this technique can be used to continue a function throughout its domain of holomorphy, starting from an arbitrary point of that domain and using all possible paths to form chains of overlapping circles. The continuation will be complete when the domain's **natural boundaries,** which are just the singular points of the function, have all been encountered. This also applies to mutivalued functions for if we analytically continue around a branch point on one Riemann sheet, we will eventually generate values appropriate to an adjacent sheet. Thus, continuing along all possible paths we will determine the behaviour of the function throughout its Riemann surface as well as the geometry of that surface.

As we have emphasized above, the only barrier this technique cannot surmount is the natural boundary of the function's domain of holomorphy. There are some functions for which this is surprisingly limiting, at least in a geometrical sense. Consider for example,

$$f(z) = 1 + z^2 + z^4 + z^8 + z^{16} + \dots = 1 + \sum_{m=1}^{\infty} z^{2^m},$$

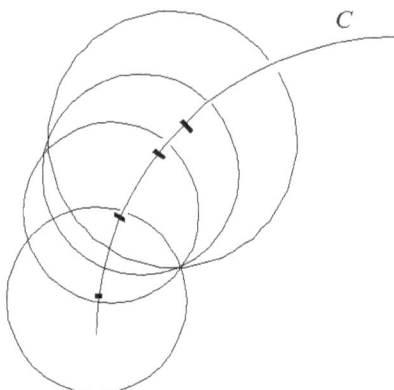

Figure 5.2: Analytic continuation around a singular point.

whose circle of convergence is $|z| = 1$. It is readily shown that any value of z that satisfies one of the equations

$$z^2 = 1, z^4 = 1, z^8 = 1, \ldots, z^{2^m} = 1, \ldots$$

is a singularity of $f(z)$. These values correspond to the points $z = e^{2\pi i k/2^m}$, where k and m are integers. On every arc of the unit circle there is an infinite number of such points. Thus, it is impossible to continue $f(z)$ outside its circle of convergence since it is also the natural boundary of the function's domain of holomorphy.

While conceptually powerful, the technique of successive Taylor expansion of a function is impractical. Fortunately, there are many alternative and more immediate methods to effect analytic continuations. Before we examine some of them, however, we need to delve a little deeper into the implications of the theorem that introduced this Section.

Suppose that we are given two analytic functions $f_1(z)$ and $f_2(z)$ whose functional forms differ from each other and are valid only in the domains D_1 and D_2, respectively. Suppose further that D_1 and D_2 overlap and that in their intersection $f_1(z)$ and $f_2(z)$ are identical. Then, our theorem tells us that the (unique) result of analytically continuing $f_1(z)$ into D_2 must coincide with $f_2(z)$ and conversely the result of analytic continuation of $f_2(z)$ into D_1 must be identical to $f_1(z)$. In fact, $f_1(z)$ and $f_2(z)$ are just local representations of the unique function

$$f(z) = \begin{cases} f_1(z), z \text{ in } D_1 \\ f_2(z), z \text{ in } D_2 \end{cases}$$

which is holomorphic throughout the combined domain $D_1 \cup D_2$.

Since the application of analytic continuation to $f_1(z)$ yields $f_2(z)$ and vice versa, one says that $f_1(z)$ and $f_2(z)$ are **analytic continuations** of each other.

Example: Consider the functions

$$f_1(z) = \sum_{m=0}^{\infty} \frac{z^m}{a^{m+1}}, \quad |z| < |a| \tag{5.1.4}$$

and

$$f_2(z) = \sum_{m=0}^{\infty} \frac{(z-b)^m}{(a-b)^{m+1}}, \quad |z-b| < |a-b| \tag{5.1.5}$$

where, to ensure that the circle of convergence of $f_2(z)$ is not interior to that of $f_1(z)$, we require that b/a not be real. Both series sum to the function $(a-z)^{-1}$. Thus, although $f_1(z)$ and $f_2(z)$ have different functional forms defined in different albeit overlapping domains, they represent the **same** analytic function, $f(z) = \frac{1}{a-z}$. Each is a unique analytic continuation of the other but, in particular, $f_2(z)$ provides the means of analytically continuing $f_1(z)$ to **any point outside** the circle $|z| = |a|$ simply by varying the value of the parameter b. However, if this is our primary interest we are by no means restricted to the use of power series representations to accomplish it. Indeed, the two **integral representations**

$$f_3(z) = \int_0^{\infty} e^{-t(a-z)}\, dt, \quad \mathrm{Re}\, z < \mathrm{Re}\, a$$

$$f_4(z) = -\int_0^{\infty} e^{t(a-z)}\, dt, \quad \mathrm{Re}\, z > \mathrm{Re}\, a$$

which converge to $(a-z)^{-1}$ in their respective domains of definition, provide a more effective means of analytically continuing $f_1(z)$ outside its circle of convergence.

In this example we have a closed-form expression for $f(z)$, namely $(a-z)^{-1}$, which is valid throughout the function's domain of holomorphy. However, this is an exception rather than a norm. The functions that can be expressed in terms of a finite number of elementary functions make up a very limited subset of the totality of analytic functions. We must get used to the idea that to know how a function behaves at widely separated points one usually requires two or more representations of very diverse appearance. A perfect example is provided by a function called the gamma function and denoted $\Gamma(z)$. Before introducing it, however, we need to complete our discussion of analytic continuation.

We have required that two functions $f_1(z)$ and $f_2(z)$ have overlapping domains of definition if they are to be analytic continuations of each other. As our next theorem shows, this is unnecessarily restrictive.

Theorem: Let $f_1(z)$ and $f_2(z)$ be holomorphic in the simply connected domains D_1 and D_2, respectively and let D_1 and D_2 have in common as part of their boundaries a simple curve C. Then, $f_1(z)$ and $f_2(z)$ are analytic continuations of each other if and

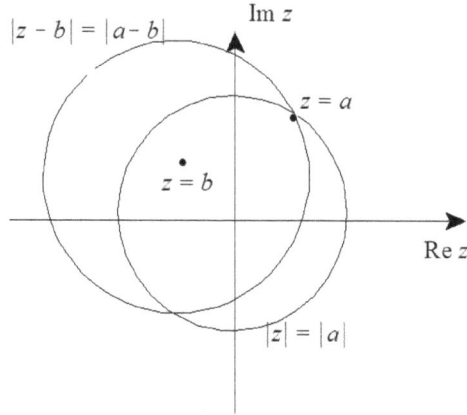

Figure 5.3: The domains of definition of the two representations of $f(z) = (a-z)^{-1}$ given in equations (5.1.4) and (5.1.5).

only if they tend uniformly to common values along C; that is, if and only if they are continuous in the regions $D_1 \cup C$ and $D_2 \cup C$, respectively and $f_1(z) = f_2(z)$ for all z on C.

Proof: That the condition is necessary is obvious. Therefore, we need only show that it is sufficient.

Define the function $f(z)$ as follows:

$$f(z) = \begin{cases} f_1(z) \text{ for } z \text{ in } D_1 \text{ or } C \\ f_2(z) \text{ for } z \text{ in } D_2 \text{ or } C. \end{cases}$$

We must now show that $f(z)$ is holomorphic throughout the entire domain

$$D = D_1 \cup D_2 \cup C.$$

Let Γ be any simple closed curve within $D_1 \cup D_2 \cup C$ and consider the integral

$$\int_\Gamma f(z)dz. \tag{5.1.6}$$

If Γ lies entirely within either D_1 or D_2, the integral vanishes by Cauchy's Theorem. If Γ lies in both D_1 and D_2 then, as shown in Figure 5.4, we can introduce two simple closed curves Γ_1 and Γ_2 separated by an infinitesimal distance so that they lie entirely within D_1 and D_2, respectively and follow a section of the boundary C in opposite directions. We can thereby write (5.1.6) as the sum of two integrals,

$$\int_\Gamma f(z)dz = \int_{\Gamma_1} f(z)dz + \int_{\Gamma_2} f(z)dz = \int_{\Gamma_1} f_1(z)dz + \int_{\Gamma_2} f_2(z)dz,$$

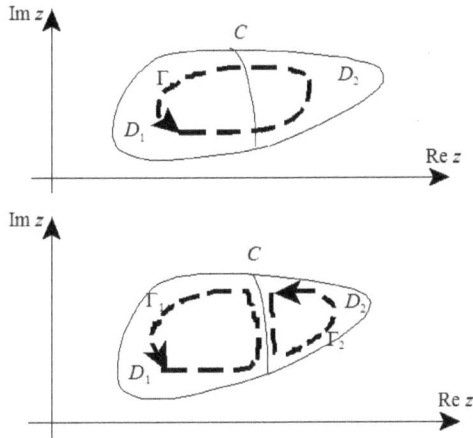

Figure 5.4: A simple closed curve $\Gamma \in D_1 \cup D_2 \cup C$ is separated at the boundary C into two simple closed curves Γ_1 and Γ_2.

both of which vanish. Thus,

$$\int_\Gamma f(z)dz = 0$$

for any closed contour Γ contained within the simply connected domain $D_1 \cup D_2 \cup C$ and so, by Morera's Theorem, $f(z)$ is holomorphic there. Therefore, $f_1(z)$ and $f_2(z)$ are analytic continuations of each other.

We are now in a position to prove a theorem which not only generalizes a common feature of the elementary functions but provides an analytic continuation technique that plays a key role in most physical applications of dispersion representations. It is known as the **Schwarz reflection principle**.

Theorem: Let $f(z)$ be holomorphic in a domain D which has, as part of its boundary, a segment C of the real axis and let D^* be the mirror image of D with respect to that axis. Then, if $f(z)$ is continuous within the region $D \cup C$ and assumes real values on C, its analytic continuation into the domain D^* exists and is given by $f^*(z^*)$ for all z in D^*.

Proof: Let Γ denote an arbitrary simple closed curve in D, described by the parametric equation $z = \zeta(t)$, $t_1 \le t \le t_2$. Since $f(z)$ is holomorphic in D, we have

$$\int_\Gamma f(z)dz = \int_{t_1}^{t_2} f[\zeta(t)]\frac{d\zeta(t)}{dt}dt = 0. \tag{5.1.7}$$

Let Γ^* be the image of Γ in D^*, (Figure 5.5). Its parametric equation must then be $z = \zeta^*(t)$, $t_1 \le t \le t_2$, with increasing t corresponding to a clockwise (counter-clockwise) traversal of Γ^* if it produces a counter-clockwise (clockwise) circuit of Γ. Thus, integrating the function $g(z) \equiv f^*(z^*)$ around Γ^*, we have

$$\int_{\Gamma^*} g(z)dz = \int_{t_1}^{t_2} g[\zeta^*(t)]\frac{d\,\zeta^*(t)}{dt}dt = \int_{t_1}^{t_2} f^*[\zeta(t)]\frac{d\,\zeta^*(t)}{dt}dt = \left[\int_{t_1}^{t_2} f[\zeta(t)]\frac{d\zeta(t)}{dt}dt\right]^* = 0$$

by equation (5.1.7) . Hence, we know from Morera's Theorem that $f^*(z^*)$ is a holomorphic function of z throughout the simply connected domain $D*$. Moreover, since $f(z)$ is real on the real axis segment C, we have $f^*(z^*) = f(z)$ for all z on C. Therefore, according to the preceding theorem, $f(z)$ and $f^*(z^*)$ are analytic continuations of each other and together define the unique function

$$F(z) = \begin{cases} f(z), z \text{ in } D \text{ or } C \\ f^*(z^*), \ z \text{ in } D^* \text{ or } C \end{cases} \tag{5.1.8}$$

which is holomorphic throughout the domain $D \cup D^* \cup C$.

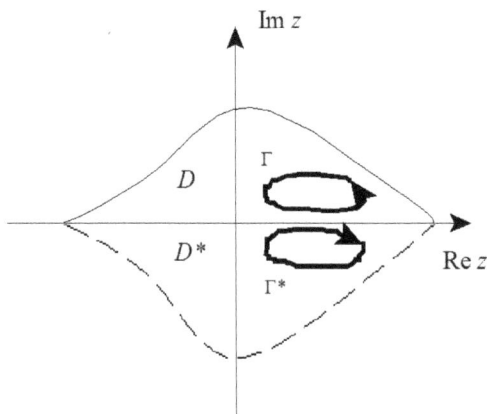

Figure 5.5: The domain D and its mirror image through the real axis $D*$.

It follows immediately from (5.1.8) that

$$F(z^*) = F^*(z) \text{ for all } z \text{ in } D \cup D^* \cup C. \tag{5.1.9}$$

Evidently, this relation must hold for any function that is holomorphic throughout a domain intersected by the real axis and that assumes real values when its argument is real. Thus, we now see why it was referred to as a **reality condition** in the preceding Chapter. More importantly, we also see that our discovery in Chapter 1 that it is satisfied by each of the single valued elementary functions was indicative of a general consequence of their definition and not mere coincidence.

5.2 The Gamma Function

In this Section we shall introduce a variety of functional forms attributable to the gamma function which is one of the more frequently encountered functions of mathematical physics. This will illustrate just how small need be our reliance on Taylor series representations to effect analytic continuations and offer further exposure to a principal alternative, the use of **integral representations**.

The usual introduction to the gamma function is via Euler's integral (of the second kind) which, for real $x > 0$, gives

$$\Gamma(x) = \int_0^\infty e^{-t}\, t^{x-1}\, dt. \tag{5.2.1}$$

Whether the introduction is made in a physics or a calculus course, it will always be pointed out that $\Gamma(x)$ is the continuous variable generalization of the factorial function, a fact that we can readily verify from equation (5.2.1). Writing t^{x-1} as $\frac{1}{x}\frac{d}{dt}\,t^x$ and integrating by parts, we obtain

$$\Gamma(x) = \left.\frac{e^{-t}}{x}\right|_0^\infty + \int_0^\infty \frac{e^{-t}}{x}\, t^x\, dt.$$

The first term vanishes and so, using (5.2.1) to identify the second term, we have

$$\Gamma(x+1) = x\Gamma(x),\ x > 0. \tag{5.2.2}$$

Thus, since

$$\Gamma(1) = \int_0^\infty e^{-t}\, dt = 1, \tag{5.2.3}$$

setting x equal to n, an integer, yields

$$\Gamma(n+1) = n!. \tag{5.2.4}$$

This makes it clear that $\Gamma(x)$ provides a smooth interpolation between the points defined by $n!,\ n = 0, 1, 2, \ldots$; explicit evaluation yields the curve shown in Figure 5.6.

The integral in (5.2.1) continues to converge when x is replaced by the complex variable z provided that $\mathrm{Re}\,z > 0$. Moreover, differentiating the function

$$\Gamma(z) = \int_0^\infty e^{-t}\, t^{z-1}\, dt, \quad \mathrm{Re}\,z > 0 \tag{5.2.5}$$

with respect to z, we obtain yet another integral that converges for $\mathrm{Re}\,z > 0$. Thus, (5.2.5) evidently defines an analytic function whose domain of holomorphy includes the right half-plane and hence, the positive real axis. Therefore, according to the theorem at the beginning of this Chapter, the integral in (5.2.5) is a representation for Re z > 0 of the only analytic function that can assume the values given by (5.2.1) when Im z = 0. Appropriately, we have denoted this function by $\Gamma(z)$. An alternative and somewhat simpler way of stating this result is to say that (5.2.5) is the analytic continuation of (5.2.1) from the positive real axis to the entire right half-plane.

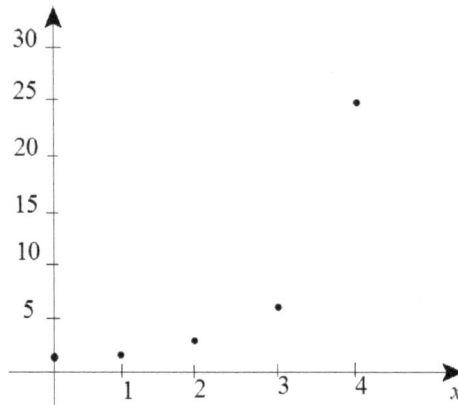

Figure 5.6: The dots indicate the discrete points defined by $n\,!$; the corresponding curve is $\Gamma(x+1)$.

With the help of the factorial property (5.2.2) we can use (5.2.5) to determine the natural boundary of the gamma function's domain of holomorphy. This will determine whether we need to seek an analytic continuation for Re z < 0 and if so, assist us with the search.

The factorial property holds for complex as well as real values of the arguments. In addition, we know that the integral representation of $\Gamma(z+1)$ converges for Re z > -1 and hence, that the right hand side of

$$z\Gamma(z) = \Gamma(z+1) \tag{5.2.6}$$

is well-defined there. Thus, dividing through by z, we immediately obtain an analytic continuation of (5.2.5) from the domain Re z > 0 to Re z > -1,

$$\Gamma(z) = \frac{1}{z}\Gamma(z+1) = \frac{1}{z}\int_0^\infty e^{-t}\, t^z\, dt, \quad z \neq 0. \tag{5.2.7}$$

Evidently, Re z = 0 is not a natural boundary of $\Gamma(z)$. The sole source of the restriction Re z > 0 that has been placed on the integral representation (5.2.5) is an isolated singularity at z = 0. Taking the limit of (5.2.7) as $z \to 0$ and recalling that $\Gamma(1) = 1$, we find

$$\Gamma(z) \to \frac{1}{z}$$

which identifies the singularity as a simple pole and the corresponding residue as unity.

This procedure may be repeated as often as one likes to obtain

$$\Gamma(z) = \frac{\Gamma(z+n+1)}{(z+n)(z+n-1)(z+n-2)...(z)} \tag{5.2.8}$$

the right hand of which is holomorphic for Re $z > -n$ except for simple poles at the points $z = 0, -1, -2, \ldots, -(n-1)$. Thus, we conclude that $\Gamma(z)$ is a meromorphic function and as such, can be continued anywhere in the left half-plane so long as we avoid its simple poles at the negative integers and zero. This allows, for example, calculation of its functional dependence on negative as well as positive real values of its argument as shown for $|x| < 5$ in Figure 5.7.

To find the residue of $\Gamma(z)$ at $z = -n$, we simply take the limit of (5.2.8) as $z \to -n$ and use $\Gamma(1) = 1$. Thus,

$$\Gamma(z) \to \frac{1}{(-1)^n n!(z+n)}$$

and hence,

$$\text{Res}[\Gamma(-n)] = \frac{(-1)^n}{n!}. \tag{5.2.9}$$

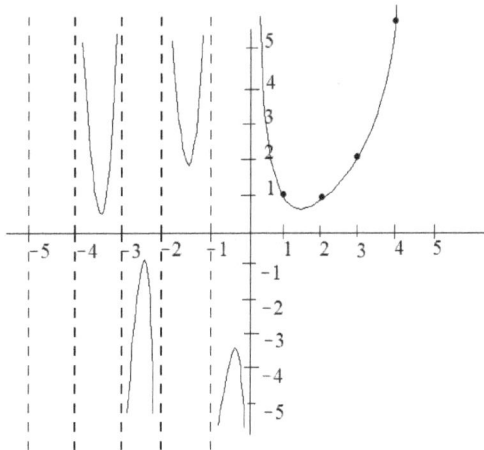

Figure 5.7: $\Gamma(z)$, as determined by (5.2.5) and (5.2.8) , plotted for $z = x$, real.

An additional means of analytically continuing $\Gamma(z)$ into the left half-plane is provided by the product $\Gamma(z)\Gamma(1-z)$. The integral representation of $\Gamma(1-z)$ converges for Re $z < 1$

and its singularities are simple poles located at the positive integers with corresponding residues $\frac{(-1)^n}{(n-1)!}$. Thus, $\Gamma(z)\Gamma(1-z)$ is a meromorphic function with simple poles at $z = n, n = 0, \pm1, \pm2, \ldots$ and residues $(-1)^n$. But these are precisely the same singularities and residues as are possessed by the function $\pi \csc \pi z$. This suggests that

$$\Gamma(z)\Gamma(1-z) = \pi \csc \pi z \tag{5.2.10}$$

which we confirm as follows.

From (5.2.5) we have, for $0 < \operatorname{Re} z < 1$,

$$\Gamma(z)\Gamma(1-z) = \int_0^\infty e^{-u} u^{z-1}\, du \int_0^\infty e^{-v} v^{-z}\, dv = 4\int_0^\infty e^{-x^2} x^{2z-1}\, dx \int_0^\infty e^{-y^2} y^{-(2z-1)}\, dy$$

where we have introduced the dummy variables of integration $u = x^2$ and $v = y^2$. Making a further change of integration variables via $r^2 = x^2 + y^2$, $\theta = \tan^{-1}(y/x)$ this becomes

$$\Gamma(z)\Gamma(1-z) = 4\int_0^\infty e^{-r^2} r\,dr \int_0^{\pi/2} [\cot\theta]^{2z-1}\, d\theta = 2\int_0^{\pi/2} [\cot\theta]^{2z-1}\, d\theta, \quad 0 < \operatorname{Re} z < 1.$$

With a final change of variable, $\cot\theta = t$, our product becomes

$$\Gamma(z)\Gamma(1-z) = 2\int_0^\infty \frac{t^{2z-1}}{t^2+1}\, dt = \pi \csc \pi z, \quad 0 < \operatorname{Re} z < 1$$

where the last equality follows from an integral evaluation that we performed in Chapter 3. By analytic continuation, this equality must hold wherever both sides are holomorphic. Thus,

$$\Gamma(z)\Gamma(1-z) = \pi \csc \pi z \text{ for all finite } z.$$

This equation permits easy calculation of $\Gamma(z)$ when $z = \frac{2n+1}{2}$, $n = 0, \pm1, \pm2, \ldots$. Setting $z = \frac{1}{2}$ we have

$$\left[\Gamma\left(\frac{1}{2}\right)\right]^2 = \pi \csc \frac{\pi}{2} = \pi$$

and hence,

$$\Gamma\left(\frac{1}{2}\right) = \sqrt{\pi}.$$

Thus, using (5.2.6), we have

$$\Gamma(n+1/2) = \frac{(2n-1)}{2}\frac{(2n-3)}{2}\cdots\frac{3}{2}\frac{1}{2}\Gamma\left(\frac{1}{2}\right) = \frac{(2n-1)!!}{2^n}\sqrt{\pi} \tag{5.2.11}$$

$$\Gamma(-n+1/2) = \frac{\Gamma\left(\frac{1}{2}\right)}{\frac{(1-2n)}{2}\frac{(3-2n)}{2}\cdots\frac{(-3)}{2}\frac{(-1)}{2}} = \frac{2^n(-1)^n\sqrt{\pi}}{(2n-1)!!} \tag{5.2.12}$$

for $n = 1, 2, 3, \ldots$.

Although we now have several ways of analytically continuing our initial integral representation of $\Gamma(z)$, we have yet to find an alternative representation whose domain of validity includes at least part of the left half-plane. A number of possibilities present themselves.

For example, since

$$\frac{1}{\Gamma(z)} = \frac{\sin \pi z}{\pi} \Gamma(1 - z)$$

is entire, we know from Chapter 3 that it must admit an infinite product representation which is valid everywhere in the finite plane. Indeed, according to equation (3.5.4) we can write

$$\frac{1}{\Gamma(z + 1)} = \frac{1}{\Gamma(1)} e^{cz} \prod_{n=1}^{\infty} \left(1 + \frac{z}{n}\right) e^{-z/n} \tag{5.2.13}$$

where $c = -\frac{d}{dz}[\ln \Gamma(z + 1)]\big|_{z=0}$ or, since $\Gamma(z + 1) = z\Gamma(z)$ and $\Gamma(1) = 1$,

$$\frac{1}{\Gamma(z)} = z e^{cz} \prod_{n=1}^{\infty} \left(1 + \frac{z}{n}\right) e^{-z/n}. \tag{5.2.14}$$

To complete the specification of this representation we need only make a more tractable identification of the constant c. Setting $z = 1$ in (5.2.13), and using $\Gamma(2) = \Gamma(1) = 1$, we find

$$e^{-c} = \prod_{n=1}^{\infty} \left(1 + \frac{1}{n}\right) e^{-1/n}$$

or,

$$c = \sum_{n=1}^{\infty} \left[\frac{1}{n} - \ln \left(1 + \frac{1}{n}\right)\right] = \lim_{m \to \infty} \left[\sum_{n=1}^{m} \frac{1}{n} - \ln m\right] \equiv \gamma \tag{5.2.15}$$

where $\gamma = 0.57721566\ldots$ is a natural number known as the Euler-Mascheroni constant. Thus, we finally obtain the representation

$$\frac{1}{\Gamma(z)} = z e^{\gamma z} \prod_{n=1}^{\infty} \left(1 + \frac{z}{n}\right) e^{-z/n} \tag{5.2.16}$$

for all finite z.

This was originally derived by Euler in a somewhat different form. Using (5.2.15) for γ we can rewrite (5.2.16) as

$$\frac{1}{\Gamma(z)} = z \lim_{m \to \infty} e^{-z \ln m} \prod_{n=1}^{m} \left(1 + \frac{z}{n}\right)$$

and thus obtain

$$\Gamma(z) = \lim_{m \to \infty} \frac{m!}{z(z + 1)\ldots(z + m)} m^z. \tag{5.2.17}$$

This is called Euler's formula for $\Gamma(z)$.

Although valid for all finite z, (5.2.16) is seldom used other than to calculate the logarithmic derivative of $\Gamma(z)$,

$$\psi(z) \equiv \frac{d}{dz} \ln \Gamma(z) = -\frac{1}{z} - \gamma + \sum_{n=1}^{\infty} \left[\frac{1}{n} - \frac{1}{z+n} \right]. \tag{5.2.18}$$

The reason for this neglect is evident: neither (5.2.16) nor its alternative form (5.2.17) admits readily to explicit evaluation for specific values of z. What we really need, if it exists, is an integral representation of $\Gamma(z)$ that is valid for all finite z.

Our starting point in trying to find one is the integral representation that we already possess

$$\Gamma(z) = \int_0^{\infty} e^{-t} t^{z-1} \, dt, \quad \operatorname{Re} z > 0. \tag{5.2.19}$$

Its integrand has branch points at $t = 0$ and $t = \infty$; indeed, it is the behaviour of the integrand in the neighbourhood of $t = 0$ that prevents convergence for $\operatorname{Re} z < 0$. However, if we join these points by a branch cut running along the positive real axis, we should be able to use the same integrand together with an integration contour that runs just above and below the cut to generate an alternative representation for $\Gamma(z)$. Therefore, let us consider

$$\int_C e^{-t} t^{z-1} \, dt \tag{5.2.20}$$

where C is the contour shown in Figure 5.8 and t^{z-1} is defined to be the branch

$$t^{z-1} = e^{(z-1)[\ln |t| + i \arg t]}, \quad 0 < \arg t < 2\pi.$$

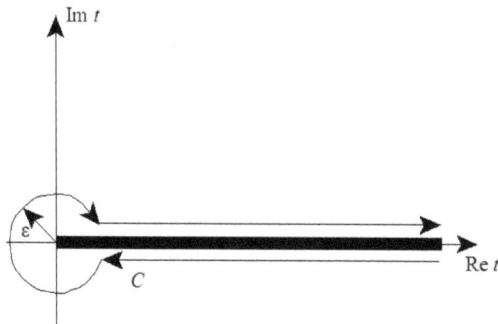

Figure 5.8: The contour in the complex t plane used in the integral (5.2.20).

With this choice of integration path we not only avoid two troublesome points, we do not cross the cut that joins them. Moreover, at the end-points $t = \infty \pm i\varepsilon$ the integrand vanishes for all finite values of z. Thus, our integral converges to a single-valued function of z for all z in the finite plane.

To find a relationship between the integral and $\Gamma(z)$ we shall temporarily restrict the domain of z to the right half-plane, $\mathrm{Re}\, z > 0$. The contribution of the semi-circular arc about $t = 0$ then vanishes in the limit as $\varepsilon \to 0$ and so,

$$\int_C e^{-t}\, t^{z-1}\, dt = \int_\infty^0 e^{-t}(e^{2\pi i}\, t)^{z-1}\, e^{2\pi i}\, dt + \int_0^\infty e^{-t}\, t^{z-1}\, dt = (1 - e^{2\pi i z})\int_0^\infty e^{-t}\, t^{z-1}\, dt$$

or, using (5.2.19),

$$\int_C e^{-t}\, t^{z-1}\, dt = 2i \sin \pi z\, e^{\pi i(z-1)}\, \Gamma(z).$$

Both sides of this equation are entire functions of z. Therefore, although derived for $\mathrm{Re}\, z > 0$, it must hold for **all** finite z and thus provides the representation

$$\Gamma(z) = \frac{1}{2i \sin \pi z} \int_C e^{-t}(-t)^{z-1}\, dt, \quad z \neq 0, -1, -2, \ldots. \tag{5.2.21}$$

This is sometimes referred to as Hankel's representation. With the help of (5.2.10) we can invert it to obtain an integral representation of $[\Gamma(z)]^{-1}$ that is valid for all finite z without exceptions:

$$\frac{1}{\Gamma(z)} = \frac{\sin \pi z \Gamma(1-z)}{\pi} = \frac{1}{2\pi i} \int_C e^{-t}(-t)^{-z}\, dt. \tag{5.2.22}$$

The utility of integral representations stems not only from their large domains of validity but also from the fact that their integration contours can be deformed at will so long as one never passes through singularities of their integrands. Thus, for example, one can often render numerical evaluation very simple by an adroit matching of the contour to the desired value(s) of the function's argument. This will be illustrated in the next Chapter in connection with evaluations for large values of $|z|$. A more immediate demonstration of this versatility is provided by the simple task of evaluating $[\Gamma(z)]^{-1}$ at $z = \pm n, n = 0, 1, 2, \ldots$.

When z assumes integer values the integrand in (5.2.22) becomes continuous across the positive real axis and hence, the integration path can be compressed to form a simple closed curve encircling $t = 0$. Evaluation of (5.2.22) then involves only a simple application of either the Residue Theorem or Cauchy's Theorem which yield, respectively

$$\frac{1}{\Gamma(z)} = \begin{cases} \frac{1}{(n-1)!} & n > 0 \\ 0 & n \leq 0 \end{cases}.$$

5.3 Integral Representations and Integral Transforms

Evidently, integral representations can play a very significant role in the definition and continuation of analytic functions. We shall conclude this Chapter with some general comments about them. And to start with, we shall determine the circumstances under which an integral representation definition is valid.

Theorem: Let $G(z, t)$ be a continuous function of both variables when z lies in a simply connected domain D and t lies on a simple curve C. Further, for each such value of t, let $G(z, t)$ be holomorphic within D. Then, the function

$$f(z) = \int_C G(z, t) dt \qquad (5.3.1)$$

is also holomorphic within D and its derivatives of all orders may be found by differentiating under the integration sign, provided that
1. C is of finite length, or
2. if C is of infinite length, the integral is uniformly convergent for z contained in any closed region interior to D.

Proof: Let Γ denote any simple closed curve in D. Then,

$$\int_\Gamma f(z) dz = \int_\Gamma \left\{ \int_C G(z, t) dt \right\} dz = \int_C \left\{ \int_\Gamma G(z, t) dz \right\} dt = 0$$

and hence, by Morera's Theorem, $f(z)$ is holomorphic in D. Here we have used the fact that the order of integration of an iterated integral with a continuous integrand can always be interchanged if it is uniformly convergent or finite. We have also used the holomorphy of $G(z, t)$ in D and Cauchy's Theorem to obtain the final equality with zero.

Similarly, using Cauchy's Differentiation Formula we have

$$\frac{df(z)}{dz} = \frac{1}{2\pi i} \int_\Gamma \frac{f(\zeta)}{(\zeta - z)^2} d\zeta = \int_C \left\{ \frac{1}{2\pi i} \int_\Gamma \frac{G(\zeta, t)}{(\zeta - z)^2} d\zeta \right\} dt = \int_C \frac{\partial G(z, t)}{\partial z} dt.$$

The most frequently encountered integral representations are of the somewhat specialized form known as **integral transforms,**

$$f(z) = \int_C K(z, t) g(t) dt. \qquad (5.3.2)$$

The function $K(z, t)$ is called the **kernel** of the transform while $g(t)$ is known as the **spectral function**. Each distinct type of transform corresponds to a specific choice of kernel and of integration contour C and each has an "existence theorem" that determines the class of spectral functions $g(t)$ and the domain of z for which the integral converges.

We were introduced to the Hilbert transform,

$$f(z) = \frac{1}{\pi} \wp \int\limits_{-\infty}^{\infty} \frac{g(t)}{t-z} dt,$$

in the course of our discussion of dispersion relations in Section 4.1. Other transforms that we will have occasion to discuss in subsequent Chapters include the Fourier transform,

$$f(z) = \frac{1}{2\pi} \int\limits_{-\infty}^{\infty} e^{itz} g(t) dt,$$

and its two close relatives, the Laplace transform

$$f(z) = \frac{1}{2\pi i} \int\limits_{c-i\infty}^{c+i\infty} e^{tz} g(t) dt$$

and the Mellin transform

$$f(z) = \frac{1}{2\pi i} \int\limits_{c-i\infty}^{c+i\infty} z^t g(t) dt.$$

As we shall see, transforms arise naturally in the solution of boundary value problems where the form of the differential equation together with the nature of the boundary conditions determines more or less uniquely which type of transform to use.

6 Asymptotic Expansions

6.1 Asymptotic Series

Physical problems often require a reasonably detailed knowledge of how particular functions behave at infinity. For example, if a function has an essential singularity at $z = \infty$ one may need a measure of how rapidly it blows up (or vanishes) as $z \to \infty$ along the real axis. Or, more detailed yet, one may actually have to evaluate the function for very large values of $|z|$. Recalling that few interesting functions are expressible in terms of the so-called elementary ones, this might appear to be a rather tall order to fill. Power series or infinite products are certainly unlikely to be helpful in most cases and integral representations would seem to offer a computational nightmare. Fortunately, the latter admit a property that obviates the need to explicitly evaluate them and hence, makes them an ideal starting point after all.

With only a modicum of manipulative effort, one can usually contrive to have an integral representation yield up an **asymptotic expansion** of the function it represents. In its simplest form an asymptotic expansion is a series in inverse powers of z which, while not convergent, has partial sums that provide an arbitrarily good approximation, for sufficiently large $|z|$, of the function to which it corresponds. Thus, by conveying detailed information about the function's large $|z|$ behaviour, it meets our analytical needs to the letter.

The idea of approximating something by a divergent series may seem a little paradoxical. Therefore, we shall assist both our intuition and our credulity by running through a simple example.

Example: Consider the function of a real variable

$$f(x) \equiv \int_x^\infty \frac{1}{t} e^{x-t}\, dt, \quad x > 0. \tag{6.1.1}$$

By making the substitution $u = t - x$, we see that

$$f(x) = \int_0^\infty \frac{e^{-u}}{u + x}\, du < \frac{1}{x} \int_0^\infty e^{-u}\, du = \frac{1}{x}, \quad x > 0$$

which immediately provides us with an upper limit on the values assumed by $f(x)$. To get a better idea of what these values may be, we now integrate (6.1.1) by parts. After n such integrations, we find

$$f(x) = \frac{1}{x} - \frac{1}{x^2} + \frac{2}{x^3} + \ldots + \frac{(-1)^{n-1}(n-1)!}{x^n} + (-1)^n\, n! \int_x^\infty \frac{e^{x-t}}{t^{n+1}}\, dt. \tag{6.1.2}$$

The form of this expression suggests that we examine the series

$$\sum_{m=0}^{\infty} u_m(x), \, u_m(x) = \frac{(-1)^m \, m!}{x^{m+1}}. \tag{6.1.3}$$

Applying the ratio test, we have

$$\lim_{n \to \infty} \left| \frac{u_{n+1}(x)}{u_n(x)} \right| = \lim_{n \to \infty} \frac{n}{x} \to \infty$$

for fixed x. Thus, the series is **divergent** for all finite values of x.

This is not the set-back it might seem. Letting $S_n(x) \equiv \sum_{m=0}^{n} u_m(x)$, we find the difference

$$|f(x) - S_n(x)| \equiv R_n(x) = (n+1)! \int_x^{\infty} \frac{e^{x-t}}{t^{n+2}} dt < (n+1)! \int_x^{\infty} \frac{dt}{t^{n+2}} = \frac{n!}{x^{n+1}}. \tag{6.1.4}$$

In other words, the error committed in approximating $f(x)$ by $S_n(x)$ is guaranteed to be less than $\frac{n!}{x^{n+1}}$. Thus, for fixed n, we can always find an x sufficiently large to make this error less than any prescribed number $\varepsilon > 0$. So, even though divergent, the series (6.1.3) has partial sums which provide an arbitrarily good approximation for $f(x)$, provided only that we restrict ourselves to sufficiently large values of x. Such a series is called an **asymptotic series** and its relationship to $f(x)$ is expressed formally by rewriting (6.1.2) to read

$$f(x) \sim \sum_{m=0}^{\infty} \frac{(-1)^m \, m!}{x^{m+1}}. \tag{6.1.5}$$

The symbol $\sim$ implies approximation rather than equality and in so doing takes cognizance of the divergent character of the series.

Notice that for a given value of x there is a value of n, N say, for which the upper bound on the error associated with the approximation is a minimum. Consequently, $N!/x^{N+1}$ is a measure of the ultimate accuracy with which $f(x)$ can be computed. To estimate N we note that

$$\frac{n!}{x^{n+1}} < \frac{n^{n-1}}{x^{n+1}} = e^{(n-1)\ln n - (n+1)\ln x}.$$

Differentiating the right hand side of this inequality with respect to n and equating the result to zero, we find

$$N \ln \frac{N}{x} + N = 1 \quad \text{or}, \quad N \simeq x \, e^{(e/x-1)}.$$

Thus, for $x = 10$, $N = 5$ which means that for this value of x, optimal accuracy is obtained with just the first five terms in the series. The associated error is then ≤ 0.00012.

Definition: A series $\sum_{m=0}^{\infty} c_m z^{-m}$, which either converges for large values of $|z|$ or diverges for all values of z, is said to be an **asymptotic series** for $f(z)$,

$$f(z) \sim \sum_{m=0}^{\infty} c_m z^{-m},\tag{6.1.6}$$

valid in a given range of values of arg z, if, for any positive integer n,

$$\lim_{|z|\to\infty}\left\{ z^n \left[f(z) - \sum_{m=0}^{\infty} c_m z^{-m} \right] \right\} = 0\tag{6.1.7}$$

for arg z in this range.

An asymptotic series will be convergent only if the function it represents is holomorphic at $z = \infty$. Thus, more often than not, they are divergent series. As we have seen, this does not adversely affect their ability to accurately represent functions for large values of $|z|$. Since the difference between $f(z)$ and the first $(n + 1)$ terms of the series (6.1.6),

$$\left| f(z) - c_0 - \frac{c_1}{z} - \ldots - \frac{c_n}{z^n} \right|,$$

is of order $1/|z|^{n+1}$, such a series is often better suited for numerical computation than a convergent representation would be. Some caution is necessary however. The addition of too many terms of a divergent series will render an approximation for fixed z worse rather than better. Indeed, as we learned from our simple example, there is always an optimum number of terms that gives rise to the best approximation for any given value of z.

If a function $f(z)$ possesses an asymptotic series

$$f(z) \sim c_0 + \frac{c_1}{z} + \frac{c_2}{z^2} + \ldots,$$

the series coefficients are uniquely determined by the equations

$$\lim_{|z|\to\infty} f(z) = c_0$$

$$\lim_{|z|\to\infty} z\left[f(z) - c_0 \right] = c_1$$

$$\lim_{|z|\to\infty} z^2 \left[f(z) - c_0 - \frac{c_1}{z} \right] = c_2$$

$$\ldots$$

$$\lim_{|z|\to\infty} z^n \left[f(z) - c_0 - \frac{c_1}{z} - \ldots - \frac{c_{n-1}}{z^{n-1}} \right] = c_n$$

$$\ldots .\tag{6.1.8}$$

This shows that a given function can only have one asymptotic series. However, knowledge of an asymptotic series does not determine a corresponding function since different functions can generate the same asymptotic series. For example, $e^{1/z}$ and $e^{1/z} + e^{-z}$

have the same asymptotic series

$$1 + \frac{1}{1!z} + \frac{1}{2!z^2} + \frac{1}{3!z^3} + \cdots,$$

valid in the range $|\arg z| < \pi/2$.

Many functions $f(z)$ do not possess an asymptotic series; e^z is an obvious example. However, even when this is the case, one can often find a second function $\varphi(z)$ such that the quotient $f(z)/\varphi(z)$ does possess a series,

$$f(z)/\varphi(z) \sim c_0 + \frac{c_1}{z} + \frac{c_2}{z^2} + \cdots,$$

for some range of $\arg z$. For such functions we shall write

$$f(z) \sim \varphi(z) \sum_{m=0}^{\infty} c_m z^{-m} \tag{6.1.9}$$

and we shall use the term **asymptotic expansion** to refer interchangably to both a representation of this form as well as the more straightforward asymptotic series representation of (6.1.6). The term $c_0\,\varphi(z)$ in (6.1.9) is often called the **dominant term** of the expansion.

Example: The exponential integral function

$$Ei(x) = \int_x^{\infty} \frac{e^{-t}}{t}\,dt, \quad x > 0$$

differs from the function (6.1.1) of our first example by a factor of e^x. Thus, without further effort we deduce that $Ei(x)$ has the asymptotic expansion =

$$Ei(x) \sim e^{-x}\left[\frac{1}{x} - \frac{1}{x^2} + \frac{2}{x^3} + \cdots + \frac{(-1)^{n-1}(n-1)!}{x^n} + \cdots \right] \sim \frac{e^{-x}}{x} \sum_{m=0}^{\infty} \frac{(-1)^m\, m!}{x^m}.$$

$$\tag{6.1.10}$$

Comparing this result with (6.1.9), we see that $Ei(x)$ has $\frac{e^{-x}}{x}$ as its dominant term.

One can show that an asymptotic series can be added, multiplied and integrated term by term. However, it is not permissable in general to perform term by term differentiation.

To determine an asymptotic expansion one almost always proceeds from an integral representation of the function in which one is interested. This is due to two exceptional results: **Watson's Lemma** and the **Method of Steepest Descents**.

6.2 Watson's Lemma

George Neville Watson (1886-1965) was noted for the application of complex analysis to the theory of special functions. His collaboration while at Trinity College, Cambridge on

the (1915) second edition of E. T. Whittaker's A Course of Modern Analysis (1902) pro-
duced an instant classic, a text that to this day is known simply as "Whittaker and Wat-
son". In 1922, he published a second classic text, Treatise on the Theory of Bessel Functions, *which is an exhaustive study of all aspects of Bessel functions including especially*
their asymptotic expansions. Watson became Professor at the University of Birmingham
in 1918, where he remained until 1951.

Watson's Lemma applies to functions $f(z)$ that can be represented by convergent
integral transforms of the form

$$f(z) \equiv \int_0^\infty e^{-zt} g(t) dt. \tag{6.2.1}$$

Although this makes it somewhat exclusive, it still covers many of the cases which
occur in practice. All we require is that the spectral function $g(t)$ be holomorphic, ex-
cept possibly for a branch-point at the origin, in the disc $|t| < T$ and admit the series
representation

$$g(t) = \sum_{m=1}^\infty c_m t^{m/p-1}, \quad |t| \leqslant \tau < T \tag{6.2.2}$$

for some $p > 0$. Also, let us suppose that when t is real and positive and $t > T$, $|g(t)| <$
$K e^{bt}, K > 0, b > 0$. One can then show that term by term integration of the series
together with the result

$$\int_0^\infty e^{-zt} t^{m/p-1} dt = \frac{1}{z^{m/p}} \int_0^\infty e^{-u} u^{m/p-1} du = \frac{\Gamma(m/p)}{z^{m/p}}, \tag{6.2.3}$$

yields the asymptotic expansion

$$f(z) \sim \sum_{m=1}^\infty c_m \Gamma(m/p) z^{-m/p} \tag{6.2.4}$$

for $|\arg z| \leqslant \pi/2 - \varepsilon$, where ε is an arbitrary positive number.

Example: To illustrate the use of the lemma, let us return to the function featured in
the first example of the previous Section,

$$f(x) = \int_x^\infty \frac{e^{x-t}}{t} dt.$$

If we let $v = \frac{1}{x}(t - x)$, we can express $f(x)$ as a transform of the appropriate type,

$$f(x) = \int_0^\infty e^{-xv} \frac{1}{v+1} dv.$$

Moreover,

$$\frac{1}{v+1} = \sum_{m=0}^{\infty}(-v)^m = \sum_{m=1}^{\infty}(-v)^{m-1}, \quad |v| < 1.$$

Therefore, using (6.2.4), we immediately obtain the by now familiar asymptotic series

$$f(x) \sim \sum_{m=1}^{\infty}(-1)^{m-1}\,\Gamma(m)\,x^{-m} = \sum_{m=1}^{\infty}(-1)^{m-1}(m-1)!\,x^{-m}.$$

Similarly, the complimentary error function

$$\mathrm{erfc}(x) = 1 - \mathrm{erf}(x) = \frac{2}{\sqrt{\pi}}\int_{x}^{\infty} e^{-t^2}\,dt$$

can be expressed as

$$\mathrm{erfc}(x) = \frac{2}{\sqrt{\pi}}e^{-x^2}\int_{x}^{\infty} e^{-(t^2 - x^2)}\,dt = \frac{2}{\sqrt{\pi}}e^{-x^2}\frac{x}{2}\int_{0}^{\infty} e^{-x^2 v}[1+v]^{-1/2}\,dv.$$

where we have set $x^2 v = t^2 - x^2$. Now,

$$(1-v)^{-1/2} = 1 - \frac{v}{2} + \frac{3v^2}{8} - \frac{5v^3}{16} + \frac{35v^4}{128} - +\dots$$

$$= \sum_{m=1}^{\infty}\frac{(-1)^{m-1}(2m-3)!!}{2^{m-1}(m-1)!}v^{m-1}, \quad |v| < 1.$$

Thus, using (6.2.4), we have

$$\mathrm{erfc}(x) \sim \frac{2}{\sqrt{\pi}}e^{-x^2}\frac{x}{2}\sum_{m=1}^{\infty}\frac{(-1)^{m-1}(2m-3)!!}{2^{m-1}}x^{-2m}$$

$$\sim \frac{2}{\sqrt{\pi}}e^{-x^2}\sum_{m=1}^{\infty}\frac{(-1)^{m-1}(2m-3)!!}{2^m}x^{-(2m-1)}.$$

The **method of steepest descent** (which is also known as the **saddle point method**) provides a much more general technique for generating asymptotic expansions. Such generality does not come about without some cost however; this method is a good deal more complicated than is Watson's Lemma.

6.3 The Method of Steepest Descent

Consider the integral

$$F(z) = \int_{C} e^{zf(t)}\,g(t)\,dt \qquad\qquad (6.3.1)$$

where $f(t)$, $g(t)$ and C have all been chosen so that the integrand is holomorphic in some domain containing C and goes to zero at either end-point of C. It is entirely reasonable to expect that the most significant contributions to $F(z)$, for large values of $|z|$, arise from those segments of C on which the real part of $[zf(t)]$ is large and positive. However, we must be mindful of the fact that the imaginary part of $[zf(t)]$ will generally increase as $|z|$ increases and that this will result in rapid oscillations of the factor $e^{i\,\text{Im}[zf(t)]}$ and hence, in a complicated pattern of cancellations among the values assumed by the integrand. Such cancellations could make the evaluation of $F(z)$ a daunting if not impossible undertaking. However, we know that the integral is path independent within the integrand's domain of holomorphy. Therefore, we shall assume that this domain permits deformation of the contour of integration C into a new contour C_0 on which $\text{Im}[zf(t)]$ is **constant** whenever $\text{Re}[zf(t)]$ assumes its largest values. Our first task then is to define C_0.

Since we want $\text{Re}[zf(t)]$ to be large, we shall require C_0 to pass through a point $t = t_0$ at which $\text{Re}[zf(t)]$ has a relative maximum and hence, at which

$$f'(t_0) \equiv \left.\frac{df(t)}{dt}\right|_{t=t_0} = 0. \tag{6.3.2}$$

As we shall see, we now need only demand that near $t = t_0$

$$\text{Im}[zf(t)] = \text{Im}[zf(t_0)] \tag{6.3.3}$$

to complete our specification of C_0.

We have already proven (the maximum modulus principle) that neither the real nor the imaginary part of a function can have an absolute maximum or minimum within its domain of holomorphy. However, we have yet to determine what does happen to a holomorphic function at a point where its first derivative vanishes. To do so now, let $w(x, y) \equiv u(x, y) + iv(x, y)$ be a holomorphic function whose derivative $\frac{dw}{dz} = 0$ at some point $z = z_0 \equiv (x_0, y_0)$. Clearly, the first partial derivatives of both $u(x, y)$ and $v(x, y)$ vanish at (x_0, y_0). Moreover, since $u(x, y)$ and $v(x, y)$ are harmonic, $\nabla^2 u = 0$ and $\nabla^2 v = 0$, we know that if, for example, $\frac{\partial^2 u}{\partial x^2} < 0$ at (x_0, y_0), then $\frac{\partial^2 u}{\partial^2 y} > 0$ there. In other words, if $u(x, y_0)$ has a maximum at $x = x_0$, then $u(x_0, y)$ has a minimum at $y = y_0$. Thus, although it is not an extremum of $u(x, y)$ and $v(x, y)$, (x_0, y_0) is a mini-max or **saddle point** of these two functions.

A further consequence of the holomorphy of $w(z)$ is that the curves $u(x, y) = constant$ and $v(x, y) = constant$ are the level curves for a conformal mapping and so are everywhere orthogonal. Thus, referring to Figure 6.1, we see that if we proceed along a curve $v(x, y) = constant$, $u(x, y)$ will vary at its maximum rate. The curves AB and CD are the two curves passing through the saddle point that correspond to a constant value of $v(x, y)$. On one of them, CD, $u(x, y)$ **increases** as rapidly as possible; on the other, AB, $u(x, y)$ **decreases** as rapidly as possible.

Returning to our integral, we now recognize that condition (6.3.2) implies that $t = t_0$ is a saddle point of $\text{Re}[zf(t)]$, while (6.3.3) and the requirement that $\text{Re}[zf(t)]$

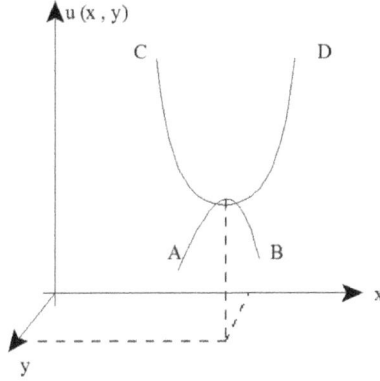

Figure 6.1: A saddle point

have a relative maximum at $t = t_0$ uniquely identifies C_0 as the counterpart of the curve AB. With this choice for C_0, $e^{zf(t)}$ goes to its end-point values by the steepest route possible, that is, by the path of **steepest descent**. Moreover, most of the value of the integral must come from the neighbourhood of $t = t_0$ since the modulus of the integrand is at a maximum there while its phase is roughly constant. (The phase is exactly constant if $g(t)$ is real.) This becomes increasingly true as $|z| \to \infty$ for the maximum becomes larger and the descent to the end-point values becomes steeper and steeper.

Assuming that we can deform our initial contour C to coincide with C_0 without encountering any of the singularities of $f(t)$ or $g(t)$, we may rewrite (6.3.1) as

$$F(z) = e^{zf(t_0)} \int_{C_0} e^{z[f(t)-f(t_0)]} g(t)dt \qquad (6.3.4)$$

where, by construction, $z[f(t)-f(t_0)]$ is real and negative for all t on C_0 except at $t = t_0$ where it vanishes. This expression can be simplified in appearance by defining a real function $\tau(t)$ via the identity

$$e^{i\arg z}[f(t) - f(t_0)] \equiv -\tau^2(t). \qquad (6.3.5)$$

We then have

$$F(z) = e^{zf(t_0)} \int_{C_0} e^{-|z|\tau^2(t)} g(t)dt$$
$$= e^{zf(t_0)} \int_{\Gamma_0} e^{-|z|\tau^2} g[t(\tau)] \frac{dt(\tau)}{d\tau} d\tau, \qquad (6.3.6)$$

where Γ_0 is the image of C_0 in the τ plane. We know that Γ_0 runs along the real axis and passes through the origin $\tau = 0$. Moreover, while its end-points will vary from case to case, this variation is no barrier to an asymptotic evaluation of (6.3.6). For, as

$|z|$ increases, the exponential in the integrand becomes sufficiently steep that the only significant contribution to $F(z)$ comes from a small segment of Γ_0 about $\tau = 0$. Thus, at the risk of committing only a negligible error we may take the end-points of Γ_0 to be $\pm\infty$ and write

$$F(z) \sim e^{zf(t_0)} \int_{-\infty}^{\infty} e^{-|z|\tau^2} g[t(\tau)] \frac{dt}{d\tau} d\tau. \tag{6.3.7}$$

To complete the asymptotic evaluation of $F(z)$ we need now only express $g(t)$ and $\frac{dt}{d\tau}$ as functions of τ. This is most appropriately done by means of power series expansions about $\tau = 0$. Thus, leaving aside for the moment the problem of determining the coefficients in the series

$$g[t(\tau)] \frac{dt}{d\tau} = \sum_{m=0}^{\infty} c_m \tau^m, \tag{6.3.8}$$

we substitute it into (6.3.7) to obtain

$$F(z) \sim e^{zf(t_0)} \sum_{m=0}^{\infty} c_m \int_{-\infty}^{\infty} e^{-|z|\tau^2} \tau^m d\tau. \tag{6.3.9}$$

The standard integral

$$I_m = \int_{-\infty}^{\infty} e^{-\alpha\tau^2} \tau^m d\tau, \quad \alpha > 0$$

is known to assume the values

$$I_m = \begin{cases} \sqrt{\pi/\alpha}, & m = 0 \\ \dfrac{(2k-1)!!}{2^k \alpha^k} \sqrt{\pi/\alpha}, & m = 2k = 2, 4, \ldots \\ 0, & m = 1, 3, 5, \ldots \end{cases}.$$

Thus, (6.3.9) can also be written as

$$F(z) \sim e^{zf(t_0)} \sqrt{\frac{\pi}{|z|}} \left[c_0 + \sum_{k=1}^{\infty} c_{2k} \frac{(2k-1)!!}{2^k} \frac{1}{|z|^k} \right] \tag{6.3.10}$$

which brings us to within a single step of our long sought-after asymptotic expansion for $F(z)$. That last step is to extract the dependence on $\arg z$ from the coefficients c_{2k} to obtain a series in z rather than $|z|$. The result is

$$F(z) \sim e^{zf(t_0)} \sqrt{\frac{\pi}{z}} \left[a_0 + \sum_{k=1}^{\infty} a_{2k} \frac{(2k-1)!!}{2^k} \frac{1}{z^k} \right] \tag{6.3.11}$$

where the coefficients

$$a_{2k} = e^{i\frac{2k+1}{2} \arg z} c_{2k}, \quad k = 0, 1, 2, \ldots . \tag{6.3.12}$$

The explicit calculation of the coefficients a_{2k} is a tedious chore because it involves expressing $t - t_0$ as a power series in τ by inversion of equation (6.3.5) and then substituting into the power series expansion of $g(t)$ about $t = t_0$. The coefficient of the leading term is the sole exception; it is readily determined as follows.

From (6.3.8) we know that

$$c_0 = \left\{ g[t(\tau)] \frac{dt(\tau)}{d\tau} \right\}\bigg|_{\tau=0} = g(t_0) \frac{dt(\tau)}{d\tau}\bigg|_{\tau=0}. \tag{6.3.13}$$

Moreover, since $f'(t_0) = 0$, we have

$$f(t) - f(t_0) = f''(t_0) \frac{(t-t_0)^2}{2!} + \ldots$$

and so

$$\tau^2 = -e^{i\theta} f''(t_0) \frac{(t-t_0)^2}{2!} + \ldots \tag{6.3.14}$$

where, for notational simplicity, we have set $\arg z = \theta$. Thus, inverting this series we find to lowest order that

$$t - t_0 = \frac{\sqrt{2}\tau}{\sqrt{e^{i(\pi+\theta)} f''(t_0)}} + \ldots$$

and hence,

$$\frac{dt}{d\tau}\bigg|_{\tau=0} = \lim_{\tau \to 0} \frac{t-t_0}{\tau} = \frac{\sqrt{2}}{\sqrt{e^{i(\pi+\theta)} f''(t_0)}}. \tag{6.3.15}$$

Substitution into (6.3.13) then yields

$$c_0 = \frac{\sqrt{2}g(t_0)}{\sqrt{e^{i(\pi+\theta)} f''(t_0)}} \quad \text{or,} \quad a_0 = \frac{\sqrt{2}g(t_0)}{\sqrt{e^{i\pi} f''(t_0)}}.$$

Therefore, retaining only the dominant term in the asymptotic expansion (6.3.11), we have

$$F(z) \sim g(t_0) \sqrt{\frac{2\pi}{e^{i\pi} f''(t_0)}} \frac{e^{zf(t_0)}}{\sqrt{z}}. \tag{6.3.16}$$

To determine higher order terms we rewrite (6.3.14) as

$$e^{-i\theta} \tau^2(t) = f(t_0) - f(t) = \sum_{m=2}^{\infty} \alpha_m (t-t_0)^m \tag{6.3.17}$$

where $\alpha_m = -\frac{f^{(m)}(t_0)}{m!}, \ m \geq 2$.

Then, since we seek a power series for $(t - t_0)$ about $\tau = 0$, we set

$$t - t_0 = \sum_{m=0}^{\infty} \beta_m \tau^{m+1} \tag{6.3.18}$$

and substitute this series into the right hand side of (6.3.17). Using

$$\left(\sum_{k=0}^{\infty} \beta_k \, \tau^{k+1}\right)^n = \sum_{m=0}^{\infty} \gamma_m \, \tau^{m+n} \tag{6.3.19}$$

where $\gamma_0 = \beta_0^n$, $\gamma_m = \dfrac{1}{m \beta_0} \displaystyle\sum_{k=1}^{m}[k(n+1) - m]\beta_k \, \gamma_{m-k}$, $m \geqslant 1$, this yields

$$
\begin{aligned}
e^{-i\theta} \tau^2 &= \alpha_2 \tau^2 [\beta_0^2 + 2\beta_1 \beta_0 \tau + (\beta_1^2 + 2\beta_2 \beta_0)\tau^2 + 2(\beta_2 \beta_1 + \beta_3 \beta_0)\tau^3 \\
&\quad + (\beta_2^2 + 2\beta_1 \beta_3 + 2\beta_0 \beta_4)\tau^4 + \ldots] \\
&\quad + \alpha_3 \tau^3 [\beta_0^3 + 3\beta_1 \beta_0^2 \tau + 3(\beta_1^2 \beta_0 + \beta_2 \beta_0^2)\tau^2 \\
&\quad + (\beta_1^3 + 6\beta_2 \beta_1 \beta_0 + 3\beta_3 \beta_0^2)\tau^3 + \ldots] \\
&\quad + \alpha_4 \tau^4 [\beta_0^4 + 4\beta_1 \beta_0^3 \tau + (6\beta_1^2 \beta_0^2 + 4\beta_2 \beta_0^3)\tau^2 + \ldots] \\
&\quad + \alpha_5 \tau^5 [\beta_0^5 + 5\beta_1 \beta_0^4 \tau + \ldots] \\
&\quad + \alpha_6 \tau^6 [\beta_0^6 + \ldots] \\
&\quad + \ldots \; .
\end{aligned} \tag{6.3.20}
$$

Thus, equating coefficients of like powers of τ we obtain the following equations expressing $\beta_0, \beta_1, \beta_2, \ldots$ in terms of the α_m, $m \geqslant 2$:

$$
\begin{aligned}
e^{-i\theta} &= \alpha_2 \beta_0^2 \\
0 &= 2\alpha_2 \beta_1 \beta_0 + \alpha_3 \beta_0^3 \\
0 &= \alpha_2(\beta_1^2 + 2\beta_2 \beta_0) + 3\alpha_3 \beta_1 \beta_0^2 + \alpha_4 \beta_0^4 \\
0 &= 2\alpha_2(\beta_2 \beta_1 + \beta_3 \beta_0) + 3\alpha_3(\beta_1^2 \beta_0 + \beta_2 \beta_0^2) + 4\alpha_4 \beta_1 \beta_0^3 + \alpha_5 \beta_0^5 \\
&\quad \vdots \quad .
\end{aligned}
$$

Solving, we find

$$
\begin{aligned}
\beta_0 &= \frac{e^{-i\theta/2}}{\sqrt{\alpha_2}} \\[2mm]
\beta_1 &= -\frac{\alpha_3}{2\alpha_2}\beta_0^2 \\[2mm]
\beta_2 &= \left[\frac{5\alpha_3^2}{8\alpha_2^2} - \frac{\alpha_4}{2\alpha_2}\right]\beta_0^3 \\[2mm]
\beta_3 &= \left[-\frac{\alpha_3^3}{\alpha_2^3} + \frac{3\alpha_3 \alpha_4}{\alpha_2^2} - \frac{\alpha_5}{2\alpha_2}\right]\beta_0^4 \\[2mm]
\beta_4 &= \left[\frac{231\alpha_3^4}{128\alpha_2^4} - \frac{63\alpha_4 \alpha_3^2}{16\alpha_2^3} + \frac{7\alpha_4^2}{8\alpha_2^2} + \frac{7\alpha_3 \alpha_5}{4\alpha_2^2} - \frac{\alpha_6}{2\alpha_2}\right]\beta_0^5 \\[2mm]
&\quad \vdots \quad .
\end{aligned} \tag{6.3.21}
$$

Substitution of these coefficients into (6.3.18) and term-wise differentiation finally produces the power series about $\tau = 0$ for $\frac{dt}{d\tau}$. To obtain the corresponding series for $g(t)$ requires still more tedious effort since we must substitute powers of (6.3.18) into

$$g(t) = \sum_{m=0}^{\infty} \frac{g^{(m)}(t_0)}{m!}(t - t_0)^m$$

and then collect coefficients of like powers of τ. Fortunately, many interesting functions have integral representations with $g(t) = 1$. In such cases the coefficients c_m of equation (6.3.8) are given by

$$c_m = (m + 1)\beta_m, \quad m = 0, 1, 2, \ldots. \tag{6.3.22}$$

Combining this result with (6.3.21) and then (6.3.12), we find that the first three of the coefficients a_{2k} in the asymptotic expansion (6.3.11) of $F(z)$ are

$$a_0 = \alpha_2^{-1/2}$$

$$a_2 = \frac{3}{2}\alpha_2^{-3/2}\left[\frac{5}{4}\frac{\alpha_3^2}{\alpha_2^2} - \frac{\alpha_4}{\alpha_2}\right]$$

$$a_4 = \frac{5}{2}\alpha_2^{-5/2}\left[\frac{231}{64}\frac{\alpha_3^4}{\alpha_2^4} - \frac{63}{8}\frac{\alpha_4\,\alpha_3^2}{\alpha_2^3} + \frac{7}{4}\frac{\alpha_4^2}{\alpha_2^2} + \frac{7}{2}\frac{\alpha_3\,\alpha_5}{\alpha_2^2} - \frac{\alpha_6}{\alpha_2}\right] \tag{6.3.23}$$

where we recall that $\alpha_m = -\frac{f^{(m)}(t_0)}{m!}$, $m \geqslant 2$.

Example: To illustrate the use of this formidable piece of mathematical machinery we shall determine the asymptotic expansion of $\Gamma(z+1)$. From Euler's definition, equation (5.2.5), we have

$$\Gamma(z + 1) = \int_0^{\infty} e^{-u}\,u^z\,du, \quad \operatorname{Re} z > -1$$

$$= z^{z+1}\int_0^{\infty} e^{z(\ln t - t)}\,dt, \tag{6.3.24}$$

where we have set $u = zt$. Thus, in this case, $f(t) = \ln t - t$ and $g(t) = 1$.

Differentiating $f(t)$ and setting the result equal to zero, $f'(t_0) = \frac{1}{t_0} - 1 = 0$, we see that $f(t)$ possesses only one saddle-point located at $t = t_0 = 1$. Hence, there is a single path of steepest descent which, because $f(t_0) = -1$, is uniquely defined by

$$\operatorname{Im}[zf(t)] = -\operatorname{Im} z \quad \operatorname{Re}[zf(t)] \leqslant -\operatorname{Re} z.$$

We do not need a more detailed identification of the path because it is already clear that the initial contour of integration, the positive real axis, can be deformed to lie along it without encountering singularities of $f(t)$. Thus, it only remains to evaluate a few derivatives of $f(t)$ at $t = t_0 = 1$ and thence, to calculate the asymptotic series coefficients $a_0, a_2, \dots$. The first of these tasks is easily accomplished: $f^{(m)}(1) = (-1)^{m-1}(m-1)!$ and so, $\alpha_m = \frac{(-1)^m}{m}$, $m \geqslant 2$. Substituting into (6.3.23) we then find

$$a_0 = 2^{1/2}, \quad a_2 = \frac{1}{12} 2^{3/2}, \quad a_4 = \frac{1}{864} 2^{5/2}.$$

Therefore, by equation (6.3.11), the asymptotic expansion of $\Gamma(z+1)$ is

$$\Gamma(z+1) \sim \sqrt{2\pi}\, z^{z+1/2}\, e^{-z} \left[1 + \frac{1}{12}\frac{1}{z} + \frac{1}{288}\frac{1}{z^2} + \dots\right] \qquad (6.3.25)$$

which is known as **Stirling's approximation.**

    Born in Scotland in 1692, James Stirling was a contemporary of and correspondent with such notable mathematicians as Euler, DeMoivre and Newton. His most important work is a treatise, Methodus Differentialis, *published in London in 1730. It contains the asymptotic formula for n! to which his name is attached. Stirling was believed to have Jacobite sympathies which was an impediment to academic advancement in Hanoverian Britain. Obliged to work as a mine manager, his mathematical output declined. Nevertheless, Euler secured his election to the Royal Academy of Berlin in 1746 just as Newton had arranged for his election to the Royal Society in 1726. Stirling died in Edinburgh in 1770.*

    We shall have occasion to use the method of steepest descent several times in subsequent Chapters, particularly when it comes time to discuss the properties of Bessel functions. However, what really commends it to physicists are its direct applications in modern physics. Important examples are the evaluation of partition functions in statistical mechanics and of generating functionals in the path integral formalism of quantum mechanics and quantum field theory.

7 Padé Approximants

7.1 From Power Series to Padé Sequences

It can often happen that one can calculate the coefficients c_m up to any order m in the power series

$$f(z) = \sum_{m=0}^{\infty} c_m z^m \qquad (7.1.1)$$

for a function of interest without being able to determine an expression for the general term and hence, without being able to determine the radius of convergence of the series. If, as also happens rather frequently, one does not possess any other representation for the function, this can pose a serious problem. One has no way of locating the singularities of $f(z)$ and may not even be able to rule out the possibility of a singularity at $z = 0$ itself. Consequently, there is no way of knowing how rapidly the series converges or even whether it converges at all for the values of z one is interested in. A good illustration is provided by perturbation theory calculations of quantum mechanical transition amplitudes. These are expressed as a power series in the "coupling strength" g of the interaction responsible for the transition and one has a formal mechanism for calculating the coefficients in the series up to any desired order. The sum of the calculated terms is then evaluated by setting g equal to its physical value. In the case of the electromagnetic interaction, for example, g is just the fine structure constant $\alpha = e^2/\hbar c = 1/137$. However, the reason why perturbation theory is used at all in such problems is the impossibility of obtaining exact solutions and the barriers to this often imply an ignorance of the analytic properties of the solutions as functions of g. In other words, one has no a priori knowledge of how or even whether the series converges at the physically relevant value of g. To make matters worse, bound and resonant states manifest themselves as singularities of transition amplitudes. Thus, a perturbation series would appear to be totally irrelevant if one's principal aim is the very important one of identifying such states. Fortunately, one can use a Taylor-like series, whether convergent or not, to define sequences of **Padé approximants** which provide much more versatile representations of the corresponding function than does the power series which is their source.

Definition: The $[L/M]$ Padé approximant to $f(z)$ is the rational function

$$f_{L/M}(z) = \frac{P_L(z)}{Q_M(z)}, \qquad Q_M(0) = 1 \qquad (7.1.2)$$

where $P_L(z)$ and $Q_M(z)$ are polynomials of degree L and M respectively, such that $f_{L/M}(z)$ and $f(z)$ have the same first $L + M$ derivatives at $z = 0$; that is, such that

$$\frac{P_L(z)}{Q_M(z)} - \sum_{m=0}^{\infty} c_m z^m = O(z^{L+M+1}) \qquad (7.1.3)$$

where $O(z^{L+M+1})$ represents terms of order $L + M + 1$ and higher: $O(z^{L+M+1}) = a\,z^{L+M+1}$ $+\,b\,z^{L+M+2} + \ldots$ for some $a, b, \ldots$.

Henri Padé (1863-1953) was born in Abbeville which is a town in the Picardy region of northern France. He graduated with his Agrégation de Mathématiques from the École Normale Supérieure in Paris in 1886 and began a career teaching in secondary schools. He also began a program of mathematical research, publishing his first paper in 1888 and commencing work on a doctoral thesis in 1890. Presented to the Sorbonne in1892, Padé's thesis provided the first systematic study of the representation of functions by rational fractions or what we now call Padé approximants. In 1897 Padé received the first of a series of university appointments culminating in that of Dean of the Faculty of Science at the University of Bordeaux in 1906. In the meantime, in 1899, he published another major work on Padé approximants and by 1908, when he left Bordeaux to become a Rector of the French Academy, had written 41 papers of which 29 were on continued fractions and Padé approximants. He remained a Rector of the Academy, first at Besançon then Dijon and finally Aix-Marseille, until he retired at age 70 in 1934.

A moment's reflection suggests that the class of functions which can be usefully approximated by rational functions is bound to be larger than the class which can be approximated by polynomials. Rational functions have poles of their own and so should be able to provide a representation even in the neighbourhood of singularities. Thus, we are predisposed to expect that the domain of convergence of a sequence of approximants $f_{L/M}(z)$ is a good deal larger than that of the corresponding Taylor series. This expectation will be confirmed; we shall find that Padé approximants provide a method of analytically continuing functions whose definition is provided solely by power series.

The coefficients of the polynomials $P_L(z)$ and $Q_M(z)$ are determined by equating coefficients of like powers of z in

$$P_L(z) - Q_M(z) \sum_{m=0}^{L+M} c_m\, z^m = O(z^{L+2M+1}), \quad Q_M(0) = 1. \tag{7.1.4}$$

The full and unique solution for $f_{L/M}(z)$ is thus found to be

$$f_{L/M}(z) = \frac{\begin{vmatrix} c_{L-M+1} & c_{L-M+2} & \cdots & c_{L+1} \\ \vdots & \vdots & \vdots & \vdots \\ c_L & c_{L+1} & \cdots & c_{L+M} \\ \sum\limits_{k=M}^{L} c_{k-M}z^k & \sum\limits_{k=M-1}^{L} c_{k-M+1}z^k & \cdots & \sum\limits_{k=0}^{L} c_k z^k \end{vmatrix}}{\begin{vmatrix} c_{L-M+1} & c_{L-M+2} & \cdots & c_{L+1} \\ \vdots & \vdots & \vdots & \vdots \\ c_L & c_{L+1} & \cdots & c_{L+M} \\ z^M & z^{M-1} & \cdots & 1 \end{vmatrix}} \tag{7.1.5}$$

where $c_k \equiv 0$ if $k < 0$ and the sums for which the lower limit is larger than the upper limit are also to be replaced by 0. Such a formal and formidable expression does little to inform one's intuition. Therefore, we list below the approximants that correspond to $L, M \leqslant 2$:

$$f_{0/0} = c_0, \quad f_{1/0} = c_0 + c_1 z, \quad f_{2/0} = c_0 + c_1 z + c_2 z^2$$

$$f_{0/1} = \frac{c_0}{1 - \frac{c_1}{c_0} z}, \quad f_{1/1} = \frac{c_0 + \frac{c_1^2 - c_0 c_2}{c_1} z}{1 - \frac{c_2}{c_1} z}, \quad f_{2/1} = \frac{c_0 + \frac{c_1 c_2 - c_0 c_3}{c_2} z + \frac{c_2^2 - c_1 c_3}{c_2} z^2}{1 - \frac{c_3}{c_2} z}$$

$$f_{0/2} = \frac{c_0}{1 - \frac{c_1}{c_0} z + \frac{c_1^2 - c_2 c_0}{c_0^2} z^2}, \quad f_{1/2} = \frac{c_0 + [c_1 + c_0 \frac{c_0 c_3 - c_1 c_2}{c_1^2 - c_0 c_2}] z}{1 + \frac{c_0 c_3 - c_1 c_2}{c_1^2 - c_0 c_2} z + \frac{c_2^2 - c_1 c_3}{c_1^2 - c_0 c_2} z^2}$$

$$f_{2/2} = \frac{c_0 + \left[c_1 + c_0 \frac{(c_1 c_4 - c_2 c_3)}{c_2^2 - c_1 c_3}\right] z + \left[c_2 + \frac{c_1(c_1 c_4 - c_2 c_3) + c_0(c_3^2 - c_2 c_4)}{c_2^2 - c_1 c_3}\right] z^2}{1 + \frac{c_1 c_4 - c_2 c_3}{c_2^2 - c_1 c_3} z + \frac{c_3^2 - c_2 c_4}{c_2^2 - c_1 c_3} z^2} \tag{7.1.6}$$

As this display presages, it is useful to group the approximants to a given function in an infinite array known as a **Padé table**:

$$
\begin{array}{cccccc}
f_{0/0} & f_{0/1} & f_{0/2} & f_{0/3} & \cdots \\
f_{1/0} & f_{1/1} & f_{1/2} & f_{1/3} & \cdots \\
f_{2/0} & f_{2/1} & f_{2/2} & f_{2/3} & \cdots \\
\vdots & \vdots & \vdots & \vdots
\end{array} \tag{7.1.7}
$$

The approximants $f_{L/M}$ with L fixed comprise a **row** of the table, while those with M fixed form a **column**. The set $\{f_{M/M}\}$ is called the **diagonal** sequence and together with the **paradiagonal** sequences $\{f_{M+j/M}\}$ with j fixed comprise the ones of most interest. The table subsumes several more specialized types of approximations. For example, the first column consists of the approximations provided by truncating the power series. Of more interest in numerical analysis, the sequence consisting alternately of members of the diagonal and first paradiagonal,

$$f_{0/0}, f_{1/0}, f_{1/1}, f_{2/1}, f_{2/2}, \ldots, f_{M/M}, f_{M+1/M}, f_{M+1/M+1}, \ldots,$$

corresponds to the approximations one would obtain by truncating the **continued fraction**

$$f(z) = a_0 + \cfrac{a_1 z}{1 + \cfrac{a_2 z}{1 + \cfrac{a_3 z}{1 + \cfrac{a_4 z}{1 + \ldots}}}}.$$

Mathematicians have yet to establish an extensive theoretical knowledge of the convergence properties of Padé approximants. Numerical experimentation indicates that the theorems that have been proven offer a limited perspective of the representational potential of Padé sequences. Consequently, following the approach used by Padé and by each successive generation of researchers, we shall examine by means of specific examples how approximants imitate the analytic structure of the functions they represent. Then, after a brief overview of the convergence theorems that have been proven, we will engage in informed speculation on just how powerful a means of analytic continuation Padé approximants really are.

7.2 Numerical Experiments

Let us commence our study of how approximants reconstruct the functions they represent by reminding ourselves that an $[L/M]$ approximant is a meromorphic function with L zeros and M poles where, for counting purposes, we are treating a zero (pole) of order n as though it were n simple zeros (poles). Thus, other meromorphic functions ought to pose a reasonably simple representational challenge. To confirm this, we look first at the trivial case provided by other rational functions.

Example: Substituting the coefficients of the geometric series

$$f(z) = \sum_{m=0}^{\infty} z^m$$

into equation (7.1.5) we see that all Padé approximants except the sequence of truncated series $\{f_{L/0}\}$ exactly reproduce the function they are supposedly approximating; that is, we have

$$f_{L/M} \equiv \frac{1}{1-z} \quad \text{for all} \quad L \geqslant 0, \ M \geqslant 1.$$

This is not a coincidental result. One can readily show that if $f(z)$ is a rational function whose numerator is of degree I and denominator degree J, then

$$f_{L/M} \equiv f(z) \quad \text{for all} \quad L \geqslant I, \ M \geqslant J.$$

Example: A more demanding task is that of reproducing the poles and zeros of a function like $\tan z$. Obviously, this is possible only in the limit that both L and $M \to \infty$. However, we should be able to get an intuitive feel for what will happen in that limit by calculating the first few approximants to $\tan z$ and identifying the location of their poles and zeros. Our starting point is the Taylor series

$$\tan z = z + \frac{1}{3} z^3 + \frac{2}{15} z^5 + \frac{17}{315} z^7 + \dots$$

whose disc of convergence, $|z| < \pi/2$, is determined by the location of the first two poles, $z = \pm\pi/2$. For calculational convenience, we shall factor out the zero at $z = 0$ and treat the remainder as a series in z^2:

$$\tan z = z f(z^2) \quad \text{where} \quad f(z^2) = 1 + \frac{1}{3} z^2 + \frac{2}{15} (z^2)^2 + \frac{17}{315} (z^2)^3 + \dots.$$

Using equations (7.1.6) we then find

$$f_{0/1}(z^2) = \frac{3}{3 - z^2}, \quad f_{1/1}(z^2) = \frac{1}{3}\frac{15 - z^2}{5 - 2z^2},$$

$$f_{2/1}(z^2) = \frac{1}{15}\frac{630 - 45z^2 - z^4}{42 - 17z^2}, \quad f_{0/2}(z^2) = \frac{45}{45 - 15z^2 - z^4},$$

$$f_{1/2}(z^2) = 5\frac{21 - 2z^2}{105 - 45z^2 + z^4},$$

which are the only approximants we can calculate in this order. Their poles and zeros are compared with those of $\tan z / z$ in Table 7.1.

Table 7.1: Zeros and Poles of Approximants of $f(z) = \tan z / z$.

Function	Zeros	Poles
$f_{0/1}$	n/a	±1.732
$f_{0/2}$	n/a	±1.601, ±4.191i
$f_{1/1}$	±3.873	±1.581
$f_{2/1}$	±3.348, ±7.497i	±1.572
$f_{1/2}$	±3.240	±1.571, ±6.522
$\frac{\tan z}{z}$	±3.142, ±6.284, ...	±1.571, ±4.712, ...

The sequence consisting of the approximants $f_{0/1}, f_{1/1}, f_{1/2}, \ldots$ acquires poles and zeros in the same alternating order as they occur for the original function. Specifically, $f_{0/1}$ has only poles and no zeros. The poles, at $z = \pm1.732$, are within about 10% of the first two poles of $\tan z / z$, at $z = \pm\pi/2$. The next approximant in the sequence, $f_{1/1}$, has poles at $z = \pm1.581$ and, in addition, has zeros at $z = \pm3.873$. In other words, we have now reproduced $z = \pm\pi/2$ to within better than 1%, as well as the first two zeros of $\tan z / z$, $z = \pm\pi$, to within about 25%. Finally, $f_{1/2}$ locates these zeros to within less than 3% and provides us with a first approximation to the location of the next pair of poles at $z = \pm3\pi/2$. Each successive approximant attempts to replicate the next pair of poles or zeros in the same order as they are encountered with increasing values of $|z|$. Moreover, each successive approximant replicates all previously located poles and zeros with ever increasing accuracy. What is perhaps most impressive is that this detailed picture of how $\tan z / z$ behaves for $|z| \geqslant \pi/2$ has been generated from a very restricted knowledge of how that function behaves for $|z| < \pi/2$, namely, a few terms of the function's Taylor expansion about $z = 0$. Thus, this Padé sequence provides a very powerful method of analytically continuing the Taylor expansion outside its circle of convergence.

The other two approximants listed in Table 7.1 are also worthy of note. While $f_{0/2}$ reproduces the poles at $z = \pm\pi/2$ with much greater accuracy than does $f_{0/1}$, it also produces extraneous poles on the imaginary axis. Similarly, $f_{2/1}$ has extraneous zeros at $z = \pm7.5i$. These features are attributable to the structural inability of these particular approximants to mimic the alternating character of the poles and zeros of the function they represent. If we were to go to higher order, we would find that the extraneous poles and zeros are very unstable with respect to position. This is because they do not reflect specific analytic features possessed by the original function in the finite plane but rather, its behaviour at infinity. In any case, this illustrates that there is a tangible advantage to be gained from being selective in one's choice of Padé sequence to represent a given function.

Example: To further illustrate the phenomenon of extraneous poles and zeros, let us consider the approximants of e^z which not only has no poles of its own, it also has no zeros. The [1/1] approximant is

$$f_{1/1} = \frac{2+z}{2-z}$$

with a zero on the negative real axis at $z = -2$ and a symmetrically placed pole on the positive real axis at $z = +2$. The [2/2] approximant is

$$f_{2/2} = \frac{12 + 6z + z^2}{12 - 6z + z^2}$$

with zeros at $z = -3 \pm \sqrt{3}i$ and poles at $z = +3 \pm \sqrt{3}i$. Thus, we perceive an interesting pattern emerging. The zeros all occur in the left half-plane with the poles in mirror-image locations in the right half-plane. Moreover, on increasing the order of the approximant, the poles and zeros have moved further away from the origin. This pattern persists as one further increases the order. The poles occur either on the real axis or on either side of it in conjugate pairs but in any case, moving ever further to the right. The zeros cluster on or about the negative real axis and move further and further to the left.

Recalling that $|e^z|$ increases without limit as $|z| \to \infty$ along the positive real axis and goes to zero as $|z| \to \infty$ along the negative real axis, this pattern becomes understandable. The poles and zeros of the approximants are simulating the exponential function's essential singularity at $z = \infty$.

This simulation evidently assists the convergence of the approximants. As one can check, an approximant gives a much better approximation of the value of e^z in the region between the poles and zeros than does the corresponding truncated Taylor series. Even at $z = -1$, a point at which the Taylor series converges fairly rapidly, the approximants improve on the accuracy of the approximation by one to two orders of magnitude. This is shown in Table 7.2 where we see that the [M/M] approximants increase their accuracy by two decimal places per unit increase in M, with the value given by $f_{4/4}$ being off by only one part in 10^8.

Example: Isolated essential singularities are not the only type of non-polar singularity that can be simulated by Padé approximants. A function with branch points is necessarily discontinuous across a cut or cuts joining the points. Simulating a discontinuity of this sort might appear to be an overly ambitious task for a sequence of functions that are continuous everywhere except for a finite set of poles. However, by clustering their poles along the cut the Padé approximants produce a representation that, very often, is convergent **everywhere** in the cut plane.

Table 7.2: $[M/M]$ Approximants of e^z Evaluated at $z = -1$.

M	$f_{M/M}$	% Deviation from e^{-1}	$\sum\limits_{m=0}^{2M} \dfrac{(-1)^m}{m!}$	% Deviation from e^{-1}
1	0.33333333	9.39	0.5	35.9
2	0.36842105	0.147	0.375	1.94
3	0.36787564	1.03×10^{-3}	0.36805555	4.79×10^{-2}
4	0.36787945	2.45×10^{-6}	0.36788194	6.79×10^{-4}
Limit	0.367879441	–	0.367879441	–

The efficacy of the simulation is perhaps best appreciated by recalling the dispersion representation for functions with branch points. For example, the function $(1+z)^{-1/2}$, with branch points at $z = -1$ and $z = \infty$, has the representation

$$(1 + z)^{-1/2} = \frac{1}{\pi} \int\limits_{1}^{\infty} \frac{d\chi}{(\chi + z)(\chi - 1)^{1/2}}, \tag{7.2.1}$$

corresponding to a cut along the real axis segment $-\infty < \chi \leqslant -1$. Using this as our starting point, we can give $(1+z)^{-1/2}$ an approximate representation by simply evaluating the integrand at a finite number of carefully selected points. Such an operation is precisely that of replacing the cut by a finite number of simple poles each located at the negative image of one of the pre-selected points and therefore, distributed along the line segment where the cut used to be. Of course, a truly discontinuous function results only in the limit as the number of such poles becomes infinite.

As we know, the choice of simple curve used as a cut joining the two branch points is arbitrary. If the choice is different from the negative real axis, it would be reflected in a corresponding change to the integration contour in (7.2.1) and hence, in the location of the poles simulating the cut were we to continue to use the integral as the basis of an approximate representation. However, there is no such flexibility with the approximate representation obtained with Padé approximants. Since the location of the branch cut has no effect on the value of $(1 + z)^{-1/2}$ and its derivatives at the origin, the approximants necessarily make an independent but specific choice of the curve to use in simulating the cut. The only question is whether this choice is predictable. The answer is an unqualified yes only in the case of a class of functions that are named after a mathematician called **Stieltjes;** they will be the subject of more detailed study a little further on. For more general functions one must rely on systematic behaviour observed in the course of numerical experimentation to justify any inference as to the location of the simulated cuts.

G.A. Baker Jr. and his collaborators have experimented with a variety of functions possessing two, three, or even four branch points. They found that the limit set of poles of diagonal Padé approximants forms circular arcs each of which joins a pair of branch points and, if extrapolated to form a complete circle, would pass through the origin. As usual we include straight lines in our definition of circle and so, in the case of $(1 + z)^{-1/2}$, the poles of the $[M/M]$ approximants conform to this rule by clustering along the real axis segment $-\infty < x < -1$.

Another interesting function studied by Baker et al. is

$$f(z) = \frac{(1 + z^2)^{1/2}}{1 + z}.$$

It has branch points at $z = \pm i$. To conform with the empirical rule, its $[M/M]$ approximants should simulate a cut which runs from $z = +i$ through $z = \infty$ to $z = -i$. As Figure 7.1 shows, this expectation is born out by actual calculation: one pole in the limit set converges to $z = -1$ and the rest cluster along the imaginary axis segments $1 < y < \infty$ and $-\infty < y < -1$.

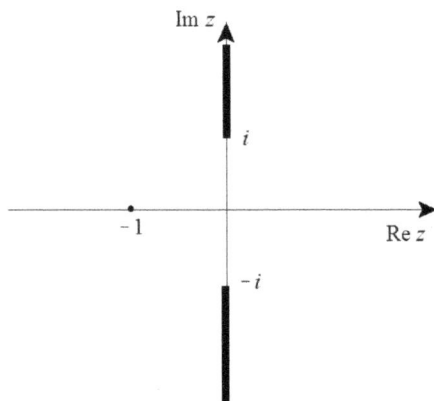

Figure 7.1: The poles of the $[M/M]$ approximants to $f(z) = (1 + z^2)^{1/2}/(1 + z)$ simulate a cut along the imaginary axis from $z = +i$ to $z = -i$, passing through $z = \infty$, as well as reproduce the polar singularity at $z = -1$.

The behaviour of these approximants at infinity is rather interesting as it reveals how cuts adversely affect convergence. From the calculated values displayed in Table 7.3 we see that the $[M/M]$ approximants converge to $\sqrt{2} \pm 1$ according as M is even or odd. Thus, this Padé sequence does not converge to anything at $z = \infty$. The sequences $[2M/2M]$ and $[2M + 1/2M + 1]$ converge separately but not to ± 1 which are the values attained by the function $(1 + z^2)^{1/2}/(1 + z)$ on alternate lips of the cut. This suggests that one should not expect convergence of Padé approximants at points located on a cut.

7.3 Stieltjes Functions

Thomas Jan Stieltjes (1856-1894) had little formal education, flunking out of the Polytechnical School of Delft in 1876, due in large part to an all-consuming passion for math-

Table 7.3: The $[M/M]$ Approximants to $(1 + z^2)^{1/2} / (1 + z)$ Evaluated at Infinity

M	$[M/M]$
1	0.333333
2	2.333333
3	0.411764
4	2.411764
5	0.414141
6	2.414141
7	0.414211
8	2.414211
Limit	$\sqrt{2} \pm 1$
Actual Value of Function	± 1

ematics. *In 1882, while employed as an assistant at the Leiden Observatory, Stieltjes began a correspondence with the distinguished mathematician Hermite. Over the remaining 12 years of Stieltjes' life, they exchanged 432 letters on his mathematical interests. Hermite became both his mentor and his sponsor, helping Stieltjes obtain academic appointments in the Netherlands and in France. Stieltjes worked on almost all branches of analysis, continued fractions and number theory but he is best remembered for the Stieltjes function defined below.*

Everything that has been empirically deduced about the behaviour of Padé approximants of functions with branch points can be rigorously proven to occur for **Stieltjes functions**. These possess either a Taylor or an asymptotic series about $z = 0$ whose coefficients can be expressed in terms of a very particular type of integral.

Definition: A function $f(z)$ which admits representation by a series of the form

$$f(z) = \sum_{m=0}^{\infty} c_m(-z)^m \tag{7.3.1}$$

is a **Stieltjes function** if and only if

$$c_m = \int_0^{\infty} t^m \, dg(t) \tag{7.3.2}$$

where $g(t)$ is a bounded non-decreasing function which takes on infinitely many values in the interval $0 \leqslant t < \infty$.

The series itself is called a **series of Stieltjes** and need not be convergent. However, if it is, with radius of convergence R, then we can interchange the order of integration and summation in (7.3.1) to obtain

$$f(z) = \int_0^{\infty} \sum_{m=0}^{\infty} (-tz)^m \, dg(t), \tag{7.3.3}$$

for $|z| < R$. We know that

$$\sum_{m=0}^{\infty}(-tz)^m = \frac{1}{1+tz} \text{ for } |tz| < 1$$

and that it diverges for $|tz| \geqslant 1$. Therefore, if (7.3.3) is to be meaningful, $f(z)$ must possess the following Stieltjes integral representation

$$f(z) = \int_0^{1/R} \frac{g'(t)dt}{1+tz}. \tag{7.3.4}$$

This provides an analytic continuation of $f(z)$ outside the circle of convergence of (7.3.1) and permits identification of the function's singularities. Indeed, if we now make the substitution $t \to x = -\frac{1}{t}$, (7.3.4) assumes the more familiar features of a dispersion representation for a real function (see equation (4.1.5):

$$f(z) = \frac{1}{\pi}\int_{-\infty}^{-R} \frac{\text{Im} f_+(\chi)}{\chi - z}d\chi \tag{7.3.5}$$

$$\text{Im}f_+(x)\Big|_{x=-1/t} = \lim_{\varepsilon \to 0}\text{Im} f[-(t+i\varepsilon)^{-1}] = -\pi t\frac{dg(t)}{dt}. \tag{7.3.6}$$

Thus, $f(z)$ is a real function, $f^*(z^*) = f(z)$, and is holomorphic everywhere in the complex plane except for a cut along the negative real axis from $z = -R$ to infinity.

One can show that these two integral representations retain their validity even in the limit of a divergent series of Stieltjes, $R \to 0$, provided that the series coefficients are such that

$$\sum_{m=1}^{\infty}(c_m)^{-1/(2m+1)}$$

also diverges. This condition is roughly equivalent to requiring that $|c_m| \leqslant (2m)!$. The cut along the negative real axis then extends from $z = 0$ to infinity and (7.3.4) becomes

$$f(z) = \int_0^{\infty} \frac{g'(t)dt}{1+tz}. \tag{7.3.7}$$

As for the series, it too retains a representational role albeit only an asymptotic one. In conformance with equation (6.1.7), we have

$$\lim_{|z|\to 0}\left\{z^{-n}\left[f(z) - \sum_{m=0}^{n}c_m z^m\right]\right\} = 0 \tag{7.3.8}$$

for any positive integer n and arg z restricted to avoid the cut joining the branch points at $z = 0$ and $z = \infty$.

Notice that the properties of $g(t)$ necessarily imply that $\text{Im} f_+(x) \leqslant 0$.

In summary, we can characterize all Stieltjes functions as satisfying the reality condition $f^*(z^*) = f(z)$, as possessing branch points at $z = \infty$ and a point on the negative real axis $z = -x_0, x_0 \geqslant 0$ and, when these points are joined by a cut along that axis, as having a negative definite, pure imaginary discontinuity across the cut. Indeed, given a function with these properties, the choice of the negative real axis as the location of its cut assures the existence of a Stieltjes integral representation since we can write

$$f(z) = \int_0^\infty \frac{g'(t)dt}{1 + tz} \tag{7.3.9}$$

with the weight function given by

$$g(t) - g(t_0) = \frac{1}{\pi} \int_{t_0}^t \lim_{\varepsilon \to 0} \text{Im}[f(-(\tau - i\varepsilon)^{-1})]\frac{d\tau}{\tau}. \tag{7.3.10}$$

Differentiation of (7.3.9) then generates a series of Stieltjes about $z = 0$.

Baker's empirical rule concerning the simulation of cuts by Padé approximants can actually be proven to be the case for Stieltjes functions. The relevant theorem also provides remarkably detailed information about the poles and zeros of successive approximants.

Theorem: If $\sum_{m=0}^\infty c_m(-z)^m$ is a series of Stieltjes, the poles and zeros of the $[M + J/M], J \geqslant -1$ Padé approximants obtained from the coefficients are on the negative real axis. Furthermore, the poles of successive approximants interlace and all the residues are positive. The roots of the numerators (the zeros) also interlace those of the corresponding denominators (the poles).

Example: Stieltjes functions are more common place than one might at first suspect. Whether by means of an adroitly chosen transformation or by more immediate measures, a very wide class of analytic functions can be cast in a Stieltjes form. Among those which are immediately recognizable as possessing properties typical of a Stieltjes function is the elementary function $f(z) = \frac{1}{z}Ln(1+z)$. It admits the following series, Stieltjes and dispersion representations:

$$\frac{1}{z}Ln(1 + z) = \begin{cases} 1 - \frac{z}{2} + \frac{z^2}{3} - \frac{z^3}{4} + - \ldots, & |z| < 1 \\ \int_0^1 \frac{dt}{1+tz}, & -\pi < \arg(z + 1) < \pi \\ \int_{-\infty}^{-1} \frac{d\chi}{\chi(\chi-z)}, & -\pi < \arg(z + 1) < \pi. \end{cases}$$

In conformance with the preceding theorem, the poles and zeros of the diagonal approximants to this function alternate along the negative real axis to the left of $z = -1$. As the order of the approximants increases, the spacing between successive poles and zeros decreases due to the interlacing effect called for by the theorem. Thus, the picture that emerges of the limiting situation as $M \to \infty$ is of an infinite sequence of very closely spaced, alternating poles and zeros extending along the negative real axis from $z = -1$ to infinity.

As we shall see, this single-minded behaviour of the poles and zeros offers the entire cut complex plane as the domain of convergence for the diagonal approximants. However, to illustrate how an alternative and, as it turns out, inappropriate choice of Padé sequence can have a much more restricted domain of convergence, we shall follow Baker's lead and explore how the $[4M/M]$ approximants to $\frac{1}{z}Ln(1+z)$ behave.

Table 7.4: $[4M/M]$ Approximants to $\frac{1}{z}Ln(1+z)$

$z \ M$	1	2	3	4	$\frac{1}{2}Ln(1+z)$
1.0	0.69242424	0.69314873	0.69314718	0.69314718	0.69314718
6.0	−0.80000000	2.1535393	−2.6374816	5.1116289	0.32431836
9.0	−4.3735294	28.911655	−176.68475	1091.7095	0.25584279

Table 7.4 compares the values of the first four of these approximants with the corresponding values of $\frac{1}{z}Ln(1+z)$ at a series of points on the positive real axis. We see that for $z = 1.0$ convergence to the exact value is almost immediate. For $z = 4.0$ (not shown in the table) convergence is slower but it still takes place. However, for $z = 6.0$ and 9.0 the approximants are oscillating in sign and, in the latter case particularly, diverging quite rapidly.

This startling change in behaviour is directly associated with the extraneous zeros possessed by these approximants. Each approximant requires only M zeros to alternate with the M poles along the cut. Since Stieltjes functions do not vanish anywhere in the cut plane, this leaves $3M$ zeros with no role to play in the reconstruction of the analytic properties of $\frac{1}{z}Ln(1+z)$. Plotting these zeros for the four approximants in Table 7.4, (Figure 7.2), one finds that they form a pattern suggestive of a closed curve surrounding a roughly heart-shaped region about the origin. Evidently, this curve will only increase in definition with increasing M and so constitutes a natural boundary for the approximants' domain of convergence.

This example illustrates a general rule for the approximation of functions possessing branch points: to enjoy a maximal domain of convergence one should restrict consideration to diagonal or para-diagonal Padé sequences.

Example: There has been speculation that the perturbation series obtained in quantum electrodynamics is an asymptotic rather that a convergent series owing to a singularity at $\alpha = 0$, where α is the electromagnetic coupling strength; (the physical value of α is $e^2/\hbar c = 1/137$). Thus, it may be useful to know something about the convergence properties of Padé sequences derived from such series.

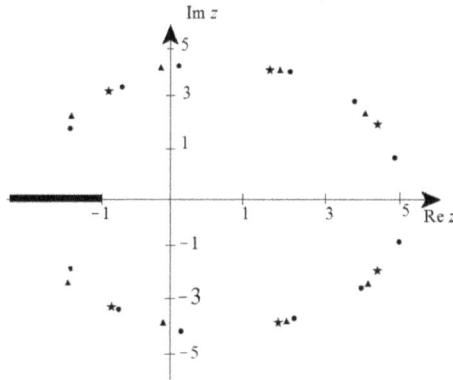

Figure 7.2: The extraneous zeros of the $[4M/M]$ Padé approximants to $[\mathrm{Ln}(1+z)]/z$. These zeros define the boundary of the domain of convergence of this sequence of approximants; (from Baker, (1975)).

Based on the last example, we might expect favourable results, for diagonal sequences at least, if the series are series of Stieltjes. This expectation is born out by both experiment and theory. An interesting case in point is provided by the Stieltjes function

$$f(z) = \int\limits_0^\infty \frac{e^{-t}}{1+zt}dt, \qquad (7.3.11)$$

which has branch points at $z = 0$ and ∞. Its asymptotic expansion about $z = 0$ is the divergent series

$$f(z) \sim 1 - (1!)z + (2!)z^2 - (3!)z^3 + \ldots = \sum_{m=0}^\infty (-1)^m \, m! \, z^m \qquad (7.3.12)$$

which makes it a simplified analogue of the electrodynamics perturbation series since the latter's general term is expected to go like $m! \, \alpha^m$ for large values of m. Calculating the $[M/M]$ approximants to $f(z)$, one finds that they do converge towards the exact values defined by (7.3.11) and do so for **all** finite z in the cut plane. Although the rate of convergence is rather slow, it is essentially unchanged as $|z|$ increases. For example, at $z = 1.0$ the $[6/6]$ approximant has the value 0.5968 compared to an exact value of 0.5963, while at infinity $f_{6/6} = \frac{1}{7}$ compared to an exact limiting value of zero. (As $x \to \infty$, $f(x) \to 0$ and $f_{M/M}(x) \to \frac{1}{M+1}$). In both cases the error is proportional to $\frac{1}{M+1}$.

This leap from a divergent series to a sequence that is convergent throughout the cut plane is a little breathtaking. It is this feature, as well as the ability to locate and identify singularities of functions defined by power series alone, that best exemplifies the analytical power of Padé approximants.

7.4 Convergence Theorems

As we have already noted, the theory of convergence of Padé sequences is still far from complete. We need be aware of only a few highlights.

The earliest theorems to be proved apply to columns of the Padé Table. For example, de Montessus de Ballore established in 1902 that, if $f(z)$ is holomorphic throughout the domain $|z| < R$, except for a finite number of poles of total multiplicity m, then the sequence $\{f_{L/m}(z)\}$ converges uniformly to $f(z)$ for $|z| \leqslant \rho < R$, except at the poles of $f(z)$. This theorem was generalized around 1930 to prove that the column sequences $\{f_{L/m+\mu}(z)\}$, $\mu \geqslant 1$ are also uniformly convergent for $|z| \leqslant \rho < R$, provided that the singularities of $f(z)$ on $|z| = R$ are no worse than a multiple pole of order greater than μ or a branch point. Moreover, the μ additional poles, not needed to represent the poles of $f(z)$ inside $|z| < R$, simulate the singularities on $|z| = R$ with the first extra pole representing the "strongest" of these singularities, the second the next strongest, and so on.

Since one can readily prove that $[f_{L/M}]^{-1}$ is the $[M/L]$ approximant to $f^{-1}(z)$ if $f_{L/M}(z)$ is an approximant to $f(z)$, theorems concerning columns of the Padé Table and poles in the z-plane translate directly into theorems concerning rows and zeros.

From this modest beginning we move on to a series of theorems established in the 1960's by Chisholm (1966) and Beardon (1968). These concern the convergence of sequences $\{f_{L/M}(z)\}$ when **both** L and $M \to \infty$. If $f(z)$ is meromorphic in a closed disc $|z| \leqslant R$, the sequence $\{f_{L/M}(z)\}$ converges uniformly to $f(z)$ in the double limit $L, M \to \infty$ provided that z is restricted to a second disc $|z| \leqslant \rho \leqslant R$ from which small neighbourhoods of the poles of $f(z)$ and of any limit points of the poles of the approximants have been removed. The radius ρ of the disc of convergence depends on the ratio L/M and, in general, is less than R. However, equality with R can be guaranteed if we are prepared to restrict ourselves to sequences for which $L \geqslant kM$ where k is a number $\geqslant 1$ that depends on both the value of R and the identity of $f(z)$.

While helpful as an aid to our intuition, these theorems do little to meet our theoretical needs. For mathematicians and physicists alike, the diagonal and paradiagonal sequences are much the most important elements of the Padé Table. This importance stems in large part from certain properties which are uniquely associated with diagonal approximants. For example, they can be shown to be invariant under mappings of the form $w = \frac{az}{cz+d}$, a, c and d constants.

Thus, it suffices to establish convergence in a restricted domain since one can immediately extend it in size by means of one of these mappings. Further, diagonal approximants can be shown to be **unitary** in the sense that if $f^*(z) = f^{-1}(z)$, then $f_{N/N}^*(z) = [f_{N/N}(z)]^{-1}$. This is a critically important property in quantum mechanical applications. So, what we really need are convergence theorems that apply specifically to diagonal approximants and cover the full range of analytic functions.

The results of numerical experimentation suggest that the boundary of the domain of convergence of a diagonal sequence should be determined only by the location of the non-polar singularities of the function it represents. However, very little progress has been made toward proving this assertion for more than a few special cases which fortunately include Stieltjes functions.

To conclude this brief survey of Padé convergence theory, we state the theorem that applies to Stieltje's functions. Given our discussion of numerical experience with approximants, it contains no surprises.

Theorem: Let $\sum\limits_{m=0}^{\infty} c_m(-z)^m$ be a series of Stieltjes. If the series is convergent with a radius of convergence R, then any sequence of $[M + J/M], J \geqslant -1$ Padé approximants to the series converges in the cut plane $(-\infty < z \leqslant -R)$ to the function $f(z)$ defined by the series. If the series is divergent, then any sequence of $[M + J/M], J \geqslant -1$ Padé approximants converges to an analytic function in the cut plane $(-\infty < z \leqslant 0)$. If, in addition, $\sum\limits_{m=0}^{\infty} (c_m)^{-1/(2m+1)}$ diverges, then all the sequences tend to a common limit $f(z)$ which possesses a Stieltjes representation of the form given by (7.3.9) and (7.3.10) and an asymptotic expansion about $z = 0$ given by the original series.

The proof of the various parts of this theorem is the subject of an entire chapter of Baker's 1975 monograph on Padé approximants.

7.5 Type II Padé Approximants

As an alternative to defining Padé approximants by means of the value of a function and its derivatives at a single point, one can build approximants that contain information at two or more points. An especially useful case arises when we use only the value of the function (and not of any of its derivatives) at an appropriate number of points. These are called Padé approximants of the second type or type II.

Definition: Let $z_1, z_2, \ldots z_N$ be N complex numbers and $f(z)$ be an analytic function which takes the values $f(z_i)$ at these points. We define the $[L/M]$ **Padé approximant of type II** to $f(z)$ to be the ratio of two polynomials in z, $P_L(z)$ and $Q_M(z)$ with $L+M = N-1$, which takes the values $f(z_i)$ at $z = z_i$,

$$f_{L/M}(z) \equiv \frac{P_L(z)}{Q_M(z)}$$

$$f_{L/M}(z_i) = f(z_i), \quad i = 1, \ldots N. \tag{7.5.1}$$

Type II approximants provide an extension of the Lagrange or polynomial method of interpolating between discrete points in the same way that their type I counterparts extend the Taylor or polynomial approximation of functions known only in the neighbourhood of $z = 0$. They are unique and one can prove (see, for example, J. Zinn-Justin, Phys. Reports 1C, No. 3 (1971)) that they possess essentially the same properties as do approximants of the first type, including convergence in the case of approximation of Stieltjes functions.

The interpolation provided by type II approximants is both more efficient and more effective at analytic continuation outside the region of the points z_i than that given by a Lagrange polynomial approach. Once again this is due to the capacity of Padé approximants to replicate the poles and simulate the non-polar singularities of the functions they approximate.

Example: If one builds the type II Padé approximants to $\tan z$ using values corresponding to z real and in the range $-\pi/2 < z < \pi/2$, one finds that the lowest orders give the nearby poles and zeros of $\tan z$ and that by increasing the order (number of points), more and more poles and zeros are successfully replicated. Moreover, this remains true even if one contracts the range of z to $-\pi/4 < z < \pi/4$ where $\tan z$ is well away from any of its poles and so is a relatively slowly varying function (in fact, it resembles a straight line there).

Another interesting application of type II approximants is in the summation of numerical series. If we have

$$S = \sum_{m=0}^{\infty} u_m,\tag{7.5.2}$$

we can treat the partial sums $S_N = \sum_{m=0}^{N} u_m$ as functions of $\frac{1}{N}$,

$$S_N = f\left(\frac{1}{N}\right) \quad \text{say,}$$

so that the sum of the series is

$$S = f(0).\tag{7.5.3}$$

We can then compute S by extrapolating from $f(1), f\left(\frac{1}{2}\right), f\left(\frac{1}{3}\right), \dots$ to $f(0)$; that is to say, by analytically continuing $f(z)$ to $z = 0$.

Example: Consider the series

$$S = \sum_{m=0}^{\infty} \frac{1}{m^2}$$

which is known to converge slowly. Using the first $N + M + 1$ partial sums as input, we can build the $[N/M]$ type II approximant to $f(z)$ and thus get an estimate of $f(0)$.

The first three partial sums are

$$S_1 = f(1) = 1 \quad S_2 = f\left(\frac{1}{2}\right) = 1.25 \quad S_3 = f\left(\frac{1}{3}\right) = 1.361.$$

These yield the approximant estimates

$$f_{[0/0]}(0) = 1 \quad f_{[0/1]}(0) = 1.5 \quad f_{[1/1]}(0) = 1.65.$$

The exact value is $f(0) = S = \pi^2/6 = 1.645\dots$. Thus, convergence has been accelerated in a quite remarkable way; while the partial sums themselves give a very crude estimate of S, using the same information as input for the [1/1] Padé approximant yields an estimate that is accurate to three significant figures.

We conclude with a reference to the use of type II approximants in the solution of algebraic equations. Suppose that we wish to solve the equation $F(z) = 0$. We begin by determining the values F_1, F_2, F_3 assumed by F at the three points z_1, z_2, z_3, respectively. Next, we build the type II approximant, $F_{[1/1]}^{(1)}(z)$, which has the same values F_i for $z = z_i$, $i = 1, 2, 3$. One can immediately read off the zero z_4 of $F_{[1/1]}^{(1)}$ which gives a first estimate of a zero of $F(z)$. Using the actual value $F_4 \equiv F(z_4)$, we repeat the process using F_2, F_3, F_4 and the points z_2, z_3, z_4 as input and determine $F_{[1/1]}^{(2)}(z)$. Its zero is a second estimate of the zero of $F(z)$. Evidently, this can be repeated until we are satisfied with the degree of convergence of our estimates yielding, if it works, a remarkably simple method of solving our original equation. The question of how well it works has been addressed by Zinn-Justin who shows that the rate of convergence from a distance is quite fast and that once we are close, it is exponential; if we are within an error ε of the zero at some point in the process, a further iteration will bring us within an error ε^3. Thus, in two steps we can go from an error of 10% to one of 1 part in 10^{-9}.

8 Fourier Series and Transforms

8.1 Trigonometrical and Fourier Series

Definition: A function $f(x)$ is said to be **periodic** if it is defined for all real x and if there is some real number T such that $f(x + T) = f(x)$. The number T is then called the **period** of $f(x)$.

Examples: The sine and cosine functions have period 2π and along with constants, which have arbitrary period, are the simplest periodic functions.

If it converges, a **trigonometrical series** of the form

$$\frac{a_0}{2} + \sum_{n=1}^{\infty}(a_n \cos nx + b_n \sin nx) \tag{8.1.1}$$

where $a_0, a_1, \ldots, a_n, \ldots, b_1, \ldots, b_n, \ldots$ are constants, has period 2π also. The representation of a function $f(x)$ by such series was first investigated by Fourier in the context of heat conduction problems. Subsequently it was found that these series have an important role to play in the theory of functions of a real variable which is the reason for our interest in them.

Joseph Fourier (1768-1830) was a French mathematician and physicist whose early career advancement resulted as much from his active participation in the French Revolution and support of Napoleon Bonaparte as it did from the success of his research. Nevertheless, he is remembered for initiating the study of Fourier series and for discovering the law of conduction that is named after him. He is also generally acknowledged to have discovered the greenhouse effect.

Suppose that the series (8.1.1) converges uniformly to the function $f(x)$:

$$f(x) = \frac{a_0}{2} + \sum_{n=1}^{\infty}(a_n \cos nx + b_n \sin nx), \quad -\pi \leqslant x \leqslant \pi. \tag{8.1.2}$$

We may multiply by $\cos mx$, where m is a positive integer, and integrate term by term from $-\pi$ to π to obtain

$$\int_{-\pi}^{\pi} f(x) \cos mx dx = \frac{a_0}{2} \int_{-\pi}^{\pi} \cos mx dx$$

$$+ \sum_{n=1}^{\infty} \left[a_n \int_{-\pi}^{\pi} \cos nx \cos mx dx + b_n \int_{-\pi}^{\pi} \sin nx \cos mx dx \right].$$

Since $\cos nx \cos mx = \frac{1}{2}\cos(n + m)x + \frac{1}{2}\cos(n - m)x$ and $\sin nx \cos mx = \frac{1}{2}\sin(n + m)x + \frac{1}{2}\sin(n - m)x$, we have the **orthogonality relations**

$$\int_{-\pi}^{\pi} \cos nx \cos mx dx = \begin{cases} 0 & \text{for } n \neq m \\ \pi & \text{for } n = m \end{cases}$$

$$\int_{-\pi}^{\pi} \sin nx \cos mx \, dx = 0 \quad \text{for all} \quad m, n$$

$$\int_{-\pi}^{\pi} \cos mx \, dx = \begin{cases} 0 & \text{for } m \neq 0 \\ 2\pi & \text{for } m = 0 \end{cases}$$

Applying these, we find

$$a_0 = \frac{1}{\pi} \int_{-\pi}^{\pi} f(x) \, dx \quad \text{and}$$

$$a_m = \frac{1}{\pi} \int_{-\pi}^{\pi} f(x) \cos mx \, dx, \quad m = 1, 2, \dots . \tag{8.1.3}$$

Similarly, multiplying (8.1.2) by $\sin mx$, integrating term by term and using the **orthogonality relation**

$$\int_{-\pi}^{\pi} \sin nx \sin mx \, dx = \begin{cases} 0 & \text{for } n \neq m \\ \pi & \text{for } n = m \end{cases},$$

we find

$$b_m = \frac{1}{\pi} \int_{-\pi}^{\pi} f(x) \sin mx \, dx, \quad m = 1, 2, \dots . \tag{8.1.4}$$

Equations (8.1.3) and (8.1.4) are called the **Euler formulae** for the coefficients and the set of numbers $\{a_m, b_m\}$ which they determine are called the **Fourier coefficients** of $f(x)$.

There is no *a priori* reason for supposing that a given function can be expanded in a uniformly convergent trigonometrical series. Therefore, the process we have just carried out is not a proof that the coefficients of a trigonometrical series representation of $f(x)$ are necessarily those determined by the Euler formulae. So, instead of starting with the series and the presumption that it has a certain property, we start from the function, calculate its Fourier coefficients, and determine the properties of the series that can be formed with them.

Definition: If we are given a function $f(x)$ that is integrable over the interval $-\pi \leqslant x \leqslant \pi$, then the integrals in (8.1.3) and (8.1.4) that define its Fourier coefficients $\{a_m, b_m\}$ exist. The trigonometrical series of the form (8.1.1) that is constructed using these coefficients is called the **Fourier series** of $f(x)$.

The question which must be addressed now is whether the Fourier series of an arbitrary function $f(x)$ converges and if it does, whether its sum is $f(x)$. Because the sum of the series, if it exists, has period 2π, we use only the values that $f(x)$ assumes in the interval $-\pi \leqslant x \leqslant \pi$; outside this interval we define it by means of periodicity, $f(x \pm 2\pi) = f(x)$, which gives us the **periodic extension** of $f(x)$.

Examples: Consider the function x^2. Its Fourier coefficients are

$$a_0 = \frac{1}{\pi} \int_{-\pi}^{\pi} x^2 \, dx = \frac{2}{3} \pi^2$$

$$a_n = \frac{1}{\pi} \int_{-\pi}^{\pi} x^2 \cos nx dx = (-1)^n \frac{4}{n^2}, \quad n > 0$$

$$b_n = \frac{1}{\pi} \int_{-\pi}^{\pi} x^2 \sin nx dx = 0.$$

Note that the sine coefficients are all zero because x^2 is an even function of x; similarly, all cosine coefficients vanish for odd functions of x.

The corresponding Fourier series

$$f(x) = \frac{\pi^2}{3} + \sum_{n=1}^{\infty} (-1)^n \frac{4}{n^2} \cos nx$$

is easily shown to be uniformly convergent for all values of x. As shown in Figure 8.1, its sum function, $f(x)$, reproduces the values assumed by x^2 on $-\pi \leqslant x \leqslant \pi$ and by its periodic extension outside that interval.

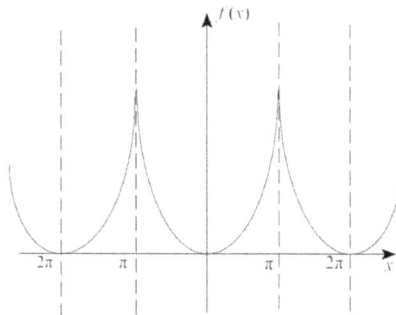

Figure 8.1: The Fourier series representation of x^2.

Next we consider the discontinuous function

$$f(x) = \begin{cases} -1, & x < 0, \\ +1, & x \geqslant 0. \end{cases}$$

This is an odd function so there are no cosine terms in its Fourier series, $a_n = 0, n \geqslant 0$. As for the sine terms, the coefficients are

$$b_n = \frac{2}{\pi} \int_{-\pi}^{\pi} \sin nx \, dx = \begin{cases} \frac{4}{n\pi}, & n = \text{odd}, \\ 0, & n = \text{even}. \end{cases}$$

The corresponding series

$$\frac{4}{\pi} \sum_{n=odd}^{\infty} \frac{1}{n} \sin nx$$

converges to
- +1 for $0 < x < \pi$,
- −1 for $-\pi < x < 0$, and
- 0 for $x = -\pi$, $x = 0$, and $x = +\pi$.

In other words, the Fourier series representation of $f(x)$ reproduces it throughout the interval **except** at the point of discontinuity, $x = 0$, and at the end-points of the interval which are points of discontinuity for the periodic extension of $f(x)$. At these three exceptional points, the series converges to the mean of the right- and left-hand limits of (the periodic extension of) $f(x)$. This behaviour at the points of discontinuity is a general feature of Fourier series.

The points of discontinuity also result in the convergence being non-uniform in any interval that includes one of them. This is exhibited in Figure 8.2 which shows successive partial sums. Notice the overshoot on either side of the points of discontinuity. This too is a general feature of Fourier series and is called Gibbs' phenomenon. The overshoot remains finite as more and more terms are added and tends to the value 0.18 in the limit of infinitely many terms.

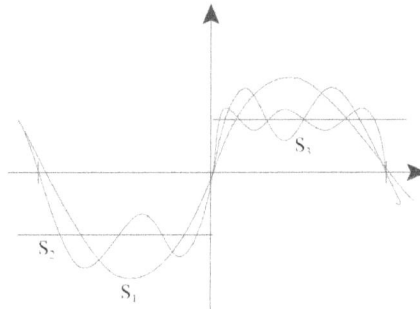

Figure 8.2: The first three partial sums of the Fourier series representation of a step function.

Finally, it is interesting to note that if we set $x = \pi/2$ in the Fourier series, we have

$$f\left(\frac{\pi}{2}\right) = 1 = \frac{4}{\pi}\left(1 - \frac{1}{3} + \frac{1}{5} - + \dots\right)$$

or,

$$\frac{\pi}{4} = 1 - \frac{1}{3} + \frac{1}{5} - \frac{1}{7} + - \ldots = \sum_{m=0}^{\infty} \frac{(-1)^m}{2m+1}$$

which is a series summation derived originally by Leibnitz using a geometrical argument.

8.2 The Convergence Question

There is a formal connection between Fourier series and Laurent series. Suppose that $f(z)$ is a single-valued analytic function, holomorphic in the annulus $R_1 < |z| < R_2$. Then

$$f(z) = \sum_{n=-\infty}^{\infty} c_n z^n,$$

where

$$c_n = \frac{1}{2\pi i} \int_C \frac{f(\zeta)}{\zeta^{n+1}} d\zeta$$

and we can choose C to be the circle $|\zeta| = r$, $R_1 < r < R_2$.

Putting $z = r e^{i\theta}$, we have

$$f(r e^{i\theta}) = \sum_{n=-\infty}^{\infty} A_n e^{in\theta},$$

where

$$A_n = \frac{1}{2\pi} \int_{-\pi}^{\pi} f(r e^{i\varphi}) e^{-in\varphi} d\varphi.$$

Combining terms pair wise for each value of $|n|$, we can rewrite the Laurent expansion as the Fourier series

$$f(r e^{i\theta}) = A_0 + \sum_{n=1}^{\infty} \left\{ (A_n + A_{-n}) \cos n\theta + i(A_n - A_{-n}) \sin n\theta \right\}, \qquad (8.2.1)$$

where

$$A_0 = \frac{1}{2\pi} \int_{-\pi}^{\pi} f(r e^{i\varphi}) d\varphi, \quad A_n + A_{-n} = \frac{1}{\pi} \int_{-\pi}^{\pi} f(r e^{i\varphi}) \cos n\varphi d\varphi, \qquad (8.2.2)$$

$$i(A_n - A_{-n}) = \frac{1}{\pi} \int_{-\pi}^{\pi} f(r e^{i\varphi}) \sin n\varphi d\varphi.$$

It follows from Laurent's Theorem that such a Fourier series converges uniformly to the function it represents. However, in general, one wants to represent a much larger class of functions than would meet the requirements of Laurent's Theorem. To see just how large a class admits such representation, we need to introduce some new concepts. The first of these is the concept of **bounded variation.**

Definition: Let $f(x)$ be defined on $a \leqslant x \leqslant b$ and let $x_1, x_2, \ldots, x_n$, be a set of points on that interval such that $a \leqslant x_1 \leqslant x_2 \leqslant \ldots \leqslant x_n \leqslant b$, as shown in Figure 8.3. Then, the sum

$$|f(a) - f(x_1)| + |f(x_1) - f(x_2)| + \ldots + |f(x_n) - f(b)|$$

is called the **variation** of $f(x)$ on the interval $a \leqslant x \leqslant b$ for the set of subdivisions $x_1, \ldots, x_n$. If the variation has an upper bound , M, independent of n, for all choices of $x_1, x_2, \ldots, x_n$, then $f(x)$ is said to be of **bounded variation** on $a \leqslant x \leqslant b$.

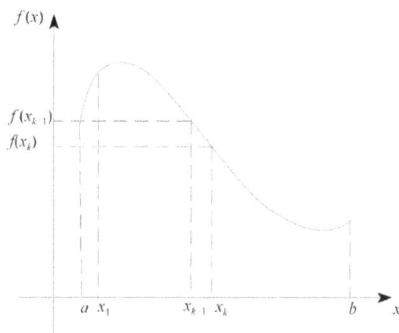

Figure 8.3: A set of points is introduced on the interval $a \leqslant x \leqslant b$ to define a corresponding variation of the function $f(x)$.

Examples: Two functions which are **not** of bounded variation are $f(x) = \frac{1}{x}$ and $f(x) = \sin\frac{\pi}{x}$ on any interval that encloses $x = 0$; see Figure 8.4. Examples of functions which **are** of bounded variation include piecewise continuous functions with a finite number of maxima and minima. A function $f(x)$ is **piecewise continuous** on a finite interval $a \leqslant x \leqslant b$ if the interval can be divided into finitely many sub-intervals, in each of which $f(x)$ is continuous and has finite limits as x approaches either endpoint of the sub-interval from the interior.

We can now state without proof the theorem that establishes the conditions for point by point convergence of a Fourier series.

Fourier's Theorem: Let $f(x)$ be defined arbitrarily on $-\pi \leqslant x \leqslant \pi$ and defined for all other x by its periodic extension, $f(x + 2\pi) = f(x)$. Also let $f(x)$ be such that $\int\limits_{-\pi}^{\pi} f(x)dx$ exists, and if this be an improper integral, let it be absolutely convergent. Then, if x is an interior point of any interval $a \leqslant x \leqslant b$ in which $f(x)$ is of bounded variation, the Fourier series of $f(x)$ converges at x to the sum $\frac{1}{2}[f(x + 0) + f(x - 0)]$. Moreover, in every closed subinterval in which $f(x)$ is continuous, the convergence is uniform.

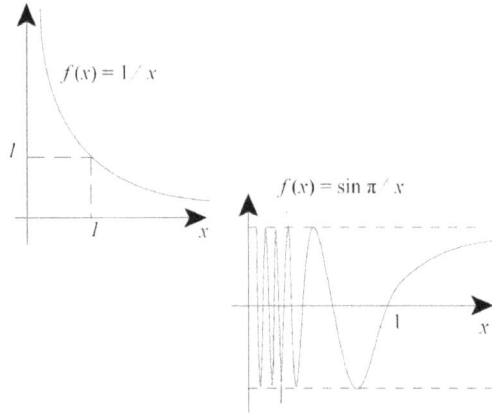

Figure 8.4: Two functions which are not of bounded variation in any interval containing $x = 0$.

A proof of the theorem can be found in E. C. Titchmarsh, Theory of Functions, Oxford University Press, New York, 1964.

As we noted with the example in the preceding Section, whenever $f(x)$ has a finite or step discontinuity the series will converge to the mean of its values on either side of the point. In particular, if $a \leqslant -\pi$ and $b \geqslant \pi$, the Fourier series will converge to $\frac{1}{2}[f(-\pi+0)+f(\pi-0)]$ at $x = \pm\pi$; thus, the series will **not** reproduce $f(x)$ at these points **unless** $f(-\pi) = f(\pi)$.

While not terribly stringent, the conditions imposed on $f(x)$ in this theorem can be relaxed even more if we replace the requirement of **pointwise convergence** by that of **convergence in the mean** which is a form of convergence that is perfectly adequate for most physical applications. To introduce the concept, however, we need an additional definition.

Definition: The integral

$$\int_a^b [f(x) - g(x)]^2 \, dx \tag{8.2.3}$$

for two functions $f(x)$ and $g(x)$ defined and square integrable on $a \leqslant x \leqslant b$ is called the **square deviation** of $f(x)$ and $g(x)$.

Evidently, a square deviation is a measure of how well one function "fits" another over the interval in question. We shall use it to meet our need for a more subtle form of convergence.

Definition: A sequence of functions $\{f_n(x)\}$ is said to **converge in the mean** to a function $f(x)$ on an interval $a \leqslant x \leqslant b$ if the corresponding square deviation of $f(x)$ and $f_n(x)$ tends to zero as $n \to \infty$:

$$\lim_{n\to\infty} \int_a^b [f(x) - f_n(x)]^2 \, dx = 0. \tag{8.2.4}$$

Convergence in the mean does **not** imply that $\lim_{n\to\infty} f_n(x) = f(x)$ at each point of $a \leqslant x \leqslant b$. On the contrary, the limit of the sequence of functions may differ widely from $f(x)$ on a discrete set of points distributed over the integration interval and still produce a zero square deviation.

We shall now make use of square deviation and convergence in the mean to further our understanding of the representations of functions that are offered by trigonometrical series. Suppose that we have a square integrable function $f(x)$ defined on $-\pi \leqslant x \leqslant \pi$ and that we want to approximate it by the partial sum

$$S_n(x) = \frac{A_0}{2} + \sum_{k=1}^{n}(A_k \cos kx + B_k \sin kx) \tag{8.2.5}$$

where the coefficients A_k, $k = 0, 1, 2, \ldots, n$, and B_k, $k = 1, 2, \ldots, n$, can be adjusted to achieve an optimal fit. In fact, we shall now try to determine what set of values for $\{A_k, B_k\}$ will minimize the square deviation

$$D_n = \int_{-\pi}^{\pi} [f(x) - S_n(x)]^2 \, dx. \tag{8.2.6}$$

Straightforward application of the orthogonality of the functions $\{\cos kx, \sin kx\}$ yields

$$D_n = \int_{-\pi}^{\pi} [f(x)]^2 \, dx + \left\{ \frac{\pi}{2} A_0^2 - A_0 \int_{-\pi}^{\pi} f(x)dx \right\}$$

$$+ \sum_{k=1}^{n} \left\{ \pi A_k^2 - 2 A_k \int_{-\pi}^{\pi} f(x) \cos kx dx \right\}$$

$$+ \sum_{k=1}^{n} \left\{ \pi B_k^2 - 2 B_k \int_{-\pi}^{\pi} f(x) \sin kx dx \right\}.$$

Using the Euler formulas for the Fourier coefficients of $f(x)$, we can rewrite this as

$$D_n = \int_{-\pi}^{\pi} [f(x)]^2 \, dx + \frac{\pi}{2}\{A_0^2 - 2 A_0 a_0\} + \pi \sum_{k=1}^{n}\{A_k^2 - 2 A_k a_k + B_k^2 - 2 B_k b_k\}.$$

But,

$$A_k^2 - 2 A_k a_k = (A_k - a_k)^2 - a_k^2 \quad \text{and} \quad B_k^2 - 2 B_k b_k = (B_k - b_k)^2 - b_k^2.$$

Thus,

$$D_n = \int_{-\pi}^{\pi} [f(x)]^2 \, dx - \pi \left\{ \frac{1}{2} a_0^2 + \sum_{k=1}^{n}[a_k^2 + b_k^2] \right\}$$

$$+ \pi \left\{ \frac{1}{2}(A_0 - a_0)^2 + \sum_{k=1}^{n} [(A_k - a_k)^2 + (B_k - b_k)^2] \right\}.$$

The last line in this expression is positive-definite:

$$\frac{1}{2}(A_0 - a_0)^2 + \sum_{k=1}^{n} [(A_k - a_k)^2 + (B_k - b_k)^2] \geqslant 0.$$

Therefore, the square deviation is minimized if the trigonometrical series coefficients A_k and B_k are chosen to be the Fourier coefficients of $f(x)$:

$$[D_n]_{\min} = \int_{-\pi}^{\pi} [f(x)]^2 \, dx - \pi \left\{ \frac{1}{2} a_0^2 + \sum_{k=1}^{n} [a_k^2 + b_k^2] \right\}. \tag{8.2.7}$$

Since $D_n \geqslant 0$, we have

$$\frac{a_0^2}{2} + \sum_{k=1}^{n} [a_k^2 + b_k^2] \leqslant \frac{1}{\pi} \int_{-\pi}^{\pi} [f(x)]^2 \, dx.$$

This holds for any n and as n increases, the sequence on the left is monotonically increasing but bounded by the integral on the right . Therefore, it possesses a limit as $n \to \infty$ and the limit satisfies

$$\frac{a_0^2}{2} + \sum_{k=1}^{\infty} [a_k^2 + b_k^2] \leqslant \frac{1}{\pi} \int_{-\pi}^{\pi} [f(x)]^2 \, dx \tag{8.2.8}$$

which is known as **Bessel's inequality**.

Suppose that we now require that the square deviation tend to zero as $n \to \infty$:

$$\lim_{n \to \infty} [D_n]_{\min} = 0.$$

By definition, this would mean that the Fourier series

$$\frac{a_0}{2} + \sum_{n=0}^{\infty} [a_n \cos nx + b_n \sin nx]$$

converges in the mean to $f(x)$. It also means that the above inequality becomes the equality

$$\frac{a_0^2}{2} + \sum_{k=1}^{\infty} [a_k^2 + b_k^2] = \frac{1}{\pi} \int_{-\pi}^{\pi} [f(x)]^2 \, dx \tag{8.2.9}$$

which is known as **Parseval's equation**.

Whenever Parseval's equation holds for a certain class of functions $f(x)$, we say that the set of trigonometrical functions $\{\cos nx, \sin nx\}$ is **complete** with respect to that class and Parseval's equation is called the **completeness relation**. The question that we now want to answer is what is the largest class of functions for which completeness has been established and hence, for which convergence in the mean of their Fourier series is assured. The formal answer, known as the **Riesz-Fischer Theorem,** makes use of a form of integration that differs somewhat from Riemann integration. Called **Lebesgue integration,** it is defined for **all** bounded functions, including those that may be discontinuous on an infinite but enumerable set of points in the interval of integration, and even for some **un**bounded functions. Thus, the class of Lebesgue integrable functions is much larger than the class of functions which are of bounded variation or which are Riemann integrable. Nevertheless, when both the Lebesgue and the Riemann integrals of a function exist, they yield identical results. Moreover, when two functions are equal **almost everywhere,** that is to say everywhere except on an enumerable set of points, they have the same Lebesgue integral. Therefore, in practice, we can usually proceed by using Riemann integration techniques without regard to discontinuous or even, in some circumstances, unbounded behaviour provided that it is restricted to an enumerable point set.

Theorem: The set of trigonometrical functions $\{\cos nx, \sin nx\}$ is complete with respect to those functions $f(x)$ that are (Lebesgue) square integrable on the interval $-\pi \leqslant x \leqslant \pi$ and hence satisfy

$$\int_{-\infty}^{\infty} [f(x)]^2 \, dx < \infty.$$

Thus, the Fourier coefficients $\{a_n, b_n\}$ for all such functions satisfy Parseval's equation and the corresponding Fourier series converge in the mean to the functions they represent.

This completes our discussion of the "convergence question".

8.3 Functions Having Arbitrary Period

Periodic functions in applications rarely have period 2π but the transition from period 2π to any period T can be effected by a simple change of scale. Suppose that $f(t)$ has period T. Then we can introduce a new variable x such that $f(t)$, as a function of x, has period 2π by setting $x = \frac{2\pi}{T} t$. Hence, if $f(t)$ has a Fourier series, it must be of the form

$$f(t) = f\left(\frac{T}{2\pi}\right) = \frac{a_0}{2} + \sum_{n=1}^{\infty} (a_n \cos nx + b_n \sin nx), \qquad (8.3.1)$$

with coefficients given by the Euler formulas:

$$a_0 = \frac{1}{2\pi} \int_{-\pi}^{\pi} f\left(\frac{T}{2\pi}\right) dx, \quad a_n = \frac{1}{2\pi} \int_{-\pi}^{\pi} f\left(\frac{T}{2\pi}\right) \cos nx dx,$$

$$b_n = \frac{1}{\pi} \int_{-\pi}^{\pi} f\left(\frac{T}{2\pi}\right) \sin nx \, dx.$$

Since $x = \frac{2\pi}{T}$, we have $dx = \frac{2\pi}{T} dt$ and the interval of integration becomes

$$-\frac{T}{2} \leqslant t \leqslant \frac{T}{2}.$$

Consequently, the Fourier coefficients of the periodic function $f(t)$ of period T must be given by

$$a_0 = \frac{2}{T} \int_{-T/2}^{T/2} f(t) dt$$

$$a_n = \frac{2}{T} \int_{-T/2}^{T/2} f(t) \cos \frac{2n\pi t}{T} dt$$

$$b_n = \frac{2}{T} \int_{-T/2}^{T/2} f(t) \sin \frac{2n\pi t}{T} dt, \quad n = 1, 2, 3, \ldots. \tag{8.3.2}$$

Furthermore, the Fourier series (8.3.1) with x expressed in terms of t becomes

$$f(t) = \frac{a_0}{2} + \sum_{n=1}^{\infty} \left(a_n \cos \frac{2n\pi t}{T} + b_n \sin \frac{2n\pi t}{T}\right). \tag{8.3.3}$$

The interval of integration in (8.3.2) may be replaced by any interval of length T: $0 \leqslant t \leqslant T$, for example.

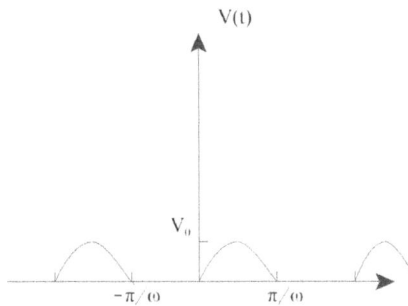

Figure 8.5: Half-Wave Rectifier

Example: Suppose that we wish to Fourier analyze the outcome of passing a sinusoidal voltage $V_0 \sin \omega t$ through a half-wave rectifier that clips the negative portion of the wave. The rectified potential will continue to have period $T = \frac{2\pi}{\omega}$ but will have the new functional form (see Figure 8.5)

$$V(t) = \begin{cases} 0 & \text{for } -T/2 < t < 0, \\ V_0 \sin \omega t & \text{for } 0 < t < T/2. \end{cases}$$

Since $V = 0$ when $-T/2 < t < 0$, we obtain from (8.3.2)

$$a_0 = \frac{\omega}{\pi} \int_0^{\pi/\omega} V_0 \sin \omega t\, dt = \frac{2\,V_0}{\pi},$$

and

$$a_n = \frac{\omega}{\pi} \int_0^{\pi/\omega} V_0 \sin \omega t \cos n\omega t\, dt = \frac{\omega\, V_0}{2\pi} \int_0^{\pi/\omega} [\sin(1+n)\omega t + \sin(1-n)\omega t]\, dt.$$

When $n = 1$, the integral on the right is zero, and when $n = 2, 3, \ldots$, we obtain

$$a_n = \frac{\omega\, V_0}{2\pi} \left[-\frac{\cos(1+n)\omega t}{(1+n)\omega} - \frac{\cos(1-n)\omega t}{(1-n)\omega} \right]_0^{\pi/\omega}$$
$$= \frac{V_0}{2\pi} \left(\frac{1 - \cos(1+n)\pi}{1+n} + \frac{1 - \cos(1-n)\pi}{1-n} \right).$$

When n is odd, this is zero; for even n it gives

$$a_n = \frac{V_0}{2\pi} \left(\frac{2}{1+n} + \frac{2}{1-n} \right) = -\frac{2\,V_0}{(n-1)(n+1)\pi}, \quad n = 2, 4, \ldots.$$

In a similar fashion we find that $b_1 = V_0/2$ and $b_n = 0$ for $n = 2, 3, \ldots$. Consequently,

$$V(t) = \frac{V_0}{\pi} + \frac{V_0}{2} \sin \omega t - \frac{2\,V_0}{\pi} \left(\frac{1}{1\cdot 3} \cos 2\omega t + \frac{1}{3\cdot 5} \cos 4\omega t + \ldots \right).$$

In many applications the independent variable is distance x and the most intuitive way to denote the period is to use $2L$. In that case the formulas read

$$f(x) = \frac{a_0}{2} + \sum_{n=1}^{\infty} \left[a_n \cos \frac{n\pi x}{L} + b_n \sin \frac{n\pi x}{L} \right], \tag{8.3.4}$$

with

$$a_n = \frac{1}{L} \int_{-L}^{L} f(x) \cos \frac{n\pi x}{L}\, dx, \quad b_n = \frac{1}{L} \int_{-L}^{L} f(x) \sin \frac{n\pi x}{L}\, dx. \tag{8.3.5}$$

8.4 Half-Range Expansions - Fourier Sine and Cosine Series

We have already seen that certain types of functions have particularly simple Fourier series. If $f(t)$ is an even function, $f(-t) = f(t)$, of period T then its Fourier sine coefficients $b_n = 0$ for all $n \geqslant 1$ and its Fourier series is a **Fourier cosine series**

$$f(t) = \frac{a_0}{2} + \sum_{n=1}^{\infty} a_n \cos \frac{2n\pi t}{T} \tag{8.4.1}$$

with coefficients

$$a_0 = \frac{4}{T} \int_0^{T/2} f(t)dt, \quad a_n = \frac{4}{T} \int_0^{T/2} f(t) \cos \frac{2n\pi t}{T} dt, \quad n = 1, 2, 3, \dots . \tag{8.4.2}$$

On the other hand, if $f(t)$ is odd, $f(-t) = -f(t)$, its Fourier cosine coefficients $a_n = 0$ for all n and its Fourier series is a **Fourier sine series**,

$$f(t) = \sum_{n=1}^{\infty} b_n \sin \frac{2n\pi t}{T} \tag{8.4.3}$$

with coefficients

$$b_n = \frac{4}{T} \int_0^{T/2} f(t) \sin \frac{2n\pi t}{T} dt. \tag{8.4.4}$$

Suppose that we have a function $f(t)$ that is defined only on some finite interval $0 \leqslant t \leqslant \tau$. If we want to represent it by a Fourier series, we now have three distinct representations to choose from:

- a Fourier cosine series of period $T = 2\tau$,
- a Fourier sine series of period $T = 2\tau$, or
- a full Fourier series of period $T = \tau$.

As before, the latter series will converge to the periodic extension of $f(t)$. The first two , however, are **half-range expansions** that converge to periodic extensions of the functions $f_1(t)$ and $f_2(t)$ respectively, where

$$f_1(t) = \begin{cases} f(t), & 0 \leqslant t \leqslant \tau \\ f(-t), & -\tau \leqslant t \leqslant 0 \end{cases}$$

and

$$f_2(t) = \begin{cases} f(t), & 0 < t \leqslant \tau \\ -f(-t), & -\tau \leqslant t < 0 \end{cases}$$

The first of these is called the **symmetric extension** and the second, the **antisymmetric extension** of $f(t)$. Thus, as shown in Figure 8.6, the Fourier cosine series

$$f(t) = \frac{a_0}{2} + \sum_{n=1}^{\infty} a_n \cos \frac{n\pi t}{\tau}, \quad 0 \leqslant t \leqslant \tau \tag{8.4.5}$$

with coefficients

$$a_n = \frac{2}{\tau} \int_0^\tau f(t) \cos \frac{n\pi t}{\tau} dt, \quad n = 0, 1, 2, \dots, \quad (8.4.6)$$

converges to the **periodic symmetric extension** of $f(t)$ of period $T = 2\tau$, while the Fourier sine series

$$f(t) = \sum_{n=1}^\infty b_n \sin \frac{n\pi t}{\tau}, \quad 0 < t < \tau \quad (8.4.7)$$

with coefficients

$$b_n = \frac{2}{\tau} \int_0^\tau f(t) \sin \frac{n\pi t}{\tau} dt, \quad n = 1, 2, \dots, \quad (8.4.8)$$

converges to the **periodic antisymmetric extension** of $f(t)$ of period $T = 2\tau$. The series (8.4.5) and (8.4.7) are called **half-range expansions** of $f(t)$.

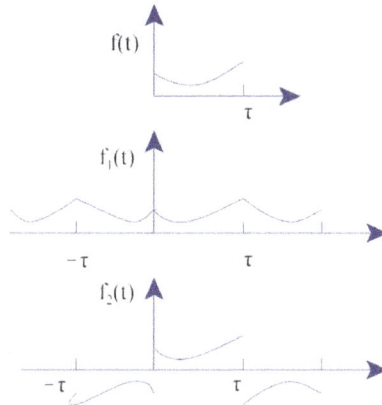

Figure 8.6: Periodic Extensions

Example: Consider the function $f(t) = \frac{t}{2\tau} + \frac{1}{2}$. Its full-range Fourier coefficients are

$$a_0 = 1, a_n = \frac{1}{\tau} \int_{-\tau}^\tau \left(\frac{t}{2\tau} + \frac{1}{2} \right) \cos \frac{n\pi t}{\tau} dt = 0,$$

$$b_n = \frac{1}{\tau} \int_{-\tau}^\tau \left(\frac{t}{2\tau} + \frac{1}{2} \right) \sin \frac{n\pi t}{\tau} dt = \frac{(-1)^{n+1}}{n\pi},$$

and the Fourier series reads

$$f(t) = \frac{1}{2} + \sum_{n=1}^\infty \frac{(-1)^{n+1}}{n\pi} \sin \frac{n\pi t}{\tau}, \quad -\tau < t < \tau.$$

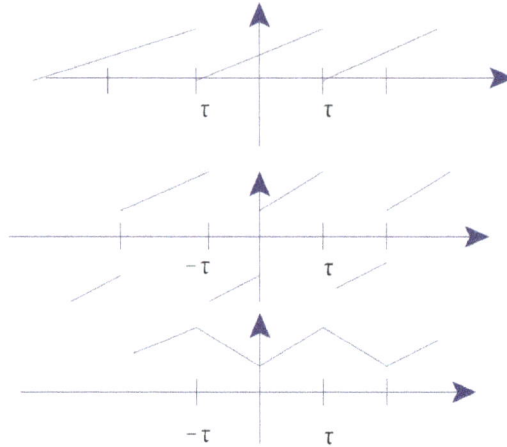

Figure 8.7: The Fourier, Fourier Sine and Fourier Cosine Series for $f(t) = \frac{t}{2\tau} + \frac{1}{2}$.

The function's half-range Fourier sine coefficients are

$$b_n = \frac{2}{\tau} \int_0^\tau \left(\frac{t}{2\tau} + \frac{1}{2} \right) \sin \frac{n\pi t}{\tau} dt = \frac{1 - 2\cos n\pi}{n\pi},$$

and its Fourier sine series reads

$$f(t) = \sum_{n=1}^\infty \frac{1 - 2(-1)^n}{n\pi} \sin \frac{n\pi t}{\tau}, \quad 0 < t < \tau.$$

Finally, the half-range Fourier cosine coefficients are

$$a_0 = \frac{3}{2}, a_n = \frac{2}{\tau} \int_0^\tau \left(\frac{t}{2\tau} + \frac{1}{2} \right) \cos \frac{n\pi t}{\tau} dt = \frac{\cos n\pi - 1}{n^2 \pi^2},$$

and so the Fourier cosine series reads

$$f(t) = \frac{3}{4} + \frac{2}{\pi^2} \sum_{n=1,3,5}^\infty \frac{1}{n^2} \cos \frac{n\pi t}{\tau}, \quad 0 \leqslant t \leqslant \tau.$$

All three series are shown in Figure 8.7 .

8.5 Complex Form of Fourier Series

As we noted at the beginning of Section 8.2, Fourier series can also be expressed in a complex form:

$$f(x) = \frac{a_0}{2} + \sum_{n=1}^{\infty} [a_n \cos nx + b_n \sin nx]$$

$$= \frac{a_0}{2} + \sum_{n=1}^{\infty} \left[a_n \frac{e^{inx} + e^{-inx}}{2} + b_n \frac{e^{inx} - e^{-inx}}{2i} \right]$$

$$= \frac{a_0}{2} + \sum_{n=1}^{\infty} \left[\frac{a_n - i b_n}{2} e^{inx} + \frac{a_n + i b_n}{2} e^{-inx} \right]$$

or,

$$f(x) = \sum_{n=-\infty}^{\infty} c_n e^{inx} \tag{8.5.1}$$

where

$$c_n = \begin{cases} \dfrac{a_0}{2}, & n = 0 \\ \dfrac{a_n - i b_n}{2}, & n > 0 \\ \dfrac{a_{|n|} + i b_{|n|}}{2}, & n < 0. \end{cases} \tag{8.5.2}$$

From the formulas for a_n and b_n we find the corresponding Euler formula for c_n:

$$c_n = \frac{1}{2\pi} \int_{-\pi}^{\pi} f(x) e^{-inx} \, dx. \tag{8.5.3}$$

Alternatively, this formula can be obtained by multiplying the series (8.5.1) by e^{-imx}, integrating, and using the (orthogonality) relation

$$\int_{-\pi}^{\pi} e^{inx} e^{-imx} \, dx = \begin{cases} 0, & n \neq m \\ 2\pi, & n = m. \end{cases}$$

For an arbitrary period $2L$ and interval $-L \leqslant x \leqslant L$, (8.5.1) and (8.5.3) become

$$f(x) = \sum_{n=-\infty}^{\infty} c_n e^{i\frac{n\pi x}{L}} \tag{8.5.4}$$

with

$$c_n = \frac{2}{L} \int_{-L}^{L} f(x) e^{-i\frac{n\pi x}{L}} \, dx. \tag{8.5.5}$$

Example: Fourier series have important applications in connection with differential equations. As an introduction we shall confine ourselves to an example involving an ordinary differential equation. Applications in connection with partial differential equations will be considered in a subsequent chapter.

The forced oscillations of a body of mass m on a spring are governed by the equation

$$m\frac{d^2 x}{d t^2} + r\frac{dx}{dt} + kx = f(t)$$

where k is the spring constant, r is the damping coefficient, and $f(t)$ is the external force. The general solution to this equation can be written (see Chapter 9) as the general solution to the associated homogeneous equation

$$m\frac{d^2 x}{d t^2} + r\frac{dx}{dt} + kx = 0$$

plus a particular solution of the non-homogeneous equation. For the homogeneous equation, substitution of the *ansatz* $x = e^{pt}$ results in the characteristic equation

$$(m p^2 + rp + k)\, e^{pt} = 0$$

with roots

$$p = \frac{-r \pm \sqrt{r^2 - 4mk}}{2m}.$$

All real physical systems have positive damping coefficients and so their homogeneous solutions are either damped sinusoidal functions or decaying exponentials:

$$c_1\, e^{-(r/2m)t} \cos \sqrt{k/m - r^2/4 m^2}\, t + c_2\, e^{-(r/2m)t} \sin \sqrt{k/em - r^2/4 m^2}\, t$$

or,

$$c_1 \exp\left[\frac{-r + \sqrt{r^2 - 4mk}}{2m} t\right] + c_2 \exp\left[\frac{-r - \sqrt{r^2 - 4mk}}{2m} t\right].$$

To determine a particular **nonhomogeneous** solution, we note that substitution of the *ansatz*

$$x(t) = X\, e^{i\omega t}$$

with an **undetermined coefficient** X gives

$$m(i\omega)^2\, X e^{i\omega t} + r(i\omega) X e^{i\omega t} + kX e^{i\omega t} = A e^{i\omega t},$$

or,

$$X = \frac{A}{-m\,\omega^2 + ir\omega + k}$$

and hence, leads immediately to the general solution

$$x(t) = \frac{A}{-m\,\omega^2 + ir\omega + k}\, e^{i\omega t} + c_1\, x_1(t) + c_2\, x_2(t)$$

where $x_1(t)$ and $x_2(t)$ are the homogeneous solutions noted above. The last two terms are called "transients" and approach zero as time increases. Thus, the solution decays to a "steady state" sinusoidal response oscillating at the same frequency as the driving term $A\,e^{i\omega t}$. This oscillation never dies out as long as the driving force is applied; its amplitude X is fixed.

Now let us suppose that the driving force $f(t)$ is not sinusoidal but that it is periodic. The steady state response of the system will again mirror the periodicity of the imposed force and to find an explicit expression for it we simply assume Fourier expansions for $f(t)$ and $x(t)$:

$$f(t) = \sum_{n=-\infty}^{\infty} A_n\, e^{in\omega t} \quad\text{and}\quad x(t) = \sum_{n=-\infty}^{\infty} c_n\, e^{in\omega t},\; \omega = 2\pi/T.$$

Assuming further that the series for $x(t)$ can be differentiated term by term the necessary number of times, we can substitute these two series plus

$$\frac{dx}{dt} = \sum_{n=-\infty}^{\infty} in\omega\, c_n\, e^{in\omega t} \quad\text{and}\quad \frac{d^2 x}{dt^2} = \sum_{n=-\infty}^{\infty} (-n^2\,\omega^2)\, c_n\, e^{in\omega t},$$

into the differential equation. Then, invoking orthogonality, we can equate the coefficients with the same exponential $e^{in\omega t}$ on both sides. The result is

$$(-n^2\,\omega^2\, m + in\omega r + k)\, c_n = A_n$$

or,

$$c_n = \frac{A_n/m}{(\omega_0^2 - n^2\,\omega^2) + 2\lambda n\omega i}$$

where $\omega_0^2 = k/m$ is the natural frequency of the oscillator and $\lambda = r/2m$ is the system's damping factor. It only remains to determine the Fourier coefficients for $f(t)$ by applying

$$A_n = \frac{1}{T} \int_{-T/2}^{T/2} f(t)\, e^{-in\omega t}\, dt.$$

Thus, we obtain the steady state solution as a superposition of sinusoidal functions with frequencies that are integral multiples of $2\pi/T$, T being the period of the driving force.

If the frequency of one of these functions is close to the natural frequency ω_0 of the system, a resonance effect occurs because of the near cancellation in the denominator of c_n; that function then becomes the dominant part of the system's response to the imposed force. To offer a concrete illustration of this, let $m = 1$ (kg), $r = 0.02$ (kg/sec), and $k = 25$ (kg /sec^2), so that the equation of motion becomes

$$\frac{d^2 x}{dt^2} + 0.02\frac{dx}{dt} + 25x = f(t)$$

where $f(t)$ is measured in $kg \cdot m/\sec^2$. Furthermore, let

$$f(t) = \begin{cases} t + \pi/2 & \text{for } -\pi < t < 0, \\ -t + \pi/2 & \text{for } 0 < t < \pi. \end{cases}$$

with $f(t + 2\pi) = f(t)$. Representing $f(t)$ by a Fourier series, we find

$$A_n = \frac{1}{2\pi} \left[\int_{-\pi}^{0} (t + \pi/2) e^{-int} \, dt + \int_{0}^{\pi} (-t + \pi/2) e^{-int} \, dt \right] = \frac{1}{2\pi} \int_{0}^{\pi} (-t + \pi/2)[e^{-int} + e^{int}] dt$$

or,

$$A_n = \frac{1}{\pi} \int_{0}^{\pi} (-t + \pi/2) \cos nt \, dt = \frac{1}{n^2 \pi}[1 - (-1)^n],$$

and thus,

$$f(t) = \frac{2}{\pi} \sum_{n=-\infty, n \text{ odd}}^{\infty} \frac{1}{n^2} e^{int}.$$

From the preceding analysis we know that the oscillator's response to this force will be a displacement $x(t)$ with Fourier coefficients

$$c_n = \frac{A_n/m}{(\omega_0^2 - n^2) + 2\lambda ni}, \quad n = \pm 1, \pm 3, \pm 5, \ldots$$

where $\omega_0^2 = k/m = 25, \lambda = r/2m = 0.01$. Thus,

$$c_n = \frac{2}{n^2 \pi} \frac{(25 - n^2) - 0.02ni}{(25 - n^2)^2 + (0.02n)^2}, \quad n = \pm 1, \pm 3, \pm 5, \ldots$$

which leads directly to

$$x(t) = \sum_{n=1,3,5,\ldots}^{\infty} [a_n \cos nt + b_n \sin nt]$$

where

$$a_n = \frac{4}{n^2 \pi} \frac{25 - n^2}{(25 - n^2)^2 + (0.02n)^2} \quad \text{and} \quad b_n = \frac{0.08}{n^2 \pi} \frac{1}{(25 - n^2)^2 + (0.02n)^2}.$$

The amplitude of each (complex) mode of oscillation is

$$|c_n| = \frac{1}{2} \sqrt{a_n^2 + b_n^2} = \frac{2}{n^2 \pi} \frac{1}{\sqrt{(25 - n^2)^2 + (0.02n)^2}}.$$

Some numerical values are

$$|c_1| = 0.0265, |c_3| = 0.0044, |c_5| = 0.2550, |c_7| = 0.0006, |c_9| = 0.0001.$$

Thus, the cancellation in the denominator of $|c_5|$ results in the $n = \pm 5$ modes dominating and in fact, since $a_5 = 0$, results in the $\sin 5t$ term dominating the Fourier series for $x(t)$. This means that the steady state response of the system is almost a pure sine wave with frequency five times that of the driving force.

8.6 Transition to Fourier Transforms

As we noted in the context of analytic continuation, integral representations of functions are often more useful than series representations. Typically, they assume the form of an **integral transform** (see Section 5.4)

$$f(z) = \int_C K(z, \zeta) g(\zeta) d\zeta$$

where the function $K(z, \zeta)$ is called the **kernel** of the representation. We noted further that a commonly used kernel is named after **Fourier**, has the functional dependence $K(z, \zeta) = \frac{1}{2\pi} e^{-iz\zeta}$ and corresponds to a choice of contour C that runs along the entire real axis, $-\infty < \zeta < \infty$. The representation that results is

$$f(z) = \frac{1}{2\pi} \int_{-\infty}^{\infty} e^{-iz\zeta} g(\zeta) d\zeta.$$

The similarity in both name and form to the representation afforded by a complex Fourier series suggests that there ought to be a connection and, at the heuristic level at least, there is.

Our starting point is equations (8.5.4) and (8.5.5):

$$f(x) = \sum_{n=-\infty}^{\infty} c_n e^{i\frac{n\pi x}{L}}, \quad -L \leqslant x \leqslant L,$$

$$c_n = \frac{2}{L} \int_{-L}^{L} f(x) e^{-i\frac{n\pi x}{L}} dx. \tag{8.6.1}$$

If we now define

$$k_n \equiv \frac{n\pi}{L} \quad \text{and} \quad \Delta k \equiv \frac{\pi}{L} = k_{n+1} - k_n,$$

these equations can be rewritten as

$$f(x) = \sum_{n=-\infty}^{\infty} c_L(k_n) e^{i k_n x} \Delta k \quad \text{with}$$

$$c_L(k_n) = \frac{L}{\pi} c_n = \frac{1}{2\pi} \int_{-L}^{L} f(x) e^{-i k_n x} dx.$$

Then, taking the limit as $L \to \infty$, we have

$$C(k) = \lim_{L \to \infty} c_L(k) = \frac{1}{2\pi} \int_{-\infty}^{\infty} f(x) e^{-ikx} dx \quad \text{and}$$

$$f(x) = \int_{-\infty}^{\infty} C(k) e^{ikx} dk$$

which comprise a **Fourier transform pair**. By modern convention, **the** Fourier transform pair is **defined** by replacing $C(k)$ by $F(k) = \sqrt{2\pi} C(-k)$; thus,

$$F(k) = \frac{1}{\sqrt{2\pi}} \int_{-\infty}^{\infty} f(x)\, e^{ikx}\, dx \quad \text{which yields the representation}$$

$$f(x) = \frac{1}{\sqrt{2\pi}} \int_{-\infty}^{\infty} F(k)\, e^{-ikx}\, dk. \tag{8.6.2}$$

The function $F(k)$ is identified as the **Fourier transform** of $f(x)$ and $f(x)$ as the **inverse Fourier transform** of $F(k)$; symbolically, we write

$$F(k) = \mathcal{F}\{f(x)\} \quad \text{and} \quad f(x) = \mathcal{F}^{-1}\{F(k)\}.$$

By construction, the representation of $f(x)$ that this affords is intended to be valid for all real x, $-\infty < x < \infty$, and periodicity is no longer a consideration.

Not surprisingly, each theorem on the convergence properties of Fourier series has a Fourier transform counterpart. Thus, the pointwise convergence of (8.6.2) is addressed by the **Fourier integral theorem**.

Theorem: Let $f(x)$ be absolutely integrable on $-\infty < x < \infty$. Then,

$$\frac{1}{2}[f(x+0) + f(x-0)] = \frac{1}{2\pi} \int_{-\infty}^{\infty} dk \int_{-\infty}^{\infty} d\xi\, f(\xi)\, e^{ik(\xi-x)} \tag{8.6.3}$$

provided that $f(\xi)$ is of bounded variation on an interval $a \leqslant \xi \leqslant b$ that includes the point $\xi = x$. Moreover, if the function is continuous on this interval, the integral on the right hand side of (8.6.3) converges uniformly to $f(x)$ for $a \leqslant x \leqslant b$.

Notice that this theorem does not reflect the striking symmetry between $f(x)$ and its Fourier transform $F(k)$; the properties of $F(k)$ are not even mentioned. While a function and its Fourier coefficients are quite different mathematical objects, a function and its Fourier transform are objects of exactly the same type and so the reciprocity implied by the equations in (8.6.2) is of considerable interest. It is addressed by **Plancherel's theorem** which, as the Fourier transform counterpart of Parseval's theorem, also addresses sufficient conditions for convergence in the mean.

Theorem: Let $f(x)$ be (Lebesgue) square integrable on $-\infty < x < \infty$. Then, the integral

$$F(k, L) = \frac{1}{\sqrt{2\pi}} \int_{-L}^{L} f(x)\, e^{ikx}\, dx$$

converges in the mean as $L \to \infty$ to a function $F(k)$ which is itself square integrable on $-\infty < k < \infty$. Furthermore, the integral

$$f(x, L) = \frac{1}{\sqrt{2\pi}} \int_{-L}^{L} F(k)\, e^{-ikx}\, dk$$

converges in the mean to $f(x)$ and

$$\int_{-\infty}^{\infty} |f(x)|^2 dx = \int_{-\infty}^{\infty} |F(k)|^2 dk. \qquad (8.6.4)$$

8.7 Examples of Fourier Transforms

Example 1. Consider the function $f(x) = \dfrac{a}{a^2 + x^2}$. From equation (8.6.2) its Fourier transform is given by

$$F(k) = \frac{a}{\sqrt{2\pi}} \int_{-\infty}^{\infty} \frac{e^{ikx}}{a^2 + x^2} dx.$$

This can be evaluated by means of the calculus of residues. The function $f(z) = \dfrac{1}{a^2 + z^2}$ satisfies the conditions specified in the theorem of Section 3.3.3 and so equations (3.3.11) and (3.3.12) apply:

$$\int_{-\infty}^{\infty} f(x) e^{ikx} dx = \begin{cases} 2\pi i \sum_{+} \text{Res}[f(z) e^{ikz}], & k > 0, \\ -2\pi i \sum_{-} \text{Res}[f(z) e^{ikz}], & k < 0. \end{cases}$$

There are simple poles at $z = \pm ia$ and the residues there are

$$\text{Res}\left[\frac{e^{ikz}}{a^2 + z^2}\right]_{z=\pm ia} = \left.\frac{e^{ikz}}{z \pm ia}\right|_{z=\pm ia} = \frac{e^{\mp ka}}{\pm 2ia}.$$

Thus,

$$F(k) = \sqrt{\frac{\pi}{2}} \cdot \begin{cases} e^{-ka}, & k > 0 \\ e^{ka}, & k < 0 \end{cases}$$

or, $F(k) = \sqrt{\frac{\pi}{2}} e^{-|k|a}$.

Verifying that equation (8.6.3)) holds, we note that

$$\frac{1}{\sqrt{2\pi}} \int_{-\infty}^{\infty} F(k) e^{-ikx} dk = \frac{1}{\sqrt{2\pi}} \sqrt{\frac{\pi}{2}} \left[\int_{-\infty}^{0} e^{ka-ikx} dk + \int_{0}^{\infty} e^{-ka-ikx} dk \right]$$

$$= \frac{1}{2} \left[\frac{1}{a - ix} + \frac{1}{a + ix} \right] = \frac{a}{a^2 + x^2} = f(x)$$

as required.

Notice that if the parameter a is small, $f(x)$ will be sharply peaked about $x = 0$; $F(k)$ on the other hand will be relatively spread out on either side of its maximum which occurs at $k = 0$. If a is large, the converse obtains: $f(x)$ is flattened while $F(k)$ is sharply peaked. This contrast in behaviour between the members of a Fourier transform pair is characteristic and is the mathematical basis for the **Heisenberg uncertainty principle** in quantum mechanics. We shall see it recur even more dramatically in each of the next two examples.

Example 2. Suppose that we now take $f(x)$ to be a Gaussian probability function: $f(x) = e^{-a x^2}$, $a = constant > 0$. Its Fourier transform is given by

$$F(k) = \frac{1}{\sqrt{2\pi}} \int\limits_{-\infty}^{\infty} e^{ikx - a x^2} \, dx$$

which, apart from the normalization factor $\frac{1}{\sqrt{2\pi}}$, is just Gauss's Integral which was evaluated in Section 3.3.6. That result gives us

$$F(k) = \frac{1}{\sqrt{2a}} e^{-k^2/4a}$$

which is **also** a Gaussian but one with a dependence on a that is the inverse of that enjoyed by $f(x)$.

In quantum mechanics it is the squared modulus of the transform pair that has physical significance; ($|f(x)|^2 dx$ and $|F(k)|^2 dk$ are the probabilities that a particle or system of particles can be localized with position x and wave number k, respectively). The function $|f(x)|^2 = e^{-2a x^2}$ decreases from its maximum of 1 at $x = 0$ to a value of $\frac{1}{e}$ at $x = \pm\frac{1}{\sqrt{2a}}$. Thus, we take its "width" to be $\Delta x = \sqrt{\frac{2}{a}}$. Similarly, the width of $|F(k)|^2 = \frac{1}{2a} e^{-k^2/2a}$ is $\Delta k = 2\sqrt{2a}$ and so,

$$\Delta x \cdot \Delta k = 4 \gtrsim O(1)$$

which, as indicated, is a number of order 1.

Another common transform pair is composed of the functions

$$f(x) = \begin{cases} \frac{1}{\sqrt{a}}, & -a/2 \leqslant x \leqslant a/2 \\ 0, & |x| > a/2. \end{cases}$$

and

$$F(k) = \frac{1}{\sqrt{2\pi}} \int\limits_{-a/2}^{a/2} \frac{1}{\sqrt{a}} e^{ikx} \, dx = \sqrt{\frac{2}{\pi a}} \frac{\sin ak/2}{k}.$$

In this case we can take Δx and Δk to be the distances between the central zeros of $|f(x)|^2$ and $|F(k)|^2$, respectively. Since the zeros are the same as the central zeros of $f(x)$ and $F(k)$, we find $\Delta x = a$ and $\Delta k = \frac{4\pi}{a}$ and so recover the relation $\Delta x \cdot \Delta k \gtrsim O(1)$. Because Δx and Δk can be identified with an uncertainty in assigning to x and k their most probable values, this has come to be called the (Heisenberg) uncertainty relation.

Example 3. As a final example, we shall consider a fairly typical function of time

$$f(t) = \begin{cases} 0, & t < 0 \\ e^{-t/T} \sin \omega_0 t, & t \geqslant 0 \end{cases}.$$

Physically, this function might represent the displacement of a damped harmonic oscillator, or the electric field in a radiated electromagnetic wave, or the current in an antenna, just to name three possibilities.

The Fourier transform of $f(t)$ is

$$F(\omega) = \frac{1}{\sqrt{2\pi}} \int\limits_{0}^{\infty} e^{-t/T} \sin \omega_0 t \, e^{i\omega t} \, dt = \frac{1}{2\sqrt{2\pi}} \left[\frac{1}{\omega + \omega_0 + i/T} - \frac{1}{\omega - \omega_0 + i/T} \right].$$

Once $f(t)$ is identified, we can use Parseval's equation (Plancherel's Theorem) to deduce a complementary physical meaning for $F(\omega)$. For example, if $f(t)$ is a radiated electric field, the radiated power will be proportional to $|f(t)|^2$ and the total energy radiated will be proportional to $\int\limits_{-\infty}^{\infty} |f(t)|^2 dt$. But, according to Parseval's equation , this is equal to $\int\limits_{-\infty}^{\infty} |F(\omega)|^2 d\omega$. Thus, to within a multiplicative constant, $|F(\omega)|^2$ must be the energy radiated per unit frequency interval .

Suppose that T is very large so that $\omega_0 T \gg 1$. Then, as happened with the resonating harmonic oscillator of Section 8.5 where one frequency or term in the Fourier series dominated over all the others, the "frequency spectrum" defined by $F(\omega)$ is sharply peaked about $\omega = \pm \omega_0$. For example, near $\omega = \omega_0$,

$$F(\omega) \approx -\frac{1}{2\sqrt{2\pi}} \frac{1}{\omega - \omega_0 + i/T},$$

and,

$$|F(\omega)|^2 \approx \frac{1}{8\pi} \frac{1}{(\omega - \omega_0)^2 + 1/T^2}.$$

When $\omega = \omega_0 \pm 1/T$, the radiated energy $|F(\omega)|^2$ is down by a factor of $\frac{1}{2}$ from its peak value. Thus, the width of the peak at half-maximum, which is a measure of the uncertainty in the frequency of the radiation, is given by $\Gamma = 2/T$. On the other hand, T is the time for the amplitude of the oscillator or of the radiated wave to "decay" by a factor of e^{-1} and so, is a measure of their mean lifetime which, in turn, is a measure of the uncertainty in the time of oscillation or of emission of the radiation. Thus, we recover another (classical) uncertainty relation:

$$\Delta t \cdot \Delta \omega = T \cdot \frac{2}{T} \gtrsim O(1).$$

8.8 The Dirac Delta Function and Transforms of Distributions

A reordering of the integrations that appear in the statement of the Fourier Integral Theorem (equation (8.6.3)) gives rise to a very suggestive result. Assuming continuity for $f(x)$, (8.6.3) reads

$$f(x) = \frac{1}{2\pi} \int\limits_{-\infty}^{\infty} dk \int\limits_{-\infty}^{\infty} d\xi \, f(\xi) \, e^{ik(\xi - x)} \tag{8.8.1}$$

which, after reordering, yields

$$f(x) = \int_{-\infty}^{\infty} d\xi \, f(\xi) \, \delta(\xi - x) \qquad (8.8.2)$$

where

$$\delta(\xi - x) = \frac{1}{2\pi} \int_{-\infty}^{\infty} dk \, e^{ik(\xi-x)} . \qquad (8.8.3)$$

At this juncture we have no idea about the legitimacy of reversing the order of integration and, given that the improper integral in (8.8.3) is not defined in any conventional sense, it does look a bit questionable. Nevertheless, let us ignore such niceties for the time being and focus on the intriguing fact that, if it does exist, $\delta(\xi - x)$ is the continuous variable generalization of the Kronecker delta function

$$\delta(m, n) = \begin{cases} 1, & m = n \\ 0, & m \neq n \end{cases}$$

That is to say, just as $\delta(m, n)$ picks out the nth term from a summation over m,

$$f_n = \sum_{m=-\infty}^{\infty} f_m \, \delta(m, n),$$

$\delta(\xi - x)$ selects and delivers the value at $\xi = x$ of the function that multiplies it in a summation over the continuous variable ξ.

A function with this property is called a **Dirac delta function** and was introduced by Paul Dirac in his landmark formulation of quantum mechanics that unified the earlier Heisenberg and Schrödinger pictures of quantum phenomena.

It is clear from (8.8.3) that the delta function depends only on the difference $\xi - x$ and not on ξ and x individually. Moreover, equation (8.8.2) tells us that the delta function is normalized for, setting $f(x) \equiv 1$, we have

$$1 = \int_{-\infty}^{\infty} \delta(\xi - x) d\xi. \qquad (8.8.4)$$

This immediately raises the question of what $\delta(\xi - x)$ "looks like" when it is plotted. From equation (8.8.2) we see that the function $f(\xi)$ can be modified anywhere except at the point $\xi = x$ without affecting the result of the integration. This implies that $\delta(\xi - x)$ must be zero everywhere except in an infinitesimal neighbourhood of $\xi = x$. Equation (8.8.4) then suggests that

$$1 = \lim_{\varepsilon \to 0} \int_{x-\varepsilon}^{x+\varepsilon} \delta(\xi - x) d\xi$$

and hence, that $\delta(\xi - x) \to \infty$ as $\xi \to x$. In other words, we can think of the delta function as having the mathematically ill-defined properties

$$\delta(\xi - x) = \begin{cases} 0 & \text{for all } \xi \neq x \\ \infty & \text{for } \xi = x. \end{cases} \tag{8.8.5}$$

Evidently, neither this "equality" nor equation (8.8.2) from which it was deduced can be used as a formal definition of the Dirac delta function. Nevertheless, they do offer an intuitive appreciation of what a delta function is as well as some idea of how to make use of them.

Another aid for our intuition and one that takes us closer to a formal definition of the Dirac delta function is to think of it as the limit of a sequence of functions that are strongly peaked about $\xi = x$ and satisfy the condition

$$\lim_{n \to \infty} \int_{-\infty}^{\infty} \delta_n(\xi - x)f(\xi)d\xi = f(x) \tag{8.8.6}$$

for all suitably behaved functions $f(x)$. Such sequences are called **delta sequences**.

For notational convenience, we are now going to reverse the roles of the symbols ξ and x and set $\xi = 0$. Equation (8.8.6) then becomes

$$\lim_{n \to \infty} \int_{-\infty}^{\infty} \delta_n(x)f(x)dx = f(0) \tag{8.8.7}$$

where $\{\delta_n(x)\}$ is a sequence of functions that are sharply peaked about $x = 0$.

Some examples of delta sequences are:

1. $\delta_n(x) = \begin{cases} 0 & \text{for } |x| \geqslant 1/n \\ n/2 & \text{for } |x| < 1/n \end{cases}$

2. $\delta_n(x) = \dfrac{n}{\sqrt{\pi}} e^{-n^2 x^2}$;

3. $\delta_n(x) = \dfrac{n}{\pi} \dfrac{1}{1 + n^2 x^2}$;

4. $\delta_n(x) = \dfrac{\sin nx}{\pi x} = \dfrac{1}{2\pi} \int_{-n}^{n} e^{ikx} dk$;

5. $\delta_n(x) = \dfrac{1}{n\pi} \dfrac{\sin^2 nx}{x^2}$.

By definition these all satisfy condition (8.8.7). We shall verify this for the first and third sequences. Notice that the fourth sequence links us back to the Fourier transform (8.8.3) that initiated our interest in the delta function.

Using the mean value theorem, substitution of sequence #1 into the integral on the left hand side of (8.8.7) yields

$$\int_{-\infty}^{\infty} \delta_n(x)f(x)dx = \int_{-1/n}^{1/n} \frac{n}{2}f(x)dx = \frac{2}{n} \cdot \frac{n}{2}f(x_m)$$

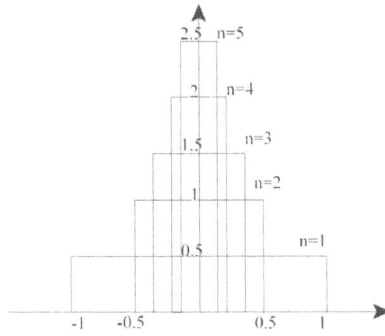

Figure 8.8: δ-sequence #1

where $-1/n \leqslant x_m \leqslant 1/n$. Clearly, as $n \to \infty$, $x_m \to 0$ and so,

$$\lim_{n\to\infty} \int_{-\infty}^{\infty} \delta_n(x)f(x)dx = f(0)$$

as required.

In the case of sequence #3 , the relevant integral is

$$\int_{-\infty}^{\infty} \frac{n}{\pi} \frac{1}{1+n^2 x^2}f(x)dx$$

which admits evaluation by residue calculus. Provided that the continuation of the test function, $f(z)$, is meromorphic in the upper half-plane and that $|f(z)|$ is bounded as $|z| \to \infty$, we can use equation (3.3.4) and write

$$\int_{-\infty}^{\infty} \frac{n}{\pi} \frac{1}{1+n^2 x^2}f(x)dx = 2\pi i \sum_{+} \text{Res}[\delta_n(z)f(z)]$$

where $\sum_{+}$ denotes the sum over all the poles of $\delta_n(z)$ and $f(z)$ in the upper half-plane. However, since $\lim_{n\to\infty} \delta_n(z) = 0$ if $z \neq 0$, the contributions to the sum from poles of $f(z)$ will vanish in the limit. Thus,

$$\lim_{n\to\infty} \int_{-\infty}^{\infty} \frac{n}{\pi} \frac{1}{1+n^2 x^2}f(x)dx = \lim_{n\to\infty} 2\pi i \text{Res}\left[\frac{n}{\pi} \frac{1}{1+n^2 z^2}f(z)\right]_{z=\frac{i}{n}}$$

or,

$$\lim_{n\to\infty} \int_{-\infty}^{\infty} \frac{n}{\pi} \frac{1}{1+n^2 x^2}f(x)dx = \lim_{n\to\infty} 2\pi i\, f\left(\frac{i}{n}\right) \frac{n}{\pi} \frac{1}{2n^2 \frac{i}{n}} = \lim_{n\to\infty} f\left(\frac{i}{n}\right) = f(0)$$

as required.

The latter sequence often appears in physical applications with n replaced by $\frac{1}{\varepsilon}$ so that $\delta_n(x)$ assumes the form

$$\delta_\varepsilon(x) = \frac{\varepsilon}{\pi}\frac{1}{x^2 + \varepsilon^2} = \frac{1}{2\pi i}\left[\frac{1}{x - i\varepsilon} - \frac{1}{x + i\varepsilon}\right] \qquad (8.8.8)$$

and the appropriate limit is now that as $\varepsilon \to 0$.

While it is helpful to **think** of the Dirac delta function, $\delta(x)$, as being the limit of any of these sequences, we cannot **define** it this way because the limits themselves are undefined at $x = 0$. Nevertheless, the delta sequences do play a role in defining the delta function not as a conventional function but as a member of a new class of mathematical entities called **generalized functions** or **distributions**. These are all defined in terms of sequences of conventional functions by means of a unique limiting process that involves the integrals of these functions "against" an appropriately well-behaved test function. In the case of $\delta(x)$ this yields a definition that recognizes that the fundamental property of the function is expressed by

$$\int_{-\infty}^{\infty} \delta(x)f(x)dx = f(0). \qquad (8.8.9)$$

Definition: Any sequence of continuous functions $g_n(x)$ defines a **distribution** $g(x)$ if, for any function $f(x)$ which is differentiable everywhere any number of times and which is non-zero only on a bounded set, the limit

$$\lim_{n \to \infty} \int_{-\infty}^{\infty} g_n(x)f(x)dx \equiv \int_{-\infty}^{\infty} g(x)f(x)dx \qquad (8.8.10)$$

exists.

The right hand side of this equation is **not** a Riemann integral but rather, it denotes the **limit** of a sequence of Riemann integrals.

Two distributions $g(x)$ and $h(x)$ are equal if the corresponding sequences satisfy

$$\lim_{n \to \infty} \int_{-\infty}^{\infty} g_n(x)f(x)dx = \lim_{n \to \infty} \int_{-\infty}^{\infty} h_n(x)f(x)dx$$

for any "test function" $f(x)$ that has the properties specified in the definition. Thus, for example, the delta-sequences $\{\delta_n(x)\}$ **all** define the same distribution $\delta(x)$:

$$\lim_{n \to \infty} \int_{-\infty}^{\infty} \delta_n(x)f(x)dx = \int_{-\infty}^{\infty} \delta(x)f(x)dx = f(0).$$

The principal result of the theory of distributions, from the perspective of a physicist, is that they admit manipulation according to the same rules of calculus as apply to conventional functions. However, one must remember that they have significance or meaning only as a multiplicative factor in an integrand.

Because of its peculiar properties, the Dirac delta function is an extremely useful artifact of mathematical physics. In addition to arising in a very natural way from a consideration of the representation of functions by means of Fourier series and transforms, they offer a functional expression of the most common idealizations of physics: that particles can be localized at points and rigid bodies can be so rigid that their elastic collisions are mediated by instantaneous impulses. The charge density for a collection of point charges and the force exerted by one perfectly rigid billiard ball on another are necessarily delta functions of position and time, respectively. This, in turn, leads to a central role for the delta function in the Green's Function approach to solving non- homogeneous boundary value problems as we shall see in Chapter 12. Thus, for all these reasons, we shall digress with an exploration of what can be learned about delta functions from an application of the δ-**calculus.**

The δ-calculus involves the treatment of $\delta(x)$ and its derivatives as though they were conventional functions albeit ones with the "unusual" properties

1. $\int_{-\infty}^{\infty} \delta(x)f(x)dx = f(0)$, and

2. $\delta(x) = \begin{cases} \infty, & x = 0 \\ 0, & x \neq 0 \end{cases}$ with $\int_{-\infty}^{\infty} \delta(x)dx = 1$.

It is a shortcut method for obtaining identities that are all derivable by a rigorous but much more onerous approach involving δ-sequences and the limiting processes that define distributions.

To begin with we shall "prove" that

$$\delta(x) = \frac{d}{dx}\theta(x) \qquad (8.8.11)$$

where θ is the step function

$$\theta(x) = \begin{cases} 1, & x > 0 \\ 0, & x < 0 \end{cases}$$

Using a continuous but otherwise arbitrary test function $f(x)$, we compose the integral

$$\int_{-\infty}^{\infty} \frac{d\theta(x)}{dx}f(x)dx$$

and proceed with integration by parts. We find immediately that

$$\int_{-\infty}^{\infty} \frac{d\theta(x)}{dx}f(x)dx = \theta(x)f(x)\Big|_{-\infty}^{\infty} - \int_{-\infty}^{\infty} \theta(x)f'(x)dx = f(\infty) - \int_{0}^{\infty} f'(x)dx = f(0) = \int_{-\infty}^{\infty} \delta(x)f(x)dx$$

which establishes the identity (8.8.11).

Similarly, by integrating

$$\int_{-\infty}^{\infty} \delta'(x)f(x)dx$$

by parts, we find

$$\int_{-\infty}^{\infty} \delta'(x)f(x)dx = \delta(x)f(x)\Big|_{-\infty}^{\infty} - \int_{-\infty}^{\infty} \delta(x)f'(x)dx$$

and hence,

$$\int_{-\infty}^{\infty} \delta'(x)f(x)dx = -f'(0). \tag{8.8.12}$$

More generally,

$$\int_{-\infty}^{\infty} \frac{d^m \delta(x)}{d x^m} f(x)dx = (-1)^m \frac{d^m f(0)}{d x^m}. \tag{8.8.13}$$

Next, we note that

$$\int_{-\infty}^{\infty} x\delta(x)f(x)dx = 0 \quad \text{and} \quad \int_{-\infty}^{\infty} x\delta'(x)f(x)dx = f(0) = \int_{-\infty}^{\infty} \delta(x)f(x)dx$$

for arbitrary continuous functions $f(x)$, and deduce the identities

$$x\delta(x) = 0 \tag{8.8.14}$$

$$x\delta'(x) = \delta(x). \tag{8.8.15}$$

A further identity is

$$\delta(ax) = \frac{1}{|a|}\delta(x), \quad a \neq 0. \tag{8.8.16}$$

To establish it, we use the usual continuous test function $f(x)$ and evaluate

$$\int_{-\infty}^{\infty} \delta(ax)f(x)dx = \begin{cases} \int_{-\infty}^{\infty} \delta(\xi)f(\xi/a)\frac{1}{a}d\xi = \frac{1}{a}f(0), & a > 0 \\ \int_{\infty}^{-\infty} \delta(\xi)f(\xi/a)\frac{1}{a}d\xi = -\frac{1}{a}f(0), & a < 0 \end{cases} = \int_{-\infty}^{\infty} \frac{1}{|a|}\delta(x)f(x)dx.$$

Evidently, $\delta(x)$ is an even function since, from (8.8.16), we have

$$\delta(-x) = \delta(x). \tag{8.8.17}$$

This comes as no great surprise given that the δ-sequences all consist of even functions.

A useful extension of (8.8.16) is the identity

$$\delta(x^2 - a^2) = \frac{1}{2a}[\delta(x - a) + \delta(x + a)], \quad a > 0. \tag{8.8.18}$$

Whereas the argument of the δ-function on the left hand side of (8.8.16) is a linear function of x, we now have a quadratic function as an argument. The derivation has the usual starting point: we compose the integral $\int_{-\infty}^{\infty} \delta(x^2 - a^2)f(x)dx$ for arbitrary continuous $f(x)$. However, we now invoke the property that $\delta(\xi) = 0$ except at $\xi = 0$ to write

$$\int_{-\infty}^{\infty} \delta(x^2 - a^2)f(x)dx = \int_{-a-\varepsilon}^{-a+\varepsilon} \delta(x^2 - a^2)f(x)dx + \int_{a-\varepsilon}^{a+\varepsilon} \delta(x^2 - a^2)f(x)dx$$

where $0 < \varepsilon < 2a$. Introducing the new variable of integration $\xi = x^2 - a^2$, we have

$$dx = \begin{cases} d\xi/2\sqrt{\xi + a^2}, & x > 0 \\ -d\xi/2\sqrt{\xi + a^2}, & x < 0 \end{cases}$$

and so,

$$\delta(x^2 - a^2)f(x)dx = \int_{2\varepsilon a}^{-2\varepsilon a} -\delta(\xi)f(-\sqrt{\xi + a^2})\frac{d\xi}{2\sqrt{\xi + a^2}} + \int_{-2\varepsilon a}^{2\varepsilon a} \delta(\xi)f(\sqrt{\xi + a^2})\frac{d\xi}{2\sqrt{\xi + a}}$$

$$= \frac{1}{2a}[f(-a) + f(a)]$$

$$= \frac{1}{2a}\int_{-\infty}^{\infty} [\delta(x + a) + \delta(x - a)]f(x)dx$$

as required.

This last identity can be extended to cover all cases of a δ-function whose argument is a differentiable function with simple zeros. If $g(x)$ is differentiable everywhere and $g(x_n) = 0$ with $g'(x_n) \neq 0$ for a countable set of points x_n then,

$$\delta[g(x)] = \sum_n \frac{1}{|g'(x_n)|}\delta(x - x_n). \tag{8.8.19}$$

The δ-calculus can also be used to determine series and integral representations for delta functions. For example, the Fourier coefficients of $\delta(x)$ are $b_n = 0$ ($\delta(x)$ is an even function) and

$$a_n = \frac{1}{L}\int_{-L}^{L} \delta(x)\cos\frac{n\pi x}{L}dx = \frac{1}{L}.$$

Thus, the Fourier series representation of $\delta(x)$ is

$$\delta(x) = \frac{1}{2L} + \frac{1}{L} \sum_{n=1}^{\infty} \cos \frac{n\pi x}{L}. \tag{8.8.20}$$

This is a divergent series, as one might expect, but it possesses the key property

$$\int_{-L}^{L} \delta(x) f(x) dx = f(0)$$

as can be seen by multiplying the right hand side of (8.8.20) by an arbitrary continuous function $f(x)$ and integrating term by term from $-L$ to L:

$$\frac{1}{2L} \int_{-L}^{L} f(x) dx + \sum_{n=1}^{\infty} \frac{1}{L} \int_{-L}^{L} f(x) \cos \frac{n\pi x}{L} dx = \frac{a_0}{2} + \sum_{n=1}^{\infty} a_n,$$

where a_n, $n = 0, 1, 2, \ldots$ are Fourier coefficients of $f(x)$. If $f(x)$ has the Fourier series

$$f(x) = \frac{a_0}{2} + \sum_{n=1}^{\infty} \left(a_n \cos \frac{n\pi x}{L} + b_n \sin \frac{n\pi x}{L} \right),$$

then

$$\frac{a_0}{2} + \sum_{n=1}^{\infty} a_n = f(0).$$

Following exactly the same steps we can determine various other series for the delta function. For instance, if $0 < \xi < L$, the Fourier sine and cosine series for $\delta(x - \xi)$ are

$$\delta(x - \xi) = \frac{2}{L} \sum_{n=1}^{\infty} \sin \frac{n\pi \xi}{L} \sin \frac{n\pi x}{L} \tag{8.8.21}$$

and

$$\delta(x - \xi) = \frac{1}{L} + \frac{2}{L} \sum_{n=1}^{\infty} \cos \frac{n\pi \xi}{L} \cos \frac{n\pi x}{L}, \tag{8.8.22}$$

respectively.

These are called the **closure relations** for the orthonormal (orthogonal and normalized) sets $\left\{ \sqrt{\frac{2}{L}} \sin \frac{n\pi x}{L} \right\}$ and $\left\{ \sqrt{\frac{2}{L}} \cos \frac{n\pi x}{L} \right\}$ on the interval $0 \leqslant x \leqslant L$. The integral representation

$$\delta(x - \xi) = \frac{1}{2\pi} \int_{-\infty}^{\infty} e^{ik(x-\xi)} dk \tag{8.8.23}$$

that was our initial introduction to the delta function is also a closure relation. In this case, the set of orthonormal functions is $\left\{ \sqrt{\frac{1}{2\pi}}\, e^{ikx} \right\}$ and their orthogonality and normalization over the interval $-\infty < x < \infty$ is expressed by

$$\frac{1}{2\pi} \int_{-\infty}^{\infty} e^{i(k-j)x}\, dx = \delta(k-j) \tag{8.8.24}$$

where, because k and j are continuous indices, the usual Kronecker delta function has been replaced by a Dirac delta function.

Equation (8.8.23) (and (8.8.24)) can also be interpreted as stating that $\delta(x - \xi)$ and $\sqrt{\frac{1}{2\pi}}\, e^{-ik\xi}$ (and $\delta(k - j)$ and $\sqrt{\frac{1}{2\pi}}\, e^{-ijx}$) comprise Fourier transform pairs. The trigonometrical exponential does not meet the integrability condition specified in either Fourier's Integral Theorem or Plancherel's Theorem. Fortunately, however, this is another apparent impediment that has been removed by appeal to distribution theory. It has been proven that every distribution has a Fourier transform which is itself a distribution. Thus, not only the trigonometrical functions but even polynomials can have well-defined transforms through the expedient of treating them as distributions or generalized functions.

We shall conclude this Section by noting that the concept of a delta function and the δ- calculus can be extended to two or more dimensions via a corresponding extension of the concept of a distribution. Using an arbitrary continuous function of position $f(\vec{r})$ we have the defining equation

$$f(\vec{r}_0) = \int_{\text{all space}} f(\vec{r})\delta(\vec{r} - \vec{r}_0)\, d^n r. \tag{8.8.25}$$

In terms of three dimensional Cartesian , spherical polar and cylindrical polar coordinates this becomes

$$f(\vec{r}_0) = \int_{-\infty}^{\infty} \int_{-\infty}^{\infty} \int_{-\infty}^{\infty} f(x, y, z)\delta(\vec{r} - \vec{r}_0)\, dx\, dy\, dz$$

$$= \int_{0}^{\infty} \int_{-1}^{1} \int_{0}^{2\pi} f(r, \cos\theta, \varphi)\delta(\vec{r} - \vec{r}_0)\, r^2\, dr\, d(\cos\theta)\, d\varphi$$

$$= \int_{0}^{\infty} \int_{0}^{2\pi} \int_{-\infty}^{\infty} f(\rho, \varphi, z)\delta(\vec{r} - \vec{r}_0)\rho\, d\rho\, d\varphi\, dz$$

from which we deduce

$$\delta(\vec{r} - \vec{r}_0) = \delta(x - x_0)\delta(y - y_0)\delta(z - z_0) = \frac{1}{r^2}\delta(r - r_0)\delta(\cos\theta - \cos\theta_0)\delta(\varphi - \varphi_0)$$

$$= \frac{1}{\rho}\delta(\rho - \rho_0)\delta(\varphi - \varphi_0)\delta(z - z_0). \tag{8.8.26}$$

An interesting representation of $\delta(\vec{r} - \vec{r}_0)$ and one that should evoke memories of Coulomb's Law is

$$\delta(\vec{r} - \vec{r}_0) = -\frac{1}{4\pi}\nabla^2\left(\frac{1}{r}\right). \tag{8.8.27}$$

To derive this relationship we shall locate the origin of our coordinate system at $\vec{r} = \vec{r}_0$ and attempt to show that

$$\int_{\text{all space}} f(\vec{r})\nabla^2\left(\frac{1}{r}\right) d^3 r = -4\pi f(0)$$

for an arbitrary but continuous function $f(\vec{r})$. The integral over all space on the left hand side of this equation can be set equal to the integral over a sphere of volume V and surface S in the limit as the radius of the sphere $R \to \infty$. Before taking that limit, however, let us apply the divergence theorem to the vector function $f(\vec{r})\vec{\nabla}\left(\frac{1}{r}\right)$:

$$\int_V \vec{\nabla}\cdot\left[f(\vec{r})\vec{\nabla}\left(\frac{1}{r}\right)\right] d^3 r = \int_S \left[f(\vec{r})\vec{\nabla}\left(\frac{1}{r}\right)\right]\cdot d\vec{S}.$$

Thus, since

$$\vec{\nabla}\cdot\left[f(\vec{r})\vec{\nabla}\left(\frac{1}{r}\right)\right] = \vec{\nabla}f(\vec{r})\cdot\vec{\nabla}\left(\frac{1}{r}\right) + f(\vec{r})\nabla^2\left(\frac{1}{r}\right),$$

we have

$$\int_V f(\vec{r})\nabla^2\left(\frac{1}{r}\right) d^3 r = \int_S \left[f(\vec{r})\vec{\nabla}\left(\frac{1}{r}\right)\right]\cdot d\vec{S} - \int_V \vec{\nabla}f(\vec{r})\cdot\vec{\nabla}\left(\frac{1}{r}\right) d^3 r.$$

But, $\vec{\nabla}\left(\frac{1}{r}\right) = -\frac{1}{r^2}\frac{\vec{r}}{r}$ and $\frac{\vec{r}}{r}\cdot\vec{\nabla}f(\vec{r}) = \frac{\partial}{\partial r}f(\vec{r})$. Therefore,

$$\int_V f(\vec{r})\nabla^2\left(\frac{1}{r}\right) d^3 r = \int_{-1}^{1}\int_0^{2\pi} f(\vec{r})\big|_{r=R}\left(-\frac{1}{r^2}\right) r^2\, d\cos\theta d\varphi$$

$$- \int_0^R \int_{-1}^1 \int_0^{2\pi}\left[-\frac{\partial}{\partial r}f(\vec{r})\right] dr d\cos\theta d\varphi$$

or,

$$\int_V f(\vec{r})\nabla^2\left(\frac{1}{r}\right) d^3 r = -\int_{-1}^1\int_0^{2\pi} f(\vec{r})\big|_{r=R}\, d\cos\theta d\varphi + \int_{-1}^1\int_0^{2\pi} f(\vec{r})\big|_{r=0}^{r=R}\, d\cos\theta d\varphi = -4\pi f(0),$$

as required.

We shall now return to the subject of Fourier transforms.

8.9 Properties of Fourier Transforms

From the defining integral,

$$\mathcal{F}\{f(x)\} \equiv F(k) = \frac{1}{\sqrt{2\pi}} \int_{-\infty}^{\infty} f(x)\, e^{ikx}\, dx, \qquad (8.9.1)$$

it follows that for real functions $f(x)$

$$F(-k) = F^*(k). \qquad (8.9.2)$$

This is called the **conjugation property** and it has two immediate corollaries:
1. $F(k)$ is real if $f(x)$ is an even function,
2. $F(k)$ is imaginary if $f(x)$ is an odd function of x.

Multiplying $f(x)$ by e^{-ax} and then calculating the transform, we have

$$\mathcal{F}\{e^{-ax} f(x)\} = \frac{1}{\sqrt{2\pi}} \int_{-\infty}^{\infty} f(x)\, e^{i(k+ia)x}\, dx = F(k + ia) \qquad (8.9.3)$$

which is called the **attenuation property**.
Similarly, a displacement of the argument of $f(x)$ results in multiplication of $F(k)$ by a phase (rather than an attenuation) factor:

$$\mathcal{F}\{f(x-a)\} = \frac{1}{\sqrt{2\pi}} \int_{-\infty}^{\infty} f(\xi)\, e^{ik(\xi+a)}\, d\xi = e^{ika} F(k) = e^{ika}\, \mathcal{F}\{f(x)\}. \qquad (8.9.4)$$

Next, let us assume that the transform of the derivative of $f(x)$, $\mathcal{F}\{f'(x)\}$, exists. Then, integrating by parts, we have

$$\frac{1}{\sqrt{2\pi}} \int_{-\infty}^{\infty} f'(x)\, e^{ikx}\, dx = \frac{1}{\sqrt{2\pi}} f(x) e^{ikx} \Big|_{-\infty}^{\infty} - \frac{ik}{\sqrt{2\pi}} \int_{-\infty}^{\infty} f(x)\, e^{ikx}\, dx.$$

The existence of its Fourier transform implies either that $f(x) \to 0$ as $x \to \pm\infty$ or that $f(x)$ is a distribution. In both cases we lose the integrated term and obtain

$$\mathcal{F}\{f'(x)\} = -ik\mathcal{F}\{f(x)\}. \qquad (8.9.5)$$

This is called the **differentiation property** and it extends in an obvious way to higher derivatives:

$$\mathcal{F}\{f''(x)\} = -k^2\, \mathcal{F}\{f(x)\} \ldots \mathcal{F}\{f^{(n)}(x)\} = (-i)^n\, \mathcal{F}\{f(x)\}. \qquad (8.9.6)$$

As we shall see, this has immediate application in the solution of differential equations.

The converse of the differentiation property arises when we transform the product of $f(x)$ and a power of x :

$$\mathcal{F}\{xf(x)\} = \frac{1}{2\pi} \int\limits_{-\infty}^{\infty} f(x)x\, e^{ikx}\, dx = \frac{1}{2\pi} \int\limits_{-\infty}^{\infty} f(x) \left(-i\frac{d}{dk}\right) e^{ikx}\, dx = -i\frac{d}{dk}\mathcal{F}\{f(x)\}, \quad (8.9.7)$$

provided that we can interchange the order of integration and differentiation. This too can be extended to read

$$\mathcal{F}\{x^n f(x)\} = (-i)^n \frac{d^n}{dx^n}\mathcal{F}\{f(x)\}. \tag{8.9.8}$$

If $F(k) = \mathcal{F}\{f(x)\}$ and $G(k) = \mathcal{F}\{g(x)\}$, then the product $H(k) = F(k)G(k)$ is the Fourier transform of the function

$$h(x) = \frac{1}{2\pi} \int\limits_{-\infty}^{\infty} f(\xi)g(x - \xi)d\xi. \tag{8.9.9}$$

The integral in (8.9.9) is of a type that yields a **convolution** of the two functions in the integrand and so this property is called the **convolution theorem**. Its proof is straightforward:

$$\mathcal{F}^{-1}\{F(k)G(k)\} = \frac{1}{\sqrt{2\pi}} \int\limits_{-\infty}^{\infty} F(k)G(k)\, e^{-ikx}\, dk = \frac{1}{2\pi} \int\limits_{-\infty}^{\infty} F(k) \int\limits_{-\infty}^{\infty} g(\xi)\, e^{ik\xi}\, d\xi\, e^{-ikx}\, dk;$$

assuming that we can interchange the order of integrations, this becomes

$$\mathcal{F}^{-1}\{F(k)G(k)\} = \frac{1}{2\pi} \int\limits_{-\infty}^{\infty} g(\xi) \int\limits_{-\infty}^{\infty} F(k)\, e^{-ik(x-\xi)}\, dkd\xi = \frac{1}{\sqrt{2\pi}} \int\limits_{-\infty}^{\infty} g(\xi)f(x - \xi)d\xi,$$

as stated .

8.10 Fourier Sine and Cosine Transforms

Suppose that we are given an even function of x, $f(x) = f(-x)$. Its Fourier transform then reduces to

$$\mathcal{F}\{f(x)\} \equiv F(k) = \frac{1}{\sqrt{2\pi}} \int\limits_{-\infty}^{\infty} f(x)\, e^{ikx}\, dx = \sqrt{\frac{2}{\pi}} \int\limits_{0}^{\infty} f(x)\cos kx dx.$$

Evidently, $F(k)$ is, in turn, an even function of k and so we also have

$$f(x) = \mathcal{F}^{-1}\{F(k)\} = \sqrt{\frac{2}{\pi}} \int\limits_{0}^{\infty} F(k)\cos kx dk.$$

Similarly, if $f(x)$ is odd, $f(-x) = -f(x)$, then

$$F(k) = \sqrt{\frac{2}{\pi}} i \int_0^\infty f(x) \sin kx \, dx$$

and,

$$f(x) = \mathcal{F}^{-1}\{F(k)\} = \sqrt{\frac{2}{\pi}}(-i) \int_0^\infty f(x) \sin kx \, dk.$$

This suggests that just as we had Fourier cosine and sine series for functions defined on the (half-) interval $0 \leqslant x \leqslant \pi$, we can introduce **Fourier cosine and sine transforms** for functions defined only on the (half-) interval $0 \leqslant x < \infty$:

$$\mathcal{F}_c\{f(x)\} \equiv F_c(k) = \sqrt{\frac{2}{\pi}} \int_0^\infty f(x) \cos kx \, dx \tag{8.10.1}$$

with

$$\mathcal{F}_c^{-1}\{F_c(k)\} \equiv \sqrt{\frac{2}{\pi}} \int_0^\infty F_c(k) \cos kx \, dk = f(x) \tag{8.10.2}$$

and,

$$\mathcal{F}_s\{f(x)\} \equiv F_s(k) = \sqrt{\frac{2}{\pi}} \int_0^\infty f(x) \sin kx \, dx \tag{8.10.3}$$

with

$$\mathcal{F}_s^{-1}\{F_s(k)\} \equiv \sqrt{\frac{2}{\pi}} \int_0^\infty F_s(k) \sin kx \, dx = f(x). \tag{8.10.4}$$

Note that

$$\mathcal{F}_c\{f(x)\} = \mathcal{F}\{f^{(+)}(x)\}$$

and

$$\mathcal{F}_s\{f(x)\} = -i\mathcal{F}\{f^{(-)}(x)\}$$

where

$$f^{(+)}(x) = \begin{cases} f(x), & x > 0 \\ f(-x), & x < 0 \end{cases}$$

is the **symmetric** extension of $f(x)$ and

$$f^{(-)}(x) = \begin{cases} f(x), & x > 0 \\ -f(-x), & x < 0 \end{cases}$$

is the **antisymmetric** extension of $f(x)$. As a result, Fourier sine and cosine transforms possess properties very similar to those of Fourier transforms. Some differences warrant our attention however. For example,

$$\mathcal{F}_c\{f'(x)\} = \sqrt{\frac{2}{\pi}} \int\limits_0^\infty f'(x) \cos kx\,dx = \sqrt{\frac{2}{\pi}} f(x) \cos kx \Big|_0^\infty + \sqrt{\frac{2}{\pi}} k \int\limits_0^\infty f(x) \sin kx\,dx.$$

As before, we can assume that $f(x) \to 0$ as $x \to \infty$ (or we can treat it as a distribution) and so obtain

$$\mathcal{F}_c\{f'(x)\} = -\sqrt{\frac{2}{\pi}} f(0) + k\,\mathcal{F}_s\{f(x)\}. \tag{8.10.5}$$

Similarly, integrating by parts twice and assuming that both $f'(x) \to 0$ and $f(x) \to 0$ as $x \to \infty$, we find

$$\mathcal{F}_c\{f''(x)\} = -\sqrt{\frac{2}{\pi}} f'(0) - k^2\,\mathcal{F}_c\{f(x)\}. \tag{8.10.6}$$

The corresponding relationships for Fourier sine transforms are

$$\mathcal{F}_s\{f'(x)\} = -k\,\mathcal{F}_c\{f(x)\} \tag{8.10.7}$$

and

$$\mathcal{F}_s\{f''(x)\} = \sqrt{\frac{2}{\pi}} kf(0) - k^2\,\mathcal{F}_s\{f(x)\}. \tag{8.10.8}$$

It should be noted that transforming derivatives of **even** order yields a transform of the undifferentiated function **and** it is a transform of the **same** type. On the other hand, transforms of derivatives of **odd** order result in a transform of the **other** type. This has immediate consequences for the application of Fourier sine and cosine transforms in the solution of differential equations: the equations must contain derivatives of only even or only odd order to avoid mixing the two types of transform. Another way in which these differentiation properties influence the application to differential equations is the "boundary condition" information they require: a knowledge of $f(0)$ in the case of sine transforms and of $f'(0)$ for cosine transforms.

The interrelation between Fourier sine and cosine transforms surfaces again in their convolution theorems. If $F_c(k) = \mathcal{F}_c\{f(x)\}$ and $G_c(k) = \mathcal{F}_c\{g(x)\}$, we have

$$\mathcal{F}_c^{-1}\{F_c(k)\,G_c(k)\} = \sqrt{\frac{2}{\pi}} \int\limits_0^\infty F_c(k)\,G_c(k) \cos kx\,dk$$

$$= \frac{2}{\pi} \int\limits_0^\infty F_c(k) \int\limits_0^\infty g(\xi) \cos k\xi \cos kx\,d\xi\,dk.$$

But, $\cos kx \cos k\xi = \frac{1}{2}[\cos k(x - \xi) + \cos k(x + \xi)]$. Thus, substituting and then interchanging the order of integration, we obtain

$$\mathcal{F}_c^{-1}\{F_c(k)\,G_c(k)\} = \frac{1}{\pi} \int_0^\infty g(\xi) \int_0^\infty F_c(k)[\cos k(x - \xi) + \cos k(x + \xi)]dkd\xi$$

$$= \frac{1}{\sqrt{2\pi}} \int_0^\infty g(\xi)[f^{(+)}(x - \xi) + f(x + \xi)]d\xi, \qquad (8.10.9)$$

where $f^{(+)}$ appears because $(x - \xi) < 0$ when $x \leqslant \xi < \infty$.

The same operations performed on the Fourier sine transform requires use of

$$\sin kx \sin k\xi = \frac{1}{2}[\cos k(x - \xi) - \cos k(x + \xi)]$$

which leads to

$$\mathcal{F}_s^{-1}\{F_s(k)\,G_s(k)\} = \frac{1}{\pi} \int_0^\infty g(\xi) \int_0^\infty F_s(k)[\cos k(x - \xi) - \cos k(x + \xi)]dkd\xi.$$

The sine transform $F_s(k)$ is now paired with cosine functions and so does not become inverted. Instead, we have to define a new function $f^{(\sim)}(x) = \mathcal{F}_c^{-1}\{F_s(k)\}$ in terms of which the convolution theorem becomes

$$\mathcal{F}_s^{-1}\{F_s(k)\,G_s(k)\} = \frac{1}{\sqrt{2\pi}} \int_0^\infty g(\xi)[f^{(\sim)}(x - \xi) - f^{(\sim)}(x + \xi)]d\xi. \qquad (8.10.10)$$

It only remains to explore applications of the three types of Fourier transforms, the most important of which are in the solution of differential equations. However, before we do so we shall introduce another and closely related integral transform, that of Laplace.

8.11 Laplace Transforms

Definition: A function $f(x)$ is said to be of **exponential order** σ if σ is the largest real number such that $|e^{-\sigma x} f(x)|$ is bounded on $0 \leqslant x < \infty$. In other words, $f(x)$ does not increase faster than $e^{\sigma x}$ as $x \to \infty$.

The Fourier transform of a function of non-zero exponential order will not exist because $|f(x)|$ will be unbounded (even in the sense of a polynomial bound) at either ∞ or $-\infty$ depending on the sign of σ. Therefore, for such functions, we form the product

$$g(x) = f(x)\,e^{-cx}\,\theta(x), \quad \text{where } c > \sigma \text{ and } \theta(x) = \begin{cases} 1, & x > 0 \\ 0, & x < 0 \end{cases}.$$

We are now assured of the convergence of the transform of this function and so define

$$G(k) \equiv \mathcal{F}\{f(x)\, e^{-cx}\, \theta(x)\} = \frac{1}{\sqrt{2\pi}} \int_0^\infty f(x)\, e^{-cx}\, e^{ikx}\, dx \qquad (8.11.1)$$

with

$$\mathcal{F}^{-1}\{G(k)\} = \frac{1}{\sqrt{2\pi}} \int_{-\infty}^\infty G(k)\, e^{-ikx}\, dk = f(x)\, e^{-cx}\, \theta(x). \qquad (8.11.2)$$

Let us introduce a new transform variable $s = c - ik$, $ds = -i\,dk$ and set

$$F(s) = \sqrt{2\pi}\,G(k).$$

Then, equations (8.11.1) and (8.11.2) become

$$F(s) = \int_0^\infty f(t)\, e^{-st}\, dt \equiv \mathcal{L}\{f(t)\} \qquad (8.11.3)$$

$$f(t)\theta(t) = \frac{1}{2\pi i} \int_{c-i\infty}^{c+i\infty} F(s)\, e^{st}\, ds \equiv \mathcal{L}^{-1}\{F(s)\} \qquad (8.11.4)$$

where, to conform with convention, we have replaced the symbol x for the independent variable with the letter t. These are the defining equations for the **Laplace transform** and its inverse. (The integral in (8.11.4) is often referred to as the **Mellin inversion integral**.) Evidently, the Laplace transform offers a means of extending the applicability of Fourier transforms to functions for which the Fourier integral is not defined. As such, they are widely used in the solution of engineering problems.

Not surprisingly, the properties of Laplace transforms are close analogs of those of Fourier transforms. Specifically, there is
–   an **attenuation property,**

$$\mathcal{L}\{e^{-at} f(t)\} = F(s + a) \quad \text{where} \quad F(s) = \mathcal{L}\{f(t)\}; \qquad (8.11.5)$$

–   a **shifting property,**

$$\mathcal{L}\{f(t - a)\theta(t - a)\} = e^{-as}\, \mathcal{L}\{f(t)\}, \quad a > 0; \qquad (8.11.6)$$

–   the **derivative property,**

$$\mathcal{L}\{f'(t)\} = s\mathcal{L}\{f(t)\} - f(0) \qquad (8.11.7)$$

which extends to

$$\mathcal{L}\{f''(t)\} = s\mathcal{L}\{f'(t)\} - f'(0) = s^2\, \mathcal{L}\{f(t)\} - sf(0) - f'(0), \qquad (8.11.8)$$

and by induction to

$$\mathcal{L}\{f^{(n)}(t)\} = s^n \, \mathcal{L}\{f(t)\} - \sum_{k=1}^{n} s^{k-1} f^{(n-k)}(0), \qquad (8.11.9)$$

and is readily established by a single integration by parts,

$$\int_0^\infty e^{-st} f(t) dt = -\frac{1}{s} e^{-st} f(t)\big|_0^\infty + \int_0^\infty e^{-st} f'(t) dt$$

$$\text{or } s \int_0^\infty e^{-st} f(t) dt = f(0) + \int_0^\infty e^{-st} f'(t) dt;$$

- multiplication by a power of t property,

$$\mathcal{L}\{tf(t)\} = -\frac{d}{ds} \mathcal{L}\{f(t)\}$$

$$\mathcal{L}\{t^n f(t)\} = (-1)^n \frac{d^n}{ds^n} \mathcal{L}\{f(t)\}; \qquad (8.11.10)$$

- and the **convolution theorem**,

$$\mathcal{L}^{-1}\{F(s)G(s)\} = \int_0^t f(\tau) g(t - \tau) d\tau \qquad (8.11.11)$$

if $F(s) = \mathcal{L}\{f(t)\}$ and $G(s) = \mathcal{L}\{g(t)\}$.

The most important application of Laplace transforms is in the solution of differential equations, especially linear differential equations. As with the use of Fourier transforms, the method consists of transforming a given differential equation to yield a **subsidiary equation** which, if the choice of method is appropriate, is an easier equation to solve. In fact, if the differential equation is linear, the subsidiary equation is **algebraic** and so is solvable by purely algebraic techniques. The choice of Laplace rather than some variety of Fourier transform hinges on the boundary conditions associated with the differential equation. Specifically, Laplace transforms are appropriate to boundaries at $t = 0$ and ∞ with a knowledge of the values of the solution **and** its first derivative at the first of these. The final step in this method is to invert the transform obtained as a solution of the subsidiary equation. For Laplace transforms this can be done by using the Mellin inversion integral. More commonly, however, one makes use of a knowledge of a few key Laplace transform pairs sufficient to invert any rational function of s, or if the solution of the subsidiary equation is more complicated, one consults a comprehensive table of Laplace transform pairs.

The inversion of a rational function proceeds as follows. Let the function be $Y(s) = P(s)/Q(s)$ where $P(s)$ and $Q(s)$ are polynomials with $\deg(P) > \deg(Q)$. If $Q(s)$ has n

simple real roots r_i, $i = 1, 2, \ldots, n$, we can express $Y(s)$ in terms of the corresponding partial fractions:

$$Y(s) = \sum_{i=1}^{n} \frac{c_i}{s - r_i} + W(s)$$

where $W(s)$ is the sum of partial fractions associated with all the **other** roots of $Q(s)$. The constants c_i can be found by multiplying both sides of this equation by $(s - r_i)$ and taking the limit as $s \to r_i$:

$$c_i = \lim_{s \to r_i} (s - r_i)Y(s) = \frac{P(r_i)}{Q'(r_i)}.$$

Thus, using $\mathcal{L}\{e^{r_i t}\} = \dfrac{1}{s - r_i}$, we have

$$y(t) = \mathcal{L}^{-1}\{Y(s)\} = \sum_{i=1}^{n} \frac{P(r_i)}{Q'(r_i)} e^{r_i t} + \mathcal{L}^{-1}\{W(s)\}. \tag{8.11.12}$$

Example: Suppose that we seek $\mathcal{L}^{-1}\left\{\dfrac{s + 1}{s^3 + s^2 - 6s}\right\}$. In terms of partial fractions, we have

$$\frac{s + 1}{s^3 + s^2 - 6s} = \frac{s + 1}{s(s - 2)(s + 3)} = \frac{c_1}{s} + \frac{c_2}{s - 2} + \frac{c_3}{s + 3}$$

where

$$c_1 = P(0)/Q'(0) = \frac{1}{3s^2 + 2s - 6)|_{s=0}} = -\frac{1}{6}$$

$$c_2 = P(2)/Q'(2) = \frac{2}{3 \cdot 4 + 4 - 6} = \frac{3}{10}$$

$$c_3 = P(-3)/Q'(-3) = \frac{-2}{3 \cdot 9 - 6 - 6} = -\frac{2}{15}.$$

Therefore,

$$y(t) = -\frac{1}{6} + \frac{3}{10} e^{2t} - \frac{2}{15} e^{-3t}.$$

If $Q(s)$ has a real root r of multiplicity m, $Y(s)$ will have a partial fraction decomposition of the form

$$Y(s) = \frac{c_m}{(s - r)^m} + \frac{c_{m-1}}{(s - r)^{m-1}} + \ldots + \frac{c_1}{s - r} + W(s)$$

where $W(s)$ again denotes the sum of the partial fractions associated with the other roots of $Q(s)$. To determine c_m we multiply both sides of this equation by $(s - r)^m$:

$$G(s) \equiv (s - r)^m Y(s) = c_m + (s - r) c_{m-1} + \ldots + (s - r)^{m-1} c_1 + (s - r)^m W(s).$$

Setting $s = r$, we obtain

$$c_m = G(r).$$

To determine c_{m-1} we differentiate $G(s)$,

$$G'(s) = c_{m-1} + 2(s - r) c_{m-2} + \ldots + m(s - r)^{m-1} W(s) + (s - r)^m W'(s),$$

and set $s = r$ to obtain

$$c_{m-1} = G'(r).$$

Continuing in this fashion up to and including the $(m - 1)^{th}$ derivative, we find

$$c_{m-2} = \frac{1}{2!} G''(r), \ldots, c_k = \frac{1}{(m-k)!} G^{(m-k)}(r), \ldots, c_1 = \frac{1}{(m-1)!} G^{(m-1)}(r).$$

Now, $\mathcal{L}^{-1} \left\{ \frac{1}{s^k} \right\} = \frac{t^{k-1}}{(k-1)!}$ and so, by the attenuation property,

$$\mathcal{L}^{-1} \left\{ \frac{1}{(s-r)^k} \right\} = \frac{e^{rt} t^{k-1}}{(k-1)!}.$$

Therefore,

$$y(t) = \mathcal{L}^{-1}\{Y(s)\} = \left[\frac{c_m t^{m-1}}{(m-1)!} + \frac{c_{m-1} t^{m-2}}{(m-2)!} + \ldots + c_2 t + c_1 \right] e^{rt} + \mathcal{L}^{-1}\{W(s)\}$$

$$= e^{rt} \sum_{k=1}^{m} \frac{G^{(m-k)}(r)}{(k-1)!(m-k)!} t^{k-1} + \mathcal{L}^{-1}\{W(s)\}. \tag{8.11.13}$$

Example: Suppose that we wish to invert $\frac{s+2}{s^5 - 2 s^4 + s^3}$. This rational function has the partial fraction decomposition

$$\frac{s+2}{s^5 - 2 s^4 + s^3} = \frac{a_3}{s^3} + \frac{a_2}{s^2} + \frac{a_1}{s} + \frac{b_2}{(s-1)^2} + \frac{b_1}{s-1}.$$

To determine the coefficients a_i, we define $G(s) = s^3 Y(s) = \frac{s+2}{(s-1)^2}$. Then,

$$a_3 = G(0) = 2, \quad a_2 = G'(0) = 5, \quad a_3 = \frac{G''(0)}{2!} = 8.$$

To determine the b_i, define $H(s) = (s-1)^2 Y(s) = \frac{s+2}{s^3}$. Then,

$$b_2 = H(1) = 3 \quad \text{and} \quad b_1 = H'(1) = -8.$$

Thus,

$$y(t) = \mathcal{L}^{-1}\{Y(s)\} = 2 t^2 + 5t + 8 + e^t(3t - 8).$$

If $Q(s)$ has real coefficients, any complex roots it may have will occur in conjugate pairs: $r = \alpha + i\beta$ and $r^* = \alpha - i\beta$. If $P(s)$ also has real coefficients, the partial fractions associated with these roots will have complex conjugate coefficients since

$$\frac{P(r^*)}{Q'(r^*)} = \left[\frac{P(r)}{Q'(r)} \right]^* = c^*.$$

Thus, using $e^{rt} = e^{\alpha t}[\cos \beta t + i \sin \beta t]$ and equation (8.11.12), we deduce for the case of n pairs of **simple** complex roots that

$$y(t) = 2 \sum_{j=1}^{n} e^{\alpha_j t} \left[\mathrm{Re} \left(\frac{P(r_j)}{Q'(r_j)} \right) \cos \beta_j t - \mathrm{Im} \left(\frac{P(r_j)}{Q'(r_j)} \right) \sin \beta_j t \right] + \mathcal{L}^{-1}\{W(s)\} \quad (8.11.14)$$

where $r_j = \alpha_j + i\beta_j$ is a member of the jth pair.

For a pair of roots with multiplicity m, we invoke (8.11.13) rather than (8.11.12) and obtain

$$y(t) = \sum_{k=1}^{m} \frac{2\,t^{k-1}}{(k-1)!} \, e^{\alpha t} [\mathrm{Re}\, a_k \cos \beta t - \mathrm{Im}\, a_k \sin \beta t] + \mathcal{L}^{-1}\{W(s)\} \quad (8.11.15)$$

where $a_k = \dfrac{1}{(m-1)!} G^{(m-k)}(r)$, $G(s) = (s-r)^m Y(s)$, and $r = \alpha + i\beta$.

Example: Consider the rational function $Y(s) = \frac{2s}{s^2 + 2s + 5}$. Its denominator has roots at $s = -1 \pm 2i$. Thus, setting $r = \alpha + i\beta = -1 + 2i$, we have

$$\frac{P(s)}{Q'(s)}\Big|_{s=r} = \frac{s}{s+1}\Big|_{s=r} = \frac{-1+2i}{2i} = 1 + \frac{i}{2}$$

and

$$y(t) = e^{-t}(2 \cos 2t - \sin 2t).$$

8.12 Application: Solving Differential Equations

One of the most important applications of integral transfoms is in the solution of **boundary value problems.** These are problems that seek the one solution of a given differential equation that satisfies certain conditions at the boundaries of the interval of variation of the independent variable. The conditions can involve specification of the value of the solution, of its first derivative, or of some linear combination of the two.

Because of the simple form assumed by their respective derivative properties, (which is due to the exponential nature of their kernels), Fourier and Laplace transforms are best suited to problems involving a differential equation with constant coefficients. Such equations become transformed into **subsidiary equations** that contain no derivatives and admit simple algebraic solutions. The latter are then subjected to an inverse transformation to produce the solutions that are appropriate to the problems in which the differential equations occur.

Since each transform requires a unique set of input or boundary condition information, one must be careful in the selection of one to use in the solution of a particular problem. It is not sufficient that the problem involves a differential equation with constant coefficients, it must have the correct range, $[0, \infty)$ or $(-\infty, \infty)$, for the independent variable and must contain boundary conditions that match the input needs of either Laplace, Fourier sine, Fourier cosine or Fourier transforms. These are summarized in the Table below (Table 8.1).

To illustrate these points we begin with the application of the Laplace transformation to the linear differential equation

$$y''(t) + ay'(t) + by(t) = r(t)$$

where a and b are known constants and $r(t)$ is a known function. Using the derivative property, equations (8.11.7) and (8.11.8), we obtain the **subsidiary equation**

$$s^2 Y(s) - sy(0) - y'(0) + a(sY(s) - y(0)) + bY(s) = R(s)$$

where $Y(s) = \mathcal{L}\{y(t)\}$ and $R(s) = \mathcal{L}\{r(t)\}$. Its solution is

$$Y(s) = \frac{(s + a)y(0) + y'(0)}{s^2 + as + b} + \frac{R(s)}{s^2 + as + b}.$$

Table 8.1: Applicability Conditions of Integral Transforms

Transformation	Range of Independent Variable	Boundary Conditions
Fourier	$-\infty < x < \infty$	$y(\pm\infty) = y'(\pm\infty) = 0$
Fourier sine	$0 \leqslant x < \infty$	$y(0) = c_1, y(\infty) = y'(\infty) = 0$
Fourier cosine	$0 \leqslant x < \infty$	$y'(0) = c_2, y(\infty) = y'(\infty) = 0$
Laplace	$0 \leqslant t < \infty$	$y(0) = c_1, y'(0) = c_2$

The next and final step is to determine the inverse transform $\mathcal{L}^{-1}\{Y(s)\} = y(t)$ and thus obtain the solution of the differential equation. The inverse of the first term in our expression for $Y(s)$ is a solution of the corresponding homogeneous equation and is called the **complementary function**. It matches the boundary conditions (or initial conditions if t is a time variable) $y(0) = c_1$ and $y'(0) = c_2$ that have to accompany the differential equation if this method is to be useful. The inverse of the second term yields a particular solution of the non-homogeneous equation corresponding to the conditions $y(0) = y'(0) = 0$. It is called a **particular integral**. Note that the first term as well as $\frac{1}{s^2 + as + b}$ are rational functions. They can be inverted by means of the techniques identified at the end of the last Section. The second term can then be inverted by application of the **convolution theorem**.

Examples: We shall start by paying a return visit to the forced, damped harmonic oscillator

$$m\frac{d^2 x}{d t^2} + r\frac{dx}{dt} + kx = f(t), \quad 0 \leqslant t < \infty$$

or,

$$\frac{d^2 x}{d t^2} + 2\lambda\frac{dx}{dt} + \omega_0^2 x = \frac{1}{m}f(t) \text{ where } \lambda = \frac{r}{2m} \text{ and } \omega_0^2 = \frac{k}{m}.$$

If the initial conditions specify both $x(0)$ and $x'(0)$, $x(0) = x_0$ and $x'(0) = v_0$ for example, it is appropriate to use Laplace transforms to solve for the motion of the oscillator. The solution of the resulting subsidiary equation is

$$X(s) = \frac{2\lambda x_0 + v_0 + s x_0}{s^2 + 2\lambda s + \omega_0^2} + \frac{F(s)}{s^2 + 2\lambda s + \omega_0^2} = X_c(s) + X_p(s).$$

The first term can be expressed as

$$X_c(s) = \frac{x_0(s + \lambda) + (v_0 + \lambda x_0)}{(s + \lambda)^2 + (\omega_0^2 - \lambda^2)}.$$

Assuming that $\omega_0^2 > \lambda^2$, which is usually the case, this has the inverse transform

$$x_c(t) = x_0 e^{-\lambda t} \cos \sigma t + \frac{v_0 + \lambda x_0}{\sigma} e^{-\lambda t} \sin \sigma t, \quad \sigma = \sqrt{\omega_0^2 - \lambda^2}.$$

This is the complementary function, or solution of the homogeneous differential equation, that satisfies the initial conditions that have been imposed in this problem.

To invert $X_p(s)$, we note that

$$\mathcal{L}^{-1}\{F(s)\} = f(t) \text{ and } \mathcal{L}^{-1}\left\{\frac{1}{s^2 + 2\lambda s + \omega_0^2}\right\} = \frac{1}{\sigma} e^{-\lambda t} \sin \sigma t, \sigma = \sqrt{\omega_0^2 - \lambda^2}.$$

Thus, using the convolution theorem, we can write

$$x_p(t) = \mathcal{L}^{-1}\left\{\frac{F(s)}{s^2 + 2\lambda s + \omega_0^2}\right\} = \int_0^t \frac{1}{\sigma} e^{-\lambda(t-\tau)} \sin \sigma(t - \tau) f(\tau) d\tau$$

so that the complete solution to our problem reads

$$x(t) = x_c(t) + x_p(t) = x_0 e^{-\lambda t} \cos \sigma t + \frac{v_0 + \lambda x_0}{\sigma} e^{-\lambda t} \sin \sigma t + \frac{1}{m\sigma} \int_0^t e^{-\lambda(t-\tau)} \sin \sigma(t-\tau) f(\tau) d\tau.$$

To illustrate what happens at resonance, let us assume for simplicity that the damping is negligible so that $\lambda \approx 0$ and $\sigma \approx \omega_0$. Then, if $f(t) = K \sin \omega_0 t$, where K is a constant, we have

$$x(t) = x_0 \cos \omega_0 t + \frac{v_0}{\omega_0} \sin \omega_0 t + \frac{K}{m \omega_0} \int_0^t \sin \omega_0(t - \tau) \sin \omega_0 \tau d\tau.$$

But,

$$\int_0^t \sin \omega_0 \tau \sin \omega_0(t - \tau) d\tau = \sin \omega_0 t \int_0^t \sin \omega_0 \tau \cos \omega_0 \tau d\tau - \cos \omega_0 t \int_0^t \sin^2 \omega_0 \tau d\tau$$

$$= \frac{1}{4 \omega_0} \left[\sin \omega_0 t(1 - \cos 2 \omega_0 t) - \cos \omega_0 t(2 \omega_0 t - \sin 2 \omega_0 t)\right]$$

$$= \frac{1}{2 \omega_0}(\sin \omega_0 t - \omega_0 t \cos \omega_0 t).$$

This means that the last term in the particular integral

$$x_p(t) = \frac{K}{2m\,\omega_0^2}(\sin \omega_0\, t - \omega_0\, t \cos \omega_0\, t)$$

has an amplitude that increases linearly with t corresponding to a resonant response by the system. The resonance is due, of course, to the frequency of the applied force coinciding with the natural frequency of the oscillator.

Suppose that instead of an harmonic force, the oscillator is subjected to an instantaneous impulse of magnitude I at time $t = t_0$. The force responsible for the impulse must be expressible as $f(t) = I\delta(t - t_0)$. Rather than use the convolution theorem, we note that the transform of $f(t)$ is particularly simple: $\mathcal{L}\{f(t)\} = e^{-s\,t_0}$. Thus,

$$X_p(s) = \frac{I}{m}\,\frac{e^{-s\,t_0}}{s^2 + 2\lambda s + \omega_0^2}.$$

But, according to equation (8.10.6), (the shifting property of Laplace transforms), the inverse of this product is just $\mathcal{L}^{-1}\{e^{s\,t_0}\,X_p(s)\} \cdot \theta(t)$ with t replaced by $t - t_0$:

$$\mathcal{L}^{-1}\{X_p(s)\} = x_p(t) = \frac{I}{m\sigma}\,e^{-\lambda(t-t_0)}\sin \sigma(t - t_0)\theta(t - t_0).$$

Therefore, the motion of the oscillator is given by

$$x(t) = x_0\, e^{-\lambda t}\cos \sigma t + \frac{v_0 + \lambda x_0}{\sigma}\, e^{-\lambda t}\sin \sigma t + \frac{I}{m\sigma}\, e^{-\lambda(t-t_0)}\sin \sigma(t - t_0)\theta(t - t_0)$$

demonstrating explicitly that whatever motion is initiated at $t = 0$ (by assignment of values to x_0 and v_0) it is modified at $t = t_0$ and thereafter by the motion caused by the impulse.

Another example along the same line is provided by an LRC-series consisting of an inductance L, resistance R and capacitance C connected in series to a switch and an emf $e(t)$. The switch is closed from $t = 0$ to $t = T$. We seek the current $i(t)$ in the circuit assuming that it is zero at $t = 0$ along with the charge $q(t)$ on the capacitor: $i(0) = 0$ and $q(0) = 0$.

The current is governed by Kirchoff's Law which requires that

$$L\frac{di(t)}{dt} + Ri(t) + \frac{q(t)}{C} = e(t)$$

where we shall assume

$$e(t) = \begin{cases} e_0, & 0 < t < T \\ 0, & t > T. \end{cases}$$

Moreover , we know that $i(t) = \dfrac{dq(t)}{dt}$.

Applying the Laplace transform to these two differential equations we obtain

$$LsI(s) - Li(0) + \frac{1}{C}Q(s) = E(s), \quad I(s) = sQ(s) - q(0).$$

Thus, using our initial conditions and solving for $I(s)$, we find

$$I(s) = \frac{E(s)}{Ls + R + 1/(sC)} = \frac{1}{L} \frac{sE(s)}{s^2 + (R/L)s + 1/(LC)}$$

where

$$E(s) = e_0 \int_0^T e^{-st}\, dt = e_0 \frac{1 - e^{-sT}}{s}.$$

Therefore,

$$I(s) = \frac{e_0}{L} \frac{1 - e^{-sT}}{s^2 + (R/L)s + 1/(LC)}.$$

The shifting property of Laplace transforms (equation (8.11.6)) takes care of the factor e^{-sT} in the numerator of the expression for $I(s)$ and so all that remains is to find the inverse of $\{s^2 + (R/L)s + 1/(LC)\}^{-1}$. Three cases arise depending on the roots of this quadratic and to classify them, we introduce the constants $\alpha = R/2L$ and $\omega^2 = 1/LC - R^2/4L^2$. Then, if $\omega^2 > 0$, the roots of the quadratic are complex,

$$\mathcal{L}^{-1}\left\{\frac{1}{s^2 + (R/L)s + 1/LC}\right\} = \frac{1}{\omega} e^{-\alpha t} \sin \omega t$$

and

$$i(t) = \frac{e_0}{\omega L} e^{-\alpha t} \sin \omega t - \frac{e_0}{\omega L} e^{-\alpha(t-T)} \sin \omega(t - T)\theta(t - T).$$

This is called the oscillatory case.

If $0 > \omega^2 = -\beta^2$, the roots are real and, replacing ω by $i\beta$, we have

$$i(t) = \frac{e_0}{\beta L} e^{-\alpha t} \sinh \beta t - \frac{e_0}{\beta L} e^{-\alpha(t-T)} \sinh \beta(t - T)\theta(t - T).$$

This is called the overdamped case.

Finally, if $\omega^2 = 0$, there is a double root and so

$$i(t) = \frac{e_0}{L} t\, e^{-\alpha t} - \frac{e_0}{L}(t - T)\, e^{-\alpha(t-T)}\, \theta(t - T).$$

This is called the critically damped case.

The next example involves a differential equation with variable coefficients and reveals some of the limitations of the Laplace transform method of solution. The differential equation is (Bessel's equation of order zero)

$$xy'' + y' + xy = 0$$

and the boundary conditions that we wish to impose are $y(0) = 1$, $y'(0) = 0$.

If $Y(s) = \mathcal{L}\{y(x)\}$, equation (8.11.10) tells us that $\mathcal{L}\{xy(x)\} = -\dfrac{dY(s)}{ds}$. Notice that if the coefficient of $y(x)$ were x^2, we would obtain the **second** derivative of $Y(s)$ and the subsidiary equation would be another second order differential equation thus doing little to advance the solution of the equation we started with.

Using the boundary conditions together with the derivative property, we also have

$$\mathcal{L}\{y'(x)\} = sY(s) - 1$$
$$\mathcal{L}\{y''(x)\} = s^2 Y(s) - s.$$

Thus, invoking (8.11.10) again, we obtain

$$\mathcal{L}\{x^2 y(x)\} = -2sY(s) - s^2 \frac{dY(s)}{ds} + 1$$

and so the subsidiary equation is

$$-2sY(s) - s^2 Y'(s) + 1 + sY(s) - 1 - Y'(s) = 0,$$

or

$$(s^2 + 1)Y'(s) + sY(s) = 0.$$

The solution of this differential equation is easy to find:

$$\ln Y(s) = -\int \frac{s}{s^2 + 1} ds = -\frac{1}{2} \ln(s^2 + 1) + \text{cnst.}$$

and so, $Y(s) = \dfrac{c}{\sqrt{s^2 + 1}}$ where c is a constant.

To find $y(x)$, we expand $Y(s)$ in inverse powers of s (a Laurent series valid for $|s| > 1$):

$$Y(s) = \frac{c}{s}\left(1 + \frac{1}{s^2}\right)^{-1/2} = c \sum_{n=0}^{\infty} \frac{(-1)^n (2n)!}{2^{2n}(n!)^2} \frac{1}{s^{2n+1}}.$$

Inverting term be term, we find

$$y(x) = c \sum_{n=0}^{\infty} \frac{(-1)^n x^{2n}}{2^{2n}(n!)^2} \quad \text{since} \quad \mathcal{L}^{-1}\left\{\frac{1}{s^{2n+1}}\right\} = \frac{x^{2n}}{(2n)!}.$$

But $y(0) = 1$. Therefore, $c = 1$, and our solution becomes

$$y(x) = \sum_{n=0}^{\infty} \frac{(-1)^n}{(n!)^2} \left(\frac{x}{2}\right)^{2n}$$

which is called the Bessel function of order zero and is conventionally denoted by $J_0(x)$.

As we shall see in the next Chapter, this differential equation has a second linearly independent solution which has an essential singularity at $x = 0$. Since the Laplace transform method requires well-defined values for both $y(0)$ and $y'(0)$, it is useless if that is the solution we seek.

To furnish examples of the application of Fourier transforms to the solution of differential equations, we return to the problem of an harmonic oscillator acted on by an external force. Using the same notation as before, the equation of motion is

$$\frac{d^2 x}{dt^2} + 2\lambda \frac{dx}{dt} + \omega_0^2 x = \frac{1}{m}f(t)$$

where t is presumed now to have the range $-\infty < t < \infty$. If we presume also that $x(t) \to 0$ as $t \to \pm\infty$ so that it possesses a Fourier transform, we can transform the differential equation to obtain the subsidiary equation

$$- \omega^2 X(\omega) - 2\lambda\omega i X(\omega) + \omega_0^2 X(\omega) = \frac{1}{m} F(\omega)$$

where $X(\omega) = \mathcal{F}\{x(t)\}$ and $F(\omega) = \mathcal{F}\{f(t)\}$. This has the solution

$$X(\omega) = \frac{1}{m} \frac{F(\omega)}{(\omega_0^2 - \omega^2) - 2\lambda\omega i}$$

and so the solution of the original problem is

$$x(t) = \frac{1}{\sqrt{2\pi}} \int_{-\infty}^{\infty} \frac{1}{m} \frac{F(\omega) e^{-i\omega t}}{(\omega_0^2 - \omega^2) - 2\lambda\omega i} d\omega.$$

In most cases this integral can be evaluated by residue calculus. To illustrate how, we shall take $\omega_0 > \lambda$, which corresponds to a weakly damped oscillator, and assume a force of the form

$$f(t) = \begin{cases} f_0, & |t| < \tau \\ 0, & |t| \geqslant \tau \end{cases}.$$

We then have

$$F(\omega) = \frac{f_0}{\sqrt{2\pi}} \int_{-\tau}^{\tau} e^{i\omega t} dt = f_0 \sqrt{\frac{2}{\pi}} \frac{\sin \omega t}{\omega}$$

and so,

$$x(t) = -\frac{f_0}{m\pi} \int_{-\infty}^{\infty} \frac{\sin \omega\tau \, e^{-i\omega t}}{\omega(\omega - \omega_1)(\omega - \omega_2)} d\omega$$

where $\omega_1 = \sigma - \lambda i$, $\omega_2 = -\sigma - \lambda i$, $\sigma = \sqrt{\omega_0^2 - \lambda^2}$.

Expressing $\sin \omega\tau$ in terms of exponentials and deforming the contour to avoid introducing an extraneous singularity at $\omega = 0$, we can rewrite $x(t)$ as

$$x(t) = -\frac{f_0}{m2\pi i} \int_{-\cup\to} \frac{e^{-i\omega(t-\tau)}}{\omega(\omega - \omega_1)(\omega - \omega_2)} d\omega + \frac{f_0}{m2\pi i} \int_{-\cup\to} \frac{e^{-i\omega(t+\tau)}}{\omega(\omega - \omega_1)(\omega - \omega_2)} d\omega$$

where the subscript on the integral signs indicates that we are going below the real axis in the neighbourhood of $\omega = 0$. We shall evaluate each term separately beginning with the first.

If $t - \tau > 0$, we must close the contour of the first integral in the lower half plane to be able to use the residue theorem. If $t - \tau < 0$, we close in the upper half plane. Thus,

$$\text{the first term} = \begin{cases} \dfrac{f_0}{m} \dfrac{e^{-i\omega_1(t-\tau)}}{\omega_1(\omega_1 - \omega_2)} - \dfrac{f_0}{m} \dfrac{e^{-i\omega_2(t-\tau)}}{\omega_2(\omega_1 - \omega_2)}, & t > \tau \\ -\dfrac{f_0}{m} \dfrac{1}{\omega_1 \omega_2}, & t < \tau. \end{cases}$$

Similarly, the second integral's contour must be closed in the lower half-plane if $t + \tau > 0$ and in the upper half-plane if $t + \tau < 0$. Thus,

$$
\text{the second term} = \begin{cases} -\dfrac{f_0}{m}\dfrac{e^{-i\omega_1(t+\tau)}}{\omega_1(\omega_1 - \omega_2)} + \dfrac{f_0}{m}\dfrac{e^{-i\omega_2(t+\tau)}}{\omega_2(\omega_1 - \omega_2)}, & t > -\tau \\ \dfrac{f_0}{m}\dfrac{1}{\omega_1\omega_2}, & t < -\tau. \end{cases}
$$

This provides us with three cases:

1. when $t < -\tau$, we have

$$
x(t) = -\frac{f_0}{m}\frac{1}{\omega_1\omega_2} + \frac{f_0}{m}\frac{1}{\omega_1\omega_2} = 0
$$

which confirms that the damped oscillator is at rest until subjected to the external force;

2. when $-\tau < t < \tau$, the displacement is

$$
x(t) = -\frac{f_0}{m}\frac{1}{\omega_1\omega_2} - \frac{f_0}{m}\frac{e^{-i\omega_1(t+\tau)}}{\omega_1(\omega_1 - \omega_2)} + \frac{f_0}{m}\frac{e^{-i\omega_2(t+\tau)}}{\omega_2(\omega_1 - \omega_2)}
$$

$$
= \frac{f_0}{m\omega_0^2} - \frac{f_0}{m\omega_0^2}[\cos\sigma(t + \tau) + \frac{\lambda}{\sigma}\sin\sigma(t + \tau)]\,e^{-\sigma(t+\tau)}
$$

3. and when $t > \tau$, it is

$$
x(t) = \frac{f_0}{m}\frac{e^{-i\omega_1(t-\tau)}}{\omega_1(\omega_1 - \omega_2)} - \frac{f_0}{m}\frac{e^{-i\omega_2(t-\tau)}}{\omega_2(\omega_1 - \omega_2)} - \frac{f_0}{m}\frac{e^{-i\omega_1(t+\tau)}}{\omega_1(\omega_1 - \omega_2)} + \frac{f_0}{m}\frac{e^{-i\omega_2(t+\tau)}}{\omega_2(\omega_1 - \omega_2)}
$$

$$
= \frac{f_0}{m\omega_0^2}[\cos\sigma(t - \tau) + \frac{\lambda}{\sigma}\sin\sigma(t - \tau)]\,e^{-\lambda(t-\tau)}
$$

$$
- \frac{f_0}{m\omega_0^2}[\cos\sigma(t + \tau) + \frac{\lambda}{\sigma}\sin\sigma(t + \tau)]\,e^{-\lambda(t+\tau)}.
$$

A rather special problem arises in the (physically unlikely) event that there is no damping. The equation of motion of the oscillator becomes

$$
\frac{d^2 x}{d t^2} + \omega_0^2 x = \frac{f(t)}{m}
$$

which, when solved by the Fourier transform procedure, yields a solution of the form

$$
x(t) = \frac{1}{m\sqrt{2\pi}}\int_{-\infty}^{\infty}\frac{F(\omega)\,e^{-i\omega t}}{\omega_0^2 - \omega^2}\,d\omega.
$$

The poles arising from the zeros of the denominator are now on the real axis at $\omega = \pm\omega_0$ and so the integral is undefined until we specify how we propose to avoid them.

Additional **physical** information is needed to resolve this ambiguity. For example, suppose that the oscillator is at rest until disturbed by a sharp blow delivered at $t = t_0$.

Representing the external force by means of a δ-function, $f(t) = f_0\, \delta(t - t_0)$ where f_0 is a constant, we have

$$F(\omega) = \frac{1}{\sqrt{2\pi}} \int_{-\infty}^{\infty} f_0\, \delta(t - t_0)\, e^{i\omega t}\, dt = \frac{f_0}{\sqrt{2\pi}}\, e^{i\omega t_0}$$

and,

$$x(t) = \frac{f_0}{m} \frac{1}{2\pi} \int_{-\infty}^{\infty} \frac{e^{-i\omega(t-t_0)}}{\omega_0^2 - \omega^2}\, d\omega.$$

Now, when $t - t_0 < 0$, this expression should yield the value $x(t) = 0$, since otherwise we would have motion occurring prior to the impulse in violation of our initial assumption and of the **principle of causality**. Moreover, when $t - t_0 < 0$, Jordan's Lemma permits us to close the contour by means of a semi-circular arc of infinite radius in the upper half-plane and evaluate the integral by means of the Residue Theorem. But the residues at the two poles $\omega = \pm\, \omega_0$ of this integrand are

$$-\frac{f_0}{m}\, \frac{e^{-i\,\omega_0(t-t_0)}}{4\pi\,\omega_0}$$

and

$$\frac{f_0}{m}\, \frac{e^{+i\,\omega_0(t-t_0)}}{4\pi\,\omega_0},$$

respectively. Thus, if one or both of the poles is included within the closed contour, the result will **not** be zero. We conclude therefore that the integral **must** be defined by deforming the contour along the real axis to pass above the poles $\omega = \pm\, \omega_0$ as shown in the diagram below. (We note that this is the equivalent of adding a vanishingly small damping force since the latter would shift the poles to $\omega = \pm\, \omega_0 - i\varepsilon$.) In other words, the **mathematical** ambiguity has been resolved by appeal to a fundamental **physical** principle; we now have an unambiguous definition of the integral and thence, can determine the oscillator's motion to be (by closing in the lower half-plane for $t - t_0 > 0$),

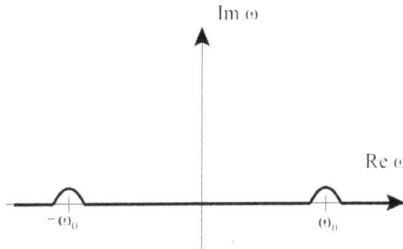

$$x(t) = \frac{f_0}{m}\, \frac{\sin \omega_0(t - t_0)}{\omega_0}\, \theta(t - t_0).$$

As a final example, we shall solve the equation

$$\frac{d^2 x}{d t^2} - \alpha^2 x = f(t), \quad 0 \leqslant t < \infty$$

subject to the boundary conditions $\frac{dx(0)}{dt} = b$ and $x(\infty) < \infty$, that is, $x(t)$ is bounded at infinity.

Since the range of the independent variable is restricted to $t \geqslant 0$, we can rule out Fourier transforms as a possible means of solution. However, that still leaves us with a choice between Fourier sine, Fourier cosine, and Laplace transforms. The fact that we have only one boundary condition at $t = 0$ means that we lack an essential piece of information for use of the latter. Moreover, we are given $x(\infty) < \infty$ which is not needed for Laplace transforms but is needed for Fourier sine and cosine transforms. To decide between these final two options, we note that sine transforms require a knowledge of $x(0)$ while cosine transforms make use of $\frac{dx(0)}{dt}$. Thus, the clear choice of method in this case is to apply Fourier cosine transforms.

Since $\mathcal{F}_C\{x''(t)\} = -\sqrt{\frac{2}{\pi}}x'(0) - \omega^2 X_C(\omega)$ where $X_C(\omega) = \mathcal{F}_C\{x(t)\}$, the differential equation transforms to the subsidiary equation

$$-\sqrt{\frac{2}{\pi}}b - \omega^2 X_C(\omega) - \alpha^2 X_C(\omega) = F_C(\omega), \ F_C(\omega) = \mathcal{F}_C\{f(t)\}$$

with solution

$$X_C(\omega) = -\sqrt{\frac{2}{\pi}}\frac{b}{\omega^2 + \alpha^2} - \frac{F_C(\omega)}{\omega^2 + \alpha^2}.$$

The inverse transform of $\frac{1}{\omega^2 + \alpha^2}$ is

$$x_1(t) = \sqrt{\frac{2}{\pi}} \int_0^\infty \frac{\cos \omega t}{\omega^2 + \alpha^2} d\omega = \frac{1}{\sqrt{2\pi}} \int_{-\infty}^\infty \frac{\cos \omega t}{\omega^2 + \alpha^2} d\omega.$$

Since $\frac{1}{\omega^2 + \alpha^2}$ is a rational function with a denominator of degree 2 and numerator of degree 0, we can evaluate this integral by using a standard formula of residue calculus where, because $t \geqslant 0$, we select the version that sums over singularities in the upper half-plane:

$$x_1(t) = -2\pi\frac{1}{\sqrt{2\pi}}\text{ImRes}\left\{\frac{e^{i\omega t}}{\omega^2 + \alpha^2}\right\}\Bigg|_{\omega=i\alpha} = -\sqrt{2\pi}\,\text{Im}\left[\frac{e^{-\alpha t}}{2\alpha i}\right] = \sqrt{\frac{\pi}{2}}\frac{e^{-\alpha t}}{\alpha}.$$

This determines the inverse of the first term in our expression for $X_C(\omega)$ as well as of the factor multiplying $F_C(\omega)$ in the second term. Thus, all we need to complete the solution is to invoke the convolution theorem for Fourier cosine transforms. Thus,

$$x(t) = -\frac{b}{\alpha} e^{-\alpha t} - \frac{1}{\sqrt{2\pi}}\sqrt{\frac{\pi}{2}}\frac{1}{\alpha} \int_0^\infty f(\tau)[e^{-\alpha|t-\tau|} + e^{-\alpha(t+\tau)}]d\tau$$

$$= -\frac{b}{\alpha} e^{-\alpha t} - \frac{1}{2\alpha} \int_0^\infty f(\tau)[e^{-\alpha|t-\tau|} + e^{-\alpha(t+\tau)}]d\tau$$

where we have used the fact that $e^{-\alpha|t|}$ is the symmetric extension of $e^{-\alpha t}$.

That completes the solution of the problem.

9 Ordinary Linear Differential Equations

9.1 Introduction

The **order** and **degree** of a differential equation (DE) are determined by the derivative of highest order after the DE has been rationalized. For example, the DE

$$\frac{d^3 y}{d x^3} + x\sqrt{\frac{dy}{dx}} + x^2 y = 0$$

is of **third order** and **second degree** since, when rationalized, it contains $\left(\frac{d^3 y}{d x^3}\right)^2$:

$$\left(\frac{d^3 y}{d x^3}\right)^2 + 2 x^2 y \frac{d^3 y}{d x^3} - x^2 \frac{dy}{dx} + x^4 y^2 = 0.$$

Thus, a **linear nth order** DE has the general form

$$\frac{d^n y}{d x^n} + a_{n-1}(x)\frac{d^{n-1} y}{d x^{n-1}} + \ldots + a_1(x)\frac{dy}{dx} + a_0(x)y = f(x) \tag{9.1.1}$$

where $f(x), a_0(x), a_1(x), \ldots, a_{n-1}(x)$ are arbitrary functions of x. If $f(x) \equiv 0$, the DE is said to be **homogeneous**. Otherwise, it is **non-homogeneous**.

For most physical applications one need not worry about orders higher than two. Moreover, first order DE's can be dispensed with by direct integration. Specifically, if the DE is

$$\frac{dy}{dx} + a(x)y = f(x), \tag{9.1.2}$$

we proceed by introducing a new function $p(x)$ whose logarithmic derivative is equal to $a(x)$:

$$\frac{1}{p(x)}\frac{dp}{dx} \equiv a(x) \ \text{ or } \ p(x) \equiv \exp\left(\int^x a(\xi)d\xi\right). \tag{9.1.3}$$

This transforms the DE into the convenient form

$$\frac{d}{dx}(p(x)y(x)) = p(x)f(x) \tag{9.1.4}$$

which integrates immediately to yield

$$y(x) = \frac{1}{p(x)}\int_{x_0}^x p(\xi)f(\xi)d\xi \tag{9.1.5}$$

where x_0 is an arbitrary initial point.

Notice that there is no lower limit for the integration in (9.1.3) but there is for the integration in (9.1.5). An arbitrary lower limit is a short hand way of adding a constant of integration to what is, in fact, an indefinite integral. It is needed in (9.1.5) to obtain the **general** solution of the DE rather than one **particular** solution. It is not needed in (9.1.3) because it would contribute an arbitrary multiplicative constant in $p(x)$ which would then cancel out in equation (9.1.5).

Example: Suppose that we wish to solve the DE

$$x^2 \frac{dy}{dx} + 2xy - x + 1 = 0.$$

To put this in the canonical form of equation (9.1.2) we divide through by x^2:

$$\frac{dy}{dx} + \frac{2}{x}y = \frac{1}{x} - \frac{1}{x^2}.$$

Comparing this with (9.1.2), we identify

$$a(x) = \frac{2}{x}, \quad f(x) = \frac{x-1}{x^2} = \frac{1}{x} - \frac{1}{x^2}.$$

Thus,

$$p(x) = \exp\left(\int^x \frac{2}{\xi} d\xi\right) = \exp(2\ln x) = x^2$$

and,

$$y(x) = \frac{1}{x^2}\int_{x_0}^x \xi^2 \frac{\xi-1}{\xi^2} d\xi = \frac{1}{x^2}\left(\frac{x^2}{2} - x + C\right) = \frac{1}{2} - \frac{1}{x} + \frac{C}{x^2}.$$

where $C = \text{constant} = x_0 - \frac{x_0^2}{2}$.

This is the **general** solution of the DE. Assigning a specific value to C will produce a **particular** solution. This is usually accomplished by imposing a particular value on $y(x)$ at some point x_0. If x_0 is an end-point of the range of variation of x, we say that we are imposing a **boundary condition**.

As an example, suppose that we require that $y(1) = 0$. Solving for C, we determine that $C = \frac{1}{2}$ and hence that the DE together with the boundary condition has the **unique** solution $y(x) = \frac{1}{2} - \frac{1}{x} + \frac{1}{2x^2}$.

9.2 Linear DE's of Second Order

Having dealt with first order DE's so readily, let us turn our attention to second order equations. The general form for a non-homogeneous, linear, second order DE is

$$\frac{d^2 y}{dx^2} + a(x)\frac{dy}{dx} + b(x)y(x) = f(x). \tag{9.2.1}$$

The corresponding homogeneous equation is

$$\frac{d^2 y}{dx^2} + a(x)\frac{dy}{dx} + b(x)y(x) = 0. \tag{9.2.2}$$

A fundamental property of homogeneous linear DE's is that **any linear combination of solutions is itself a solution**. Thus, if $y_1(x)$ and $y_2(x)$ are solutions of (9.2.2) so is $y(x) = c_1 y_1(x) + c_2 y_2(x)$, c_1, c_2 = constants.

Two solutions are **linearly independent** if the algebraic equation

$$c_1 y_1(x) + c_2 y_2(x) = 0 \tag{9.2.3}$$

can be satisfied for all x only if $c_1 = c_2 = 0$. In other words, they are linearly independent if they are not constant multiples of each other.

Differentiating (9.2.3) produces a second linear algebraic equation for c_1 and c_2,

$$c_1 \frac{dy_1}{dx} + c_2 \frac{dy_2}{dx} = 0. \tag{9.2.4}$$

The simultaneous equations (9.2.3) and (9.2.4) imply that y_1 and y_2 will be **linearly independent**, $c_1 = c_2 = 0$, if the determinant of the coefficients of c_1 and c_2 is non-zero:

$$y_1(x)\frac{dy_2}{dx} - y_2(x)\frac{dy_1}{dx} \equiv W[y_1, y_2] \neq 0. \tag{9.2.5}$$

$W[y_1, y_2]$ is called the **Wronskian** of y_1 and y_2. If $W[y_1, y_2] \equiv 0$, $y_1(x)$ and $y_2(x)$ are necessarily linearly dependent since

$$y_1 \frac{dy_2}{dx} - y_2 \frac{dy_1}{dx} = 0$$

integrates immediately to give $y_2(x)$ = constant $\times y_1(x)$ for all x.

The following theorem extends and formalizes these conclusions.

Theorem: The Wronskian of two solutions of a linear, homogeneous, second order DE is either identically zero or never zero and hence, a necessary and sufficient condition for linear independence is that the Wronskian be non-zero at any point x_0.

Proof: From (9.2.5),

$$\frac{dW}{dx} = y_1(x)\frac{d^2 y_2}{dx^2} - y_2(x)\frac{d^2 y_1}{dx^2}.$$

But, y_1 and y_2 are known to satisfy

$$\frac{d^2 y_1}{dx^2} + a(x)\frac{dy_1}{dx} + b(x)y_1 = 0$$

$$\frac{d^2 y_2}{dx^2} + a(x)\frac{dy_2}{dx} + b(x)y_2 = 0.$$

Multiplying the first of these by $(-y_2(x))$ and the second by $y_1(x)$ and adding, we obtain

$$y_1(x)\frac{d^2 y_2}{dx^2} - y_2(x)\frac{d^2 y_1}{dx^2} + a(x)[y_1(x)\frac{dy_2}{dx} - y_2(x)\frac{dy_1}{dx}] = 0$$

or,

$$\frac{dW}{dx} + a(x)W = 0.$$

Integrating, we find that

$$W(x) = W(x_0)\exp\left\{-\int_{x_0}^{x} a(\xi)d\xi\right\} \qquad (9.2.6)$$

where x_0 is an arbitrary point. If $W(x_0) = 0$, $W(x) \equiv 0$ and if $W(x_0) \neq 0$, $W(x)$ is never zero and the theorem is proved.

The relevance of linear independence is brought out by the next theorem.

Theorem: If $y_1(x)$ and $y_2(x)$ are any pair of linearly independent solutions of a homogeneous linear DE of second order, then any **other** solution of that equation can be expressed as a linear combination of y_1 and y_2,

$$y(x) = c_1 y_1(x) + c_2 y_2(x) \qquad (9.2.7)$$

where c_1 and c_2 are constants. The pair y_1 and y_2 are called a **fundamental set of solutions** and the linear combination (9.2.7), with c_1 and c_2 arbitrary, is the **general solution** of the DE.

Proof: We write the DE (9.2.2) in the modified but equivalent form

$$p(x)\frac{d^2 y}{dx^2} + q(x)\frac{dy}{dx} + r(x)y = 0, \quad p(x) \neq 0.$$

Thus, if $y(x), y_1(x), y_2(x)$ are **any** three solutions of the DE then

$$p(x)s\frac{d^2 y}{dx^2} + q(x)\frac{dy}{dx} + r(x)y = 0$$

$$p(x)\frac{d^2 y_1}{dx^2} + q(x)\frac{dy_1}{dx} + r(x)y_1 = 0$$

$$p(x)\frac{d^2 y_2}{dx^2} + q(x)\frac{dy_2}{dx} + r(x)y_2 = 0.$$

Considered as simultaneous, linear, algebraic equations for p, q, and r, these will admit a non-trivial solution if and only if the determinant of their coefficients is zero. Interchanging rows and columns in that determinant we see that this implies that the algebraic equations

$$ay + a_1 y_1 + a_2 y_2 = 0$$

$$a\frac{dy}{dx} + a_1\frac{dy_1}{dx} + a_2\frac{dy_2}{dx} = 0$$

$$a\frac{d^2 y}{dx^2} + a_1\frac{d^2 y_1}{dx^2} + a_2\frac{d^2 y_2}{dx^2} = 0$$

must also admit a non-trivial solution for the constants a, a_1 and a_2. Moreover, if y_1 and y_2 are linearly independent, a cannot be zero. Therefore,

$$y(x) = c_1 y_1(x) + c_2 y_2(x)$$

where $c_1 = -a_1/a$ and $c_2 = -a_2/a$ are constants. Since y, y_1 and y_2 are **any** three solutions of (9.2.2), subject to y_1 and y_2 being linearly independent, we conclude that (9.2.7) is the most general solution of (9.2.2).

Is there a similarly simple expression for the most general solution of a non-homogeneous, linear, second order DE ? The answer is yes and is addressed in detail by the next theorem.

Theoem: If $y_p(x)$ is a **particular solution** of the non-homogeneous DE

$$\frac{d^2 y}{d x^2} + a(x)\frac{dy}{dx} + b(x)y = f(x)$$

and $y_1(x)$ and $y_2(x)$ are a fundamental set of solutions of the corresponding homogeneous equation

$$\frac{d^2 y}{d x^2} + a(x)\frac{dy}{dx} + b(x)y = 0,$$

then **every** other solution of the non-homogeneous equation may be expressed as the linear combination

$$y(x) = y_p(x) + y_c(x) = y_p(x) + c_1 y_1(x) + c_2 y_2(x) \qquad (9.2.8)$$

where c_1 and c_2 are constants. The functions $y_p(x)$ and $y_c(x)$ are called the **particular integral** and the **complementary function** of the non-homogeneous DE, respectively.

Proof: Substitution of $y(x) = y_p(x) + y_c(x)$ into the non-homogeneous DE yields

$$\frac{d^2 y_c}{d x^2} + a(x)\frac{d y_c}{dx} + b(x) y_c = 0$$

whose general solution is $y_c = c_1 y_1(x) + c_2 y_2(x)$.

9.3 Given One Solution, Find the Others

Suppose that we know one solution of the homogeneous DE (9.2.2). Can we use this information to find a second linearly independent solution? Not only can we do that, we can find the solution to any non-homogeneous counterpart as well. To see how this comes about, we return to a consideration of the Wronskian which, according to (9.2.6), is determined by the coefficient of the first derivative in the DE. And, of course, it relates a first solution $y_1(x)$ to a linearly independent second solution $y_2(x)$. In fact,

$$\frac{d}{dx}\left(\frac{y_2}{y_1}\right) = \frac{y_1 y_2' - y_2 y_1'}{y_1^2} = \frac{W(x)}{y_1^2}.$$

So, using (9.2.6), we find

$$\frac{y_2(x)}{y_1(x)} = W(x_0) \int^x \frac{\exp[-\int_{x_0}^{\xi} a(\zeta)d\zeta]}{[y_1(\xi)]^2} d\xi + C$$

where C is a constant of integration. Adding a constant times $y_1(x)$ to $y_2(x)$ does not provide new information and so we shall set $C = 0$. Similarly, including the multiplicative constant $W(x_0)$ is unnecessary and so we arbitrarily set it equal to one. Thus, our final expression for a second linearly independent solution is

$$y_2(x) = y_1(x) \cdot \int^x \frac{\exp[-\int^\xi a(\zeta)d\zeta]}{[y_1(\xi)]^2} d\xi. \tag{9.3.1}$$

This remarkably simple expression is obtainable by another approach which has the advantage of applying to both the homogeneous and non-homogeneous cases. Called the **method of variation of constants**, it starts from the premise that a second linearly independent solution is necessarily related to a first solution by a multiplicative function or **"variable constant"**.

Thus, we set $y_2(x) = u(x)y_1(x)$, $u(x) \neq$ a constant and substitute into the homogeneous DE (9.2.2). The result is a DE for $u(x)$:

$$\frac{d^2 u}{dx^2} + \left[\frac{2y_1' + a(x)y_1}{y_1} \right] \frac{du}{dx} = 0$$

or,

$$\frac{d}{dx} \ln \left(\frac{du}{dx} \right) = -a(x) - 2\frac{d}{dx} \ln y_1(x). \tag{9.3.2}$$

Integrating, we find

$$\frac{du}{dx} = \frac{c_1}{[y_1(x)]^2} \exp \left[-\int^x a(\zeta)d\zeta \right]$$

whence

$$u(x) = c_1 \int^x \frac{\exp[-\int^\xi a(\zeta)d\zeta]}{[y_1(\xi)]^2} d\xi + c_2$$

where c_1 and c_2 are constants of integration. Since any solution for $u(x)$ will do, we set $c_1 = 1$ and $c_2 = 0$. Then, multiplying $u(x)$ by $y_1(x)$ we recover (9.3.1) as expected.

Similarly, a particular integral of (9.2.1)

$$\frac{d^2 y}{dx^2} + a(x)\frac{dy}{dx} + b(x)y = f(x)$$

cannot be a constant multiplier of a solution of its homogeneous counterpart (9.2.2). Therefore, we set $y_p(x) = v(x)y_1(x)$ where $y_1(x)$ is again a known solution of (9.2.2). Substituting into (9.2.1) we obtain a DE for $v(x)$:

$$v'' y_1 + 2v' y_1' + v y_1'' + a(x)(v' y_1 + v y_1') + b(x)v y_1 = f(x),$$

or

$$v'' y_1 + [2 y_1' + a(x) y_1] v' + [y_1'' + a(x) y_1' + b(x) y_1] v = f(x),$$

or

$$\frac{d^2 v}{dx^2} + \left[a(x) + \frac{2}{y_1(x)} \frac{dy_1}{dx} \right] \frac{dv}{dx} = \frac{f(x)}{y_1(x)}. \tag{9.3.3}$$

We note from (9.3.2) that $a(x) + \dfrac{2}{y_1} \dfrac{dy_1}{dx}$ can be set equal to the logarithmic derivative $\dfrac{1}{R(x)} \dfrac{dR}{dx}$ where

$$R(x) = \left[\frac{d}{dx} \left(\frac{y_2}{y_1} \right) \right]^{-1} = \frac{y_1^2}{W[y_1, y_2]} \tag{9.3.4}$$

and $y_2(x)$ is a second linearly independent solution of the homogeneous DE. Thus, (9.3.3) becomes

$$\frac{d}{dx} \left(\frac{dv}{dx} \right) + \frac{1}{R} \frac{dR}{dx} \frac{dv}{dx} = \frac{f(x)}{y_1(x)},$$

a first order, non-homogeneous DE for $\dfrac{dv}{dx}$. Using (9.1.5) for its solution, we find

$$\frac{dv}{dx} = \frac{1}{R(x)} \int_{x_0}^{x} \frac{R(\xi) f(\xi)}{y_1(\xi)} d\xi$$

$$= \frac{d}{dx} \left(\frac{y_2}{y_1} \right) \cdot \int_{x_0}^{x} \frac{y_1(\xi) f(\xi)}{W[y_1(\xi), y_2(\xi)]} d\xi$$

$$= \frac{d}{dx} \left[\frac{y_2}{y_1} \int_{x_0}^{x} \frac{y_1(\xi) f(\xi)}{W[y_1(\xi), y_2(\xi)]} d\xi \right] - \frac{y_2(x) f(x)}{W[y_1(x), y_2(x)]}.$$

Therefore, integrating to obtain $v(x)$ and multiplying the result by $y_1(x)$, we conclude that a particular integral of (9.2.1) is

$$y_p(x) = y_2(x) \int^{x} \frac{y_1(\xi) f(\xi)}{W[y_1(\xi), y_2(\xi)]} d\xi - y_1(x) \int^{x} \frac{y_2(\xi) f(\xi)}{W[y_1(\xi), y_2(\xi)]} d\xi. \tag{9.3.5}$$

Notice that we have omitted the lower limits on the integrals which means that we have omitted the two constants of integration. Were they included, we would have the **general solution**, $y(x) = y_p(x) + c_1 y_1(x) + c_2 y_2(x)$, whereas our original objective was simply to find a **particular integral**, $y_p(x)$.

Thus, knowledge of just one solution of a homogeneous, linear, second order DE is sufficient to determine all other solutions of that DE **and** of all of its non-homogeneous counterparts!

Example: Consider the DE

$$x^2 \frac{d^2 y}{d x^2} - 2y = x.$$

It is clear from inspection that a particular solution of the corresponding homogeneous DE,

$$x^2 \frac{d^2 y}{d x^2} - 2y = 0,$$

is $y_1(x) = x^2$. Thus, noting that $a(x) = 0$ in this case, we obtain a second linearly independent solution by performing the integration

$$y_2(x) = x^2 \int^x \frac{e^0}{\xi^4} d\xi$$

$$= x^2 \left(\frac{-1}{3} \right) \frac{1}{x^3}$$

$$= -\frac{1}{3x}.$$

The Wronskian of $y_1(x)$ and $y_2(x)$ is

$$W[y_1, y_2] = y_1 y_2' - y_2 y_1' = x^2 \left(\frac{1}{3 x^2} \right) - \left(\frac{-1}{3x} \right) 2x = 1.$$

Therefore, using (9.3.5) with $f(x) = \frac{1}{x}$, we find

$$y_p(x) = \left(\frac{-1}{3x} \right) \int^x \xi^2 \frac{1}{\xi} d\xi - x^2 \int^x \left(\frac{-1}{3\xi} \right) \frac{1}{\xi} d\xi$$

$$= \left(\frac{-1}{3x} \right) \frac{x^2}{2} - x^2 \left(\frac{1}{3x} \right)$$

$$= -\frac{x}{2}$$

as a particular integral of the non-homogeneous DE. The general solution of the non-homogeneous DE is thus

$$y(x) = -\frac{x}{2} + c_1 x^2 + c_2 \frac{1}{x}.$$

9.4 Finding a Particular Solution for Homogeneous DE's

We can now focus on homogeneous DE's and in particular, on the problem of finding a first solution. We might as well start with the simplest class of such DE's, those with constant coefficients:

$$\frac{d^2 y}{d x^2} + a \frac{dy}{dx} + by = 0 \qquad (9.4.1)$$

where a and b are real constants.

The only function that enjoys a linear relationship with its derivatives is the exponential. Therefore, let us try a solution of the form $y(x) = e^{\lambda x}$. Substituting into (9.4.1) we find

$$(\lambda^2 + a\lambda + b)\, e^{\lambda x} = 0. \qquad (9.4.2)$$

This means that $e^{\lambda x}$ is indeed a solution if and only if λ is a solution of the quadratic equation

$$\lambda^2 + a\lambda + b = 0$$

which is called the **characteristic equation** of the original DE (9.4.1).

The roots of the characteristic equation are

$$\lambda_1 = \frac{1}{2}(-a + \sqrt{a^2 - 4b}) \quad \text{and} \quad \lambda_2 = \frac{1}{2}(-a - \sqrt{a^2 - 4b}). \qquad (9.4.3)$$

If $a^2 \neq 4b$, $\lambda_1 \neq \lambda_2$ and we have two linearly independent solutions

$$y_1(x) = e^{\lambda_1 x} \quad \text{and} \quad y_2(x) = e^{\lambda_2 x}. \qquad (9.4.4)$$

If $a^2 < 4b$, the roots will be complex. Should an explicitly real solution be required, the two solutions in (9.4.4) can be combined to give

$$y_1(x) = e^{-ax/2} \cos\left(\sqrt{4b - a^2}\,\frac{x}{2}\right) \quad \text{and} \quad y_2(x) = e^{-ax/2} \sin\left(\sqrt{4b - a^2}\,\frac{x}{2}\right). \qquad (9.4.5)$$

If $a^2 = 4b$, $\lambda_1 = \lambda_2 = -\frac{a}{2}$ and we have only one solution $y_1(x) = e^{-ax/2}$. A second, linearly independent solution is then obtained from an application of (9.3.1):

$$y_2(x) = e^{-ax/2} \int^x \frac{\exp[-\int^\xi a\,d\zeta]}{e^{-a\xi}} d\xi$$

$$= e^{-ax/2} \int^x d\xi$$

$$= x\,e^{-ax/2}. \qquad (9.4.6)$$

Examples: Consider $y'' + y' - 2y = 0$. The characteristic equation is $\lambda^2 + \lambda - 2 = 0$ with roots $\lambda_1 = 1$ and $\lambda_2 = -2$. Therefore, this DE has the general solution $y(x) = c_1 e^x + c_2 e^{-2x}$.

Next, suppose that we wish to solve $y'' - 2y' + 10y = 0$. Its characteristic equation is $\lambda^2 - 2\lambda + 10 = 0$ with roots $\lambda_1 = 1 + 3i$ and $\lambda_2 = 1 - 3i$. So, the general solution in this case is $y(x) = e^x(c_1 \cos 3x + c_2 \sin 3x)$.

Finally, consider the DE $y'' + 8y' + 16y = 0$. The characteristic equation is $\lambda^2 + 8\lambda + 16 = 0$ which has the double root $\lambda = -4$. This means that the general solution must be $y(x) = e^{-4x}(c_1 + c_2 x)$.

When we make the transition to DE's with variable coefficients, the task of finding a solution becomes more complicated as do the solutions themselves. In fact, the solutions are seldom expressible in terms of elementary functions. Experience tells us that we should therefore seek solutions in the form of power series or integral representations. We shall examine both approaches in due course but will discover that power series provide sufficient insight into the analytical properties of solutions that we can make them a principal focus of our attention.

The power series approach is called the **method of Frobenius** and it is fairly straight forward to apply. Therefore, before launching into an exposition of the underlying theory, we shall illustrate its practical content by means of some simple examples.

Ferdinand Georg Frobenius (1849-1917) was a student of Weierstrass. Best known for his contributions to the theory of differential equations and to group theory,he taught at the University of Berlin and at ETH Zurich.

As we can quickly verify by checking its characteristic equation, the DE

$$\frac{d^2 y}{d x^2} + \omega^2 y = 0, \quad \omega = a \text{ real constant} \tag{9.4.7}$$

has the general solution $y(x) = c_1 \cos \omega x + c_2 \sin \omega x$. Our challenge is to re-derive this result starting from the assumption that the DE admits a solution that can be represented by a Taylor series about $x = 0$:

$$y(x) = a_0 + a_1 x + a_2 x^2 + \ldots = \sum_{m=0}^{\infty} a_m x^m . \tag{9.4.8}$$

We begin by differentiating (9.4.8) term by term (which means we are assuming uniform convergence) to obtain

$$y'' = 2 \cdot 1 \, a_2 + 3 \cdot 2 \, a_3 x + 4 \cdot 3 \, a_4 x^2 + \ldots = \sum_{m=2}^{\infty} m(m-1) x^{m-2} .$$

Inserting this together with (9.4.8) into the DE, we have

$$\sum_{m=2}^{\infty} m(m-1) a_m x^{m-2} + \sum_{m=0}^{\infty} \omega^2 a_m x^m = 0.$$

Remembering that power series representations are unique, we recognize that this equality of series implies equality on a term by term basis. Simply put, this means that we can equate the coefficient of each power that appears on the left hand side to zero:

$$2 \cdot 1 \, a_2 + \omega^2 \, a_0 = 0,$$
$$3 \cdot 2 \, a_3 + \omega^2 \, a_1 = 0,$$
$$4 \cdot 3 \, a_4 + \omega^2 \, a_2 = 0, \quad \ldots,$$
$$m(m-1) a_m + \omega^2 \, a_{m-2} = 0, \quad \ldots.$$

This means that $a_2 = -\frac{\omega^2}{2 \cdot 1} a_0, a_3 = -\frac{\omega^2}{3 \cdot 2} a_1, a_4 = -\frac{\omega^2}{4 \cdot 3} a_2 = (-1)^2 \frac{(\omega^2)^2}{4 \cdot 3 \cdot 2 \cdot 1} a_0$, and in general,

$$a_m = -\frac{\omega^2}{m(m-1)} a_{m-2} = \ldots = (-1)^k \frac{(\omega^2)^k}{m(m-1)\ldots(m-k+1)} a_{m-2k} . \tag{9.4.9}$$

This is a **recurrence relation** for the coefficients. It expresses them all in terms of either a_0 or a_1 depending on whether m is even or odd. This splits the solution into two linearly independent parts, one which is an even function and the other an odd function of x. More specifically, we find

$$y(x) = a_0 \left(1 - \frac{\omega^2 x^2}{2!} + \frac{\omega^4 x^4}{4!} - + \ldots\right) + \frac{a_1}{\omega} \left(\omega x - \frac{\omega^3 x^3}{3!} + \frac{\omega^5 x^5}{5!} - + \ldots\right)$$

or,

$$y(x) = c_1 \cos \omega x + c_2 \sin \omega x$$

where c_1 and c_2 are arbitrary constants. Thus, as we hoped we would, we have recovered the general solution of the DE.

Suppose that we up the ante by using the same approach to solve the (non-homogeneous) DE with variable coefficients

$$\frac{d^2 y}{dx^2} + xy = x^3 . \tag{9.4.10}$$

Proceeding in a tentative fashion, we start with the homogeneous counterpart

$$\frac{d^2 y}{dx^2} + xy = 0 \tag{9.4.11}$$

and substitute into it a Taylor series representation $y(x) = \sum_{m=0}^{\infty} a_m x^m$. We find

$$\sum_{m=2}^{\infty} m(m-1) a_m x^{m-2} + \sum_{m=0}^{\infty} a_m x^{m+1} = 0.$$

Next, we set $m - 2 = n$ in the first series and $m + 1 = n$ in the second series to obtain

$$\sum_{n=0}^{\infty} (n+2)(n+1) a_{n+2} x^n + \sum_{n=1}^{\infty} a_{n-1} x^n = 0$$

or,

$$2 \cdot 1 a_2 + \sum_{n=1}^{\infty} \{(n+2)(n+1) a_{n+2} + a_{n-1}\} x^n = 0.$$

Then, equating the coefficients of successive powers of x to zero, we have $a_2 = 0$ and

$$a_{n+2} = -\frac{a_{n-1}}{(n+2)(n+1)}, \quad n \geq 1. \tag{9.4.12}$$

Applying this recurrence relation a few times gives

$$a_2 = 0, a_3 = -\frac{a_0}{3 \cdot 2}, a_4 = -\frac{a_1}{4 \cdot 3}, a_5 = -\frac{a_2}{5 \cdot 4} = 0, a_6 = -\frac{a_3}{6 \cdot 5} = \frac{a_0}{6 \cdot 5 \cdot 3 \cdot 2},$$

$$a_7 = -\frac{a_4}{7 \cdot 6} = \frac{a_1}{7 \cdot 6 \cdot 4 \cdot 3}, a_8 = -\frac{a_5}{8 \cdot 7} = 0, a_9 = -\frac{a_6}{9 \cdot 8} = -\frac{a_0}{9 \cdot 8 \cdot 6 \cdot 5 \cdot 3 \cdot 2}, \dots$$

and hence,

$$y(x) = a_0 \left(1 - \frac{x^3}{3 \cdot 2} + \frac{x^6}{6 \cdot 5 \cdot 3 \cdot 2} - + \dots \right) + a_1 \left(x - \frac{x^4}{4 \cdot 3} + \frac{x^7}{7 \cdot 6 \cdot 4 \cdot 3} - + \dots \right)$$

$$(9.4.13)$$

where a_0 and a_1 are arbitrary. Each of the two terms in brackets in (9.4.13) is a particular solution of the homogeneous DE. Since they are also linearly independent, their linear combination is the general solution. Thus, once again, the assumption of a Taylor series representation about $x = 0$ has lead directly to the general solution. The only apparent difference from the previous example being that we do not recognize the series as ones that sum to some combination of elementary functions.

Using the same approach to solve the non-homogeneous DE (9.4.10) looks like a reasonable proposition because the non-homogeneous term is a monomial. If it were anything more complicated than a polynomial, we would have to expand it in a power series and the solution of the DE, while still feasible, would become much more complicated. Proceeding as we did for the homogeneous case, we substitute $y(x) = \sum_{m=0}^{\infty} a_m x^m$ into (9.4.10) and obtain

$$\sum_{m=2}^{\infty} m(m-1) a_m x^{m-2} + \sum_{m=0}^{\infty} a_m x^{m+1} = x^3,$$

or

$$2 \cdot 1 a_2 + \sum_{n=1}^{\infty} \{(n+2)(n+1) a_{n+2} + a_{n-1}\} x^n = x^3.$$

Equating the coefficients of successive powers of x on the left hand side of the equality to their counterparts on the right, we have

$$a_2 = 0, 3 \cdot 2 a_3 + a_0 = 0, 4 \cdot 3 a_4 + a_1 = 0, 5 \cdot 4 a_5 + a_2 = 1,$$

and

$$m(m-1) a_m + a_{m-3} = 0, \quad m \geq 6.$$

Thus,

$$a_3 = -\frac{a_0}{3 \cdot 2}, a_4 = -\frac{a_1}{4 \cdot 3}, a_5 = 1, a_6 = \frac{a_0}{6 \cdot 5 \cdot 3 \cdot 2}, a_7 = \frac{a_1}{7 \cdot 6 \cdot 4 \cdot 3}, a_8 = -\frac{1}{8 \cdot 7}, \dots$$

and so,

$$y(x) = a_0 \left(1 - \frac{x^3}{3 \cdot 2} + \frac{x^6}{6 \cdot 5 \cdot 3 \cdot 2} - + \dots \right) + a_1 \left(x - \frac{x^4}{4 \cdot 3} + \frac{x^7}{7 \cdot 6 \cdot 4 \cdot 3} - + \dots \right) + y_p(x)$$

where $y_p(x)$ is the **particular integral**

$$y_p(x) = x^5 - \frac{1}{8 \cdot 7} x^8 + \frac{1}{11 \cdot 10 \cdot 8 \cdot 7} x^{11} - + \dots . \tag{9.4.14}$$

Thus, regardless of whether the DE is homogeneous or non-homogeneous, the power series method appears to yield not just one solution but the general solution. But wait. There must be DE's whose solutions do not admit a Taylor series expansion about $x = 0$. In fact, invoking equation (9.3.1), we see that if $a(x) = \frac{1}{x}$, the relationship between any two linearly independent solutions is

$$y_2(x) = y_1(x) \int^x \frac{d\xi}{\xi [y_1(\xi)]^2} .$$

Therefore, if $y_1(x)$ is non-singular and non-zero at $x = 0$ then $y_2(x)$ will have a logarithmic singularity there and if $y_1(x)$ has a zero at $x = 0$ then $y_2(x)$ will have a pole of the same order there. In other words, one can have a DE that has a particular but not a general solution that is expressible as a Taylor series about $x = 0$. There is a physically important DE whose solutions illustrate this point.

The DE

$$\frac{d^2 y}{d x^2} + \frac{1}{x} \frac{dy}{dx} - \frac{m^2}{x^2} y = 0, \quad m = \text{ an integer}, \tag{9.4.15}$$

is a special case of what is known as Cauchy's DE of order 2, $x^2 \frac{d^2 y}{d x^2} + ax \frac{dy}{dx} + by = 0$ with a and b held constant. The standard approach to solving this class of DE's is to attempt a solution of the general power $y(x) = x^s$. However, because we want to illustrate the power series method of solution, we shall attempt a Taylor series, $y(x) = \sum\limits_{n=0}^{\infty} a_n x^n$. Substituting into (9.4.15), we have

$$\sum_{n=2}^{\infty} n(n-1) a_n x^{n-2} + \sum_{n=1}^{\infty} n a_n x^{n-2} - \sum_{n=0}^{\infty} m^2 a_n x^{n-2} = 0.$$

Assuming $m \neq 0$ and equating coefficients of successive powers of x to zero, we find

$$m^2 a_0 = 0, \ 1 a_1 - m^2 a_1 = 0, \ 2 \cdot 1 a_2 + 2 a_2 - m^2 a_2 = 0, \ \dots,$$

$$m(m-1) a_m + m a_m - m^2 a_m = 0, \ \dots, n(n-1) a_n + n a_n - m^2 a_n = 0, \ \dots .$$

Thus, $a_n = 0$ for all $n \neq m$, and a_m is arbitrary. In other words, $y(x) = x^m$ is the only non-trivial solution with a Taylor series representation about $x = 0$. To find a second linearly independent solution we are obliged to resort to (9.3.1):

$$y_2(x) = y_1(x) \int^x \frac{d\xi}{\xi [y_1(\xi)]^2} = x^m \int^x \frac{d\xi}{\xi^{2m+1}} = \frac{-1}{2m+1} x^{-m} .$$

So, the general solution of this DE is

$$y(x) = c_1 x^m + c_2 x^{-m}, \quad m \neq 0. \tag{9.4.16}$$

If $m = 0$, a_0 will be the only non-zero coefficient implying that $y_1 = 1$ is now the non-trivial solution with a Taylor series representation about $x = 0$. Using (9.3.1) again, we find that

$$y_2(x) = \int^x \frac{d\xi}{\xi} = \ln x$$

is the second linearly independent solution and so the general solution becomes

$$y(x) = c_1 + c_2 \ln x, \quad m = 0. \tag{9.4.17}$$

Evidently, we need some answers to questions regarding when and where the power series method can be used and what sort of solutions we can expect when a Taylor series is no longer valid for the general solution. As we shall now discover, the answers are provided by a sequence of theorems due to a nineteenth century mathematician called Frobenius.

9.5 Method of Frobenius

As we know, the most appropriate language for a discussion of power series representations is that of complex analysis. Therefore, we replace the real variable x by the complex variable z and rewrite the canonical DE (9.2.2) in the format

$$\frac{d^2 y}{d z^2} + a(z)\frac{dy}{dz} + b(z)y = 0, \tag{9.5.1}$$

where $a(z)$, $b(z)$ and $y(z)$ are complex functions of the complex variable z that satisfy the reality conditions $a^*(z) = a(z^*)$, $b^*(z) = b(z^*)$, $y^*(z) = y(z^*)$. Next, we **define** as **ordinary points** of the DE all points at which both $a(z)$ and $b(z)$ are holomorphic. **Theorem:** If $z = z_0$ is an ordinary point of (9.5.1) then every solution of the DE is holomorphic there.

The **proof** is quite straight forward. The holomorphy of $a(z)$ and b (z) implies that they have Taylor series about $z = z_0$. Substituting these as well as an assumed Taylor series for $y(z)$,

$$y(z) = \sum_{m=0}^{\infty} c_m(z - z_0)^m, \tag{9.5.2}$$

into (9.5.1), we determine a consistent set of equations for the c_m by equating the coefficients of successive powers of $(z - z_0)$ to zero. The radius of convergence of the series (9.5.2) will be the distance from $z = z_0$ to the nearest point which is not ordinary.

This confirms and extends the experience we acquired via the examples of the preceding Section. It also brings us to the question of what happens at "non-ordinary" or singular points.

If $a(z)$ and /or $b(z)$ have poles at $z = z_0$ but

$$(z - z_0)a(z) \quad \text{and} \quad (z - z_0)^2\, b(z)$$

are holomorphic there, $z = z_0$ is **defined** to be a **regular singular point** of the DE (9.5.1). If one or both of these functions has an isolated singularity at $z = z_0$, the DE has as an **irregular singular point** there.

Theorem: If $z = z_0$ is a regular singular point then **at least one** solution of the DE (9.5.1) can be expressed as a **Frobenius series**

$$y(z) = (z - z_0)^s \sum_{m=0}^{\infty} c_m(z - z_0)^m, \quad c_0 \neq 0 \text{ and } s \text{ is real or complex}, \qquad (9.5.3)$$

which converges in any circle about $z = z_0$ that contains no other singularities.

Note that if

1. $s = n$, $n =$ an integer, $y(z)$ has a zero of order n at $z = z_0$;
2. $s = 0$, $y(z)$ is holomorphic and non-zero at $z = z_0$;
3. $s = -n$, $n =$ an integer, $y(z)$ has a pole of order n at $z = z_0$;
4. $s \neq 0, \pm 1, \pm 2, \ldots$, $z = z_0$ is a branch point of $y(z)$.

To prove this theorem, we rewrite (9.5.1) in the form

$$(z - z_0)^2 \frac{d^2 y}{d z^2} + (z - z_0)A(z)\frac{dy}{dz} + B(z)y = 0 \qquad (9.5.4)$$

where

$$A(z) = (z - z_0)a(z) = \sum_{m=0}^{\infty} a_m(z - z_0)^m = a_0 + a_1(z - z_0) + a_2(z - z_0)^2 + \ldots \qquad (9.5.5)$$

and

$$B(z) = (z - z_0)^2\, b(z) = \sum_{m=0}^{\infty} b_m(z - z_0)^m = b_0 + b_1(z - z_0) + b_2(z - z_0)^2 + \ldots \qquad (9.5.6)$$

are holomorphic at $z = z_0$ and so have Taylor series expansions about that point. Substituting the series (9.5.3), (9.5.5) and (9.5.6) into the DE (9.5.4), we find

$$(z - z_0)^s \left[s(s - 1)c_0 + \sum_{m=1}^{\infty}(m + s)(m + s - 1)c_m(z - z_0)^m \right]$$

$$+ (z - z_0)^s \left[s\, c_0 + \sum_{m=1}^{\infty}(m + s)c_m(z - z_0)^m \right] \times \sum_{m=0}^{\infty} a_M(z - z_0)^m$$

$$+ (z - z_0)^s \left[c_0 + \sum_{m=1}^{\infty} c_m(z - z_0)^m \right] \times \sum_{m=0}^{\infty} b_m(z - z_0)^n = 0. \qquad (9.5.7)$$

Equating coefficients of successive powers of $(z - z_0)$ to zero, we obtain

$$c_0[s(s - 1) + a_0 s + b_0] = 0$$
$$c_1[(s + 1)s + a_0(s + 1) + b_0] + c_0[s a_1 + b_1] = 0$$
$$c_2[(s + 2)(s + 1) + a_0(s + 2) + b_0] + c_1[(s+!) a_1 + b_1] + c_0[s a_2 + b_2] = 0$$
$$\vdots$$
$$c_m[(s + m)(s + m - 1) + a_0(s + m) + b_0] + \ldots = 0$$
$$\vdots \tag{9.5.8}$$

Since $c_0 \neq 0$, the first of these equations becomes the **indicial equation**

$$s(s - 1) + a_0 s + b_0 = 0 \tag{9.5.9}$$

whose roots, s_1 and s_2, Re $s_1 \geq$ Re s_2, are the only permissable values for the index s.

Substituting the value s_1 for s in the remaining equations of (9.5.8), we can solve successively for $c_1, c_2, \ldots c_n, \ldots$ in terms of c_0. The latter becomes an arbitrary multiplicative constant which can be assigned any value but zero. Often but certainly not invariably, the value assigned to it is one.

Thus, we have generated the one Frobenius solution guaranteed by our theorem. Note that this first solution corresponds to the root of the indicial equation with the largest real part, s_1. Does the other root, s_2, generate a second linearly independent solution? The answer is obviously no if $s_1 = s_2$, that is, if the indicial equation has a **double root**. In that case, we have to rely on equation (9.3.1) to produce a second solution from our knowledge of the first. The result is novel within the context of our current level of experience. This is because

$$a(z) = \frac{a_0}{z - z_0} + a_1 + a_2(z - z_0) + \ldots,$$

and so,

$$\int^z a(\zeta)d\zeta = a_0 \ln(z - z_0) + a_1 z + \frac{a_2}{2}(z - z_0)^2 + \ldots.$$

Thus,

$$\exp\left[-\int^z a(\zeta)d\zeta\right] = \frac{1}{(z - z_0)^{a_0}} \exp[-a_1 z - \frac{a_2}{2}(z - z_0)^2 - \ldots] \tag{9.5.10}$$

which, when multiplied by $\frac{1}{[y_1(z)]^2}$ where $y_1(z) = (z - z_0)^{s_1} \sum_{m=0}^{\infty} c_m(z - z_0)^m$, yields

$$y_2(z) = y_1(z) \int^z \frac{1}{(\zeta - z_0)^{a_0 + 2 s_1}} f(\zeta)d\zeta \tag{9.5.11}$$

where $f(z)$ has a Taylor series expansion about $z = z_0$ and is non-vanishing there: $f(z) = \sum_{m=0}^{\infty} f_m(z - z_0)^m$, $f_0 \neq 0$. But, when we examine the indicial equation, we see that a double root occurs when $b_0 = \frac{(a_0-1)^2}{4}$ in which case the root is $s_1 = \frac{(1-a_0)}{2}$. This means that (9.5.11) becomes

$$y_2(z) = y_1(z) \int^z \frac{f(\zeta)}{(\zeta - z_0)} d\zeta = y_1(z) \times \left[f_0 \ln(z - z_0) + \sum_{m=1}^{\infty} \frac{f_m}{m}(z - z_0)^m \right].$$

In other words, the second linearly independent solution is of a form that is **defined** to be a **generalized Frobenius series.** Specifically,

$$y_2(z) = y_1(z) \cdot \ln(z - z_0) + (z - z_0)^{s_1} \sum_{m=1}^{\infty} d_m(z - z_0)^m. \tag{9.5.12}$$

Therefore, when the indicial equation has a double root, a first solution of the Frobenius form (9.5.3) is determined by substituting into the DE, determining the indicial equation and its root, and then solving the recurrence equation(s) (9.5.8) for the coefficients c_m. A second, linearly independent solution of the generalized Frobenius form is then determined by substituting (9.5.12) into the DE and solving for the coefficients d_m.

A similar kind of phenomenon occurs when the two roots of the indicial equation differ by an integer, $s_1 - s_2 = N$. Because s_1 is a root of the indicial equation, we know that

$$(s_2 + N)(s_2 + N - 1) + a_0(s_2 + N) + b_0 = 0. \tag{9.5.13}$$

But the left hand side of this equation is the coefficient of c_N in the Nth of the equations (9.5.8). This means that we cannot solve for c_N. If the other terms in the Nth equation are non-zero, c_N as well as all subsequent coefficients is undefined and we cannot determine a second solution of the Frobenius form. If the other terms in the equation are zero, c_N is arbitrary. The result is a second solution consisting of a superposition that contains c_N multiplied by the **first** solution or, put another way, the result is the **general solution** of the DE. Curiously, expanding about an ordinary point is one instance of this situation. If $z = z_0$ is an ordinary point, $a_0 = b_0 = b_1 = 0$ and the indicial equation becomes $s(s - 1) = 0$ with roots $s_1 = 1$ and $s_2 = 0$.

To cover either situation, c_N undefined or c_N arbitrary, we can again use (9.3.1) leading to equation (9.5.11). However, now we have $s_1 + s_2 = 1 - a_0$ and so $a_0 + 2s_1 = 1 + N$. Therefore, in this case,

$$y_2(z) = y_1(z) \int^z \frac{f(\zeta)}{(\zeta - z_0)^{N+1}} d\zeta$$

$$= y_1(z) \left[\frac{1}{N} \frac{-f_0}{(z - z_0)^N} + \ldots + \frac{-f_{N-1}}{z - z_0} + f_N \ln(z - z_0) + \sum_{m=1}^{\infty} f_{N+m}(z - z_0)^m \right].$$

$$\tag{9.5.14}$$

We have no a priori information about f_N. Depending on the DE it may be either zero or non- zero corresponding to either a Frobenius or a generalized Frobenius represen- tation for $y_2(z)$.

We can allow for both cases by using a multiplicative constant and setting

$$y_2(z) = c\, y_1(z) \ln(z - z_0) + (z - z_0)^{s_2} \sum_{m=0}^{\infty} d_m (z - z_0)^m. \qquad (9.5.15)$$

Notice that, consistent with (9.5.14), we have chosen
1. the index for the Frobenius series in the second term of (9.5.15) to be s_2, and
2. the summation to begin with $m = 0$.

Thus, should the multiplicative constant c turn out to be zero, not only will the DE have **two** linearly independent solutions of the Frobenius form but they will correspond respectively to the **two** distinct roots of the indicial equation. This happy outcome always obtains when the distinct roots differ by a non-integer.

When the roots are distinct and do not differ by an integer, the equations (9.5.8) yield two distinct and well-defined sets of solutions for the coefficients c_m correspond- ing to the two allowed values of s, s_1 and s_2. Denoting these to sets by $\{c_m\}$ and $\{d_m\}$ respectively, we again have two linearly independent solutions of the Frobenius form,

$$y_1(z) = (z - z_0)^{s_1} \sum_{m=0}^{\infty} c_m (z - z_0)^m, \quad c_0 \neq 0 \quad \text{and}$$

$$y_2(z) = (z - z_0)^{s_2} \sum_{m=0}^{\infty} d_m (z - z_0)^m, \quad d_0 \neq 0. \qquad (9.5.16)$$

We will illustrate all of these possibilities with some examples. But first, a comment about irregular singular points is in order. It is easy to verify that if $a(z)$ and $b(z)$ are more singular than we have assumed, the indicial equation will have at most one root and so the DE may have no solution of the Frobenius form. In that case, often corre- sponding to solutions with an essential singularity at $z = z_0$, other techniques are required.

9.6 The Legendre Differential Equation

Adrien-Marie Legendre (1752-1833), a Parisian from a wealthy background, taught more or less continuously at the École Militaire and the École Normale despite the many, often turbulent, regime changes of that period. Although best known as a geometer, he also made important contributions to classical mechancs, mathematical analysis, number theory and statistics. He was made an officer of the Légion d'Honneur in 1831.

A DE that arises in a great many physical applications that require the use of spherical coordinates is named after Legendre. Using real variable notation again, it is

$$(1 - x^2)\frac{d^2 y}{d x^2} - 2x\frac{dy}{dx} + \lambda y = 0, \quad \lambda = \text{ a real constant.} \tag{9.6.1}$$

In applications, the variable x is actually the cosine of the polar angle θ and so it has the range $-1 \leq x \leq 1$. Recognizing that $x = \pm 1$ are regular singular points of the DE we begin to worry that its solutions will be singular there. This is something we cannot permit and so this DE is always accompanied in applications by the **boundary conditions** $|y(\pm 1)| < \infty$.

The point $x = 0$ is an ordinary point of the DE. Therefore, we know that its general solution has a Taylor series representation $y(x) = \sum_{m=0}^{\infty} c_m x^m$, $|x| < 1$. Given our observation about $x = \pm 1$ we expect that the series will diverge there, an expectation that can be confirmed explicitly. Thus, while seeking the coefficients c_m we will also be interested in finding some means of modifying the representation so that its range of validity is extended to include ± 1.

Rather than start with the normal assumption of a Taylor series, we shall assume a Frobenius series

$$y(x) = \sum_{m=0}^{\infty} c_m x^{s+m}, c_0 \neq 0 \tag{9.6.2}$$

and then confirm that the roots of the indicial equation give rise to a Taylor series for the general solution.

Differentiating (9.6.2) and substituting into (9.6.1) we have

$$\sum_{m=0}^{\infty} c_m(s + m)(s + m - 1) x^{s+m-2} - \sum_{m=0}^{\infty} c_m[(s + m)(s + m - 1) + 2(s + m) - \lambda] x^{s+m} = 0. \tag{9.6.3}$$

The first term corresponds to y'', the second to $- x^2 y''$, the third to $-2xy'$, and the fourth to λy.

The lowest power of x in this equation is x^{s-2} (from the $m = 0$ term in the first sum). Its coefficient $c_0 s(s - 1)$ and the constraint $c_0 \neq 0$ gives us the indicial equation:

$$s(s - 1) = 0 \tag{9.6.4}$$

whose roots are $s_1 = 1$ and $s_2 = 0$. We note that the roots differ by an integer.

The next power of x is x^{s-1} (from the $m = 1$ term in the first sum). Its coefficient

$$c_1(s + 1)s$$

must also be zero. Therefore,

1. if $s = 1, c_1 = 0$ and

2. if $s = 0$, c_1 is arbitrary.

So far, everything is working out as our analysis in the previous section suggested it would.

Now we consider the coefficient of the general power x^{s+m-2}. Equating it to zero we have

$$c_m(s + m)(s + m - 1) - c_{m-2}[(s + m - 2)(s + m - 3) + 2(s + m - 2) - \lambda] = 0.$$

This becomes a recurrence relation for the coefficients. In fact, inserting the value $s = 1$, we find

$$c_m = \frac{[(m - 1)m - \lambda]}{m(m + 1)} c_{m-2}, \quad m \geq 2, \tag{9.6.5}$$

which expresses all of the even coefficients in terms of c_0 and all of the odd coefficients in terms of c_1. But, since c_1 is zero for this value of s, this means that we obtain a single solution, multiplied by the arbitrary constant c_0:

$$y_1(x) = x + \frac{1 \cdot 2 - \lambda}{3!} x^3 + \frac{(3 \cdot 4 - \lambda)(1 \cdot 2 - \lambda)}{5!} x^5 + \frac{(5 \cdot 6 - \lambda)(3 \cdot 4 - \lambda)(1 \cdot 2 - \lambda)}{7!} x^7 + \dots$$
$$\tag{9.6.6}$$

where we have set $c_0 = 1$. Note that the solution is an odd function of x.

If we use the other root $s = 0$, the recurrence relation becomes

$$c_m = \frac{[(m - 2)(m - 1) - \lambda]}{m(m - 1)} c_{m-2}, m \geq 2. \tag{9.6.7}$$

Again, all of the even coefficients relate back to c_0 and the odd coefficients to c_1 but this time c_1 is arbitrary. Thus, as expected for an expansion about an ordinary point, (9.6.7) generates the general solution in the form of a linear combination of two linearly independent Taylor series. Moreover, as predicted in the commentary following equation (9.5.13), the particular solution multiplying c_1 is just the $y_1(x)$ in equation (9.6.6). The solution multiplying c_0 is

$$y_0(x) = 1 + \frac{(-\lambda)}{2!} x^2 + \frac{(3 \cdot 2 - \lambda)(-\lambda)}{4!} x^4 + \frac{(5 \cdot 4 - \lambda)(3 \cdot 2 - \lambda)(-\lambda)}{6!} x^6 + \dots. \tag{9.6.8}$$

Neither series converges at $x = \pm 1$. Therefore, the only way we can be assured of having a solution that is well-defined for all x in the range $-1 \leq x \leq 1$ is to take advantage of the fact that the numerator of (9.6.7) can vanish. In fact, $c_{l+2} = 0$ when $\lambda = l(l + 1)$, $l = 0, 1, 2, \dots$, and one of the two linearly independent solutions becomes a **polynomial** of degree l. When normalized to have the value 1 at $x = 1$, these solutions are called **Legendre polynomials** and are denoted by $P_l(x)$. They occur in a wide range of physical applications and will be studied in some detail in Chapter 11. For now it suffices to note that our normalization requirement can be met by setting

$$c_l = \frac{(2l)!}{2^l (l!)^2}. \tag{9.6.9}$$

Then, using (9.6.7) in the form

$$c_{m-2} = \frac{(m-1)m}{(m-2)(m-1) - l(l+1)} c_m,$$

we find

$$c_{l-2k} = (-1)^k \frac{(2l-2k)!}{2^l \, k!(l-k)!(l-2k)!}. \tag{9.6.10}$$

Thus, for example,

$$P_0(x) = 1,$$
$$P_1(x) = x,$$
$$P_2(x) = \frac{1}{2}(3\,x^2 - 1),$$
$$P_3(x) = \frac{1}{2}(5\,x^3 - 3x),$$
$$P_4(x) = \frac{1}{8}(35\,x^4 - 30\,x^2 + 3), \ldots .$$

The linearly independent solution for $\lambda = l(l+1)$ is denoted by $Q_l(x)$ and is singular at $x = \pm 1$. In fact, it has branch points there. For the special case of $l = 0$, equation (9.6.7) gives

$$Q_o(x) = x + \frac{x^3}{3} + \frac{x^5}{5} + \frac{x^7}{7} + \ldots = \frac{1}{2} \ln \frac{1-x}{1+x}.$$

9.7 Bessel's Differential Equation

Friedrich Wilhelm Bessel (1784-1846) was a German astronomer who, in the course of studying the dynamics of many body systems, systematized the functions that now bear his name. Although this would be a sufficient accomplishment to rank him as one of the more important mathematicians of this period, his contributions to astronomy were even more important. In particular, he was the first to use parallax to calculate the distance to a star.

Another DE that occurs in many, many physical guises is called **Bessel's equation.** In its most general (real variable) form it is

$$x^2 \frac{d^2 y}{d x^2} + x \frac{dy}{dx} + (x^2 - \mu^2)y = 0. \tag{9.7.1}$$

Here μ is a non-negative real parameter called the **order** of the equation.

We notice immediately that $x = 0$ is a regular singular point. As we shall see, everything that can happen with expansions about a regular singular point do happen for the solutions of (9.7.1) as we let the order parameter vary. However, we are always assured of the existence of at least one solution with a Frobenius expansion about $x =$

0:

$$y(x) = \sum_{m=0}^{\infty} c_m x^{s+m}, \quad c_0 \neq 0. \tag{9.7.2}$$

Substituting this into (9.7.1) we find

$$\sum_{m=0}^{\infty}(s+m)(s+m-1)c_m x^{s+m} + \sum_{m=0}^{\infty}(s+m)c_m x^{s+m} + \sum_{m=0}^{\infty} c_m x^{s+m+2} - \sum_{m=0}^{\infty} \mu^2 c_m x^{s+m} = 0$$

or,

$$\sum_{m=0}^{\infty}[(s+m)^2 - \mu^2]c_m x^{s+m} + \sum_{m=0}^{\infty} c_m x^{s+m+2} = 0. \tag{9.7.3}$$

Equating the coefficients of successive powers of x in (9.7.3) to zero yields the following equations

$$c_0(s^2 - \mu^2) = 0$$
$$c_1[(s+1)^2 - \mu^2] = 0 \tag{9.7.4}$$
$$c_m[(s+m)^2 - \mu^2] + c_{m-2} = 0, \quad m \geq 2.$$

From the first of these we obtain the indicial equation

$$s^2 - \mu^2 = 0 \tag{9.7.5}$$

which has the roots $s = \pm\mu$. Recalling the conclusions of our theoretical analysis in Section 9.5, we note that there is a need to consider four cases based on possible values of μ:

$\mu = 0; \mu = $ an integer; $\mu = a$ half – integer; $\mu = $ anything else.

From the second equation we see that c_1 must be zero unless $\mu = \frac{1}{2}$ and we choose the root $s = -\mu$. In that one exceptional case c_1 is arbitrary and so we are free to set $c_1 = 0$. As we learned in Section 9.5, the terms that we lose by exercising this freedom sum to the solution obtained with the larger root $s = \mu = +\frac{1}{2}$.

From the third equation we obtain the recurrence relation

$$c_m = \frac{-1}{(s+\mu+m)(s-\mu+m)}c_{m-2}, \quad m \geq 2. \tag{9.7.6}$$

This relates all even coefficients to c_0 and all odd coefficients to c_1. Thus, since $c_1 = 0$, **all** of the odd coefficients must be zero also.

The solution corresponding to the largest root of the indicial equation, $s = \mu$, will have a Frobenius expansion about $x = 0$ regardless of the value of μ and we are now in a position to determine what it is. Setting $s = \mu$ in (9.7.6) we have

$$c_m = \frac{-1}{(2\mu+m)m}c_{m-2}$$

or, since m is even, $m = 2k$, $k = 0, 1, 2, \ldots,$

$$c_{2k} = \frac{-1}{2^2 k(\mu + k)} c_{2k-2} . \tag{9.7.7}$$

Thus, starting with c_{2k} and applying (9.7.7) k times, we have

$$c_{2k} = \frac{(-1)^k}{2^{2k} k!(\mu + 1)(\mu + 2)\ldots(\mu + k)} c_0 . \tag{9.7.8}$$

At this point it is conventional to choose

$$c_0 = \frac{1}{2^\mu \, \Gamma(\mu + 1)}$$

so that (9.7.8) becomes

$$c_{2k} = \frac{(-1)^k}{2^{\mu+2k} k!\Gamma(\mu + k + 1)} . \tag{9.7.9}$$

This completes the determination of the solution corresponding to $s = \mu$ which, following convention, we will denote $J_\mu(x)$:

$$J_\mu(x) = \sum_{k=0}^{\infty} \frac{(-1)^k}{k!\Gamma(\mu + k + 1)} \left(\frac{x}{2}\right)^{\mu+2k} . \tag{9.7.10}$$

The rather lengthy name that is attached to this series is **Bessel function of the first kind of order** μ. It looks a good deal more friendly when we assign μ integer or half-integer values.

Specifically, if $\mu = m$, an integer or zero, we can replace $\Gamma(m + k + 1)$ by $(m + k)!$ and obtain

$$J_m(x) = \sum_{k=0}^{\infty} \frac{(-1)^k}{k!(m + k)!} \left(\frac{x}{2}\right)^{m+2k}, \quad m \geq 0. \tag{9.7.11}$$

And, if $\mu = \frac{1}{2}$,

$$J_{\frac{1}{2}}(x) = \sqrt{\frac{x}{2}} \sum_{k=0}^{\infty} \frac{(-1)^k}{k!\Gamma(\frac{1}{2} + k + 1)} \left(\frac{x}{2}\right)^{2k}$$

$$= \sqrt{\frac{2x}{\pi}} \sum_{k=0}^{\infty} \frac{(-1)^k}{(2k + 1)!} x^{2k+1}$$

$$= \sqrt{\frac{2}{\pi x}} \sin x \tag{9.7.12}$$

where we have used the Taylor series for $\sin x$ and the identity

$$\Gamma(k + 1 + \frac{1}{2}) = \frac{(2k + 1)!}{2^{2k+1} k!} \sqrt{\pi}.$$

We shall now turn our attention to finding a second, linearly independent solution to Bessel's equation. The first case we will consider is μ not equal to zero, an integer or a half- integer so that the two roots of the indicial equation will be distinct and will not differ by an integer. According to our analysis in Section 9.5, the second solution will then be the Frobenius series obtained with the root $s = -\mu$. Thus, making that substitution in (9.7.6) we obtain the recurrence relation

$$c_{2k} = \frac{-1}{2^2 k(k - \mu)} c_{2k-2}, \quad k \geq 1. \tag{9.7.13}$$

Applying this k times we find

$$c_{2k} = \frac{(-1)^k}{2^{2k} k!(1 - \mu)(2 - \mu)\ldots(k - \mu)} c_0. \tag{9.7.14}$$

Therefore, choosing

$$c_0 = \frac{1}{2^{-\mu} \Gamma(1 - \mu)}$$

in analogy to what was done for the first solution, we obtain as a second, linearly independent solution the Bessel function of the first kind of order $-\mu$:

$$J_{-\mu}(x) = \sum_{k=0}^{\infty} \frac{(-1)^k}{k!\Gamma(k - \mu + 1)} \left(\frac{x}{2}\right)^{2k-\mu}. \tag{9.7.15}$$

This continues to be a well-defined, independent solution of Bessel's equation when μ has half-integer value. In fact,

$$J_{-\frac{1}{2}}(x) = \sqrt{\frac{2}{\pi x}} \cos x. \tag{9.7.16}$$

Evidently, this is an instance of both solutions having Frobenius representations even though the roots of the indicial equation differ by an integer. Will our luck hold with the same being true when $\mu = m$, an integer? The answer is no. Since $\Gamma(k - m + 1)$ is infinite for $k = 0, 1, 2, \ldots, m - 1$, the coefficients of the first m terms in the series for $J_{-m}(x)$ vanish and the summation starts with $k = m$:

$$J_{-m}(x) = \sum_{k=m}^{\infty} \frac{(-1)^k}{k!(k - m)!} \left(\frac{x}{2}\right)^{2k-m}.$$

But, changing the summation index to $j = k - m$, this becomes

$$J_{-m}(x) = \sum_{j=0}^{\infty} \frac{(-1)^{m+j}}{j!(j + m)!} \left(\frac{x}{2}\right)^{2j+m} = (-1)^m J_m(x), \quad m = 1, 2, \ldots. \tag{9.7.17}$$

Thus, for the case of $\mu = m$, an integer, the second, linearly independent solution can only be expanded about $x = 0$ in a **generalized** Frobenius series. And, of course, the same is true for $\mu = 0$. These solutions are called **Bessel functions of the**

second kind and are denoted by $Y_m(x)$, $m = 0, 1, 2, \ldots$. Their generalized Frobenius representations are

$$Y_m(x) = J_m(x) \ln x + x^{-m} \sum_{k=0}^{\infty} d_k\, x^k, \quad m = 1, 2, \ldots, \tag{9.7.18}$$

and

$$Y_0(x) = J_0(x) \ln x + \sum_{k=1}^{\infty} d_k\, x^k. \tag{9.7.19}$$

We shall solve explicitly for the coefficients $\{d_k\}$ for the case $m = 0$ and state the outcome of such solution for integer values of m.

Substituting (9.7.19) into Bessel's equation of order zero,

$$x \frac{d^2 y}{d x^2} + \frac{dy}{dx} + xy = 0, \tag{9.7.20}$$

and using the fact that $J_0(x)$ is known to be a solution of this DE, we find

$$2 \frac{dJ_0}{dx} + \sum_{k=1}^{\infty} k(k-1)\, d_k\, x^{k-1} + \sum_{k=1}^{\infty} k\, d_k\, x^{k-1} + \sum_{k=1}^{\infty} d_k\, x^{k+1} = 0. \tag{9.7.21}$$

From (9.7.11) we know that

$$\frac{dJ_0}{dx} = \sum_{k=1}^{\infty} \frac{(-1)^k}{k!(k-1)!} \left(\frac{x}{2} \right)^{2k-1}$$

and so inserting this series in (9.7.21), we have

$$2 \sum_{k=1}^{\infty} \frac{(-1)^k}{k!(k-1)!} \left(\frac{x}{2} \right)^{2k-1} + \sum_{k=1}^{\infty} k^2\, d_k\, x^{k-1} + \sum_{k=1}^{\infty} d_k\, x^{k-1} = 0.$$

The coefficient of the lowest power of $x(x^0)$ is just d_1 and so we immediately obtain

$$d_1 = 0.$$

Equating the coefficient of any even power of $x(x^{2k})$ to zero, we have

$$(2k+1)^2\, d_{2k+1} + d_{2k-1} = 0, \quad k = 1, 2, \ldots.$$

Therefore, since $d_1 = 0$, so must $d_3 = 0$, $d_5 = 0, \ldots$, successively.
Equating the coefficient of any odd power of $x(x^{2k+1})$ to zero, we have

$$-1 + 4\, d_2 = 0, \quad k = 0, \text{ and}$$

$$\frac{(-1)^{k+1}}{2^{2k}(k+1)!k!} + (2k+2)^2\, d_{2k+2} + d_{2k} = 0, \quad k = 1, 2, \ldots. \tag{9.7.22}$$

Thus, $d_2 = \frac{1}{4}$, $\frac{1}{8} + 16 d_4 + d_2 = 0$ or, $d_4 = \frac{-3}{128}$ and, in general,

$$d_{2k} = \frac{(-1)^{k-1}}{2^{2k}(k!)^2}\left\{1 + \frac{1}{2} + \frac{1}{3} + \frac{1}{4} + \ldots + \frac{1}{k}\right\}, \quad k = 1, 2, \ldots. \tag{9.7.23}$$

Collecting all of these results and applying them to (9.7.19) we can express the Bessel function of the second kind of order zero as

$$Y_0(x) = J_0(x)\ln x + \sum_{k=1}^{\infty} \frac{(-1)^{k-1}}{(k!)^2}\left\{1 + \frac{1}{2} + \frac{1}{3} + \ldots + \frac{1}{k}\right\}\left(\frac{x}{2}\right)^{2k}$$

$$= J_0(x)\ln x + \frac{1}{4}x^2 - \frac{3}{128}x^4 + - \ldots. \tag{9.7.24}$$

Exactly the same procedure can be followed to determine the second linearly independent solution of

$$x^2 \frac{d^2 y}{dx^2} + x\frac{dy}{dx} + (x^2 - m^2)y = 0, \quad m = 1, 2, \ldots. \tag{9.7.25}$$

That is to say, we can assume a solution of the form

$$Y_m(x) = J_m(x)\ln x + x^{-m}\sum_{k=0}^{\infty} d_k x^k, \tag{9.7.26}$$

substitute it into (9.7.25) and solve for the coefficients $\{d_k\}$. This is precisely what we would do if (9.7.25) were just any old DE. However, because of the physical relevance of Bessel's equation, it is conventional to take an approach that yields a second solution that is defined in an order-independent way.

The definition of the second, linearly independent solution that is used for **all** values of the order parameter μ is

$$N_\mu(x) = \frac{J_\mu(x)\cos\mu\pi - J_{-\mu}(x)}{\sin\mu\pi}. \tag{9.7.27}$$

This is called the **Neumann function** of order μ. For $\mu \neq$ an integer or zero it is a well-defined linear combination of $J_\mu(x)$ and $J_{-\mu}(x)$ and in particular, for $\mu = \frac{2l+1}{2}, l = 0, 1, 2, \ldots,$

$$N_{\frac{2l+1}{2}}(x) = (-1)^l J_{-\frac{2l+1}{2}}(x). \tag{9.7.28}$$

For $\mu =$ an integer or zero (9.7.27) produces $\frac{0}{0}$ by dint of the identity (9.7.17). For these cases the Neumann function is defined by an application of L'Hospital's rule:

$$N_m(x) = \lim_{\mu\to\infty} \frac{J_\mu(x)\cos\mu\pi - J_{-\mu}(x)}{\sin\mu\pi} = \lim_{\mu\to\infty} \frac{\frac{dJ_\mu}{d\mu}\cos\mu\pi - \pi\sin\mu\pi J_\mu(x) - \frac{dJ_{-\mu}}{d\mu}}{\pi\cos\mu\pi}$$

$$= \frac{1}{\pi}\left\{\frac{dJ_\mu(x)}{d\mu}\bigg|_{\mu=m} - (-1)^m \frac{dJ_\mu(x)}{d\mu}\bigg|_{\mu=m}\right\}, \quad m = 0., 1, 2\ldots. \tag{9.7.29}$$

To calculate the derivatives with respect to μ we use the series expansions (9.7.10) and (9.7.15):

$$\frac{dJ_\mu(x)}{d\mu}\bigg|_{\mu=m} = J_m(x)\ln\frac{x}{2} + \sum_{k=0}^\infty \frac{(-1)^k}{k!}\left(\frac{x}{2}\right)^{m+2k}\frac{d}{dz}\frac{1}{\Gamma(z)}\bigg|_{z=m+k+1}; \qquad (9.7.30)$$

$$\frac{dJ_{-\mu}(x)}{d\mu}\bigg|_{\mu=m} = -J_{-m}(x)\ln\frac{x}{2} - \sum_{k=0}^\infty \frac{(-1)^k}{k!}\left(\frac{x}{2}\right)^{-m+2k}\frac{d}{dz}\frac{1}{\Gamma(z)}\bigg|_{z=-m+k+1}. \qquad (9.7.31)$$

We know from Section 5.2 that

$$\frac{d}{dz}\frac{1}{\Gamma(z)}\bigg|_{z=n} = \begin{cases} -\frac{1}{(n-1)!}\left[\frac{1}{n-1} + \frac{1}{n-2} + \ldots + \frac{1}{1} - \gamma\right], & n \geq 2 \\ \gamma, & n = 1 \\ (-1)^{|n|}\,|n|!, & n = 0, -1, -2, \ldots \end{cases}$$

$$(9.7.32)$$

where γ is the Euler-Mascheroni constant,

$$\gamma = \lim_{n\to\infty}\left[\frac{1}{1} + \frac{1}{2} + \ldots + \frac{1}{n} - \ln n\right] = 0.5772\ldots.$$

Thus, on substituting (9.7.30) through (9.7.32) in (9.7.29), we obtain finally

$$N_m(x) = \frac{2}{\pi}J_m(x)\ln\frac{x}{2}$$

$$+ \frac{1}{\pi}\sum_{k=0}^\infty \frac{(-1)^k}{k!}\left(\frac{x}{2}\right)^{m+2k}\frac{-1}{(m+k)!}\left\{\frac{1}{m+k} + \frac{1}{m+k-1} + \ldots + \frac{1}{1} - \gamma\right\}$$

$$+ \frac{1}{\pi}\sum_{k=0}^{m-1}\frac{(-1)^{k+m}}{k!}\left(\frac{x}{2}\right)^{-m+2k}(-1)^{m-k-1}(m-k-1)!$$

$$+ \frac{1}{\pi}\sum_{k=m}^\infty \frac{(-1)^{k+m}}{k!}\left(\frac{x}{2}\right)^{-m+2k}\frac{(-1)}{(k-m)!}\left\{\frac{1}{k-m} + \frac{1}{k-m-1} + \ldots + \frac{1}{1} - \gamma\right\}.$$

$$(9.7.33)$$

Setting $m = 0$ and simplifying we recognize the series that emerges for $N_0(x)$ as the linear combination

$$N_0(x) = \frac{2}{\pi}[Y_0(x) + (\gamma - \ln 2)J_0(x)]. \qquad (9.7.34)$$

Had we solved explicitly for $Y_m(x)$, m = an integer, we would find that this relationship between Neumann functions and Bessel functions of the second kind obtains for non-zero values of the order as well:

$$N_m(x) = \frac{2}{\pi}[Y_m(x) + (\gamma - \ln 2)J_m(x)], \quad m = 0, 1, 2, \ldots. \qquad (9.7.35)$$

9.8 Some Other Tricks of the Trade

Representation of the solution by a series about the origin is a sufficiently powerful approach that we can undertake the solution of most physical problems with confidence. Nevertheless, the method does have its limitations. Therefore, it is useful to be aware of the supplementary techniques discussed in this Section. Each is designed to circumvent one or more of the limitations.

9.8.1 Expansion About the Point at Infinity

Expansion about the origin is possible when the DE

$$\frac{d^2 y}{d z^2} + a(z)\frac{dy}{dz} + b(z)y = 0 \qquad (9.8.1)$$

has an ordinary or regular singular point at $z = 0$. Two immediate limitations are
1. it yields a solution whose domain of definition is restricted to $|z| < |\zeta|$ where $z = \zeta$ is the location of the next or nearest singularity, and
2. it cannot be used at all if $z = 0$ is an irregular singular point.

These can be circumvented if $z = \infty$ is an ordinary or regular singular point because that means that the DE admits at least one solution with a representation of the form

$$y(z) = \left(\frac{1}{z}\right)^s \sum_{m=0}^{\infty} c_m \left(\frac{1}{z}\right)^m \qquad (9.8.2)$$

for $|z| > |\zeta|$ where $z = \zeta$ is the singularity furthest from the origin.

As usual, we test the status of $z = \infty$ by making the substitution $z = \frac{1}{w}$ and determining whether $w = 0$ is an ordinary, regular singular or irregular singular point. Since

$$\frac{dy}{dz} = \frac{dy}{dw}\frac{dw}{dz} = \frac{-1}{z^2}\frac{dy}{dw} = -w^2\frac{dy}{dw}$$

$$\frac{d^2 y}{d z^2} = \frac{d}{dw}\left(\frac{dy}{dz}\right)\frac{dw}{dz} = -w^2\left[-2w\frac{dy}{dw} - z^2\frac{d^2 y}{d z^2}\right] = 2w^3\frac{dy}{dw} + w^4\frac{d^2 y}{d z^2},$$

the DE (9.8.1) becomes

$$w^4\frac{d^2 y}{d w^2} + [2w^3 - w^2 a(w^{-1})]\frac{dy}{dw} + b(w^{-1})y = 0. \qquad (9.8.3)$$

Thus, if $\frac{2}{w} - \frac{a(w^{-1})}{w^2}$ has no worse than a first order pole and $\frac{b(w^{-1})}{w^4}$ has no worse than a second order pole at $w = 0$, our original DE (9.8.1) will admit a solution that can be expanded in a Frobenius series about $z = \infty$ like that in (9.8.2).

Examples: Consider the Legendre DE of order zero,

$$(1 - x^2)\frac{d^2 y}{d x^2} - 2x\frac{dy}{dx} = 0$$

We know that it has the two linearly independent solutions

$$P_0(x) = 1 \quad \text{and} \quad Q_0(x) = x + \frac{x^3}{3} + \frac{x^5}{5} + \dots, \quad |x| < 1.$$

Since we know that the series sums to $\ln\left(\frac{1+x}{1-x}\right)$ we have an analytic continuation for $|x| > 1$. However, we shall ignore this knowledge and seek a series solution that is valid in the latter domain.

Setting $x = \frac{1}{u}$ and noting that

$$a(u^{-1}) = \frac{-2/u}{1 - 1/u^2} = \frac{-2u}{u^2 - 1} \quad \text{and} \quad b(u^{-1}) = 0,$$

we find that

$$\frac{2}{u} - \frac{1}{u^2} a(u^{-1}) = \frac{2u}{u^2 - 1} \quad \text{and} \quad \frac{b(u^{-1})}{u^4} = 0 \quad \text{and so the DE becomes}$$

$$\frac{d^2 y}{d u^2} + \frac{2u}{u^2 - 1} \frac{dy}{du} = 0.$$

Recognizing that the transformed DE is still Legendre's equation of order zero, we can write down two linearly independent solutions without further ado:

$$y_1(u^{-1}) = 1 \quad \text{and} \quad y_2(u^{-1}) = u + \frac{u^3}{3} + \frac{u^5}{5} + \dots \quad \text{for } |u| < 1.$$

Transforming back, this means that the original DE has the solutions

$$y_1(x) = 1 \quad \text{and} \quad y_2(x) = \frac{1}{x} + \frac{1}{3x} + \frac{1}{5x} + \dots = \sum_{m=0}^{\infty} \frac{x^{-(2m+1)}}{2m+1} \quad \text{for } |x| > 1.$$

Evidently, $y_1(x)$ is just $P_0(x)$ while $y_2(x)$ may be identified with $Q_0(x)$. Thus, we have found the representation we were seeking and, of course, it sums to $\frac{1}{2} \ln\left(\frac{x+1}{x-1}\right)$.

As a second **example**, consider the DE

$$x^4 \frac{d^2 y}{d x^2} + 2 x^3 \frac{dy}{dx} - y = 0$$

which has an **irregular** singular point at $x = 0$. Substituting $u = \frac{1}{x}$, we find

$$\frac{2}{u} - \frac{1}{u^2} a(u^{-1}) = \frac{2}{u} - \frac{1}{u^2} \cdot 2u = 0,$$

and

$$\frac{1}{u^4} b(u^{-1}) = \frac{1}{u^4}(-u^4) = -1$$

so that the transformed DE reads

$$\frac{d^2 y}{d u^2} - y = 0.$$

This has constant coefficients and can be readily solved. The result is

$$y_1(u^{-1}) = e^u = \sum_{m=0}^{\infty} \frac{u^m}{m!} \quad \text{and} \quad y_2(u^{-1}) = e^{-u} = \sum_{m=0}^{\infty} \frac{(-1)^m u^m}{m!}.$$

Thus, transforming back to $x = u^{-1}$, the solutions of our original DE are

$$y_1(x) = e^{\frac{1}{x}} = \sum_{m=0}^{\infty} \frac{x^{-m}}{m!} \quad \text{and} \quad y_2(x) = e^{-\frac{1}{x}} = \sum_{m=0}^{\infty} \frac{(-1)^m x^{-m}}{m!}, \quad |x| > 0.$$

9.8.2 Factorization of the Behaviour at Infinity

In appropriate units, the Schrodinger equation for a one-dimensional harmonic oscillator reduces to

$$\frac{d^2 y}{d x^2} + (\lambda - x^2)y = 0, \quad \lambda = \text{a real constant}. \tag{9.8.4}$$

This equation has an irregular singular point at infinity and its solutions have essential singularities there. In fact, for large values of x and all values of λ, the DE is approximated by the equations

$$\frac{d^2 y}{d x^2} - (x^2 \pm 1)y = 0$$

whose solutions are $y(x) = e^{\pm\frac{x^2}{2}}$. Therefore, it is plausible to expect the solutions of (9.8.4) to behave like either $e^{\frac{x^2}{2}}$ or $e^{-\frac{x^2}{2}}$ as $x \to \infty$. However, to be physically acceptable, solutions must be bounded everywhere. To ensure that we comply with this (boundary) condition, we **factor** in the desired **behaviour at infinity** and set $y(x) = v(x)\, e^{-\frac{x^2}{2}}$. The DE is then transformed into a simpler DE for $v(x)$:

$$\frac{d^2 v}{d x^2} - 2x\frac{dv}{dx} + (\lambda - 1)v = 0. \tag{9.8.5}$$

This is known as the **Hermite equation**. It has an ordinary point at the origin and so both of its linearly independent solutions can be represented by Taylor series about $x = 0$. Both of these series behaves like e^{x^2} for large x. However, as with the Legendre DE, we can arrange to have one of the series terminate and become a polynomial of degree n by restricting λ to the the integer values $\lambda = 2n + 1, n = 0, 1, 2, \ldots$. With appropriate normalization, this solution defines the **Hermite polynomial of order n** and is denoted by $H_n(x)$. Thus, the physically acceptable solution of (9.8.4), (the wavefunction of a one-dimensional harmonic oscillator in its nth energy level), is

$$y(x) = e^{-\frac{x^2}{2}} H_n(x).$$

Explicitly solving (9.8.5) provides a useful exercise. We start with the usual assumption for an expansion about an ordinary point:

$$v(x) = \sum_{m=0}^{\infty} c_m x^m, \quad c_0 \neq 0.$$

Substitution into the DE then yields

$$\sum_{m=0}^{\infty} c_m \, m(m-1) \, x^{m-2} + \sum_{m=0}^{\infty} c_m (\lambda - (2m+1)) \, x^m = 0.$$

Equating coefficients of successive powers of x to zero, we find that c_0 and c_1 are arbitrary and that all coefficients c_m are linked by the recurrence relation

$$c_{m+2} = \frac{(2m+1) - \lambda}{(m+2)(m+1)} \, c_m \, .$$

This yields the general solution

$$v(x) = c_0 \left[1 + \frac{1-\lambda}{2!} \, x^2 + \frac{(1-\lambda)(5-\lambda)}{4!} \, x^4 + \ldots \right] + c_1 \left[x + \frac{3-\lambda}{3!} \, x^3 + \frac{(3-\lambda)(7-\lambda)}{5!} \, x^5 + \ldots \right].$$

For a specific set of values of λ, $\lambda = 2n + 1$, $n = 0, 1, 2, \ldots$, one of these series terminates after the x^n term and, suitably normalized, yields the **Hermite polynomial, $H_n(x)$**. The large x behaviour of the other series is determined by the higher order terms, those with $m \gg n$. For such large m, the recurrence relation is approximately

$$\frac{c_{m+2}}{c_m} \approx \frac{2}{m}$$

which is the relation satisfied by the coefficients in the Taylor series expansion of e^{x^2}. Thus, the corresponding solution of (9.8.4) behaves like $y(x) \approx e^{\frac{x^2}{2}}$ as $x \to \infty$.

9.8.3 Changing the Independent Variable

If we make the substitution $x \to t = T(x)$, the DE

$$\frac{d^2 y}{d x^2} + a(x)\frac{dy}{dx} + b(x)y = 0 \tag{9.8.6}$$

becomes

$$[T'(x)]^2 \, \frac{d^2 y}{d t^2} + [T''(x) + a(x)T'(x)]\frac{dy}{dt} + b(x)y = 0. \tag{9.8.7}$$

One can now choose $T(x)$ to simplify the transformed equation and thereby obtain a DE that (one hopes) is easier to solve. The obvious choice is to require

$$\frac{d^2 T}{d x^2} + a(x)\frac{dT}{dx} = 0,$$

or

$$\frac{dT}{dx} = \exp\left\{ -\int^x a(\xi)d\xi \right\},$$

or

$$T(x) = \int^{x} \exp\left\{ -\int^{\zeta} a(\xi)d\xi \right\} d\zeta \tag{9.8.8}$$

so that we eliminate the second term in (9.8.7).

Example: The **Euler equation**

$$x^2 \frac{d^2 y}{dx^2} + x\frac{dy}{dx} + y = 0 \tag{9.8.9}$$

can be simplified by setting

$$t = T(x) = \int^{x} \exp\left\{ -\int^{\zeta} \frac{1}{\xi}d\xi \right\} d\zeta = \int^{x} \frac{d\zeta}{\zeta} = \ln x$$

or, $x = e^t$. Under this transformation the DE becomes

$$\left(\frac{1}{x}\right)^2 \frac{d^2 y}{dt^2} + \frac{1}{x^2}y = 0 \ \text{ or } \ \frac{d^2 y}{dt^2} + y = 0$$

whose solutions are $y_1(t) = \cos t$ and $y_2(t) = \sin t$. Therefore, two linearly independent solutions of the Euler equation are

$$y_1(x) = \cos(\ln x) \ \text{ and } \ y_2(x) = \sin(\ln x). \tag{9.8.10}$$

The Euler equation can be solved almost as quickly using the Frobenius method. As an exercise, show that this yields an equivalent linearly independent pair, $y_1(x) = x^i$ and $y_2(x) = x^{-i}$.

9.8.4 Changing the Dependent Variable

Factoring the behaviour at infinity illustrated the utility of replacing the dependent variable $y(x)$ by a product $u(x) \cdot v(x)$ where $u(x)$ is a known function or a function that can be selected with the express purpose of simplifying the resulting DE for $v(x)$. In fact, making the substitution $y(x) = u(x) \cdot v(x)$ in

$$\frac{d^2 y}{dx^2} + a(x)\frac{dy}{dx} + b(x)y = 0$$

we obtain

$$uv'' + (2u' + a(x)u)v' + (u'' + a(x)u' + b(x)u)v = 0. \tag{9.8.11}$$

At this point we need to know the functional dependence of $a(x)$ and $b(x)$ if we are to make the optimal choice for the function $u(x)$. However, one simplification that can always be achieved is to eliminate the second term in (9.8.11) because that simply requires

$$2\frac{du}{dx} + a(x)u = 0 \quad \text{or} \quad u(x) = \exp\left\{-\int^x a(\xi)d\xi\right\}. \tag{9.8.12}$$

The DE for $v(x)$ then becomes

$$\frac{d^2 v}{d x^2} - \left(b(x) - \frac{1}{2}\frac{da}{dx} - \frac{1}{4}[a(x)]^2\right)v = 0. \tag{9.8.13}$$

With just two terms in this DE, one can more readily estimate the behaviour of its solutions at infinity which turns out to be the principal application of this technique.
Example: The equation $x^2 y'' + 2xy' + (x^2 - 2)y = 0$ is a variant of Bessel's DE called the **spherical** Bessel equation of order two. If we divide through by x^2 it becomes

$$\frac{d^2 y}{d x^2} + \frac{2}{x}\frac{dy}{dx} + \left(1 - \frac{2}{x^2}\right)y = 0.$$

The large x behaviour of the solutions of this equation is not apparent. However, if we make the substitution $y(x) = u(x) \cdot v(x)$ where $u(x)$ is defined by (9.8.12),

$$u(x) = \exp\left\{-\frac{1}{2}\int^x \frac{2}{\xi}d\xi\right\} = \exp\{-\ln x\} = \frac{1}{x},$$

then $v(x)$ is a solution of

$$\frac{d^2 v}{d x^2} + \left(1 - \frac{2}{x^2}\right)v = 0$$

which is approximated by $v'' + v = 0$ for large values of $|x|$. This implies that $v(x)$ behaves like $\cos x$ or $\sin x$ asymptotically and hence that $y(x) \approx \frac{\cos x}{x}$ or $\frac{\sin x}{x}$ as $|x| \rightarrow \infty$. The exact solutions for $v(x)$ can be found by a straightforward application of the Frobenius method. The result is $v_1(x) = \cos x - \frac{\sin x}{x}$ and $v_2(x) = \sin x + \frac{\cos x}{x}$ which do indeed behave like $\cos x$ and $\sin x$ for large $|x|$.

9.9 Solution by Definite Integrals

This Section will pull together a couple of loose ends from earlier chapters. We have seen that integral representations offer an alternative to representation by power series and that they are often more useful because of a larger domain of definition and because they provide a basis from which still other representations such as asymptotic expansions can be derived. We have also seen that Fourier and Laplace transforms offer a particularly effective means of solving a specialized class of differential equations accompanied by appropriate boundary conditions. And of course the result of their application is solutions expressed as Fourier and Laplace integrals. All of this suggests that we should explore the possibility of a general method of solving DE's based on the representation of the solutions as definite integrals.

We start by introducing an **operator** notation that will turn out to be very convenient both here and in Chapter 10 where we address boundary value problems in a formal way. In this notation the DE

$$a_2(x)\frac{d^2 y}{d x^2} + a_1(x)\frac{dy}{dx} + a_0(x)y = 0 \qquad (9.9.1)$$

becomes

$$\mathfrak{L}_x\, y(x) = 0 \quad \text{where} \quad \mathfrak{L}_x \equiv a_2(x)\frac{d^2}{d x^2} + a_1(x)\frac{d}{dx} + a_0(x) \qquad (9.9.2)$$

and is called a **differential operator**. It indicates the operations involved in obtaining the differential equation but is itself only symbolic. Notice that we have abandoned our usual convention of taking the coefficient of the highest derivative to be one. This is for ease of discussion later on and has no fundamental significance.

For any such operator $\mathfrak{L}_x$ and a specific interval $\alpha \le x \le \beta$, one can **define** an **adjoint** $\mathfrak{L}_x^+$ with respect to a **weight function** $w(x)$ by the requirement that for any sufficiently differentiable functions $u(x)$ and $v(x)$,

$$w(x)[v(x)\, \mathfrak{L}_x\, u(x) - u(x)\, \mathfrak{L}_x^+\, v(x)] = \frac{d}{dx} Q(u, v) \qquad (9.9.3)$$

where $Q(u, v)$ is a bilinear combination of $u(x)$, $v(x)$, $\frac{du}{dx}$ and $\frac{dv}{dx}$ and $w(x)$ is some function that is positive definite on the interval in question. The preceding sentence is quite a mouthful. Put more succinctly, it simply requires the left hand side of (9.9.3) be a perfect differential so that $\mathfrak{L}_x^+$ can be determined by a process of partial integration. We will illustrate with an example as soon as we introduce some more nomenclature. Equation (9.9.3) is called the **Lagrange identity** and if we integrate it over the interval $\alpha \le x \le \beta$, we obtain the **generalized Green's identity**

$$\int_{\alpha}^{\beta} [v(x)\, \mathfrak{L}_x\, u(x)]w(x)dx - \int_{\alpha}^{\beta} [u(x)\, \mathfrak{L}_x^+\, v(x)]w(x)dx = Q(u, v)\big|_{x=\beta} - Q(u, v)\big|_{x=\alpha}. \quad (9.9.4)$$

The right hand side of this equation is called the **boundary** or **surface** term.

Joseph-Louis Lagrange (1736-1813) was born Giuseppe Lodovico Lagrangia in Turin. Although he did not take up residence in France until 1787 at the age of 51, he is generally considered to have been a French mathematician and physicist. Indeed, after surviving the French Revolution, his accomplishments were recognized by Napoleon who made him a Count of the Empire. When he died at age 77, he was buried in the Panthéon in Paris. Lagrange made significant contributions to analysis and number theory but he is most noted for his work in classical mechanics. His two volume monograph on analytical mechanics, published in 1788, was the most comprehensive presentation of classical mechanics since Newton. It provided a foundation for the development of mathematical physics over the next century.

George Green (1793-1841) was an English miller and a self-taught mathematician and physicist. In 1828, four years prior to his admission as a mature undergraduate at Cambridge, he published a treatise that contained the first mathematical theory of electricity and magnetism. Like Lagrange's famous treatise, this provided a foundation for the work of subsequent phyicists such as Maxwell and Lord Kelvin.

Example: The simplest differential operator possible is just $\mathfrak{L}_x = \frac{d}{dx}$. A single partial integration applied to the product $u(x)\,\mathfrak{L}_x\,v(x) = u(x)\frac{dv}{dx}$ gives us

$$\int_\alpha^\beta v(x)\frac{du}{dx}\,dx - \int_\alpha^\beta u(x)\left[-\frac{dv}{dx}\right]dx = u(\beta)v(\beta) - u(\alpha)v(\alpha).$$

Comparing with (9.9.4), we deduce that the adjoint of $\mathfrak{L}_x = \frac{d}{dx}$ with respect to a weight $w(x) = 1$ is $\mathfrak{L}_x^+ = -\frac{d}{dx}$ and that the function $Q(u, v) = u(x)v(x)$. Notice that $\mathfrak{L}_x^+ \neq \mathfrak{L}_x$. If they had been equal, we would say that the operator $\mathfrak{L}_x$ is **self-adjoint.** As it turns out one can convert $\frac{d}{dx}$ into a self-adjoint operator by the simple expedient of multiplication by the pure imaginary i.

When complex functions are involved, the products in the Lagrange identity have to be modified accordingly and (9.9.4) becomes

$$\int_\alpha^\beta [v^*(\mathfrak{L}_x u)]w\,dx - \int_\alpha^\beta [u(\mathfrak{L}_x^+ v)^*]w\,dx = Q(u, v^*)\Big|_{x=\beta} - Q(u, v^*)\Big|_{x=\alpha}. \tag{9.9.5}$$

Applying this to $\mathfrak{L}_x = i\frac{d}{dx}$, we have

$$\int_\alpha^\beta v^*\left[i\frac{du}{dx}\right]dx - \int_\alpha^\beta u\left[i\frac{dv}{dx}\right]^* dx = i[u(\beta)\,v^*(\beta) - u(\alpha)\,v^*(\alpha)].$$

Thus, $\mathfrak{L}_x^+ = i\frac{d}{dx} = \mathfrak{L}_x$, which confirms that this operator is self-adjoint with respect to the weight $w(x) = 1$.

Finding the adjoint is just as straight forward but somewhat more tedious as one increases the order of the differentials in the operator $\mathfrak{L}_x$. In the case of the second order operator with real coefficients,

$$\mathfrak{L}_x = a_2(x)\frac{d^2}{d\,x^2} + a_1(x)\frac{d}{dx} + a_0(x), \tag{9.9.6}$$

one can show that the adjoint, with respect to weight $w(x) = 1$, is

$$\mathfrak{L}_x^+ = a_2(x)\frac{d^2}{d\,x^2} + (2\,a_2'(x) - a_1(x))\frac{d}{dx} + (a_2''(x) - a_1'(x) + a_0(x)). \tag{9.9.7}$$

We note in passing that the $\mathfrak{L}_x$ of (9.9.6) becomes self-adjoint with respect to the weight function

$$w(x) = \frac{1}{a_2(x)} \exp\left\{ \int^x \frac{a_1(\xi)}{a_2(\xi)} d\xi \right\}. \tag{9.9.8}$$

As we saw in Section 5.3, an integral representation of a solution of our DE would take the form (cf. equation (5.3.2))

$$y(z) = \int_C K(z, t)v(t)dt$$

for some kernel $K(z, t)$ and spectral function $v(t)$. Specializing to real variables and choosing C to be the real line segment $\alpha \le t \le \beta$, this becomes

$$y(x) = \int_\alpha^\beta K(x, t)v(t)dt. \tag{9.9.9}$$

Substitution into the DE (9.9.1) results in

$$\mathfrak{L}_x y(x) = \int_\alpha^\beta [\mathfrak{L}_x K(x, t)]v(t)dt \tag{9.9.10}$$

and so the first question to address is what can we do with $\mathfrak{L}_x K(x, t)$ short of assuming an explicit functional form for the kernel $K(x, t)$? The answer is to assume the existence of a differential operator in t, M_t, such that

$$\mathrm{M}_t K(x, t) = \mathfrak{L}_x K(x, t). \tag{9.9.11}$$

Then, applying the Lagrange identity to M_t and its adjoint, we have

$$v(t)[\mathrm{M}_t K(x, t)] - K(x, t)[\mathrm{M}_t^+ v(t)] = \frac{\partial}{\partial t} Q(K, v) \tag{9.9.12}$$

where $Q(K, v)$ is a bilinear function of $K(x, t)$, $v(t)$, and their derivatives. This means that

$$\mathfrak{L}_x y(x) = \int_\alpha^\beta [\mathrm{M}_t K(x, t)]v(t)dt = \int_\alpha^\beta K(x, t)\, \mathrm{M}_t^+ v(t)dt + Q(K, v)\big|_{t=a}^{t=\beta}. \tag{9.9.13}$$

Therefore, $y(x)$ will be a solution of $\mathfrak{L}_x y(x) = 0$ if
1. $Q(K, v)\big|_{t=b} - Q(K, v)\big|_{t=a} = 0$, and
2. $v(t)$ is a solution of the (adjoint) equation

$$M_t^+ v(t) = 0. \tag{9.9.14}$$

Evidently, the success of this approach rests on the relative ease of solving (9.9.14) in comparison with (9.9.1). This in turn depends on a judicious matching of the kernel $K(x, t)$ to the differential operator $\mathcal{L}_x$. We shall illustrate this point with a few **examples.**

As we saw in the last chapter, Laplace Transforms are an effective means of solving DE's with coefficients that are either constant or at worst, linear functions of x. In the present context, this suggests that choosing the kernel

$$K(x, t) = e^{xt}$$

should be useful. Not surprisingly, this is called the **Laplace kernel**.

Because of the unique properties of the exponential, it is particularly easy to match up any $\mathcal{L}_x$ with its M_t counterpart. In fact, one merely replaces each power x^i with the differential $\frac{d^i}{dt^i}$ and each differential $\frac{d^j}{dx^j}$ by the power t^j. For specificity, let's consider the DE

$$\mathcal{L}_x y(x) \equiv x\frac{d^2 y}{dx^2} + (a + b + x)\frac{dy}{dx} + by = 0 \tag{9.9.15}$$

where a and b are constants. Making the prescribed replacements, we have

$$M_t v(t) = t^2 \frac{dv}{dt} + (a + b)tv + t\frac{dv}{dt} + bv = t(t + 1)\frac{dv}{dt} + [(a + b)t + b]v. \tag{9.9.16}$$

Therefore, from (9.9.7), the adjoint equation that we must solve is

$$M_t^+ v(t) = -t(t + 1)\frac{dv}{dt} + [(a + b - 2)t + b - 1]v = 0. \tag{9.9.17}$$

In addition, the Lagrange identity (9.9.12) is

$$v M_t K - K M_t^+ v = \frac{\partial}{\partial t}[t(t + 1)vK]. \tag{9.9.18}$$

The first order DE (9.9.17) can be rewritten as

$$\frac{1}{v(t)}\frac{dv}{dt} = \frac{(a + b - 2)t + b - 1}{t(t + 1)} = \frac{b - 1}{t} + \frac{a - 1}{t + 1}.$$

Thus, integrating and exponentiating, we find that it has the solution

$$v(t) = t^{b-1}(t + 1)^{a-1}. \tag{9.9.19}$$

This means that the right hand side of the Lagrange identity is

$$\frac{\partial}{\partial t}Q(v, K) = \frac{\partial}{\partial t}[t^b(t + 1)^a e^{xt}]$$

and hence, the solution of the original DE (9.9.15) is

$$y(x) = \int_\alpha^\beta e^{xt} t^{b-1}(t + 1)^{a-1} dt \tag{9.9.20}$$

where α and β are chosen so that

$$e^{xt} t^b (t+1)^a \Big|_{t=\beta} - e^{xt} t^b (t+1)^a \Big|_{t=\alpha} = 0. \tag{9.9.21}$$

Rather than complete the solution by making an appropriate choice for α and β, we shall pause to consider what would happen if we just used the Laplace Transform techniques of the preceding chapter. Using the notation of that chapter, Laplace transformation of the DE (9.9.15) gives us

$$-s(s+1)\frac{dY}{ds} + [(a+b-2)s + b - 1]Y(s) - (a+b-1)y(0) = 0. \tag{9.9.22}$$

We do not kow the value of $y(0)$ and so we shall set it to zero arbitrarily. The result is a DE that is identical to the adjoint equation (9.9.17) but with t replaced by s and $v(t)$ by $Y(s)$. Therefore, we know that it has the solution

$$Y(s) = s^{b-1}(s+1)^{a-1}.$$

The next and last step is to invert the transform which, as we have seen can be done with the Mellin inversion integral. Thus, we find

$$y(x)\theta(x) = \frac{1}{2\pi i} \int_{c-i\infty}^{c+i\infty} e^{sx} s^{b-1}(s+1)^{a-1} ds \tag{9.9.23}$$

which, apart from the multiplicative constant, is the same as our solution (9.9.20) but with a specific choice for the integration limits α and β. Recall that c is any real number greater than the exponential order of $y(x)$. We do not know what that is of course but the requirement that the integral vanish for $x < 0$ implies that $c > 0$. By inspection we can see that the surface term (9.9.21) will vanish with these limits only if $a + b < 0$. Presumably, the same constraint applies to the domain of definition of the Mellin integral (9.9.23). Thus, Laplace transformation, while similar, is less general than the assumption of an integral representation with a Laplace kernel. The latter requires less input knowledge and is subject to fewer restrictions on both the independent variable x and the parameters a and b.

Continuing where we left off in the solution by integral representation, we now list pairs of values for α and β that will result in the vanishing of the surface term (9.9.21):

1. $\alpha = -1$ and $\beta = 0$ ($a > 0$, $b > 0$),
2. $\alpha = -\infty$ and $\beta = 0$ ($x > 0$, $b > 0$),
3. $\alpha = -\infty$ and $\beta = -1$ ($x > 0$, $a > 0$),
4. $\alpha = 0$ and $\beta = \infty$ ($x < 0$, $b > 0$),
5. $\alpha = -1$ and $\beta = \infty$ ($x < 0$, $a > 0$).

Thus, for example, when a, b and x are all positive, the general solution of (9.9.15) can be written

$$y(x) = c_1 \int_{-1}^{0} e^{xt} t^{b-1}(t+1)^{a-1} dt + c_2 \int_{-\infty}^{-1} e^{xt} t^{b-1}(t+1)^{a-1} dt \tag{9.9.24}$$

where c_1 and c_2 are arbitrary constants.

Our second **example** involves DE's in which the coefficient of $\frac{d^m y}{dx^m}$ is a polynomial in x of degree m. Such equations can always be expressed as, (this generalizes in an obvious way for DE's of order higher than 2),

$$\mathcal{L}_x\, y(x) \equiv G_0(x)\frac{d^2 y}{dx^2} - \mu\, G_0'(x)\frac{dy}{dx} + \frac{\mu(\mu+1)}{1\cdot 2}\, G_0''(x)y - G_1(x)\frac{dy}{dx} + (\mu+1)\, G_1'(x)y$$
$$+ G_2(x)y = 0 \tag{9.9.25}$$

where μ is a constant and $G_j(x)$ is a polynomial of degree $2 - j$. Such equations can be solved by using the **Euler kernel**, $K(x, t) = (x - t)^{\mu+1}$ since this results in

$$\mathcal{L}_x[(x - t)^{\mu+1}] = C\, M_t[(x - t)^{\mu+p-1}] \tag{9.9.26}$$

where C is a constant, p is the largest value for which $G_p(x)$ is non-zero, and

$$M_t \equiv G_0(t)\frac{d^p}{dt^p} + G_1(t)\frac{d^{p-1}}{dt^{p-1}} + \ldots + G_p(t). \tag{9.9.27}$$

Thus, if $G_2(x) = 0$, M_t will be of first order and so will the adjoint equation

$$M_t^+\, v(t) = 0.$$

Notice that in this case the kernel $K(x, t)$ does not satisfy equation (9.9.11), $\mathcal{L}_x\, K(x, t) = M_t\, K(x, t)$, but rather a generalization of it: $\mathcal{L}_x\, K(x, t) = M_t\, \kappa(x, t)$ where $\kappa(x, t) = C(x - t)^{\mu+p-1}$. Thus, assuming a representation of the form

$$y(x) = \int_{\alpha}^{\beta} (x - t)^{\mu+1}\, v(t)dt, \tag{9.9.28}$$

we obtain

$$\mathcal{L}_x\, y(x) = C\int_{\alpha}^{\beta} M_t[(x - t)^{\mu+p-1}]v(t)dt = C\int_{\alpha}^{\beta}(x - t)^{\mu+p-1}\, M_t^+\, v(t)dt + \left[Q(\kappa, v)\right]_{t=\alpha}^{t=\beta}. \tag{9.9.29}$$

Therefore, (9.9.28) will be a solution of $\mathcal{L}_x\, y(x) = 0$ if $v(t)$ is a solution of the adjoint equation $M_t^+\, v(t) = 0$ and if α and β are so chosen that $\left[Q(\kappa, v)\right]_{t=\alpha}^{t=\beta} = 0$.

To illustrate, we recall that the Legendre DE fits the description prescribed for this type of kernel. It reads

$$\mathcal{L}_x\, y(x) = (1 - x^2)\frac{d^2 y}{dx^2} - 2x\frac{dy}{dx} + l(l+1)y = 0.$$

Thus, using the notation introduced above, we have

$$G_0(x) = 1 - x^2$$

$$\mu \, G_0'(x) + G_1(x) = 2x$$

$$\frac{1}{2}\mu(\mu + 1) \, G_0''(x) + (\mu + 1) \, G_1'(x) + G_2(x) = l(l + 1).$$

Since $G_0'(x) = -2x$, $G_1(x) = 2(\mu + 1)x$. This means $G_0''(x) = -2$ and $G_1'(x) = 2(\mu + 1)$ and so the third equation becomes

$$(\mu + 1)(\mu + 2) + G_2(x) = l(l + 1)$$

which is consistent with $G_2(x) = 0$ provided that $\mu = l - 1$ or $\mu = -l - 2$. Therefore, p is indeed equal to one and the differential operator M_t is

$$M_t \equiv (1 - t^2)\frac{d}{dt} + 2(\mu + 1)t.$$

This yields an adjoint equation of the form

$$M_t^+ v(t) = (1 - t^2)\frac{dv}{dt} - 2(\mu + 2)tv = 0.$$

Rewriting this as

$$\frac{1}{v(t)}\frac{dv}{dt} = \frac{2(\mu + 2)t}{1 - t^2},$$

we integrate and exponentiate to find $v(t) = (1 - t^2)^{-\mu-2}$.

The surface term can be found from

$$\frac{\partial}{\partial t}Q(\kappa, x) = v(t)\, M_t\, \kappa(x, t) - \kappa(x, t)\, M_t^+\, v(t) = v(t)\, M_t\, \kappa(x, t) = C\frac{\partial}{\partial t}[(x - t)^\mu(1 - t^2)^{-\mu-1}].$$

Thus, we require values of α and β such that

$$\left[(x - t)^\mu(1 - t^2)^{-\mu-1}\right]_{t=\alpha}^{t=\beta} = 0.$$

Using the value $\mu = -l - 2$ with $l \geq 0$ and assuming $|x| \neq 1$, $Q(\kappa, x) = 0$ when $t = \pm 1$. Therefore, we set $\alpha = -1$ and $\beta = 1$ and obtain the function

$$y(x) = \int_{-1}^{1} (x - t)^{-l-1}(1 - t^2)^l \, dt$$

as a solution of Legendre's equation. As a matter of fact, this is an integral representation of the Legendre function of the second kind, $Q_l(x)$:

$$Q_l(x) = \frac{1}{2^{l+1}} \int_{-1}^{1} (x - t)^{-l-1}(1 - t^2)^l \, dt.$$

As a final **example** we return to Bessel's equation which reads

$$\mathcal{L}_x \, y(x) \equiv \frac{d^2 y}{d x^2} + \frac{1}{x}\frac{dy}{dx} + \left(1 - \frac{\mu^2}{x^2}\right) y = 0.$$

This does not fit into a class of DE's associated with any of the standard kernels or transformations. However, it has been found by experiment that the kernel

$$K(x, t) = \left(\frac{x}{2}\right)^{\mu} \exp\left(t - \frac{x^2}{4t}\right)$$

is effective in producing a solution. In fact, one can readily show that

$$\mathfrak{L}_x K(x, t) = \left(\frac{\partial}{\partial t} + \frac{\mu + 1}{t}\right) K(x, t)$$

and so, we identify

$$M_t \equiv \frac{d}{dt} - \frac{\mu + 1}{t}.$$

This means the adjoint equation is

$$M_t^{+} v(t) = -\frac{dv}{dt} - \frac{\mu + 1}{t} v = 0$$

which has the solution

$$v(t) = t^{-\mu - 1}.$$

Therefore, a solution of Bessel's equation is

$$y(x) = \left(\frac{x}{2}\right)^{\mu} \int_{\alpha}^{\beta} t^{-\mu - 1} e^{(t - x^2/4t)}\, dt \qquad (9.9.30)$$

where α and β are chosen so that

$$Q(K, v)\Big|_{t=\alpha}^{t=\beta} = \left[t^{-\mu-1} e^{(t-x^2/4t)} \left(\frac{x}{2}\right)^{\mu}\right]_{t=\alpha}^{t=\beta} = 0. \qquad (9.9.31)$$

The only pair of points that qualifies is $t = 0$ and $t \to -\infty$. Thus, we find as a solution,

$$y(x) = \left(\frac{x}{2}\right)^{\mu} \int_{-\infty}^{0} t^{-\mu - 1} \exp\left(t - \frac{x^2}{4t}\right) dt. \qquad (9.9.32)$$

This is not quite equal to the Bessel function $J_{\mu}(x)$. As we will show in Chapter 11, the latter can be represented by the closely related contour integral

$$J_{\mu}(z) = \frac{1}{2\pi i} \left(\frac{z}{2}\right)^{\mu} \int_{C} t^{-\mu - 1} \exp\left(t - \frac{z^2}{4t}\right) dt \qquad (9.9.33)$$

where $-\pi < \arg t < \pi$ and C is a contour that encloses the cut along the negative real axis.

When μ is an integer, $\mu = m$, the contour C can be closed around the origin. This is an obvious generalization of the real definite integral (9.9.30) subject to the constraint (9.9.31) since, by using a contour that does not cross the cut, we are assured that

$$\int_{C} \frac{d}{dt} Q(K, v) dt = \int_{C} \frac{d}{dt}\left[t^{-\mu-1} \exp\left(t - \frac{z^2}{4t}\right)\right] dt = 0.$$

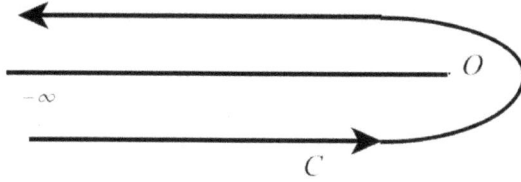

9.10 The Hypergeometric Equation

Gauss's hypergeometric differential equation is

$$x(1-x)y'' + [c - (a+b+1)x]y' - aby = 0 \qquad (9.10.1)$$

where a, b, c are constants. It has regular singular points at $x = 0, 1$ and ∞.

Attempting a series solution about the origin produces an indicial equation with roots 0 and $1 - c$. Thus there is a solution y_1 that has a Taylor series expansion about the origin and which can be normalized to unity. It is

$$y_1(x) = 1 + \frac{ab}{1!c}x + \frac{a(a+1)b(b+1)}{2!c(c+1)}x^2 + \dots$$

or

$$y_1(x) = \frac{\Gamma(c)}{\Gamma(a)\Gamma(b)} \sum_{n=0}^{\infty} \frac{\Gamma(a+n)\Gamma(b+n)}{\Gamma(c+n)\Gamma(n+1)}x^n, \, c \neq 0, -1, -2, \dots . \qquad (9.10.2)$$

This series is called the **hypergeometric series**. Its sum is denoted by $F(a, b; c; x)$ and is called the **hypergeometric function**. Note that the expansion of $F(1, b; b; x)$ is just the geometric series which explains the use of the term "hypergeometric".

If $1 - c \neq$ an integer, a second solution of (9.10.1) is of the form

$$x^{1-c}u(x)$$

where $u(x)$ has a Taylor series expansion about the origin. Substituting into (9.10.1) yields a differential equation for $u(x)$:

$$x(x-1)u'' + [(a+b-2c+3)x + c - 2]u' + (a-c+1)(b-c+1)u = 0.$$

Comparing this with (9.10.1) shows that $u(x)$ is itself the hypergeometric function

$$F(b-c+1, a-c+1; 2-c; x).$$

Therefore, when $c - 1$ is not an integer, the general solution of (9.10.1) is a linear combination of $F(a, b; c; x)$ and $x^{1-c}F(b-c+1, a-c+1; 2-c; x)$.

The importance of the hypergeometric function $F(a, b; c; x)$ stems from its generality. A great many functions can be written in terms of it. For example,

$$F(-a, b; b; -x) = (1+x)^a \text{ and } F(1, 1; 2; -x) = \frac{\ln(1+x)}{x}.$$

Also, by making the substitutions $a = -l$, $b = l + 1$, $c = 1$ and $x \to \frac{1-x}{2}$ we see that (9.10.1) becomes Legendre's equation

$$\left(1 - x^2\right)\frac{d^2 y}{dx^2} - 2x\frac{dy}{dx} + l(l+1)y = 0.$$

Thus, the Legendre polynomial $P_l(x) = F\left(-l, l+1; 1; \frac{1-x}{2}\right)$.

To relate Bessel functions to the hypergeometric function we must accompany a substitution with a limiting process that results in the singularity at $x = 1$ coalescing with that at infinity. Specifically, we make the substitution $x \to \frac{x}{b}$ so that (9.10.1) becomes

$$x(x - b)y'' + [cb - (a + b + 1)x]y' - aby = 0.$$

Dividing through by $x - b$ and taking the limit as $b \to \infty$, this simplifies to

$$x\frac{d^2 y}{dx^2} + (c - x)\frac{dy}{dx} - ay = 0 \qquad (9.10.3)$$

which is called the **confluent hypergeometric equation**. It is important to note that the point at infinity is now an irregular singular point.

The roots of the indicial equation corresponding to the regular singular point at the origin continue to be 0 and $1 - c$. Thus, there is a solution that has a Taylor series expansion about the origin and can be normalized to unity. By convention, this solution is denoted by $\Phi(a, c; x)$ and called the **confluent hypergeometric function**.

We shall use this as an opportunity to illustrate the method of definite integrals and in so doing obtain an integral representation of $\Phi(a, c; x)$.

Comparison of (9.10.3) with (9.9.15) suggests that we again try the Laplace kernel $K(x, t) = e^{xt}$ and seek a solution of (9.10.3) of the form

$$y(x) = \int_\alpha^\beta e^{xt} v(t)\, dt.$$

Making the replacements prescribed in Section 9.9 we find the differential operator

$$M_t = t(t - 1)\frac{d}{dt} + (ct - a)$$

and so $v(t)$ must be a solution of the adjoint equation

$$M_t^+ v(t) = -t(t - 1)\frac{dv}{dt}(2t - 1)v(t) + (ct - a)v(t) = 0$$

where M_t^+ was obtained by an application of (9.9.7). This gives us

$$\frac{1}{v(t)}\frac{dv}{dt} = \frac{(c - 2)t + 1 - a}{t(t - 1)} \quad \text{or}$$

$$\frac{1}{v(t)}\frac{dv}{dt} = \frac{a - 1}{t} + \frac{c - a - 1}{t - 1}.$$

Integrating, we have

$$v(t) = kt^{a-1}(1-t)^{c-a-1}$$

where k is an integration constant and we have anticipated that $0 \leq t \leq 1$.

The end-points of the integral must be chosen so that the surface term

$$Q[K, v]\big|_\alpha^\beta = t(t-1)v(t)e^{xt}\big|_\alpha^\beta = -kt^a(1-t)^{c-a}e^{xt}\big|_\alpha^\beta = 0.$$

Here we have used the Lagrange identity to obtain $Q[K, v]$:

$$\frac{\partial}{\partial t}Q[K, v] = vM_tK - KM_t^+v = \frac{\partial}{\partial t}\left[t(t-1)v(t)e^{xt}\right].$$

If $c > a > 0$, this will happen with $\alpha = 0$ and $\beta = 1$. Therefore, we obtain the solution

$$y(x) = k\int_0^1 e^{xt}t^{a-1}(1-t)^{c-a-1}dt.$$

At $x = 0$ this becomes

$$y(0) = k\int_0^1 t^{a-1}(1-t)^{c-a-1}dt = k\frac{\Gamma(a)\Gamma(c-a)}{\Gamma(c)}$$

where we have used Euler's Integral of the first kind. Therefore, the solution that is normalized to unity has

$$k = \frac{\Gamma(c)}{\Gamma(a)\Gamma(c-a)}$$

and is the confluent hypergeometric function

$$\Phi(a, c; x) = \frac{\Gamma(c)}{\Gamma(a)\Gamma(c-a)}\int_0^1 e^{xt}t^{a-1}(1-t)^{c-a-1}dt, \, c > a > 0. \tag{9.10.4}$$

Replacing e^{xt} by its Taylor series, (9.10.4) becomes

$$\Phi(a, c; x) = \frac{\Gamma(c)}{\Gamma(c-a)\Gamma(a)}\sum_{n=0}^\infty \frac{x^n}{\Gamma(n+1)}\int_0^1 t^{a+n-1}(1-t)^{c-a-1}dt.$$

But a second application of Euler's Integral gives us

$$\int_0^1 t^{a+n-1}(1-t)^{c-a-1}dt = \frac{\Gamma(a+n)\Gamma(c-a)}{\Gamma(c+n)}.$$

Therefore,

$$\Phi(a, c; x) = \frac{\Gamma(c)}{\Gamma(a)}\sum_{n=0}^\infty \frac{\Gamma(a+n)}{\Gamma(c+n)\Gamma(n+1)}x^n \tag{9.10.5}$$

which is called the **confluent hypergeometric series**. As one would expect, this can be obtained from the hypergeometric series (9.10.2) by making the substitution $x \to x/b$ and taking the limit $b \to \infty$,

$$\Phi(a, c; x) = \lim_{b \to \infty} F(a, b; c; \frac{x}{b}).$$

An interesting special case of (9.10.4) occurs for $a = \frac{1}{2}$, $c = \frac{3}{2}$ and $x \to -x^2$:

$$\Phi\left(\frac{1}{2}, \frac{3}{2}; -x^2\right) = \frac{1}{2} \int_0^1 e^{-x^2 t} t^{-\frac{1}{2}} dt.$$

Changing the variable of integration to u where $u^2 = x^2 t$, this becomes

$$\Phi\left(\frac{1}{2}, \frac{3}{2}; -x^2\right) = \frac{1}{x} \int_0^x e^{-u^2} du = \frac{1}{x} \operatorname{erf}(x)$$

where erf (x) is the *error function*.

But what is the connection with Bessel's equation and Bessel functions? A little work is required to transform

$$\frac{d^2 y}{dx^2} + \frac{1}{x}\frac{dy}{dx} + \left(1 - \frac{\mu^2}{x^2}\right) y = 0$$

into the confluent hypergeometric form. Specifically, we have to set

$$y(x) = x^\mu e^{-ix} u(x)$$

and substitute into Bessel's equation to obtain

$$x\frac{d^2 u}{dx^2} + [(2\mu + 1) - 2ix]\frac{du}{dx} - i(2\mu + 1)u = 0.$$

Comparing this with (9.10.3) we see that a solution for $u(x)$ is

$$\Phi\left(\frac{2\mu + 1}{2}, 2\mu + 1; 2ix\right).$$

Therefore, a solution of Bessel's equation must be

$$x^\mu e^{-ix}\Phi(\frac{2\mu + 1}{2}, 2\mu + 1; 2ix)$$

which, when multiplied by $\frac{1}{2^\mu}\frac{1}{\Gamma(\mu+1)}$, yields the same power series as we obtained for $J_\mu(x)$. Thus,

$$J_\mu(x) = \frac{1}{2^\mu \Gamma(\mu + 1)} x^\mu e^{-ix}\Phi\left(\frac{2\mu + 1}{2}, 2\mu + 1; 2ix\right). \tag{9.10.6}$$

Other special functions that can be expressed in terms of confluent hypergeometric functions include both the Hermite and the Laguerre polynomials.

10 Partial Differential Equations and Boundary Value Problems

10.1 The Partial Differential Equations of Mathematical Physics

Almost all of classical physics and a significant part of quantum physics involves only three types of partial differential equation (or PDE). These are
- Laplace's / Poisson's equation $\nabla^2 \psi(r) = \sigma(r)$,
- the diffusion / heat conduction equation $D \nabla^2 \psi(r, t) - \frac{\partial \psi}{\partial t} = \sigma(r, t)$, and
- the wave equation $\nabla^2 \psi(r, t) - \frac{1}{c^2} \frac{\partial^2 \psi}{\partial t^2} = \sigma(r, t)$.

In each case σ represents a "source" or "sink" of the scalar field ψ; if it is zero, which happens in many applications, the equations become **homogeneous.** Familiar examples are provided by Maxwell's equations expressed in terms of the potentials A and Φ. In SI units these reduce to the wave equation

$$\nabla^2 A - \frac{1}{c^2} \frac{\partial^2 A}{\partial t^2} = -\mu_0 j \quad \text{and} \quad \nabla^2 \Phi - \frac{1}{c^2} \frac{\partial^2 \Phi}{\partial t^2} = -\frac{\rho}{\varepsilon_0}$$

or, in the event of time independence, to Poisson's equation.

Pierre-Simon Laplace (1749-1827) was a French mathematician whose work was central to the development of both astronomy and statistics. Among his many accomplishments, he derived the equation and introduced the transform that bear his name. The Poisson equation (and distribution) are named after Simeon Denis Poisson (1781-1840) who was one of Laplace's students. Bonaparte made Laplace a count of the Empire in 1806. Demonstrating that he too could recognize genius, Louis-Phillipe made him a marquis in 1817, after the restoration of the monarchy.

What distinguishes the different physical phenomena that are described by any one of these equations are the identity of the scalar field ψ (i.e. whether it is an electrostatic potential, a temperature, a density, a transverse displacement of a vibrating medium, or what have you) and the **boundary conditions** and, where time is an independent variable, the **initial conditions** that are imposed on it. Indeed, it is only when a partial differential equation is accompanied by such conditions that it will admit a unique solution.

The impact of boundary conditions or more precisely of the geometrical character of the boundaries is first experienced in the choice of coordinate system to use when ψ is defined on a multi-dimensional space. It is enormously convenient to be able to specify the boundary of the domain or region of definition by means of fixed values of one or more of the coordinates. Thus, for example, if the boundary is a rectangular box, which can be specified by $x = a$ and b, $y = c$ and d, $z = e$ and f, where a, b, c, d, e and f are constants, one should choose Cartesian coordinates; if it is a sphere, which can be specified by $r = a$ constant, $0 \leqslant \theta \leqslant \pi$, $0 \leqslant \varphi \leqslant 2\pi$, use spherical polars;

and if it is a cylinder, corresponding to $\rho = R$, $z = a$ and b, $0 \leqslant \varphi \leqslant 2\pi$, where R, a and b are constants, cylindrical polars are the obvious choice. Consequently, we need to be able to express the ∇^2 differential operator in any type of curvilinear coordinate system.

10.2 Curvilinear Coordinates

A point in space can be described by any three independent parameters (u_1, u_2, u_3). To move in either direction between such a system of coordinates and the Cartesian system, there must be some definite functional relationship relating the two sets of coordinates at any point:

$$x = f_1(u_1, u_2, u_3), \quad y = f_2(u_1, u_2, u_3), \quad z = f_3(u_1, u_2, u_3) \tag{10.2.1}$$

and

$$u_1 = F_1(x, y, z), \quad u_2 = F_2(x, y, z), \quad u_3 = F_3(x, y, z). \tag{10.2.2}$$

The three coordinate surfaces $u_i = a$ constant can be drawn. If the orientations of these surfaces change from point to point, the u_i are called **curvilinear coordinates** and if the three surfaces are mutually perpendicular everywhere, they are called **orthogonal curvilinear coordinates**.

At any point, specified by the radius vector from the origin $\boldsymbol{r} = x\boldsymbol{i} + y\boldsymbol{j} + z\boldsymbol{k}$, we can construct unit vectors $\boldsymbol{e}_i$ normal to the surfaces $u_i = a$ constant by means of

$$\boldsymbol{e}_i \equiv \frac{\partial \boldsymbol{r}/\partial u_i}{|\partial \boldsymbol{r}/\partial u_i|}. \tag{10.2.3}$$

These clearly form an orthogonal system when the coordinates are orthogonal. The quantities

$$h_i = |\partial \boldsymbol{r}/\partial u_i| = \sqrt{\left(\frac{\partial x}{\partial u_i}\right)^2 + \left(\frac{\partial y}{\partial u_i}\right)^2 + \left(\frac{\partial z}{\partial u_i}\right)^2} \tag{10.2.4}$$

are called **scale factors** and depend upon the position of $\boldsymbol{r}$ in space.

Consider a small displacement $d\boldsymbol{r} = dx\boldsymbol{i} + dy\boldsymbol{j} + dz\boldsymbol{k}$. Its curvilinear components can be read off from

$$d\boldsymbol{r} = \frac{\partial \boldsymbol{r}}{\partial u_1} d u_i + \frac{\partial \boldsymbol{r}}{\partial u_2} d u_2 + \frac{\partial \boldsymbol{r}}{\partial u_3} d u_3 = h_1 d u_i \, \boldsymbol{e}_1 + h_2 d u_2 \, \boldsymbol{e}_2 + h_3 d u_3 \, \boldsymbol{e}_3. \tag{10.2.5}$$

Let us assume that the curvilinear system is orthogonal. The **line element** or element of arc length is then given by the square root of

$$ds^2 = d\boldsymbol{r} \cdot d\boldsymbol{r} = h_1^2 (d u_1)^2 + h_2^2 (d u_2)^2 + h_3^2 (d u_3)^2. \tag{10.2.6}$$

In addition, the **volume element,** the volume of the parallelepiped formed by the surfaces

$$u_1 = c_1, u_1 = c_1 + d u_1, u_2 = c_2, u_2 = c_2 + d u_2, u_3 = c_3, u_3 = c_3 + d u_3,$$

is

$$dV = h_1 h_2 h_3 \, d u_1 \, d u_2 \, d u_3 . \tag{10.2.7}$$

To express grad ψ in terms of curvilinear coordinates we start with

$$(\text{grad } \psi) \cdot d\mathbf{r} = d\psi \equiv \frac{\partial \psi}{\partial u_1} d u_1 + \frac{\partial \psi}{\partial u_2} d u_2 + \frac{\partial \psi}{\partial u_3} d u_3$$

and rewrite it in the form

$$(\text{grad } \psi) \cdot d\mathbf{r} = \left(\frac{1}{h_1} \frac{\partial \psi}{\partial u_1} \right) h_1 \, d u_1 + \left(\frac{1}{h_2} \frac{\partial \psi}{\partial u_2} \right) h_2 \, d u_2 + \left(\frac{1}{h_3} \frac{\partial \psi}{\partial u_3} \right) h_3 \, d u_3 .$$

It follows immediately that

$$\text{grad } \psi = \frac{1}{h_1} \frac{\partial \psi}{\partial u_1} \mathbf{e}_1 + \frac{1}{h_2} \frac{\partial \psi}{\partial u_2} \mathbf{e}_2 + \frac{1}{h_3} \frac{\partial \psi}{\partial u_3} \mathbf{e}_3 . \tag{10.2.8}$$

To determine the divergence of a vector field $\mathbf{A}$ we make use of the definition

$$\nabla \cdot \mathbf{A} \equiv \lim_{\Delta V \to 0} \frac{\int_{\Delta S} \mathbf{A} \cdot d\mathbf{s}}{\Delta V} = \lim_{\Delta V \to 0} \frac{\text{net flux of } \mathbf{A} \text{ through surface } \Delta \, S \text{ bounding } \Delta V}{\Delta V}$$

and note that the flux through an elementary area oriented perpendicular to the $\mathbf{e}_1$ direction is $A_1 h_2 \, d u_2 \, h_3 \, d u_3$. Thus, the net flux of $\mathbf{A}$ through two such areas separated by a distance $h_1 \, d u_1$ is

$$h_1 \, d u_1 \frac{1}{h_1} \frac{\partial}{\partial u_1} (A_1 h_2 h_3) d u_2 \, d u_3 .$$

Adding the corresponding contributions from the other four faces of a volume element and dividing by its volume, we obtain

$$\text{div } \mathbf{A} = \frac{1}{h_1 h_2 h_3} \left[\frac{\partial}{\partial u_1} (A_1 h_2 h_3) + \frac{\partial}{\partial u_2} (A_2 h_3 h_1) + \frac{\partial}{\partial u_3} (A_3 h_1 h_2) \right]. \tag{10.2.9}$$

The expression for the Laplacian is obtained by combining the formulas for gradient and divergence:

$$\nabla^2 \psi = \frac{1}{h_1 h_2 h_3} \left[\frac{\partial}{\partial u_1} \left(\frac{h_2 h_3}{h_1} \frac{\partial \psi}{\partial u_1} \right) + \frac{\partial}{\partial u_2} \left(\frac{h_3 h_1}{h_2} \frac{\partial \psi}{\partial u_2} \right) + \frac{\partial}{\partial u_3} \left(\frac{h_1 h_2}{h_3} \right) \right].$$
$$\tag{10.2.10}$$

Examples: The most common curvilinear coordinates in physics are
- plane polar: $x = r \cos \theta, \quad y = r \sin \theta$

- cylindrical polar: $x = r \cos \theta, \quad y = r \sin \theta, z = z$
- spherical polar: $x = r \sin \theta \cos \varphi, \quad y = r \sin \theta \sin \varphi, \quad z = r \cos \theta$.

The scale factors h_i are defined for any set of coordinates by

$$h_i = \sqrt{\left(\frac{\partial x}{\partial u_i}\right)^2 + \left(\frac{\partial y}{\partial u_i}\right)^2 + \left(\frac{\partial z}{\partial u_i}\right)^2}.$$

Thus, for spherical polars, we have

$$h_r = 1, \quad h_\theta = r, \quad h_\varphi = r \sin \theta$$

and so,

$$d s^2 = d r^2 + r^2 d \theta^2 + r^2 \sin^2 \theta d \varphi^2,$$

$$dV = r^2 \sin \theta dr d\theta d\varphi,$$

$$\operatorname{div} \boldsymbol{A} = \frac{1}{r^2 \sin \theta} \left[\frac{\partial}{\partial r}(r^2 \sin \theta A_r) + \frac{\partial}{\partial \theta}(r \sin \theta A_\theta) + \frac{\partial}{\partial \varphi}(r A_\varphi) \right]$$

or,

$$\operatorname{div} \boldsymbol{A} = \frac{\partial A_r}{\partial r} + \frac{2}{r} A_r + \frac{1}{r} \frac{\partial A_\theta}{\partial \theta} + \frac{\cot \theta}{r} A_\theta + \frac{1}{r \sin \theta} \frac{\partial A_\varphi}{\partial \varphi},$$

and,

$$\nabla^2 \psi = \frac{1}{r^2} \frac{\partial}{\partial r}\left(r^2 \frac{\partial \psi}{\partial r}\right) + \frac{1}{r^2 \sin \theta} \frac{\partial}{\partial \theta}\left(\sin \theta \frac{\partial \psi}{\partial \theta}\right) + \frac{1}{r^2 \sin^2 \theta} \frac{\partial^2 \psi}{\partial \varphi^2}.$$

An analogous determination in cylindrical polars can be easily done and one finds in particular that the Laplacian is

$$\nabla^2 \psi = \frac{\partial^2 \psi}{\partial r^2} + \frac{1}{r} \frac{\partial \psi}{\partial r} + \frac{1}{r^2} \frac{\partial^2 \psi}{\partial \theta^2} + \frac{\partial^2 \psi}{\partial z^2}.$$

10.3 Separation of Variables

Various methods have been devised for the solution of partial differential equations corresponding to different kinds of boundaries (whether finite or at infinity) and different kinds of boundary and initial conditions. What we shall come to recognize is that they all involve representation of the solution as an expansion in terms of **eigenfunctions** of one or more of the partial differential operators in the equation which achieves a separation of the dependence on the individual coordinate variables involved. Eigenfunction is a term that we have not encountered before. Therefore, to provide some context, we shall present a method of solution that is actually called the **separation of variables** method. We will then abstract from it the key elements that are common to all methods of solution.

The "laboratory" we will use to investigate separation of variables is the **vibrating string problem**. Suppose that we have a string of mass per unit length ρ stretched under a tension T along the line $x = 0$ to $x = L$ and fixed at both ends. The string is set in motion at time $t = 0$ by means of some combination of plucking and striking it. Denoting the transverse displacement of the string by $\psi(x, t)$, the string's equation of motion (obtained by an application of Newton's 2nd law) is the one-dimensional wave equation

$$\frac{\partial^2 \psi}{\partial x^2} = \frac{1}{c^2} \frac{\partial^2 \psi}{\partial t^2} \text{ where } c = \sqrt{\frac{T}{\rho}}. \tag{10.3.1}$$

The manner in which motion is initiated is described by the **initial conditions** $\psi(x, 0) = u_0(x)$ and $\left. \frac{\partial \psi(x,t)}{\partial t} \right|_{t=0} = v_0(x)$, where $u_0(x)$ and $v_0(x)$ are known functions. Finally, the fact that the string has fixed end-points is captured by the **boundary conditions** $\psi(0, t) = 0$ and $\psi(L, t) = 0$.

One can prove that the solution of a linear partial differential equation accompanied by a complete set of boundary/initial conditions is unique. Thus, if we find a solution, no matter by what means, we are assured that it is the only solution to the problem. The means we shall employ here begins with the assumption that the motion at any point $0 \leqslant x \leqslant L$, and time $t \geqslant 0$, can be expressed in the form

$$\psi(x, t) = X(x)T(t). \tag{10.3.2}$$

If our method works, if we obtain a solution, this assumption will be justified a posteriori.

Substituting into the wave equation and dividing through by $\psi = XT$ we find

$$\frac{1}{X(x)} \frac{d^2 X}{d x^2} = -\lambda = \frac{1}{c^2} \frac{1}{T(t)} \frac{d^2 T}{d t^2}, \tag{10.3.3}$$

where λ must be a constant since the first equality implies it is independent of t while the second equality implies it is independent of x. The two equalities yield the same ordinary differential equation (ODE) with constant coefficients. The general solutions are

$$X_\lambda(x) = \begin{cases} A \cos \sqrt{\lambda}x + B \sin \sqrt{\lambda}x & \text{if } \lambda > 0 \\ A e^{\sqrt{-\lambda}x} + B e^{-\sqrt{-\lambda}x} & \text{if } \lambda < 0 \\ Ax + B & \text{if } \lambda = 0 \end{cases} \tag{10.3.4}$$

and

$$T_\lambda(t) = \begin{cases} A \cos \sqrt{\lambda}ct + B \sin \sqrt{\lambda}ct & \text{if } \lambda > 0 \\ A e^{\sqrt{-\lambda}ct} + B e^{-\sqrt{-\lambda}ct} & \text{if } \lambda < 0 . \\ Act + B & \text{if } \lambda = 0 \end{cases} \tag{10.3.5}$$

The initial conditions on $\psi(x, t)$ involve functions of x and so place no restrictions on $T_\lambda(t)$ other than a general requirement of boundedness on $0 \leqslant t < \infty$. Such conditions are called **non-homogeneous.** In contrast, the boundary conditions on $\psi(x, t)$

are **homogeneous**: they require that $\psi(x, t)$ and hence each $X_\lambda(x)$ vanish at $x = 0$ and $x = L$. A quick check tells us that these conditions cannot be satisfied for any values of $\lambda \leqslant 0$ (except in the trivial case of $A = B = 0$). On the other hand, there is an infinite set of values of $\lambda > 0$, $\lambda_n = \frac{n^2 \pi^2}{L^2}$, $n = 1, 2, 3, \ldots$, which admit these conditions and the corresponding solutions are, to within an arbitrary multiplicative constant, $X_n(x) = \sin \frac{n\pi x}{L}$. The latter are called the **eigenfunctions** of the differential operator $\frac{d^2}{dx^2}$ appropriate to these boundary conditions and the λ_n are its **eigenvalues** (characteristic values). This means that of those functions that vanish at $x = 0$ and $x = L$, there is a unique subset ,$\{X_n\}$, with the property that when operated on by $\frac{d^2}{dx^2}$ they are reproduced multiplied by a characteristic constant (an eigenvalue).

With λ determined, so is T_λ. In fact, we now have an infinite set of factored solutions

$$\psi_n(x, t) = X_n(x)\, T_n(t) = \sin \frac{n\pi x}{L} \left(A_n \cos \frac{n\pi ct}{L} + B_n \sin \frac{n\pi ct}{L} \right), \quad n = 1, 2, 3, \ldots,$$

each of which satisfies both the wave equation and the boundary conditions. Moreover, because the equation is linear, every linear combination of solutions satisfies both the wave equation and the boundary conditions. But, what about the initial conditions? Evidently, unless $u_0(x)$ and $v_0(x)$ are themselves sinusoidal with period $2L$, we will not be able to reproduce them with one or even a linear combination of several of the $\psi_n(x, t)$. Therefore, we shall use **all** of them. We form the superposition

$$\psi(x, t) = \sum_{n=1}^{\infty} \sin \frac{n\pi x}{L} \left(A_n \cos \frac{n\pi ct}{L} + B_n \sin \frac{n\pi ct}{L} \right) \tag{10.3.6}$$

and impose the initial conditions via

$$u_0(x) = \sum_{n=1}^{\infty} \sin \frac{n\pi x}{L} A_n, \quad \text{and} \tag{10.3.7}$$

$$v_0(x) - \sum_{n=1}^{\infty} \sin \frac{n\pi x}{L} B_n \frac{n\pi c}{L}. \tag{10.3.8}$$

We recognize the summations in these three equations as Fourier sine series. Thus, so long as $u(x, t)$, $u_0(x)$ and $v_0(x)$ are continuous functions of x, the series will converge uniformly to these functions when A_n and $\frac{n\pi c}{L} B_n$ are replaced by the Fourier sine coefficients of $u_0(x)$ and $v_0(x)$, respectively; that is, when

$$A_n = \frac{2}{L} \int_0^L u_0(x) \sin \frac{n\pi x}{L} dx, \quad \text{and} \tag{10.3.9}$$

$$B_n = \frac{2}{n\pi c} \int_0^L v_0(x) \sin \frac{n\pi x}{L} dx. \tag{10.3.10}$$

This completes the solution of the problem.

Evidently, the method of separation of variables works. Let us be certain we understand why.

Working backwards, we see that a critical element is the implicit representation of $\psi(x, t)$ by the **eigenfunction expansion**

$$\psi(x, t) = \sum_{n=1}^{\infty} b_n(t) \sin \frac{n\pi x}{L} \qquad (10.3.11)$$

which is a Fourier sine series, every term of which satisfies the boundary conditions. Thus, it must converge uniformly to $\psi(x, t)$ and so, when we substitute it into the wave equation, we can interchange the order of summation and integration. The result is

$$\sum_{n=1}^{\infty} b_n(t) \frac{d^2}{dx^2} \sin \frac{n\pi x}{L} = \sum_{n=1}^{\infty} b_n(t) \left(\frac{-n^2 \pi^2}{L^2} \right) \sin \frac{n\pi x}{L} = \sum_{n=1}^{\infty} \frac{1}{c^2} \frac{d^2 b_n(t)}{dt^2} \sin \frac{n\pi x}{L}.$$

Because the sine functions are orthogonal, the second equality implies that

$$\frac{d^2 b_n}{dt^2} + \left(\frac{n\pi c}{L} \right)^2 b_n(t) = 0$$

and hence, that

$$b_n(t) = A_n \cos \frac{n\pi ct}{L} + B_n \sin \frac{n\pi ct}{L}.$$

The solution is then completed by relating A_n and B_n to the Fourier sine coefficients of $u_0(x)$ and $v_0(x)$.

The separation of variables method is successful because it amounts to an expansion of $\psi(x, t)$ in terms of the eigenfunctions of the differential operator associated with homogeneous boundary conditions. The differential operator is replaced by its eigenvalues and thereby eliminated from the partial differential equation. The PDE is replaced by a series of ODE's with constant coefficients.

That being said, we shall now perform a practical inventory of the steps that comprise this method and do so in the course of solving another boundary value problem. The problem is to find the electrostatic potential everywhere inside a conducting rectangular box of dimensions $a \times b \times c$ which has all of its walls grounded except for the top which is separated from the other walls by thin insulating strips and maintained at a potential V.

The PDE to be solved is Laplace's equation

$$\nabla^2 \psi = 0.$$

Here is how we proceed with its solution.

Step 1. Choose an appropriate coordinate system.

We choose Cartesian coordinates with the origin at one corner of the box so that its interior and boundaries are defined by $0 \leqslant x \leqslant a, 0 \leqslant y \leqslant b, 0 \leqslant z \leqslant c$. The boundary conditions then become

$$\psi(0, y, z) = \psi(a, y, z) = \psi(x, 0, z) = \psi(x, b, z) = \psi(x, y, 0) = 0 \text{ and } \psi(x, y, c) = V.$$

Step 2. Separate the PDE into ODE's.

Substitute $\psi(x, y, z) = X(x)Y(y)Z(z)$ into the PDE and then divide through by ψ to obtain

$$\frac{1}{X}\frac{d^2 X}{d x^2} + \frac{1}{Y}\frac{d^2 Y}{d^2 y} + \frac{1}{Z}\frac{d^2 Z}{d z^2} = 0.$$

This can hold for all x, y and z if and only if each term is separately equal to a constant with the three constants summing to zero:

$$\frac{d^2 X}{d x^2} = \lambda_1 X, \; \frac{d^2 Y}{d y^2} = \lambda_2 Y \text{ and } \frac{d^2 Z}{d z^2} = \lambda_3 Z \text{ with } \lambda_1 + \lambda_2 + \lambda_3 = 0.$$

These three DE's are identical to each other and to the separated DE's of the stretched string problem. Therefore, they have the same three sets of solutions (10.3.4) corresponding to positive, negative and null values of the separation constants.

Step 3. Impose the single-coordinate boundary conditions that are homogeneous at **both** *boundaries and solve the corresponding eigenvalue equations for the functions of those coordinates.*

In this case, the homogeneous boundary conditions require that

$$X(0) = X(a) = 0, \; Y(0) = Y(b) = 0.$$

We know from the stretched string problem that this implies eigenvalues

$$\lambda_1 = -\left(\frac{n\pi}{a}\right)^2, n = 1, 2, \ldots \text{ and } \lambda_2 = -\left(\frac{m\pi}{b}\right)^2, m = 1, 2, \ldots,$$

corresponding to the eigenfunctions

$$X_n(x) = \sin\frac{n\pi x}{a} \quad \text{and} \quad Y_m(y) = \sin\frac{m\pi y}{b}.$$

Notice that while we know that $Z(0) = 0$ we have no information bearing directly on $Z(c)$.

Step 4. Solve for the remaining function(s).

We now know that

$$\lambda_3 = \left(\frac{n\pi}{a}\right)^2 + \left(\frac{m\pi}{b}\right)^2.$$

Since this is always positive, the corresponding solution for $Z(z)$ is a linear combination of $e^{\sqrt{\lambda_3}z}$ and $e^{-\sqrt{\lambda_3}z}$. But we must also satisfy $Z(0) = 0$. Therefore, an appropriate linear combination is

$$Z_{nm}(z) = \sinh\left(\sqrt{\left(\frac{n\pi}{a}\right)^2 + \left(\frac{m\pi}{b}\right)^2}\,z\right).$$

Step 5. Form a linear superposition of all factored solutions.

We now have a doubly infinite set of factored solutions

$$X_n(x)\, Y_m(y)\, Z_{nm}(z), \quad n, m = 1, 2, \ldots$$

each of which satisfies the PDE as well as the homogeneous boundary conditions. To prepare for the imposition of the non-homogeneous boundary condition, we take advantage of the linearity of the PDE and form the superposition

$$\psi(x, y, z) = \sum_{n=1}^{\infty} \sum_{m=1}^{\infty} A_{nm} \sin \frac{n\pi x}{a} \sin \frac{m\pi y}{b} \sinh\left(\sqrt{\left(\frac{n\pi}{a}\right)^2 + \left(\frac{m\pi}{b}\right)^2}\, z\right).$$

Step 6. Impose the remaining boundary condition(s).

In this case there remains only one condition: $\psi(x, y, c) = V$. Imposing it on our superposition we have the requirement that

$$V = \sum_{n=1}^{\infty} \sum_{m=1}^{\infty} A_{nm} \sin \frac{n\pi x}{a} \sin \frac{m\pi y}{b} \sinh\left(\sqrt{\left(\frac{n\pi}{a}\right)^2 + \left(\frac{m\pi}{b}\right)^2}\, c\right).$$

This is a double Fourier sine series and so we can use the Euler formula for the coefficients of such series to determine A_{nm}. Thus,

$$A_{nm} = \frac{V}{\sinh\left(\sqrt{\left(\frac{n\pi}{a}\right)^2 + \left(\frac{m\pi}{b}\right)^2}\, c\right)} \frac{4}{ab} \int_0^a \int_0^b \sin \frac{n\pi x}{a} \sin \frac{m\pi y}{b}\, dx\, dy$$

or,

$$A_{nm} = \frac{V}{\sinh\left(\sqrt{\left(\frac{n\pi}{a}\right)^2 + \left(\frac{m\pi}{b}\right)^2}\, c\right)} \frac{4}{nm\pi^2}(1 - (-1)^n)(1 - (-1)^m).$$

Substituting back into the superposition we obtain as our solution

$$\psi(x, y, z) = \frac{16V}{\pi^2} \sum_{n=1,3,5,\ldots}^{\infty} \sum_{m=1,3,5,\ldots}^{\infty} \frac{1}{nm} \sin \frac{n\pi x}{a} \sin \frac{m\pi y}{b} \frac{\sinh\left(\sqrt{\left(\frac{n\pi}{a}\right)^2 + \left(\frac{m\pi}{b}\right)^2}\, z\right)}{\sinh\left(\sqrt{\left(\frac{n\pi}{a}\right)^2 + \left(\frac{m\pi}{b}\right)^2}\, c\right)}.$$

10.4 What a Difference the Choice of Coordinate System Makes!

We shall now investigate the solution of the homogeneous versions of the partial differential equations we introduced in Section 10.1 when applied to a three dimensional medium with either rectangular, spherical or cylindrical symmetry.

Recall that the PDE's are

$$\nabla^2 \psi = 0,$$

$$\nabla^2 \psi - \frac{1}{D} \frac{\partial \psi}{\partial t} = 0,$$

$$\nabla^2 \psi - \frac{1}{c^2} \frac{\partial^2 \psi}{\partial t^2} = 0. \tag{10.4.1}$$

We start by separating off the time dependence in the case of the last two of these. We do this by assuming a solution of the form

$$\psi(\mathbf{r}, t) = u(\mathbf{r})T(t). \tag{10.4.2}$$

Substituting into the diffusion equation and dividing by $\psi = uT$ we find

$$\frac{1}{u(\mathbf{r})}\nabla^2 u(\mathbf{r}) = -\lambda = \frac{1}{D}\frac{1}{T(t)}\frac{dT}{dt}, \tag{10.4.3}$$

where λ must be a constant since the first equality implies it is independent of t and the second equality implies it is independent of $\mathbf{r}$. The second equality is a simple differential equation for $T(t)$ with solution

$$T(t) = A\,e^{-\lambda D t}.$$

Unlike the vibrating string problem, we will not assume an initial condition but require only that $T(t)$ be bounded for $0 \leqslant t < \infty$, (i.e. that it satisfy the homogeneous **boundary condition** $|T(\infty)| < \infty$). This means that λ must be positive and so we set $\lambda = k^2$. Thus,

$$T(t) = A\,e^{-k^2 D t} \tag{10.4.4}$$

and the first part of (10.4.3) is

$$\nabla^2 u(\mathbf{r}) + k^2 u(\mathbf{r}) = 0 \tag{10.4.5}$$

which is called **Helmholtz' equation**. Note that Laplace's equation is the special case of Helmholtz' equation corresponding to $k^2 = 0$.

The Helmholtz' equation is named for Hermann von Helmholtz (1821-1894), a German physician and physicist. He made significant contributions in neurophysiology, physics (electrodynamics and thermodynamics), and philosophy. The Helmholtz Association of German research centres is named after him.

We can do a similar separation of the time dependence in the case of the wave equation. Assuming a solution of the form (10.4.2), substituting into the equation, and dividing by $\psi = uT$, we find

$$\frac{1}{u}\nabla^2 u = -\lambda = \frac{1}{c^2}\frac{1}{T}\frac{d^2 T}{dt^2}, \tag{10.4.6}$$

with λ a constant. The second part of (10.4.6) is the same differential equation for T that we encountered in the vibrating string problem. This time we will accompany it with the homogeneous boundary condition $|T(\pm\infty)| < \infty$ which makes it an eigenvalue equation with solution $\lambda = k^2$, $0 \leqslant k < \infty$ and

$$T_\lambda(t) = T_k(t) = A_k\,e^{ikct} + B_k\,e^{-ikct} = \begin{Bmatrix} e^{ikct} \\ e^{-ikct} \end{Bmatrix} = \begin{Bmatrix} \sin kct \\ \cos kct \end{Bmatrix}. \tag{10.4.7}$$

The use of curly braces in (10.4.7) is a convenient short-hand for a linear combination of the functions that appear between them.

This means that we obtain the Helmholtz equation

$$\nabla^2 u + k^2 u = 0$$

once again to describe the space dependence.

To proceed further with the separation of variables we must now adopt one or another type of coordinate system.

In Cartesian coordinates Helmholtz' equation is

$$\frac{\partial^2 u}{\partial x^2} + \frac{\partial^2 u}{\partial y^2} + \frac{\partial^2 u}{\partial z^2} + k^2 u = 0 \tag{10.4.8}$$

and so, assuming a separated solution of the form $u = X(x)Y(y)Z(z)$, substituting into (10.4.8), and dividing through by $u = XYZ$, we have

$$\frac{1}{X}\frac{d^2 X}{dx^2} + \frac{1}{Y}\frac{d^2 Y}{dy^2} + \frac{1}{Z}\frac{d^2 Z}{dz^2} + k^2 = 0. \tag{10.4.9}$$

Thus,

$$\frac{1}{X}\frac{d^2 X}{dx^2} = -\lambda_1 = -k^2 - \frac{1}{Y}\frac{d^2 Y}{dy^2} - \frac{1}{Z}\frac{d^2 Z}{dz^2}, \tag{10.4.10}$$

where λ_1 must be a (separation) constant. Then, as in the potential problem of the last Section, we find

$$\frac{1}{Y}\frac{d^2 Y}{dy^2} = -\lambda_2 \tag{10.4.11}$$

and,

$$\frac{1}{Z}\frac{d^2 Z}{dz^2} = -\lambda_3 \tag{10.4.12}$$

but now

$$\lambda_1 + \lambda_2 + \lambda_3 = k^2. \tag{10.4.13}$$

We now need some information about the spatial boundaries. Rather than confine the medium to a finite box as was the case in the potential problem, we shall assume that the medium is infinite and that the boundary conditions require X, Y, and Z to be bounded for all x, y, and z. The differential equations for X, Y, and Z are the same as the differential equation for $T(t)$ and so we know that they admit bounded solutions if and only if $\lambda_i \geqslant 0$ for $i = 1, 2$, and 3. Thus, we set $\lambda_1 = k_1^2$, $\lambda_2 = k_2^2$, $\lambda_3 = k_3^2$ with $-\infty < k_1, k_2, k_3 < \infty$ to obtain

$$X(x) \propto e^{i k_1 x}$$

$$Y(y) \propto e^{i k_2 y}$$

$$Z(z) \propto e^{i k_3 z}, \tag{10.4.14}$$

with

$$k_1^2 + k_2^2 + k_3^2 = \boldsymbol{k}^2. \tag{10.4.15}$$

Multiplying these together we get (plane wave) solutions for $u(\boldsymbol{r})$ of the form

$$u(\boldsymbol{r}) = A_k \, e^{i \boldsymbol{k} \cdot \boldsymbol{r}} \tag{10.4.16}$$

where $\boldsymbol{k}$ is a three-dimensional vector with norm $\boldsymbol{k} \cdot \boldsymbol{k} = k^2$.

This is as far as we can go without having information about how the wave motion or diffusion was initiated, that is, without having initial conditions to impose. Therefore, let us turn instead to the question of the kind of waves we would obtain if there was cylindrical geometry.

In cylindrical coordinates Helmholtz' equation is

$$\frac{\partial^2 u}{\partial r^2} + \frac{1}{r} \frac{\partial u}{\partial r} + \frac{1}{r^2} \frac{\partial^2 u}{\partial \theta^2} + \frac{\partial^2 u}{\partial z^2} + k^2 u = 0. \tag{10.4.17}$$

Assuming a separated solution of the form $u = R(r)\Theta(\theta)Z(z)$, substituting in (10.4.17), and dividing through by $u = R\Theta Z$, we find

$$\frac{1}{R} \left[\frac{d^2 R}{d r^2} + \frac{1}{r} \frac{dR}{dr} \right] + \frac{1}{r^2} \frac{1}{\Theta} \frac{d^2 \Theta}{d \theta^2} + \frac{1}{Z} \frac{d^2 Z}{d z^2} + k^2 = 0. \tag{10.4.18}$$

Separating variables yields the equations

$$\frac{1}{Z} \frac{d^2 Z}{d z^2} = -\lambda_2 \tag{10.4.19}$$

$$\frac{1}{\Theta} \frac{d^2 \Theta}{d \theta^2} = -\lambda_1 \tag{10.4.20}$$

$$\frac{1}{R} \left[\frac{d^2 R}{d r^2} + \frac{1}{r} \frac{dR}{dr} \right] - \frac{\lambda_1}{r^2} - \lambda_2 + k^2 = 0. \tag{10.4.21}$$

Most physical applications involve the **boundary condition** $\Theta(\theta + 2\pi) = \Theta(\theta)$ to ensure that Θ is a single valued function. It then follows that $\lambda_1 = m^2, m = 0, 1, 2, \ldots$ and

$$\Theta(\theta) \equiv \Theta_m(\theta) = A_m \cos m\theta + B_m \sin m\theta \tag{10.4.22}$$

are the **eigensolutions** (or characteristic solutions) of equation (10.4.19) .

There is no common boundary condition that can be applied to the solutions of (10.4.20) and so we write them for now as

$$Z(z) = C \exp(\sqrt{-\lambda_2} z) + D \exp(-\sqrt{-\lambda_2} z) \tag{10.4.23}$$

and note that this will be a trigonometric function if $\lambda_2 > 0$ and a hyperbolic function if $\lambda_2 < 0$.

Setting $k^2 - \lambda_2 = \alpha^2$ and $\rho = \alpha r$, equation (10.4.21) becomes

$$\frac{d^2 R}{d\rho^2} + \frac{1}{\rho}\frac{dR}{d\rho} + \left(1 - \frac{m^2}{\rho^2}\right) R = 0 \tag{10.4.24}$$

which is **Bessel's equation**. As we have seen, its general solution can be expressed as the linear combination

$$R(r) = E J_m(\alpha r) + F N_m(\alpha r) \tag{10.4.25}$$

where $J_m(x)$ and $N_m(x)$ are the Bessel and Neumann functions of order m, respectively. These have an oscillatory dependence on x with an infinite number of zeros. Moreover, the Neumann function, $N_m(x)$, is singular at $x = 0$. Thus, if there are homogeneous boundary conditions such as $R(0) = R(a) = 0$, we would require $F = 0$ for all m and determine **eigensolutions** $J_m(\alpha_{m,n} r)$ where $\alpha_{m,n} = x_{m,n}/a$ and $x_{m,n}$ is the nth zero of $J_m(x)$.

If $\lambda_2 > k^2$, α will be a pure imaginary. In that case it is conventional to replace (10.4.25) by the linear combination

$$R(r) = G I_m(|\alpha|r) + H K_m(|\alpha|r) \tag{10.4.26}$$

where $I_m(x)$ and $K_m(x)$ are called **modified** Bessel functions. The modified Bessel functions are not oscillatory in behaviour but rather behave exponentially for large x. Specifically, $I_m \to \infty$ and $K_m \to 0$ as $x \to \infty$. At the other end of the scale , as $x \to 0$, $K_m \to \infty$ while $I_m \to 0$ if $m \neq 0$ and $I_0 \to 1$. Note that the modified Bessel functions arise when the z- dependence is given by oscillatory sine and cosine functions. The converse is true also: if $Z(z)$ is non-oscillatory, $\lambda_2 < 0$ and $R(r)$ is given by the oscillatory form (10.4.25).

In the event that $\lambda_2 = k^2$, $\alpha = 0$ and one must require that $m = 0$ and $F = 0$ in (10.4.25) to obtain a bounded but non-null solution. The overall solution is then the plane wave

$$u(r, \theta, z) \sim e^{\pm ikz} .$$

A second special case involving $\alpha = 0$ arises when $k^2 = 0$ (so the partial differential equation is Laplace's equation) and $\lambda_2 = 0$ (so there is no z dependence). The equation for R becomes

$$\frac{d^2 R}{dr^2} + \frac{1}{r}\frac{dR}{dr} - \frac{m^2}{r^2} R = 0 \tag{10.4.27}$$

which has the general solutions

$$R(r) = \begin{cases} G r^m + H r^{-m}, & m \neq 0 \\ G + H \ln r, & m = 0. \end{cases} \tag{10.4.28}$$

Forming a superposition of solutions as we did in the examples of Section 10.3 gives us the potential

$$\psi(r, \theta) = A_0 + B_0 \ln r + \sum_{m=1}^{\infty}(A_m \, r^m + B_m \, r^{-m})(C_m \cos m\theta + D_m \sin m\theta) \qquad (10.4.29)$$

which we recognize as a full Fourier series in θ. The coefficients A_m, B_m, C_m and D_m, $m \geqslant 0$ can be determined by imposing non-homogeneous boundary conditions at two fixed values of r, $\psi(a, \theta) = V_1(\theta)$ and $\psi(b, \theta) = V_2(\theta)$ for example, and then using the Euler formulae for the Fourier coefficients of the functions $V_1(\theta)$ and $V_2(\theta)$.

As if (10.4.29) is not complicated enough, a superposition of solutions of the Helmholtz equation with homogeneous boundary conditions at $r = 0$ and $r = a$ has the form

$$u(r, \theta, z) = \sum_{n=1}^{\infty}\sum_{m=0}^{\infty} J_m(\alpha_{mn} \, r)(A_{mn} \cos m\theta + B_{mn} \sin m\theta)$$

$$\times (C_{mn} \cosh \sqrt{\alpha_{mn}^2 - k^2}\,z + D_{mn} \sinh \sqrt{\alpha_{mn}^2 - k^2}\,z).$$
$$(10.4.30)$$

This is a Fourier series in the θ coordinate as well as a series unlike anything we have seen thus far: an expansion in terms of an infinite set of Bessel functions. Evidently, our knowledge of series representations requires extension if we are to feel comfortable working with cylindrical polars.

What further complications await us when we switch to spherical polars? In spherical coordinates Helmholtz' equation assumes the form

$$\frac{1}{r}\frac{\partial^2}{\partial r^2}(ru) + \frac{1}{r^2 \sin \theta}\left[\frac{\partial}{\partial \theta}(\sin \theta \frac{\partial u}{\partial \theta}) + \frac{1}{\sin \theta}\frac{\partial^2 u}{\partial \varphi^2}\right] + k^2 u = 0. \qquad (10.4.31)$$

Assuming a separated solution $u = R(r)Y(\theta, \varphi)$, substituting into (10.4.31), and dividing by $u = RY$, we obtain

$$\frac{1}{R}\frac{1}{r}\frac{d^2}{dr^2}(rR) + \frac{1}{r^2}\frac{1}{Y \sin \theta}\left[\frac{\partial}{\partial \theta}\left(\sin \theta \frac{\partial Y}{\partial \theta}\right) + \frac{1}{\sin \theta}\frac{\partial^2 Y}{\partial \varphi^2}\right] + k^2 = 0. \qquad (10.4.32)$$

Separating variables yields the equation

$$\frac{1}{Y}\frac{1}{\sin \theta}\left[\frac{\partial}{\partial \theta}(\sin \theta \frac{\partial Y}{\partial \theta}) + \frac{1}{\sin \theta}\frac{\partial^2 Y}{\partial \varphi^2}\right] = -\lambda \qquad (10.4.33)$$

for the angular dependence of $u(r, \theta, \varphi)$ plus the radial equation

$$\frac{1}{R}\frac{1}{r}\frac{d^2}{dr^2}(rR) + k^2 - \frac{\lambda}{r^2} = 0 \qquad (10.4.34)$$

where λ is the separation constant.

We shall start with the angular equation (10.4.33) which we subject to a further separation of variables. Setting $Y = \Theta(\theta)\Phi(\varphi)$, we obtain

$$\frac{1}{\Theta}\frac{1}{\sin\theta}\frac{d}{d\theta}(\sin\theta\frac{d\Theta}{d\theta}) + \frac{1}{\sin^2\theta}\frac{1}{\Phi}\frac{d^2\Phi}{d\varphi^2} + \lambda = 0$$

and hence,

$$\frac{1}{\Phi}\frac{d^2\Phi}{d\varphi^2} = -m^2, \quad m = 0, \pm1, \pm2, \ldots, \tag{10.4.35}$$

and

$$\frac{1}{\sin\theta}\frac{d}{d\theta}(\sin\theta\frac{d\Theta}{d\theta}) + \left(\lambda - \frac{m^2}{\sin^2\theta}\right)\Theta = 0, \tag{10.4.36}$$

where we have invoked the **boundary condition** $\Phi(\varphi + 2\pi) = \Phi(\varphi)$ to ensure single-valued solutions and determine the second separation constant. The corresponding eigensolutions are a linear combination of $\cos m\varphi$ and $\sin m\varphi$ or of $e^{\pm im\varphi}$. In this instance we will choose the latter and write

$$\Phi \equiv \Phi_m(\varphi) = A_m\, e^{im\varphi} + B_m\, e^{-im\varphi}. \tag{10.4.37}$$

To identify solutions of (10.4.36), we introduce the new variable $x = \cos\theta$ which transforms the equation into a version of **Legendre's equation**:

$$(1 - x^2)\frac{d^2 P}{dx^2} - 2x\frac{dP}{dx} + \left[\lambda - \frac{m^2}{1 - x^2}\right]P = 0 \text{ where } P(x) = \Theta(\cos^{-1}x). \tag{10.4.38}$$

Since θ varies over the range $0 \leqslant \theta \leqslant \pi$, x has the range $-1 \leqslant x \leqslant 1$. But as we know, the boundary points $x = \pm1$ are regular singular points of Legendre's equation. Therefore, an obvious **boundary condition** to impose on the solutions of (10.4.38) is that they be bounded at $x = \pm1$. This can be satisfied if and only if λ is assigned one of the discrete eigenvalues $\lambda = l(l + 1)$, $l = 0, 1, 2, \ldots$ with $l \geqslant |m|$; the corresponding eigensolutions are the **associated Legendre polynomials**, denoted $P_l^m(x)$. How do these relate to the Legendre polynomials whose acquaintance we made in Chapter 9? The answer will be derived in Chapter 11 but here is a preview:

$$P_l^m(x) = (1 - x^2)^{\frac{m}{2}}\frac{d^m P_l(x)}{dx^m}, \quad m = 0, 1, 2, \ldots l. \tag{10.4.39}$$

The product
$$Y = \Theta(\theta)\Phi(\varphi) = P_l^m(\cos\theta)\, e^{im\varphi},$$

with appropriate normalization to be defined later, is called a **spherical harmonic**. We will have occasion in Chapter 11 to study its properties in some detail. It suffices at present to note that it is an eigenfunction solution of (10.4.33) combined with the periodicity and boundedness conditions and hence, an eigenfunction of the angular part of the partial differential operator ∇^2.

We can now turn our attention to the radial equation (10.4.34) with λ set equal to $l(l+1)$:

$$\frac{d^2 R}{d r^2} + \frac{2}{r}\frac{dR}{dr} + \left[k^2 - \frac{l(l+1)}{r^2}\right] R = 0. \tag{10.4.40}$$

If $k^2 \neq 0$, we set $r = \rho/k$ and $R = \frac{1}{\sqrt{\rho}}S$ to transform this equation into

$$\frac{d^2 S}{d\rho^2} + \frac{1}{\rho}\frac{dS}{d\rho} + \left[1 - \frac{(l+1/2)^2}{\rho^2}\right] S = 0 \tag{10.4.41}$$

which we recognize as Bessel's equation of order $l + 1/2$. Thus, we conclude that the general solution of the radial equation is

$$R(r) = A\frac{1}{\sqrt{kr}} J_{l+1/2}(kr) + B\frac{1}{\sqrt{kr}} N_{l+1/2}(kr) = A' j_l(kr) + B' n_l(kr), \tag{10.4.42}$$

where $j_l(x) \equiv \sqrt{\frac{\pi}{2x}} J_{l+1/2}(x)$ and $n_l(x) \equiv \sqrt{\frac{\pi}{2x}} N_{l+1/2}(x)$ are called spherical Bessel and Neumann functions of order l, respectively.

If $k^2 = 0$, which corresponds to Laplace's equation, the radial equation is

$$\frac{d^2 R}{d r^2} + \frac{2}{r}\frac{dR}{dr} - \frac{l(l+1)}{r^2}R = 0 \tag{10.4.43}$$

which has the general solution

$$R = A r^l + B\frac{1}{r^{l+1}}. \tag{10.4.44}$$

Summarizing, the sort of superpositions we can expect in problems with spherical geometry are potentials of the form

$$\psi(r, \theta, \varphi) = \sum_{l=0}^{\infty} \sum_{m=-l}^{l} [A_{lm} r^l + B_{lm} r^{-l-1}] Y_l^m(\theta, \varphi) \tag{10.4.45}$$

and waves like

$$u(r, \theta, \varphi) = \sum_{l=0}^{\infty} \sum_{m=-l}^{l} [A_{lm} j_l(kr) + B_{lm} n_l(kr)] Y_l^m(\theta, \varphi). \tag{10.4.46}$$

Thus, Fourier series do not figure in the solutions at all when spherical coordinates are used. Rather, we have a double series expansion in terms of the spherical harmonics and so yet another type of series representation to become familiar with. Fortunately, all of these representations are special cases of a **Sturm-Liouville** eigenfunction expansion and so we can acquire a comprehensive understanding by considering a single eigenvalue problem.

A student of Poisson at the École Polytechnique, Joseph Liouville (1809-1882) contributed widely to mathematics, mathematical physics and astronomy. The Liouville theorem of complex analysis and the Liouville theorem of classical mechanics are both named after him as is the Liouville crater on the moon. He developed Sturm-Liouville theory in collaboration with a colleague at the École Polytecnique, Jacques Sturm (1803-1855).

10.5 The Sturm-Liouville Eigenvalue Problem

To review, the separation of variables in a partial differential equation results in two or more ordinary differential equations which, when combined with homogeneous boundary conditions, become eigenvalue equations whose solutions correspond to characteristic values of the separation constant(s). Once all of these are known, we use them to express the solution of the original partial differential equation as an eigen-function expansion and then impose whatever non-homogeneous boundary conditions may be associated with the problem. In the vibrating string problem the eigen-function expansion is a Fourier sine series which is a type of series that is reasonably familiar to us. But what are the convergence properties of series involving Legendre polynomials or Bessel functions? How do we determine their coefficients? Is there a connection with the theory of Fourier representations?

The most general way of answering these questions is to study the **Sturm-Liouville eigenvalue problem.** It consists of solving a differential equation of the form

$$\mathfrak{L}u(x) \equiv \frac{d}{dx}\left[p(x)\frac{du(x)}{dx}\right] - q(x)u(x) = -\lambda\rho(x)u(x) \qquad (10.5.1)$$

where $\rho(x) \geqslant 0$ on the interval $a \leqslant x \leqslant b$ of the real line and the solution $u(x)$ is subject to (homogeneous) boundary conditions such as $u(a) = u(b)$ and $u'(a) = u'(b)$, (the periodicity condition is an example of this), or

$$\alpha_1 u + \beta_1 \frac{du}{dx} = 0 \text{ at } x = a \text{ and}$$

$$\alpha_2 u + \beta_2 \frac{du}{dx} = 0 \text{ at } x = b \qquad (10.5.2)$$

where $\alpha_1, \beta_1, \alpha_2$, and β_2 are given constants. The form of the differential operator $\mathfrak{L}$ in (10.5.1) is quite general since after multiplication by a suitable factor any second order linear differential operator can be expressed this way.

The differential equations obtained by separating variables in the preceding Sec-tion are all of the Sturm-Liouville type, the separation constants being the eigenvalue parameters λ. The boundary conditions to go with them, such as boundedness or pe-riodicity, were determined by the requirements of the physics problem in which the equations arise and this is invariably the case.

Changing the boundary conditions can result in a profound change to the eigen-value spectrum of a differential operator. To illustrate, we shall consider the simple operator $\mathfrak{L} \equiv \frac{d^2}{dx^2}$. Its Sturm-Liouville equation is

$$\mathfrak{L}u(x) \equiv \frac{d^2}{dx^2}u(x) = -\lambda u(x), \qquad (10.5.3)$$

corresponding to $p(x) \equiv 1, q(x) \equiv 0, \rho(x) \equiv 1$. We know already that this equation has solutions

$$u(x) = \begin{cases} A \cos \sqrt{\lambda}x + B \sin \sqrt{\lambda}x & \text{if } \lambda > 0 \\ A \cosh \sqrt{-\lambda}x + B \sinh \sqrt{-\lambda}x & \text{if } \lambda < 0 \\ Ax + B & \text{if } \lambda = 0 \end{cases} \qquad (10.5.4)$$

So, if the boundary conditions are
- $u(x + 2\pi) = u(x)$ then,

$$\lambda = m^2 \text{ and } u_m(x) = A_m \cos mx + B_m \sin mx, \qquad m = 0, 1, 2, \ldots, \qquad (10.5.5)$$

- $u(0) = 0$ and $u(b) = 0$ then,

$$\lambda = \frac{n^2 \pi^2}{b^2} \text{ and } u_n(x) = A_n \sin \frac{n\pi x}{b}, \qquad n = 1, 2, 3, \ldots, \qquad (10.5.6)$$

- $u'(0) = 0$ and $|u(\infty)| < \infty$ then,

$$\lambda = k^2 \text{ and } u_k(x) = A_k \cos kx, \qquad 0 \leqslant k < \infty, \qquad (10.5.7)$$

- $|u(\pm\infty)| < \infty$ then,

$$\lambda = k^2 \text{ and } u_k(x) = A_k e^{ikx}, \qquad -\infty < k < \infty. \qquad (10.5.8)$$

The multiplicative constants in these expressions are determined by some nonlinear normalization condition such as

$$\int_0^b |u_n(x)|^2 \, dx = 1. \qquad (10.5.9)$$

Some other Sturm-Liouville problems encountered in Section 10.4 are reviewed in the following table (Table 10.1).

Table 10.1: Sturm-Liouville Problems in Section 10.4

Equation	Boundary Conditions	Eigenfunctions	Eigenvalues	$\rho(x)$	$p(x)$	$q(x)$
Legendre	$|u(\pm 1)|^2 < \infty$	$P_l(x)$ $l=0,1,2,\ldots$	$l(l+1)$	1	$1-x^2$	0
Associated Legendre	ditto	$P_l^m(x)$	$l(l+1)$	1	$1-x^2$	$\frac{m^2}{1-x^2}$
Bessel	$|u(0)| < \infty$	$J_m(x)$	1	x	x	$\frac{m^2}{x}$
Spherical Bessel	ditto	$j_l(x)$	1	x^2	x^2	$l(l+1)$

The Sturm-Liouville eigenvalue problem is an infinite-dimensional analogue of the matrix eigenvalue problem

$$Mu = \lambda u, \tag{10.5.10}$$

where M is an $n \times n$ matrix and u is an n–dimensional column vector, encountered in connection with finite dimensional vector spaces. In both cases there are solutions u_n only for certain values of the **eigenvalue** λ_n. These are called the **eigenvectors** of M while in the case of $\mathfrak{L}$, they are the **eigenfunctions** corresponding to the particular choice of boundary conditions that accompany the equation. Significantly, like the eigenvectors of matrices, the eigenfunctions of $\mathfrak{L}$ can be used as **basis vectors** spanning a type of vector space in which the vectors are functions. Such **function spaces** are generally infinite dimensional corresponding either to a countable infinity or to a continuum of eigenfunctions and eigenvalues. An example of the former is the space consisting of all square integrable functions defined on a finite interval $a \leqslant x \leqslant b$. A continuum normally arises when one or both of the end-points is at infinity.

In a conventional vector space each vector is an ordered n-tuple of numbers,

$$\boldsymbol{a} \equiv |a> \equiv (a_1, a_2, \ldots, a_n). \tag{10.5.11}$$

The numbers can be real or imaginary. The number of dimensions, n, can be finite or infinite. Various operations such as addition, subtraction and multiplication by a scalar are defined as is the operation of scalar product,

$$\boldsymbol{a} \cdot \boldsymbol{b} \equiv < a|b> \equiv \sum_{i=1} a_i^* b_i. \tag{10.5.12}$$

The ordering is discrete and even if the number of dimensions is infinite, it is a "countable infinity".

Functions also provide ordered sets of numbers although now the ordering is continuous: $f(x)$, $a \leqslant x \leqslant b$, denotes an ordered continuum of numbers. Thus, the set of all functions which satisfy certain behavioural conditions on an interval of the real line, $a \leqslant x \leqslant b$, can define a vector space called a **function space**. An example is the set of functions which are square integrable. The scalar product is defined in analogy with the definition for a conventional vector space,

$$< u|v> \equiv \sum_{\text{all components}} u^*(x)v(x) \equiv \int_a^b u^*(x)v(x)\rho(x)dx. \tag{10.5.13}$$

Here, $\rho(x)$ is a weight function that determines how one counts "components" as x varies along the real line from a to b : $\rho(x)dx$ = the number of "components" in the interval dx about x.

In conventional vector spaces it is convenient to define a basis (or bases) of orthogonal unit vectors $\boldsymbol{e}_i$, $i = 1, 2, \ldots n$ with $\boldsymbol{e}_i \cdot \boldsymbol{e}_j = \delta_{i,j}$ so that each vector $\boldsymbol{a}$ can be expressed as

$$a = \sum_{i=1} a_i \, e_i, \quad a_j = e_j \cdot a \quad (j = 1, 2, \ldots). \tag{10.5.14}$$

Notice that $e_i \cdot e_j = \delta_{i,j}$ captures both orthogonality and normalization. The corresponding expressions in "ket" notation are

$$|a> = \sum_{i=1} a_i \, |e_i>, \quad a_j = <e_j|a> \text{ and } <e_i|e_j> = \delta_{i,j}. \tag{10.5.15}$$

When **all** vectors in a space can be so expressed the basis is said to be **complete** with respect to the space. (The adjective **complete** should have a familiar ring to it.)

The same is true of function spaces. One can determine basis vectors (functions) $u_n(x)$ which are orthonormal,

$$< u_m \, | \, u_n > = \int_a^b u_m^*(x) \, u_n(x) \rho(x) dx = \delta_{m,n}, \tag{10.5.16}$$

and which are complete with respect to the space. That means that each $f(x)$ in the space can be expanded in the series

$$f(x) = \sum_{m=1}^{\infty} c_m \, u_m(x) \text{ where } c_m = \int_a^b u_m^*(x) f(x) \rho(x) dx. \tag{10.5.17}$$

Now we remember where we have encountered the term **complete** before. It was in connection with Parseval's equation and the representation of functions that are square integrable on $-\pi \leqslant x \leqslant \pi$ in terms of the Fourier functions $\{\cos nx, \sin nx\}$.

Having digressed into the algebraic perspective on series representations, let us return to the analysis of the Sturm-Liouville problem. Its solutions have some general properties of key importance. These follow in large part from general properties possessed by the Sturm-Liouville operator $\mathfrak{L}$. Specifically, suppose that $u(x)$ and $v(x)$ are arbitrary twice differentiable functions. For increased generality, we shall take them to be complex. We write

$$v^*(x)\mathfrak{L}u(x) \equiv v^*(x)\tfrac{d}{dx}[p(x)\tfrac{du(x)}{dx}] - v^*(x)q(x)u(x),$$

$$u(x)(\mathfrak{L}v(x))^* \equiv u(x)\tfrac{d}{dx}[p(x)\tfrac{dv^*(x)}{dx}] - u(x)q(x)\,v^*(x),$$

take the difference, and then integrate by parts to obtain

$$\int_a^b v^*(\mathfrak{L}u)dx - \int_a^b u(\mathfrak{L}v)^* dx = p(x)\left[v^*(x)\frac{du(x)}{dx} - u(x)\frac{dv^*(x)}{dx}\right]\Bigg|_{x=a}^{x=b}. \tag{10.5.18}$$

This is an instance of the **generalized Green's identity** that we encountered in Section 9.9 and its appearance here tells us that the Sturm-Liouville operator $\mathfrak{L}$ is **self-adjoint** wth respect to the weight function $w(x) = 1$.

Note that if the functions $u(x)$ and $v(x)$ both satisfy the homogeneous boundary conditions

$$\alpha_1 y(a) + \beta_1 y'(a) = 0 \qquad (10.5.19)$$
$$\alpha_2 y(b) + \beta_2 y'(b) = 0 \qquad (10.5.20)$$

or the conditions

$$y(a) = y(b) \text{ and } y'(a) = y'(b) \text{ together with } p(a) = p(b), \qquad (10.5.21)$$

the **surface term** on the right hand side of (10.5.18) vanishes and we obtain the **Green's identity** for self-adjoint operators

$$\int_a^b v^*(x)(\mathfrak{L}u(x))dx = \int_a^b u(x)(\mathfrak{L}v(x))^* dx. \qquad (10.5.22)$$

In algebraic terms, imposing homogeneous boundary conditions on a set of functions $u(x)$ defines a function space; a self-adjoint differential operator $\mathfrak{L}$ then defines a **Hermitian** operator on that space.

The fact that the Sturm-Liouville operator is self-adjoint has important consequences for its eigenfunctions and eigenvalues. Suppose that we have two different eigenfunctions $u_n(x)$ and $u_m(x)$ corresponding to the eigenvalues λ_n and λ_m, $\lambda_n \neq \lambda_m$:

$$\mathfrak{L}u_n(x) = -\lambda_n \rho(x) u_n(x), \qquad (10.5.23)$$

$$\mathfrak{L}u_m(x) = -\lambda_m \rho(x) u_m(x). \qquad (10.5.24)$$

We shall allow for the possibility of complex eigenfunctions and even complex eigenvalues but, by definition, $\rho(x)$ and $\mathfrak{L}$ are real. Multiplying (10.5.23) by $u_m^*(x)$ and the complex conjugate of (10.5.24) by $u_n(x)$, subtracting and integrating, we find

$$\int_a^b [u_m^*(x)\mathfrak{L}u_n(x) - u_n(x)\mathfrak{L}u_m^*(x)]dx = -(\lambda_n - \lambda_m^*) \int_a^b u_m^*(x) u_n(x)\rho(x)dx. \qquad (10.5.25)$$

Since $u_n(x)$ and $u_m(x)$ are eigenfunctions, they satisfy homogeneous boundary conditions and, as we have seen, that means that the left hand side of (10.5.25) must vanish. Thus,

$$(\lambda_n - \lambda_m^*) \int_a^b u_m^*(x) u_n(x)\rho(x)dx = 0. \qquad (10.5.26)$$

If $n = m$, the integral cannot vanish because both $\rho(x)$ and $|u_m(x)|^2$ are non-negative. Therefore, we conclude that $\lambda_m^* = \lambda_m$; **all the eigenvalues of the Sturm-Liouville operator are real**.

If $n \neq m$ and $\lambda_n \neq \lambda_m$, we conclude that

$$\int_a^b u_m^*(x)\, u_n(x)\rho(x)dx = 0. \qquad (10.5.27)$$

Two functions $u_n(x)$ and $u_m(x)$ satisfying a condition like (10.5.27) are said to be **orthogonal with respect to the weight function** $\rho(x)$. In other words, **the functions** $\{u_m(x)\}$ **comprise an orthogonal set of vectors** in a function space where the scalar product between two vectors $u(x)$ and $v(x)$ is **defined** to be

$$\boldsymbol{v}^* \cdot \boldsymbol{u} \equiv\; < v|u > \; \equiv \int_a^b v^*(x)u(x)\rho(x)dx. \qquad (10.5.28)$$

If we normalize the eigenfunctions by requiring

$$\int_a^b |u_m(x)|^2\, \rho(x)dx \equiv ||\, u_m\, ||^2 = 1, \qquad (10.5.29)$$

we obtain an **orthonormal** set and (10.5.27) and (10.5.29) combine to read

$$\int_a^b u_m^*(x)\, u_n(x)\rho(x)dx = \delta_{m,n}\,. \qquad (10.5.30)$$

It is also possible to have $n \neq m$, but $\lambda_n = \lambda_m = \lambda$. If this happens, we say that λ is **degenerate** and equation (10.5.26) no longer requires the corresponding eigenfunctions to be orthogonal.. However, we can always choose or construct them to be orthogonal by forming orthogonal linear combinations.

The most important consequence of the self-adjoint character of the Sturm-Liouville operator is one that we shall state without proof. (The proof can be found in a variety of analysis texts such as Courant and Hilbert or E.C. Titchmarsh.) Its statement is as follows: **the eigenfunctions of a Sturm-Liouville operator comprise a complete set of functions.** Algebraically, this means that they **span** the function space on which they are defined and can be used as **basis vectors** for that space. Thus, any other function (vector) $f(x)$ in the space can be expanded in terms of them,

$$f(x) = \sum_m c_m\, u_m(x) \qquad (10.5.31)$$

where the coefficients c_m are the "components of $f(x)$ along the 'unit' vectors $u_m(x)$",

$$c_m = \; < u_m\, |f> \; = \int_a^b u_m^*(x')f(x')\rho(x')dx'. \qquad (10.5.32)$$

Keep in mind that we have normalized the functions $\{u_m(x)\}$. If that were not the case, (10.5.32) would become

$$c_m = \frac{\int_a^b u_m^*(x')f(x')\rho(x')dx'}{\int_a^b |u_m(x')|^2 \rho(x')dx'}. \tag{10.5.33}$$

In general, the **eigenfunction expansion** in (10.5.31) is an infinite series and so the statement of **completeness** implies a statement about the convergence of the series. Since the Fourier functions $\left\{ \frac{1}{\sqrt{T}} \exp\left[i\frac{2\pi m}{T}x\right] \right\}$ are Sturm-Liouville eigenfunctions, it is not surprising that the convergence properties of Fourier series characterize those of all such eigenfunction expansions. In fact, (10.5.31) is sometimes referred to as a **generalized Fourier series** and the c_m as **generalized Fourier coefficients**. In particular, if $f(x)$ is square integrable with respect to $\rho(x)$ over $a \leqslant x \leqslant b$, then we are assured that the series

$$\sum_{m=1}^{\infty} c_m u_m(x) \text{ with } c_m = \int_a^b u_m^*(x')f(x')\rho(x')dx' \tag{10.5.34}$$

must at least converge in the mean to $f(x)$ and therefore,

$$< f|f > = \int_a^b |f(x)|^2 \rho(x)dx = \sum_{m=1}^{\infty} |c_m|^2 = \sum_{m=1}^{\infty} < f|u_m >< u_m|f > . \tag{10.5.35}$$

Equation (10.5.35) is called a **completeness relation**. Having it hold for all vectors $f(x)$ in a function space defined over $a \leqslant x \leqslant b$ is a necessary and sufficient condition for the set $\{u_m(x)\}$ to be complete with respect to that space.

We encountered convergence in the mean in connection with Fourier series. To remind, it means that if S_N is the Nth partial sum of the series,

$$S_N = \sum_{m=1}^{N} c_m u_m(x),$$

then,

$$\lim_{N\to\infty} \int_a^b |f(x) - S_N(x)|^2 \rho(x)dx = 0. \tag{10.5.36}$$

This does not imply point-wise convergence let alone uniform convergence of the series. However, one can prove that if we further restrict $f(x)$ so that it is piecewise continuous with a square integrable first derivative over $a \leqslant x \leqslant b$, the eigenfunction expansion (10.5.34) converges absolutely and uniformly to $f(x)$ in all sub-intervals free of points of discontinuity and at the points of discontinuity it converges to the arithmetic mean of the right and left hand limits of $f(x)$. If there are no points of discontinuity and in addition, if $f(x)$ satisfies the boundary conditions imposed on $\{u_m(x)\}$, the expansion converges uniformly throughout $a \leqslant x \leqslant b$.

In physics texts the completeness relation is often complemented by and confused with an equation called the **closure relation**. Substituting (10.5.32) into (10.5.31) and reversing the order of summation and integration, we have

$$f(x) = \int_a^b [\rho(x') \sum_m u_m(x)\, u_m^*(x')] f(x') dx'$$

for an arbitrary function $f(x)$. Comparing this with the defining property of Dirac delta functions,

$$\int_a^b \delta(x' - x) f(x') dx' = f(x),\ a \leqslant x \leqslant b,$$

we conclude that

$$\rho(x') \sum_m u_m(x)\, u_m^*(x') = \delta(x' - x). \tag{10.5.37}$$

10.6 A Convenient Notation (And Another Algebraic Digression)

We have defined the scalar product in a complex function space as

$$\mathbf{v}^* \cdot \mathbf{u} =< v|u> = \int_a^b v^*(x)u(x)\rho(x)dx, \tag{10.6.1}$$

where $\rho(x)$ is a suitable weight function. This is a generalization of the expression

$$< a|b> = \mathbf{a}^* \cdot \mathbf{b} = \sum_{j=1}^N a_j^*\, b_j \tag{10.6.2}$$

for the scalar product in an N-dimensional complex number vector space in which an orthonormal basis has been chosen. In a function space, the vector $|u>$ corresponds to the entire set (or continuum) of values assumed by a function $u(x)$ for $a \leqslant x \leqslant b$. Therefore, it is convenient to consider the number $u(x)$ for a specific value of x to be the xth component of the vector $|u>$. This implies the existence of a set of basis vectors $|x>, a \leqslant x \leqslant b$, such that

$$u(x) \equiv < x|u>. \tag{10.6.3}$$

The continuity of x gives rise to difficulties in defining the normalization of $|x>$. We assume, of course, that two distinct basis vectors $|x>$ and $|x'>$ are orthogonal,

$$< x'|x >= 0 \text{ for } x' \neq x.$$

Moreover, we assume that the analogue of the familiar decomposition of a vector in terms of an orthonormal basis,

$$|a >= \sum_{j}^{N} a_j |e_j>, \quad a_k =< e_k|a> (k = 1, 2, \ldots, N),$$

is

$$|u >= \int_{a}^{b} dx \rho(x) u(x) |x>. \tag{10.6.4}$$

This means the scalar product of $|u>$ with $|x'>$ can be written

$$< x'|u >= u(x') = \int_{a}^{b} dx \rho(x) u(x) < x'|x >$$

which implies that $\rho(x) < x'|x >$ has the properties of a Dirac δ-function. Evidently, we cannot normalize $|x>$ to unity. Rather, the analogue of $< e_j|e_k >= \delta_{j,k} = \begin{cases} 1 \text{ if } j = k \\ 0 \text{ if } j \neq k \end{cases}$ is

$$< x|x' >= \frac{1}{\sqrt{\rho(x)\rho(x')}} \delta(x - x') = \frac{1}{\rho(x)} \delta(x - x') = \frac{1}{\rho(x')} \delta(x - x'), \tag{10.6.5}$$

which is not so surprising once we remember that the Dirac δ-function is a continuous variable analogue of the Kronecker δ-function $\delta_{j,k}$.

As we saw in the preceding Section, our function space can also have an enumerable orthonormal basis consisting of vectors $|u_m>$ represented by the functions $u_m(x)$,

$$u_m(x) =< x|u_m >, \quad m = 1, 2, \ldots.$$

The closure relation satisfied by these functions is (see (10.5.37))

$$\frac{1}{\rho(x')} \delta(x' - x) = \sum_{m=1}^{\infty} u_m(x) u_m^*(x').$$

Using the normalization equation (10.6.5), we can rewrite this as

$$< x|x' >= \sum_{m=1}^{\infty} < x|u_m >< u_m|x' >=< x| \left(\sum_{m=1}^{\infty} |u_m >< u_m| \right) |x' >.$$

Since $|x >$ and $|x' >$ are arbitrary basis vectors, the object in brackets must be the identity operator $\mathbb{I}$. Thus, an alternative expression of closure is

$$\mathbb{I} = \sum_{m=1}^{\infty} |u_m >< u_m|. \qquad (10.6.6)$$

The analogous relation for the basis $|x >$, $a \leqslant x \leqslant b$, is

$$\mathbb{I} = \int_a^b dx \rho(x)|x >< x|. \qquad (10.6.7)$$

10.7 Fourier Series and Transforms as Eigenfunction Expansions

The most familiar examples of complete sets are the Fourier functions

$$\frac{1}{\sqrt{T}} \exp\left[i\frac{2\pi m}{T}x \right], \quad m = 0, \pm 1, \pm 2, \ldots, a \leqslant x \leqslant a + T \qquad (10.7.1)$$

and

$$\frac{1}{\sqrt{2\pi}} e^{-ikx}, \quad -\infty < k < \infty, -\infty < x < \infty. \qquad (10.7.2)$$

As we have seen already, the first of these is comprised of the eigenfunction solutions of the Sturm-Liouville equation

$$\frac{d^2}{dx^2}u(x) = -\lambda u(x), \quad a \leqslant x \leqslant a + T \qquad (10.7.3)$$

subject to the periodic boundary condition $u(x) = u(x + T)$. The corresponding eigenvalues are $\lambda = \left(\frac{2\pi m}{T}\right)^2$. This discrete set is called the **spectrum** of $\mathcal{L} \equiv \frac{d^2}{dx^2}$ when applied to functions which satisfy this boundary condition. The orthogonality relation satisfied by these eigenfunctions is

$$\int_a^{a+T} u_m^*(x)\, u_n(x)dx = \delta_{m,n}. \qquad (10.7.4)$$

which also tells us that they are normalized to unity.

An eigenfunction expansion of a function $f(x)$ in terms of this basis provides a Fourier series representation:

$$f(x) = \sum_{m=-\infty}^{\infty} c_m \frac{1}{\sqrt{T}} \exp\left[i\frac{2\pi m}{T}x \right] = \frac{a_0}{2} + \sum_{m=0}^{\infty}\left[a_m \cos\left(\frac{2\pi m}{T}\right) + b_m \sin\left(\frac{2\pi m}{T}\right) \right],$$
$$(10.7.5)$$

with

$$c_m = <u_m | f> = \frac{1}{\sqrt{T}} \int_a^{a+T} \exp\left[-i\frac{2\pi m}{T}x\right] f(x)dx. \qquad (10.7.6)$$

The completeness relation for the Fourier functions is

$$\int_a^{a+T} |f(x)|^2 \, dx = \sum_{m=-\infty}^{\infty} |c_m|^2 \qquad (10.7.7)$$

or,

$$\frac{a_0^2}{2} + \sum_{m=1}^{\infty} [a_m^2 + b_m^2] = \frac{2}{T} \int_a^{a+T} |f(x)|^2 \, dx. \qquad (10.7.8)$$

The latter expression is known as Parseval's equation in the theory of Fourier series.

At this point the reader may wish to return to our analysis of the solution of the stretched string problem since it centres upon the identification of a Fourier sine series as an eigenfunction expansion.

Suppose that the range of x is the entire real line so that (10.7.3) is replaced by

$$\frac{d^2}{dx^2}u(x) = -\lambda u(x), \qquad -\infty < x < \infty \qquad (10.7.9)$$

and the periodic boundary condition is replaced by

$$|u(\pm\infty)| < \infty. \qquad (10.7.10)$$

As we have seen, the eigenfunctions are now

$$u_k(x) = e^{-ikx}, \qquad -\infty < k < \infty$$

and the corresponding eigenvalues are $\lambda = k^2$. Notice that $\mathfrak{L} \equiv \frac{d^2}{dx^2}$ now has a continuous spectrum. The eigenfunctions' orthogonality relation is

$$\int_{-\infty}^{\infty} e^{ikx} e^{-ik'x} \, dx = 2\pi\delta(k - k'). \qquad (10.7.11)$$

Thus, normalizing the eigenfunctions, we arrive at the form given in (10.7.2):

$$u_k(x) = \frac{1}{\sqrt{2\pi}} e^{-ikx}, \qquad -\infty < k < \infty.$$

An eigenfunction expansion of a function $f(x)$ defined on $-\infty < x < \infty$ in terms of this basis is given by the continuous sum

$$f(x) = \int_{-\infty}^{\infty} F(k) u_k(x)dk \qquad (10.7.12)$$

where

$$F(k) = <u_k|f> = \int\limits_{-\infty}^{\infty} u_k^*(x')f(x')dx'. \tag{10.7.13}$$

Substituting for $u_k(x)$, we see that $F(k)$ is the Fourier transform of $f(x)$,

$$F(k) = \frac{1}{\sqrt{2\pi}}\int\limits_{-\infty}^{\infty} e^{ikx'} f(x')dx' \equiv \mathcal{F}\{f(x)\}, \tag{10.7.14}$$

and (10.7.12) is a Fourier integral representation of $f(x)$:

$$f(x) = \frac{1}{2\pi}\int\limits_{-\infty}^{\infty} e^{-ikx} F(k)dk \equiv \mathcal{F}^{-1}\{F(k)\}. \tag{10.7.15}$$

The completeness relation in this case reads

$$\int\limits_{-\infty}^{\infty} |f(x)|^2\, dx = \int\limits_{-\infty}^{\infty} |F(k)|^2\, dk \tag{10.7.16}$$

which is a result known as Plancherel's Theorem in the theory of Fourier transforms. The closure relation is

$$\int\limits_{-\infty}^{\infty} u_k(x)\, u_k^*(x')dk = \frac{1}{2\pi}\int\limits_{-\infty}^{\infty} e^{-ik(x-x')}\, dk = \delta(x - x'). \tag{10.7.17}$$

We shall now consider a concrete problem involving a continuous eigenvalue spectrum. Four large conducting plates are arranged with the electrostatic potentials shown in the diagram below.

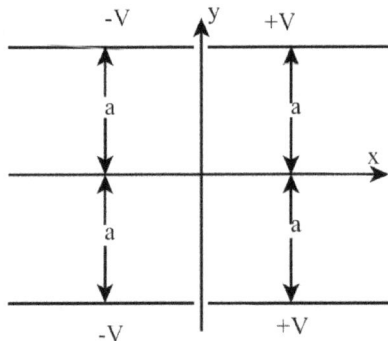

The size of the plates is much larger than the separation $2a$ and so they can be treated as though they extend to infinity in the x- and z-directions. We wish to find the electrostatic potential in the region between the plates. That means we wish to solve Laplace's equation

$$\frac{\partial^2 \psi}{\partial x^2} + \frac{\partial^2 \psi}{\partial y^2} + \frac{\partial^2 \psi}{\partial z^2} = 0 \tag{10.7.18}$$

subject to boundary conditions at the x- and y-boundaries but not at the z-boundaries. In fact, the symmetry of the problem tells us that there is no z-dependence at all.

Evidently, the conditions at the y-boundaries are

$$\psi(x, \pm a) = V \text{ for } x > 0 \text{ and } \psi(x, \pm a) = -V \text{ for } x < 0 \tag{10.7.19}$$

which implies, among other things, that $\psi(x, y)$ is an even function of y.

Formulating the conditions at the x-boundaries requires a little more thought. We note that the conditions in (10.7.19) are odd with respect to $x \to -x$:

$$\psi(-x, \pm a) = -\psi(x, \pm a).$$

This must also be true for all other values of y, that is $\psi(-x, y) = -\psi(x, y)$ for $-a \leqslant y \leqslant a$. Therefore, we must have the following condition at $x = 0$:

$$\psi(0, y) = 0 \text{ for all } y \text{ in } -a \leqslant y \leqslant a. \tag{10.7.20}$$

The other x-boundaries are at infinity where we can require

$$\lim_{x \to \pm \infty} |\psi(x, y)| < \infty. \tag{10.7.21}$$

Since the conditions at the x-boundaries are homogeneous, we will begin our solution of Laplace's equation by eliminating the derivative with respect to x . This means expanding $\psi(x, y)$ in terms of the eigenfunctions of $\mathfrak{L} = \frac{d^2}{dx^2}$ that satisfy the boundary conditions $X(0) = 0$ and $|X(\pm\infty)| < \infty$. The boundedness requirement means that we have to rule out the possibility that $\lambda < 0$ since that corresponds to the exponential solutions $\exp(\pm\sqrt{-\lambda}x)$ of the equation

$$\frac{d^2 X}{dx^2} = -\lambda X.$$

The condition at $x = 0$ further eliminates the possibility of $\lambda = 0$ and of the cosine solution when $\lambda > 0$. This leaves us with the eigensolutions $X_k(x) = \sin kx, 0 \leqslant k < \infty$ and eigenvalues $\lambda = k^2$.

Since the only restriction placed on k is that it be real and positive-definite, we have obtained a **continuous** eigenvalue spectrum and the eigenfunctions $X_k(x)$ comprise a non- denumerably infinite set. Of course we know that when normalized to become $\left\{ \sqrt{\frac{2}{\pi}} \sin kx \right\}$ this set of eigenfunctions provides the complete, orthonormal basis for Fourier sine transform representations. Therefore, we can represent $\psi(x, y)$ by the uniformly convergent eigenfunction expansion

$$\psi(x, y) = \sqrt{\frac{2}{\pi}} \int_0^\infty \Psi(k, y) \sin kx\, dk \tag{10.7.22}$$

where the expansion coefficients $\Psi(k, y)$ comprise the Fourier sine transform of $\psi(x, y)$:

$$\Psi(k, y) = \mathcal{F}_S\{\psi(x, y)\}. \tag{10.7.23}$$

Substituting the representation (10.7.22) into Laplace's equation and interchanging the order of differentiation and integration gives us

$$\sqrt{\frac{2}{\pi}} \int_0^\infty \left[-k^2 \, \Psi(k, y) + \frac{\partial^2 \Psi}{\partial y^2} \right] \sin kx \, dk = 0. \tag{10.7.24}$$

In other words, the PDE has been reduced to its Fourier sine transform

$$\frac{\partial^2 \Psi}{\partial y^2} - k^2 \, \Psi(k, y) = 0. \tag{10.7.25}$$

Solving this equation and using the fact that $\Psi(k, y)$ is an even function of y (because $\psi(x, y)$ is even) we obtain

$$\Psi(k, y) = C(k) \cosh ky.$$

To find the remaining unknown $C(k)$ we impose the boundary condition $\psi(x, a) = V$; that is, we require that

$$\Psi(k, a) = \mathcal{F}_S\{V\} = \sqrt{\frac{2}{\pi}} \frac{V}{k}.$$

Thus,

$$C(k) = \sqrt{\frac{2}{\pi}} \frac{V}{k} \frac{1}{\cosh ka}$$

and our final solution is

$$\psi(x, y) = \frac{2V}{\pi} \int_0^\infty \frac{\cosh ky}{\cosh ka} \frac{\sin kx}{k} dk. \tag{10.7.26}$$

This demonstrates that using integral transforms is completely equivalent to performing an eigenfunction expansion when the eigenfunctions correspond to a **continuous** eigenvalue spectrum.

10.8 Normal Mode (or Initial Value) Problems

Having started our discussion of boundary value problems with a vibrating string, we shall conclude with a vibrating membrane (or drum head). But first, we shall pursue some theoretical considerations that are relevant to any system that is set in motion via an action that is expressible by means of non-homogeneous **initial conditions**.

The equation of motion of such a system will typically be one of

$$\nabla^2 \psi(\mathbf{r}, t) = \frac{1}{D} \frac{\partial \psi}{\partial t} \quad \text{or} \quad \nabla^2 \psi(\mathbf{r}, t) = \frac{1}{c^2} \frac{\partial^2 \psi}{\partial t^2} \tag{10.8.1}$$

and the problem will be to solve it within a region V bounded by a surface S subject to time independent boundary conditions on S **and** to **initial conditions** that specify ψ and in the case of the wave equation $\frac{\partial \psi}{\partial t}$ throughout V at $t = 0$.

The most efficacious way of proceeding is to set

$$\psi(\boldsymbol{r}, t) = \psi_1(\boldsymbol{r}) + \psi_2(\boldsymbol{r}, t) \tag{10.8.2}$$

where

1. $\nabla^2 \psi_1(\boldsymbol{r}) = 0$ and $\psi_1(\boldsymbol{r})$ satisfies the **same boundary conditions on** S as does $\psi(\boldsymbol{r}, t)$
2. $\nabla^2 \psi_2(\boldsymbol{r}, t) = \frac{1}{D}\frac{\partial \psi_2}{\partial t}$ or $\nabla^2 \psi_2(\boldsymbol{r}, t) = \frac{1}{c^2}\frac{\partial^2 \psi_2}{\partial t^2}$ and $\psi_2(\boldsymbol{r}, t)$ satisfies **homogeneous boundary conditions** on S and the **same initial conditions** as does $\psi(\boldsymbol{r}, t)$.

We have seen how one goes about solving for $\psi_1(\boldsymbol{r})$ in either Cartesian, cylindrical or spherical coordinates. Therefore, we can focus on the initial value problem associated with $\psi_2(\boldsymbol{r}, t)$.

We already know from Section 10.3 how to separate the time dependence. It results in separated solutions of the form

$$\psi_2(\boldsymbol{r}, t) = e^{-Dk^2 t}\, u(\boldsymbol{r}) \text{ or } \psi_2(\boldsymbol{r}, t) = \begin{Bmatrix} \cos kct \\ \sin kct \end{Bmatrix} u(\boldsymbol{r}) \tag{10.8.3}$$

depending on which PDE we are solving. Moreover, in **both** cases, the time-independent function $u(\boldsymbol{r})$ is required to be a solution of

$$\nabla^2 u(\boldsymbol{r}) + k^2 u(\boldsymbol{r}) = 0 \tag{10.8.4}$$

which when accompanied by homogeneous boundary conditions on S is a multi-dimensional Sturm-Liouville **eigenvalue problem**. Denoting its eigenfunctions and eigenvalues by $u_n(\boldsymbol{r})$ and k_n^2 respectively, we can assert that the former comprise a **complete, orthogonal** set of functions. In other words,

$$\int_V u_n^*(\boldsymbol{r})\, u_m(\boldsymbol{r}) dV = 0 \quad \text{if } n \neq m \tag{10.8.5}$$

and, any function $f(\boldsymbol{r})$ that is square integrable over V can be represented by the (convergent) series

$$f(\boldsymbol{r}) = \sum_n c_n u(\boldsymbol{r}) \tag{10.8.6}$$

where

$$c_n = \frac{\int_V u_n^*(\boldsymbol{r}) f(\boldsymbol{r}) dV}{\int_V |u_n(\boldsymbol{r})|^2 \, dV}. \tag{10.8.7}$$

This means that there is an infinite set of solutions of the diffusion and wave equations which satisfy homogeneous boundary conditions on S. Each solution has a **characteristic time dependence** and collectively, they are called the **normal modes** of the system in question. Forming superpositions of them, we can express any other solution of the diffusion or wave equation as

$$\psi_2(\mathbf{r}, t) = \sum_n c_n\, e^{-D k_n^2 t}\, u_n(\mathbf{r}) \quad \text{or} \quad \sum_n [a_n \cos k_n\, ct + b_n \sin k_n\, ct]\, u_n(\mathbf{r}) \qquad (10.8.8)$$

respectively. A complete determination of the solution is then made by imposing the initial conditions via an application of (10.8.7):

$$c_n = \frac{\int\limits_V u_n^*(\mathbf{r})\, \psi_2(\mathbf{r}, 0)\, dV}{\int\limits_V |u_n(\mathbf{r})|^2\, dV}, \quad \text{or} \qquad (10.8.9)$$

$$a_n = \frac{\int\limits_V u_n^*(\mathbf{r})\, \psi_2(\mathbf{r}, 0)\, dV}{\int\limits_V |u_n(\mathbf{r})|^2\, dV} \quad \text{and} \qquad (10.8.10)$$

$$b_n = \frac{1}{k_n c} \frac{\int_V u^*(\mathbf{r}) \frac{\partial \psi_2}{\partial t}\big|_{t=0}\, dV}{\int_V |u_n(\mathbf{r})|^2\, dV}. \qquad (10.8.11)$$

As advertised at the beginning of this Section, we shall illustrate the use of this machinery by trying it out on a vibrating membrane. The transverse vibrations of a horizontal membrane of rectangular shape that is stretched equally with a tension τ in all directions will satisfy the two- dimensional wave equation

$$\frac{\partial^2 \psi}{\partial x^2} + \frac{\partial^2 \psi}{\partial y^2} = \frac{1}{c^2} \frac{\partial^2 \psi}{\partial t^2}, \quad c^2 = \frac{\tau}{\mu} \qquad (10.8.12)$$

where μ is the mass per unit area and $\psi(x, y; t)$ is the vertical displacement of the membrane at any point (x, y) and time t. We shall set up our coordinate axes along two of the edges of the membrane. Then, assuming that it is fixed along all four of its edges and that it has sides of length a and b, the boundary conditions for this problem are

$$\psi(0, y; t) = \psi(a, y; t) = 0$$
$$\psi(x, 0; t) = \psi(x, b; t) = 0 \qquad \text{for all } t. \qquad (10.8.13)$$

As we have just seen, the solution can be expressed as

$$\psi(x, y; t) = \sum_v [a_v \cos k_v\, ct + b_v \sin k_v\, ct]\, u_v(x, y) \qquad (10.8.14)$$

with

$$\frac{\partial^2 u_v}{\partial x^2} + \frac{\partial^2 u_v}{\partial y^2} + k_v^2 \, u_v(x, y) = 0 \qquad (10.8.15)$$

and

$$u_v(0, y) = u_v(a, y) = u_v(x, 0) = u_v(x, b) = 0.$$

Setting $u_v(x, y) = X(x)Y(y)$ the Helmholtz equation in (10.8.15) separates into the eigenvalue equations

$$\frac{d^2 X}{d x^2} = -\lambda_1 \, X(x) \text{ and } \frac{d^2 Y}{d y^2} = -\lambda_2 \, Y(y) \qquad (10.8.16)$$

subject to

$$X(0) = X(a) = 0 \text{ and } Y(0) = Y(b) = 0,$$

where $\lambda_1 + \lambda_2 = k_v^2$. These are identical to the eigenvalue equation in the stretched string problem and so we know that the eigenvalues are

$$\lambda_1 = \frac{m^2 \pi^2}{a^2}, m = 1, 2, \ldots \text{ and } \lambda_2 = \frac{n^2 \pi^2}{b^2}, n = 1, 2, \ldots \qquad (10.8.17)$$

and the corresponding eigenfunctions are

$$u_v(x, y) = u_{m,n}(x, y) = X_m(x) \, Y_n(y) = \sin \frac{m\pi x}{a} \sin \frac{n\pi y}{b}. \qquad (10.8.18)$$

Further, since $k_v^2 = \lambda_1 + \lambda_2 = \pi^2 \left(\frac{m^2}{a^2} \right)$, the corresponding time dependence is given by

$$T_{m,n}(t) = a_{m,n} \cos \omega_{m,n} t + b_{m,n} \sin \omega_{m,n} t \qquad (10.8.19)$$

with $\omega_{m,n} = \pi c \sqrt{\frac{m^2}{a^2} + \frac{n^2}{b^2}}$, m and $n = 1, 2, \ldots$. Thus, **each pair** of integers (m, n) defines a distinct **normal mode** of vibration of the membrane and the complete solution of the two-dimensional wave equation with homogeneous conditions at rectangular boundaries is the superposition

$$\psi(x, y; t) = \sum_{m=1}^{\infty} \sum_{n=1}^{\infty} [a_{m,n} \cos \omega_{m,n} t + b_{m,n} \sin \omega_{m,n} t] \sin \frac{m\pi x}{a} \sin \frac{n\pi y}{b}. \qquad (10.8.20)$$

This is a double Fourier sine series and so equations (10.8.10) and (10.8.11) for the coefficients reproduce the familiar Euler formulae. Specifically, if we impose initial conditions

$$\psi(x, y; 0) = u_0(x, y) \text{ and } \left. \frac{\partial \psi}{\partial t} \right|_{t=0} = v_0(x, y).$$

we have

$$a_{m,n} = \frac{4}{ab} \int_0^a \int_0^b u_0(x, y) \sin m\pi \frac{x}{a} \sin \frac{n\pi y}{b} dx dy \qquad (10.8.21)$$

and

$$b_{m,n} = \frac{4}{ab\,\omega_{m,n}} \int_0^a \int_0^b v_0(x, y) \sin \frac{m\pi x}{a} \sin \frac{n\pi y}{b} dx dy. \tag{10.8.22}$$

It is interesting to explore the properties of the individual normal modes. Because

$$\sin \frac{m\pi x}{a} = 0 \quad \text{at} \quad x = \frac{a}{m}, \frac{2a}{m}, \dots, (m-1)\frac{a}{m}$$

and

$$\sin \frac{n\pi y}{b} = 0 \quad \text{at} \quad y = \frac{b}{n}, \frac{2b}{n}, \dots, (n-1)\frac{b}{n},$$

the (m, n) normal mode,

$$\psi_{m,n}(x, y; t) = \sin \frac{m\pi x}{a} \sin \frac{n\pi y}{b} [a_{m,n} \cos \omega_{m,n} t + b_{m,n} \sin \omega_{m,n} t],$$

has $(m - 1)$ **nodal lines** parallel to the y-axis and $(n - 1)$ **nodal lines** parallel to the x-axis. Every point on each of these lines remains at rest for all t.

Two modes can possess the same frequency if $\frac{a}{b}$ is a rational number. When that happens we say that the frequency is **degenerate** because it is associated with more than one eigenfunction. A simple example is afforded by a square membrane since then every pair of transposed integers defines a pair of normal modes with the same frequency. For instance, the $(2, 1)$ and $(1, 2)$ modes both have frequency $\frac{\sqrt{5}\pi c}{a}$. More-over, any linear combination of the $(2, 1)$ and $(1, 2)$ modes,

$$\psi(x, y; t) = \left(A \sin \frac{2\pi x}{a} \sin \frac{\pi y}{a} + B \sin \frac{\pi x}{a} \sin \frac{2\pi y}{a} \right) \cos \frac{\sqrt{5}c\pi}{a} t,$$

represents a harmonic motion with the same frequency. These solutions are called **hybrid modes** and are vectors in the space spanned by the normal modes. The hybrid modes have **nodal curves** whose location depends on the relative value of the coefficients A and B.

11 Special Functions

11.1 Introduction

The "special functions" of mathematical physics are simply functions that occur so frequently in the solution of physical problems that they have been studied exhaustively resulting in an unusually complete knowledge of their properties. We made the acquaintance of a number of these functions in the last Chapter. Now what we need to do is learn enough about them that acquaintance waxes into friendship or at least into that level of familiarity needed to feel comfortable when they arise in the solution of boundary value problems. And, to promote that sense of comfort, we shall solve problems drawn from several fields of physics. More often than not, the problems will be classified not by their physical origin but by their spatial symmetry.

We shall commence with a study of spherical harmonics.

11.2 Spherical Harmonics: Problems Possessing Spherical Symmetry

11.2.1 Introduction

As we learned in Section 10.4, spherical harmonics are eigenfunctions of the angular part of the Laplacian differential operator ∇^2 when it is expressed in spherical coordinates. Thus, they arise in descriptions of electromagnetic phenomena and of classical and quantum mechanical wave motion. This also means that they are eigenfunctions of the orbital angular momentum operator in quantum mechanics and so they figure in the description of molecules, atoms and nuclei and even in some models of sub-nuclear or quark matter. All of which is to say, spherical harmonics warrant our attention.

We start by reviewing a few lines from Section 10.4. Substitution of a separated solution $u(r, \theta, \varphi)$ or $\psi(r, \theta, \varphi) = R(r)Y(\theta, \varphi)$ into the Helmholtz equation or Laplace's equation resulted in the following equation for Y:

$$\frac{1}{\sin\theta} \frac{\partial}{\partial\theta} \left(\sin\theta \frac{\partial Y}{\partial\theta} \right) + \frac{1}{\sin^2\theta} \frac{\partial^2 Y}{\partial\varphi^2} = -\lambda_1 Y(\theta, \varphi). \tag{11.2.1}$$

Accompanied by suitable boundary conditions, this is an eigenvalue equation and the eigenfunctions are found by performing a second separation of variables.

Specifically, we set $Y(\theta, \varphi) = \Theta(\theta)\Phi(\varphi)$ and obtain

$$\frac{d^2 \Phi}{d\varphi^2} = -\lambda_2 \Phi(\varphi), \tag{11.2.2}$$

and

$$\frac{1}{\sin\theta}\frac{d}{d\theta}\left(\sin\theta\frac{d\Theta}{d\theta}\right) + \left(\lambda_1 - \frac{\lambda_2}{\sin^2\theta}\right)\Theta = 0 \qquad (11.2.3)$$

where $0 \le \varphi \le 2\pi$ and $0 \le \theta \le \pi$.

Since we want single-valued solutions we impose the boundary condition $\Phi(\varphi + 2\pi) = \Phi(\varphi)$. This implies that

$$\lambda_2 = m^2 \quad \text{and} \quad \Phi = \Phi_m(\varphi) = \frac{1}{\sqrt{2}}e^{im\varphi}, \quad m = 0, \pm 1, \pm 2, \ldots \qquad (11.2.4)$$

where we have included a normalization factor.

We now turn our attention to equation (11.2.3) which can be rendered more familiar by a transformation of the independent variable. In terms of the new variable $x = \cos\theta$, (11.2.3) becomes

$$(1 - x^2)\frac{d^2 P}{dx^2} - 2x\frac{dP}{dx} + \left(\lambda - \frac{m^2}{1 - x^2}\right)P = 0 \qquad (11.2.5)$$

where $P(x) = \Theta(\cos^{-1} x)$ and $-1 \le x \le 1$. Note that we have suppressed the subscript on λ_1. This is a variant of Legendre's equation called the **associated Legendre equation**. Like the original Legendre DE, which corresponds to setting $m = 0$ in (11.2.5), it has regular singular points at $x = \pm 1$. Therefore, the boundary condition that we must impose on its solutions is that they be bounded, $|P(\pm 1)| < \infty$. We shall now use the analytical tools of Chapter 9 to find the eigenfunctions that result from that imposition.

11.2.2 Associated Legendre Polynomials

Our first tentative move will be to expand $P(x)$ in Frobenius series about the regular singular points $x = \pm 1$. Our object is to determine its leading behaviour there.

Substituting $P(x) = \sum_{k=0}^{\infty} c_k(x-1)^{s+k}$, $c_0 \ne 0$ into (11.2.5) and equating the coefficient of the lowest power of $(x - 1)$ to zero, we obtain the indicial equation

$$4s(s - 1) + 4s - m^2 = 0 \quad \text{with roots} \quad s = \pm\frac{m}{2}.$$

To remove a source of ambiguity, we shall restrict $m \ge 0$ for the time being.

The root $s = -\frac{m}{2}$ can be ruled out immediately because $P(x)$ has to be bounded at $x = 1$. Therefore, we must be able to write $P(x) = (1 - x)^{\frac{m}{2}}f(x)$ where $f(x)$ is bounded and non-vanishing at $x = 1$.

Next, we substitute $P(x) = \sum_{k=0}^{\infty} c_k(x+1)^{s+k}$, $c_0 \ne 0$ into (11.2.5) and again determine the indicial equation. The result is the same as before and so s must again be set equal to $\frac{m}{2}$. This means that $P(x) = (1 + x)^{\frac{m}{2}}g(x)$ where $g(x)$ is bounded and non-vanishing at $x = -1$.

Combining these two results, we conclude that $P(x)$ can be expressed in the factored form

$$P(x) = (1 - x^2)^{\frac{m}{2}} u(x) \tag{11.2.6}$$

where $u(x)$ must be bounded and non-vanishing at $x = \pm 1$. A differential equation for $u(x)$ can be found by substituting (11.2.6) into the associated Legendre equation. The result is

$$(1 - x^2)\frac{d^2 u}{d x^2} - 2(m + 1)x\frac{du}{dx} + (\lambda - m - m^2)u = 0. \tag{11.2.7}$$

Since $x = 0$ is an ordinary point, the solutions of this DE will have the Taylor series representation $u(x) = \sum\limits_{k=0}^{\infty} c_k x^k$. Substitution into (11.2.7) and equation of coefficients of successive powers of x to zero results in the recurrence relation

$$c_{k+2} = \frac{k(k - 1) + 2(m + 1)k - \lambda + m(m + 1)}{(k + 1)(k + 2)} c_k, \quad k \geq 0. \tag{11.2.8}$$

This relates the coefficients of all even powers in $u(x)$ back to c_0 and of all odd powers back to c_1 . Thus, as we learned to expect in Chapter 9, we have obtained two linearly independent solutions.

Applying standard convergence tests (the ratio test for example), one finds that the series diverge at $x = \pm 1$. However, there is a remedy at hand and it is one we have invoked before.

If we choose λ so that $c_{k+2} = 0$ for some k, one of the series will terminate to become a **polynomial** of degree k. Therefore, the requirement that $|u(\pm 1)| < \infty$ implies that

$$k(k - 1) + 2(m + 1)k - \lambda + m(m + 1) = 0,$$

or

$$\lambda = (m + k)(m + k + 1) \quad \text{for some} \quad k.$$

This has a familiar ring to it and to reinforce the familiarity we set $m + k = l$. The eigenvalues λ are then specified as

$$\lambda = l(l + 1), \quad l \geq m, \, m = 0, 1, 2, \ldots . \tag{11.2.9}$$

The corresponding solutions for $u(x)$ are **polynomials of degree** $l - m$.

Normally, the next step would be to use the recurrence relation (11.2.8) to determine an explicit expression for $u(x)$. However, it is less messy as well as more instructive to make the determination in a somewhat different way. As we know from Chapter 9, Legendre polynomials satisfy the equation

$$(1 - x^2)\frac{d^2 P_l}{d x^2} - 2x\frac{d P_l}{dx} + l(l + 1) P_l = 0. \tag{11.2.10}$$

Differentiating m times we obtain

$$(1 - x^2)\frac{d^2}{dx^2}\left(\frac{d^m P_l}{dx^m}\right) - 2(m+1)x\frac{d}{dx}\left(\frac{d^m P_l}{dx^m}\right) + [l(l+1) - m - m^2]\left(\frac{d^m P_l}{dx^m}\right) = 0$$

which is identical to the DE for $u(x)$. This enables us to make the identification $u(x) = \frac{d^m}{dx^m} P_l(x)$ and conclude that the **eigenfunction solutions of the associated Legendre equation are**

$$P_l^m(x) = (1 - x^2)^{\frac{m}{2}}\frac{d^m}{dx^m} P_l(x), \quad l = 0, 1, 2, \ldots \text{ and } 0 \leq m \leq l \tag{11.2.11}$$

corresponding to the **eigenvalues** $\lambda = l(l+1)$. These functions are called **associated Legendre polynomials** (but are polynomials only when m is even). We shall begin an exploration of their properties by focussing on the subset we have met before, the ($m = 0$) Legendre polynomials.

11.2.3 Properties of Legendre Polynomials

The explicit polynomial expression for $P_l(x)$ that we found in Section 9.6 reads

$$P_l(x) = \frac{1}{2^l}\sum_{k=0}^{[l/2]}(-1)^k \frac{(2l-2k)!}{k!(l-2k)!(l-k)!} x^{l-2k} \tag{11.2.12}$$

where $[l/2] = \frac{l}{2}$ if l is even and $[l/2] = \frac{l-1}{2}$ if l is odd . This can be recast in a form that has a variety of uses, both practical and theoretical, by noticing first that

$$P_l(x) = \frac{1}{2^l}\frac{d^l}{dx^l}\left[\sum_{k=0}^{[l/2]}(-1)^k \frac{1}{k!(l-k)!} x^{2l-2k}\right]$$

and then that

$$(x^2-1)^l = \sum_{k=0}^{[l/2]}(-1)^k \frac{l!}{k!(l-k)!} x^{2l-2k}.$$

Combined, these two identities give us **Rodrigues' formula** for $P_l(x)$,

$$P_l(x) = \frac{1}{2^l l!}\frac{d^l}{dx^l}(x^2-1)^l. \tag{11.2.13}$$

This formula was derived by Olinde Rodrigues (1795-1851) and appears in his doctoral thesis. After graduation from Université de Paris in 1815, Rodrigues became a banker, a not uncommon fate for mathematicians then as now.

One can generate the lowest order polynomials fairly easily from Rodrigues' formula and thus confirm what we found in Section 7.6:

$$P_0(x) = 1, \quad P_1(x) = x, \quad P_2(x) = \frac{1}{2}(3x^2 - 1), \quad P_2(x) = \frac{1}{2}(5x^3 - 3x), \ldots.$$

Yet another way of generating the polynomials and more importantly, of deducing many of their properties is to make use of the so-called **generating function**

$$G(x, t) \equiv \frac{1}{\sqrt{1 - 2xt + t^2}} = \sum_{l=0}^{\infty} t^l P_l(x), \quad |t| < 1, |x| \le 1. \tag{11.2.14}$$

The proof of this identity can be established by a "brute-force" method that begins with the expansion

$$(1 - 2xt + t^2)^{-\frac{1}{2}} = [1 + t(t - 2x)]^{-\frac{1}{2}} = \sum_{m=0}^{\infty} \frac{(-\frac{1}{2})(-\frac{3}{2})\dots(\frac{1}{2} - m)}{m!} t^m (t - 2x)^m,$$

uses the binomial theorem to expand $(t - 2x)^m$, and then performs an "inspired" change of summation index to obtain

$$G(x, t) = \sum_{l=0}^{\infty} \sum_{k=0}^{[l/2]} \left(-\frac{1}{2}\right)^{l-k} \frac{1 \cdot 3 \cdot \dots (2l - 2k - 3)(2l - 2k - 1)}{k!(l - 2k)!} t^l (-2x)^{l-2k}$$

which is recognizable, after some simplification, as the left hand side of (11.2.14). We shall establish it by a more elegant approach that employs integral representations of $P_l(x)$.

A contour integral representation follows immediately from Rodrigues' formula and the Cauchy differentiation formula:

$$P_l(z) = \frac{1}{2^l} \frac{1}{2\pi i} \int_C \frac{(\zeta^2 - 1)^l}{(\zeta - z)^{l+1}} d\zeta \tag{11.2.15}$$

where C is any simple closed contour enclosing the point $\zeta = z$. This is called **Schläfli's integral representation.**

Ludwig Schläfli (1814-1895) was a Swiss geometer and complex analyst.

We shall choose C to be a circle about z with radius $|\sqrt{z^2 - 1}|$ in which case any point on C is defined by

$$\zeta = z + \sqrt{z^2 - 1} \, e^{i\theta}, \quad 0 \le \theta \le 2\pi.$$

It does not matter which branch of $\sqrt{z^2 - 1}$ is used here so long as we are consistent. A little algebraic manipulation then gives us

$$\zeta^2 - 1 = 2(\zeta - z)(z + \sqrt{z^2 - 1} \cos \theta) \quad \text{and} \quad d\zeta = i(\zeta - z)d\theta.$$

Therefore, substituting into (11.2.15), we obtain **Laplace's integral representation,**

$$P_l(z) = \frac{1}{\pi} \int_0^{\pi} (z + \sqrt{z^2 - 1} \cos \theta)^l \, d\theta. \tag{11.2.16}$$

This is the representation we need for the generating function identity. Substituting (11.2.16) into the right hand side of (11.2.14) and interchanging the order of summation and integration, we have

$$\sum_{l=0}^{\infty} t^l P_l(x) = \frac{1}{\pi} \int_0^{\pi} \sum_{l=0}^{\infty} t^l (x + \sqrt{x^2 - 1} \cos \theta)^l \, d\theta$$

$$= \frac{1}{\pi} \int_0^{\pi} \frac{d\theta}{1 - tx - t\sqrt{x^2 - 1} \cos \theta}$$

$$= \frac{1}{\sqrt{1 - 2tx + t^2}} = G(x, t),$$

where the evaluation of the integral over θ is done by residue calculus (or by reference to a set of integral tables). This is just (11.2.14) written in reverse order and so our derivation is complete.

The generating function readily yields the values assumed by $P_l(x)$ at a number of special points. For example, at $x = 1$, we have

$$\sum_{l=0}^{\infty} t^l P_l(1) = G(1, t) = \frac{1}{1 - t} = \sum_{l=0}^{\infty} t^l, \quad |t| < 1.$$

Therefore,

$$P_l(1) = 1 \quad \text{for all} \ \ l \geq 0. \tag{11.2.17}$$

At $x = 0$, we have

$$\sum_{l=0}^{\infty} t^l P_l(0) = (1 + t^2)^{-\frac{1}{2}} = 1 - \frac{1}{2} t^2 + \left(-\frac{1}{2}\right) \left(-\frac{3}{2}\right) \frac{t^4}{2} + \dots .$$

Therefore,

$$P_l(0) = \begin{cases} 0 & \text{if } l \text{ is odd} \\ \dfrac{(-1)^{\frac{l}{2}} \, l!}{2^l \left(\frac{l}{2}\right)^2} & \text{if } l \text{ is even} \end{cases} . \tag{11.2.18}$$

At $x = -1$, we have

$$\sum_{l=0}^{\infty} t^l P_l(-1) = \frac{1}{1 + t} = \sum_{l=0}^{\infty} (-t)^l$$

and so,

$$P_l(-1) = (-1)^l \quad \text{for all} \ \ l \geq 0. \tag{11.2.19}$$

More generally,

$$P_l(-x) = (-1)^l P_l(x) \tag{11.2.20}$$

which follows from

$$G(-x, -t) = \frac{1}{\sqrt{1 - 2(-x)(-t) + (-t)^2}} = G(x, t)$$

or

$$\sum_{l=0}^{\infty} P_l(-x)(-t)^l = \sum_{l=0}^{\infty} P_l(x) t^l.$$

The generating function is also the source of a number of useful identities connecting Legendre polynomials of different orders. These are found by differentiating $G(x, t)$.

Differentiating with respect to t, we obtain

$$\frac{\partial G}{\partial t} = \frac{x - t}{(1 - 2xt + t^2)^{\frac{3}{2}}} = \frac{x - t}{(1 - 2tx + t^2)} G(x, t) = \sum_{l=0}^{\infty} l t^{l-1} P_l(x).$$

After cross-multiplying and then substituting for $G(x, t)$, this yields

$$(x - t) \sum_{l=0}^{\infty} t^l P_l(x) = (1 - 2xt + t^2) \sum_{l=0}^{\infty} l t^{l-1} P_l(x)$$

or

$$\sum_{l=0}^{\infty} l t^{l-1} P_l(x) - \sum_{l=0}^{\infty} (2l + 1)x t^l P_l(x) + \sum_{l=0}^{\infty} (l + 1) t^{l+1} P_l(x) = 0.$$

Equating coefficients of like powers of t, we find that this implies

$$(2l + 1)x P_l(x) = (l + 1) P_{l+1}(x) + l P_{l-1}(x), \quad l = 1, 2, 3, \dots. \tag{11.2.21}$$

This is called a **recursion relation.** An immediate application of it is to determine all Legendre polynomials from a knowledge of $P_0(x) = 1$ and $P_1(x) = x$.

If we differentiate $G(x, t)$ with respect to x rather than t, we obtain

$$\frac{\partial G}{\partial x} = \frac{t}{(1 - 2xt + t^2)^{\frac{3}{2}}} = \frac{t}{1 - 2xt + t^2} G(x, t) = \sum_{l=0}^{\infty} t^l P_l'(x), \quad P_l'(x) \equiv \frac{d P_l}{dx}.$$

Cross-multiplication followed by substitution for $G(x, t)$ makes this read

$$\sum_{l=0}^{\infty} t^{l+1} P_l(x) = (1 - 2xt + t^2) \sum_{l=0}^{\infty} t^l P_l'(x).$$

Equating coefficients of like powers of t, we find the second recursion relation:

$$P_l(x) = P_{l+1}'(x) - 2x P_l'(x) + P_{l-1}'(x), \quad l = 1, 2, 3, \dots. \tag{11.2.22}$$

Many other recursion relations can be derived from linear combinations of these two basic ones including, for example,

$$(2l + 1) P_l(x) = P'_{l+1}(x) - P'_{l-1}(x)$$
$$l P_l(x) = x P'_l(x) - P'_{l-1}(x)$$
$$P'_{l+1}(x) = x P'_l(x) + (l + 1) P_l(x)$$
$$(x^2 - 1) P'_l(x) = lx P_l(x) - l P_{l-1}(x). \tag{11.2.23}$$

We now turn our attention to those properties of Legendre polynomials which bear directly on their relevance in the solution of boundary value problems. These are orthogonality, normalization and completeness. Since the polynomials are solutions of the Sturm-Liouville eigenvalue problem

$$\frac{d}{dx}\left[(1 - x^2)\frac{d P_l}{dx}\right] = -l(l + 1) P_l(x) \text{ with } |P_l(\pm 1)| < \infty, \tag{11.2.24}$$

they are mutually orthogonal with respect to the weight function $\rho(x) = 1$:

$$\int_{-1}^{1} P_l(x) P_m(x)dx = 0 \text{ if } l \neq m. \tag{11.2.25}$$

Verifying this provides an instructive example of the utility of Rodrigue's formula. Taking $l < m$, we have

$$\int_{-1}^{1} P_l(x) P_m(x)dx = \frac{1}{2^{l+m}}\frac{1}{l!m!}\int_{-1}^{1}\left[\frac{d^l}{d x^l}(x^2 - 1)^l\right]\left[\frac{d^m}{d x^m}(x^2 - 1)^m\right]dx. \tag{11.2.26}$$

Integrating by parts, we obtain

$$\int_{-1}^{1}\frac{d^l}{d x^l}(x^2 - 1)^l\frac{d^m}{d x^m}(x^2 - 1)^m dx$$

$$= \frac{d^{l-1}}{d x^{l-1}}(x^2 - 1)^l\frac{d^m}{d x^m}(x^2 - 1)^m\bigg|_{x=-1}^{x=1}$$

$$- \int_{-1}^{1}\frac{d^l}{d x^l}(x^2 - 1)^l\frac{d^{m+1}}{d x^{m+1}}(x^2 - 1)^m dx.$$

The integrated term vanishes because $\frac{d^{l-1}}{d x^{l-1}}(x^2 - 1)^l$ has simple zeros at the end-points $x = \pm 1$. We now repeat the integration by parts $l - 1$ times. In each of these integrations the integrated term vanishes and we are left with

$$\int_{-1}^{1}\frac{d^l}{d x^l}(x^2 - 1)^l\frac{d^m}{d x^m}(x^2 - 1)^m dx = (-1)^l\int_{-1}^{1}(x^2 - 1)\frac{d^{l+m}}{d x^{l+m}}(x^2 - 1)^m dx. \tag{11.2.27}$$

The function $(x^2 - 1)^m$ is a polynomial of degree $2m$. Thus, if $l > m$, we are differentiating it more than $2m$ times and so the result is zero. Therefore,

$$\int_{-1}^{1} P_l(x)\, P_m(x)\, dx = 0, \quad l \ne m.$$

If l does equal m, this integral becomes the **normalization integral**

$$N_l = \int_{-1}^{1} [P_l(x)]^2\, dx = \frac{1}{2^{2l}} \frac{1}{(l!)^2} (-1)^l \int_{-1}^{1} (x^2 - 1)^l \frac{d^{2l}}{d\,x^{2l}} (x^2 - 1)^l\, dx$$

where we have used (11.2.26) and (11.2.27) with $m = l$. We know that

$$\frac{d^{2l}}{d\,x^{2l}} (x^2 - 1)^l = (2l)!$$

and so,

$$N_l = (-1)^l \frac{(2l)!}{2^{2l}(l!)^2} \int_{-1}^{1} (x^2 - 1)^l\, dx.$$

The latter integral can be evaluated by repeated integration by parts. One finds

$$\int_{-1}^{1} (x^2 - 1)^l\, dx = (-1)^l \frac{2^{l+1}\, l!}{1 \cdot 3 \cdot 5 \cdots (2l+1)}.$$

Thus,

$$\int_{-1}^{1} [P_l(x)]^2\, dx = \frac{2}{2l + 1}. \tag{11.2.28}$$

The normalization integral can also be derived from the generating function. Specifically, we can square $G(x, t)$,

$$[G(x, t)]^2 = \frac{1}{1 - 2xt + t^2} = \left[\sum_{l=0}^{\infty} t^l P_l(x)\right]^2 = \sum_{l=0}^{\infty}\sum_{m=0}^{\infty} t^{l+m} P_l(x)\, P_m(x),$$

and then integrate from -1 to 1 to obtain

$$\int_{-1}^{1} \frac{dx}{1 - 2xt + t^2} = \sum_{l=0}^{\infty}\sum_{m=0}^{\infty} t^{l+m} \int_{-1}^{1} P_l(x)\, P_m(x)\,dx = \sum_{l=0}^{\infty} t^{2l} \int_{-1}^{1} [P_l(x)]^2\, dx \tag{11.2.29}$$

where we have used the orthogonality of the Legendre polynomials to eliminate all but the terms with $m = l$ on the right hand side of the equation. Introducing a new variable of integration $y = 1 - 2tx + t^2$ the left hand side of the equation is found to be

$$\int_{-1}^{1} \frac{dx}{1 - 2tx + t^2} = \frac{1}{2t} \int_{(1-t)^2}^{(1+t)^2} \frac{dy}{y} = \frac{1}{t} \ln\left(\frac{1+t}{1-t}\right)$$

which has the power series expansion

$$\frac{1}{t} \ln \left(\frac{1+t}{1-t} \right) = 2 \sum_{l=0}^{\infty} \frac{t^{2l}}{2l+1}. \tag{11.2.30}$$

Equating coefficients of like powers of t in (11.2.29) and (11.2.30), we recover

$$N_l = \int_{-1}^{1} [P_l(x)]^2 \, dx = \frac{2}{2l+1}.$$

Orthogonality and normalization are conveniently combined in the single equation

$$\int_{-1}^{1} P_l(x) \, P_m(x) dx = \frac{2}{2l+1} \, \delta_{l,m}. \tag{11.2.31}$$

As eigenfunctions of a Sturm-Liouville problem, Legendre polynomials form a **complete set**: any function $f(x)$ that is square-integrable with respect to the weight $\rho(x) = 1$ on the interval $-1 \le x \le 1$ can be represented on that interval by the expansion

$$f(x) = \sum_{l=0}^{\infty} c_l P_l(x) \quad \text{where} \quad c_l = \frac{2l+1}{2} \int_{-1}^{1} P_l(x') f(x') dx'. \tag{11.2.32}$$

This is sometimes called a **Fourier-Legendre series**. The **completeness relation** for Legendre polynomials is

$$\sum_{l=0}^{\infty} \frac{2}{2l+1} c_l^2 = \int_{-1}^{1} [f(x)]^2 \, dx \tag{11.2.33}$$

and the **closure relation** is

$$\sum_{l=0}^{\infty} \frac{2l+1}{2} P_l(x) P_l(x') = \delta(x - x'). \tag{11.2.34}$$

In theory at least, the coefficients of like powers in an expansion over an infinite set of polynomials can always be summed so that the expansion is converted to a power series. This raises the question of how power series fit within what is in fact an algebraic picture. The answer is straightforward but instructive. A Taylor series about $x = 0$ is an expansion over the monomials $1, x, x^2, x^3, \ldots$ and the algebraic counterpart of Taylor's Theorem is a theorem due to Weierstrass that this set is complete with respect to any space of square integrable functions. However, they do not form an orthogonal set. Rather, once a weight function and interval of definition is given, one has to form orthogonal linear combinations of the monomials. There are well-defined methods for doing so, the most general being the Gram-Schmidt orthogonalization procedure which constructs successively polynomials of degree l that are orthogonal to the polynomials of degree $0, 1, 2, \ldots l - 1$. The outcome is a set of orthogonal poly-nomials that is unique to the choice of interval and weight function used. We will dis-cuss this in more detail in a subsequent section. For now it suffices to know that the orthogonalization process in the case of the interval $-1 \le x \le 1$ and weight $\rho(x) = 1$ results in the Legendre polynomials.

11.2.4 Problems Possessing Azimuthal Symmetry

Returning to the discussion in Section 11.2.1 of what happens with separation of vari-ables in problems with spherical symmetry, we now know that the angular depen-dence will be governed by the complete set of functions

$$P_l^m(\cos\theta) \left\{ \begin{array}{c} \cos m\varphi \\ \sin m\varphi \end{array} \right\} l = 0, 1, 2, \ldots \quad \text{and} \quad 0 \le m \le l \qquad (11.2.35)$$

where we have used an explicitly real form for $\Phi_m(\varphi)$.

Many applications involve symmetry about the z-axis. This means that there will be no dependence on the azimuthal angle φ and so it is often referred to as azimuthal symmetry. A glance at (11.2.35) tells us that no φ -dependence necessarily implies $m = 0$ and hence, a θ - dependence expressible in terms of Legendre polynomials rather than *associated* Legendre polynomials. In the case of Laplace's equation, for example, a solution with azimuthal symmetry will have the Fourier-Legendre expansion

$$\psi(r, \theta) = \sum_{l=0}^{\infty} R_l(r)\, P_l(\cos\theta) \qquad (11.2.36)$$

where, as we saw in Section 10.4, $R_l(r)$ is a solution of

$$\frac{d^2\,R_l}{d\,r^2} + \frac{2}{r}\frac{d\,R_l}{dr} - \frac{l(l+1)}{r^2}\,R_l = 0. \qquad (11.2.37)$$

The general solution of this equation is

$$R_l(r) = A_l\, r^l + B_l\, r^{-l-1}. \qquad (11.2.38)$$

As a concrete example, suppose that we have two conducting hemispherical shells of radius b, insulated from each other by a thin strip along their circle of contact and maintained at potentials $+V$ and $-V$, respectively. We seek the potential everywhere inside the composite sphere.

This is an example of an **interior problem**: the lower limit of the range for r is zero and, since we require a solution that is bounded there, we can set $B_l = 0$ for all l. (The corresponding **exterior problem** would be to find the potential everywhere **outside**

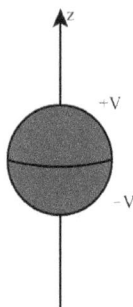

the sphere. In that case, the requirement of a solution that is zero at infinity results in $A_l = 0$ for all l and the B_l have to be determined from the potential at the surface of the sphere.) Thus, it only remains to impose the boundary condition

$$\psi(b, \theta) = \begin{cases} +V, & 0 \le \theta < \frac{\pi}{2} \\ -V, & \frac{\pi}{2} < \theta \le \pi \end{cases}$$

on the Fourier-Legendre series

$$\psi(r, \theta) = \sum_{l=0}^{\infty} A_l\, r^l\, P_l(\cos \theta).$$

Using (11.2.32), we find that this implies

$$A_l = \frac{1}{b^l} \frac{2l+1}{2} \int_0^\pi \psi(b, \theta)\, P_l(\cos \theta) \sin \theta d\theta$$

and, since $\psi(b, \theta)$ is an **odd** function of $\cos \theta$ while $P_l(\cos \theta)$ has **parity** $(-1)^l$, this becomes

$$A_l = \begin{cases} \dfrac{1}{b^l} \dfrac{2l+1}{2} 2V \displaystyle\int_0^1 P_l(\cos \theta)d(\cos \theta) & \text{if } l \text{ is odd} \\ 0 & \text{if } l \text{ is even} \end{cases}$$

Recalling (from (11.2.23)) that

$$(2l+1)\, P_l(x) = P'_{l+1}(x) - P'_{l-1}(x),$$

we find

$$\int_0^1 P_l(\cos \theta)d(\cos \theta) = \frac{1}{2l+1} \left[P_{l+1}(x) - P_{l-1}(x) \right]_{x=0}^{x=1}.$$

But, $P_l(1) = 1$ for all l and $P_l(0) = \frac{(-1)^{\frac{l}{2}} l!}{2^l \left(\frac{l}{2}\right)^2}$ if l is even and is zero otherwise. Thus,

$$\int_0^1 P_l(\cos \theta)d(\cos \theta) = \frac{(-1)^{\frac{l+1}{2}}(l+1)!}{2^{l+1} \left(\frac{l+1}{2}\right)^2 l} \qquad (l \text{ odd}).$$

Therefore, the electrostatic potential at any point inside the sphere is

$$\psi(r, \theta) = V \sum_{l=1,3,5,\ldots}^{\infty} \frac{(2l+1)(-1)^{\frac{l-1}{2}}(l+1)!}{2^{l+1}\left(\frac{l+1}{2}\right)^2 l} \left(\frac{r}{b}\right)^l P_l(\cos\theta).$$

As a further example we shall consider the problem of a conducting sphere in a uniform electric field E_0. What we seek is the new, perturbed electrostatic potential ψ.

Because there are no charges present, $\nabla^2 \psi = 0$. And, choosing our z-axis to be in the direction of E_0, we have azimuthal symmetry. Thus,

$$\psi(r, \theta) = \sum_{l=0}^{\infty} [A_l r^l + B_l r^{-l-1}] P_l(\cos\theta).$$

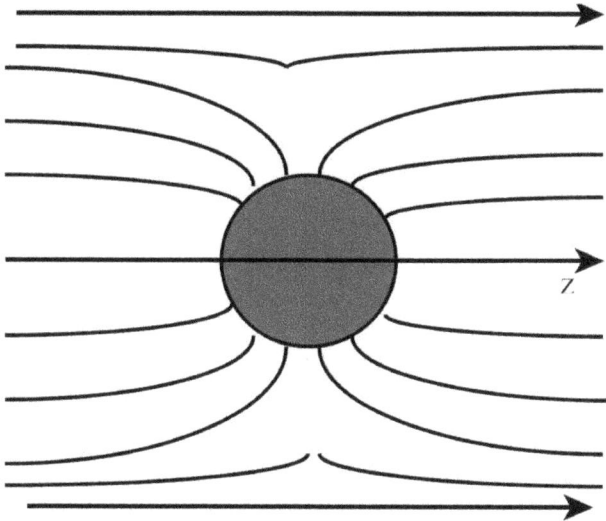

Taking the origin to be at the centre of the sphere, the effect of the perturbing sphere should go to zero as $r \to \infty$. Therefore, we require

$$\lim_{r \to \infty} \psi(r, \theta) = -E_0 z = -E_0 r \cos\theta = -E_0 r P_1(\cos\theta).$$

It then follows that $A_l = 0$ for all $l \geq 2$ (as one would expect for an **exterior problem**) and $A_1 = -E_0$.

The conducting sphere must be at a constant value of potential and so, denoting the sphere's radius by a, we have the boundary condition

$$\psi(a, \theta) = a \text{ constant} = A_0 + \frac{B_0}{a} + \left(\frac{B_1}{a^2}\right) P_1(\cos\theta) + \sum_{l=2}^{\infty} B_l \frac{P_l(\cos\theta)}{a^{l+1}}.$$

In order that this hold for all θ, each coefficient of a $P_l(\cos\theta)$ with $l \neq 0$ must vanish. Thus, $B_l = 0$ for $l \geq 2$ and $B_1 = E_0\, a^3$ which gives us

$$\psi(r, \theta) = A_0 + \frac{B_0}{r} - E_0\, r \left(1 - \frac{a^3}{r^3}\right) P_1(\cos\theta).$$

Since it is $\mathbf{E} = \nabla\psi$ rather than ψ itself that has physical significance, the constant A_0 can be dropped. In addition, we know from Gauss' Law that B_0 is determined by the net charge Q on the sphere: $B_0 = \frac{Q}{4\pi\varepsilon_0}$. Therefore, our final answer is

$$\psi(r, \theta) = \frac{1}{4\pi\varepsilon_0} \frac{Q}{r} - E_0\, r \left(1 - \frac{a^3}{r^3}\right) \cos\theta.$$

In effect, the sphere has perturbed the external field by adding both a monopole term, $\frac{1}{4\pi\varepsilon_0}\frac{Q}{r}$, and a dipole term, $E_0\frac{a^3}{r^2}\cos\theta$, corresponding to an induced dipole moment

$$p = a^3\, E_0.$$

11.2.5 Properties of the Associated Legendre Polynomials

From (11.2.11) and the Rodrigues' formula for $P_l(x)$, we can write down the corresponding formula for $P_l^m(x)$ immediately:

$$P_l^m(x) = \frac{(1 - x^2)^{\frac{m}{2}}}{2^l\, l!} \frac{d^{l+m}}{dx^{l+m}}(x^2 - 1)^l, \quad m = 0, 1, 2, \ldots, l, \quad l = 0, 1, 2, \ldots. \quad (11.2.39)$$

This identity yields well-defined functions even if **m is negative**, provided that $|m| \leq l$. However, they are not independent of their positive m counterparts since one can show that

$$P_l^{-m}(x) = (-1)^m \frac{(l - m)!}{(l + m)!} P_l^m(x), \quad 0 \leq m \leq l. \quad (11.2.40)$$

From the recursion formulas for $P_l(x)$ one can readily obtain formulas for $P_l^m(x)$. Of particular use are the **m-raising and m-lowering relations**,

$$m x\, P_l^m(x) + (1 - x^2)\frac{d P_l^m}{dx} = (1 - x^2)^{\frac{1}{2}} P_l^{m+1}(x)$$

$$m x\, P_l^m(x) - (1 - x^2)\frac{d P_l^m}{dx} = (l + m)(l - m + 1)(1 - x^2)^{\frac{1}{2}} P_l^{m-1}(x), \quad (11.2.41)$$

and the **l-raising and l-lowering relations**,

$$(l + 1)x\, P_l^m(x) - (1 - x^2)\frac{d P_l^m}{dx} = (l - m + 1) P_{l+1}^m(x)$$

$$l x\, P_l^m(x) + (1 - x^2)\frac{d P_l^m}{dx} = (l + m) P_{l-1}^m(x). \quad (11.2.42)$$

As for **parity**, we know already that $P_l(-x) = (-1)^l P_l(x)$. Therefore, since

$$\frac{d^m}{d(-x)^m} = (-1)^m \frac{d^m}{dx^m},$$

$$P_l^m(-x) = (-1)^{l+m} P_l^m(x). \tag{11.2.43}$$

And, in regard to special values, it is obvious that $P_l^m(\pm 1) = 0$.

Of most interest to us are the properties the associated Legendre polynomials have by dint of being eigenfunction solutions of a Sturm-Liouville problem. As we have seen, the equation

$$\frac{d}{dx}\left[(1-x^2)\frac{dP}{dx}\right] + \left[\lambda - \frac{m^2}{1-x^2}\right]P = 0$$

together with the boundary condition $|P(\pm 1)| < \infty$ has the eigensolutions $P(x) = P_l^m(x)$, $0 \le m \le l$, corresponding to eigenvalues $\lambda = l(l+1)$, $l = 0, 1, 2, \ldots$. Thus, the set $P_l^m(x)$, $l = m, m+1, m+2, \ldots$ for fixed m is complete and orthogonal with respect to the weight function $\rho(x) = 1$ on the interval $-1 \le x \le 1$.

The statement of orthogonality is

$$\int_{-1}^{1} P_l^m(x) P_k^m(x)dx = 0 \quad \text{for all } l \ne k. \tag{11.2.44}$$

The associated Legendre polynomials satisfy a second orthogonality relation. It arises because

$$\frac{d}{dx}\left[(1-x^2)\frac{d^2 P_l^m}{dx}\right] + l(l+1) P_l^m(x) = \frac{m^2}{1-x^2} P_l^m(x), \quad |P_l^m(\pm 1)| < \infty$$

is also an eigenvalue problem with eigenvalues, for fixed l, $\lambda = -m^2$, $0 \le m \le l$, and weight function $\rho(x) = \frac{1}{1-x^2}$. Thus,

$$\int_{-1}^{1} \frac{P_l^m(x) P_l^n(x)}{1-x^2}dx = 0 \quad \text{for all } m \ne n. \tag{11.2.45}$$

Orthogonality needs to be accompanied by knowledge of the corresponding normalization integral. For the physically relevant case, this means that we need to evaluate

$$N_{l,m} = \int_{-1}^{1} [P_l^m(x)]^2\, dx = \int_{-1}^{1}(1-x^2)^m \frac{d^m P_l}{dx^m}\frac{d^m P_l}{dx^m}dx. \tag{11.2.46}$$

Integrating by parts, we have

$$N_{l,m} = (1-x^2)^m \frac{d^m P_l}{dx^m}\frac{d^{m-1} P_l}{dx^{m-1}}\Bigg|_{-1}^{1} - \int_{-1}^{1}\frac{d}{dx}\left[(1-x^2)^m \frac{d^m P_l}{dx^m}\right]\frac{d^{m-1} P_l}{dx^{m-1}}dx.$$

The integrated term is zero and the derivative in the integral is

$$\frac{d}{dx}\left[(1-x^2)^m\frac{d^m P_l}{dx^m}\right] = -2mx(1-x^2)^{m-1}\frac{d^m P_l}{dx^m} + (1-x^2)^m\frac{d^{m+1} P_l}{dx^{m+1}}$$

which we can simplify by using Legendre's equation. Differentiating (11.2.10) $m-1$ times we have

$$(1-x^2)\frac{d^{m+1} P_l}{dx^{m+1}} - 2mx\frac{d^m P_l}{dx^m} = -[l(l+1)-m(m-1)]\frac{d^{m-1} P_l}{dx^{m-1}}.$$

Thus,

$$N_{l,m} = [l(l+1)-m(m-1)]\int_{-1}^{1}(1-x^2)^{m-1}\left[\frac{d^{m-1} P_l}{dx^{m-1}}\right]^2 dx. \tag{11.2.47}$$

The numerical factor in front of the integral can be rearranged to read $(l+m)(l-m+1)$ while the integral itself is recognizable as

$$N_{l,m-1} = \int_{-1}^{1}[P_l^{m-1}(x)]^2\, dx.$$

In other words,

$$N_{l,m} = (l+m)(l-m+1)\, N_{l,m-1}$$

and if we apply the same procedure m times, this becomes

$$N_{l,m} = (l+m)(l-m+1)(l+m-1)(l-m+2)\, N_{l,m-2}$$
$$= \ldots =$$
$$= (l+m)(l+m+1)\ldots(l+1)l\ldots(l-m+2)(l-m+1)\, N_{l,0}$$
$$= \frac{(l+m)!}{(l-m)!}\int_{-1}^{1}[P_l(x)]^2\, dx. \tag{11.2.48}$$

But the latter integral is just the normalization integral for the Legendre polynomials and is given in (11.2.31). Therefore, our final result, combining the statement of orthogonality and normalization, is

$$\int_{-1}^{1}P_l^m(x)P_k^m(x)dx = \frac{2}{2l+1}\frac{(l+m)!}{(l-m)!}\delta_{l,k}. \tag{11.2.49}$$

The statement of completeness is

$$\int_{-1}^{1}[f(x)]^2\, dx = \sum_{l=m}^{\infty}\frac{2}{2l+1}\frac{(l+m)!}{(l-m)!}[c_l^m]^2 \tag{11.2.50}$$

where

$$c_l^m = \frac{2l+1}{2} \frac{(l-m)!}{(l+m)!} \int\limits_{-1}^{1} P_l^m(x) f(x) dx,$$

m is a fixed positive integer, and $f(x)$ is any function that is square integrable over the interval $-1 \leq x \leq 1$. Seen in isolation, this result does not appear to add a significant new tool to our box of problem solving techniques. After all, no one would choose to expand functions in terms of associated Legendre functions rather than the much simpler Legendre polynomials unless, of course, some other aspect of the problem gives rise to a compelling reason. This is exactly what happens in problems with spherical but not azimuthal symmetry and it is of such importance that we will devote a separate sub-section to it.

11.2.6 Completeness and the Spherical Harmonics

Separation of variables using spherical coordinates has taught us that any problem with spherical symmetry has a solution whose angular dependence can be expressed in terms of the solutions of the eigenvalue problem

$$\frac{1}{\sin\theta} \frac{\partial}{\partial\theta}\left(\sin\theta \frac{\partial Y}{\partial\theta}\right) + \frac{1}{\sin^2\theta} \frac{\partial^2 Y}{\partial\varphi^2} = -\lambda Y(\theta, \varphi)$$

where $Y(\theta, \varphi)$ is required to be single-valued and finite over (the sphere) $0 \leq \varphi \leq 2\pi$, $-1 \leq \cos\theta \leq 1$. These solutions are

$$Y(\theta, \varphi) = P_l^m(\cos\theta) e^{\pm im\theta}, \quad m = 0, 1, 2, \dots, l \tag{11.2.51}$$

corresponding to eigenvalues $\lambda = l(l+1), l = 0, 1, 2, \dots$. Rather than carry the plus-minus sign in the exponent, we shall allow m to assume both positive and negative values. A further convenience is provided by inclusion of normalizing factors and a phase that is useful in quantum mechanical applications. Making all of these modifications, we obtain

$$Y_l^m(\theta, \varphi) = (-1)^m \sqrt{\frac{2l+1}{4\pi} \frac{(l-m)!}{(l+m)!}} P_l^m(\cos\theta) e^{im\varphi}, \quad l = 0, 1, 2, \dots,$$

$$m = -l, -l+1, \dots, 0, \dots, l-1, l. \tag{11.2.52}$$

This set of functions is called **spherical harmonics**.

Notice that

$$(Y_l^m(\theta, \varphi))^* = (-1)^m Y_l^{-m}(\theta, \varphi), \tag{11.2.53}$$

and

$$\int\limits_0^{2\pi} \int\limits_0^{\pi} (Y_l^m(\theta, \varphi))^* Y_k^n(\theta, \varphi) \sin\theta \, d\theta \, d\varphi = \delta_{l,k} \, \delta_{m,n}. \tag{11.2.54}$$

	$m = 0$	$m = 1$	$m = 2$
$l = 0$	$\frac{1}{\sqrt{4\pi}}$		
$l = 1$	$\sqrt{\frac{3}{4\pi}}\cos\theta$	$\sqrt{\frac{3}{8\pi}}\sin\theta\, e^{i\varphi}$	
$l = 2$	$\sqrt{\frac{5}{4\pi}}\left(\frac{3}{2}\right)$	$-\frac{1}{4}\sqrt{\frac{30}{\pi}}\cos\theta\sin\theta\, e^{i\varphi}$	$\frac{1}{8}\sqrt{\frac{30}{\pi}}\sin^2\theta\, e^{2i\varphi}$

Since for each value of l there are $2l + 1$ allowed values of m, each eigenvalue $\lambda = l(l + 1)$ corresponds to $2l + 1$ eigenfunctions $Y_l^m(\theta, \varphi)$ and so we say that it is $(2l + 1)$– fold **degenerate.**

As eigenfunctions, the spherical harmonics are complete with respect to the space of square-integrable functions defined on the surface of a sphere. Thus, any function $f(\theta.\varphi)$ that is square integrable over the sphere can be represented by the convergent series

$$f(\theta, \varphi) = \sum_{l=0}^{\infty}\sum_{m=-l}^{l} c_{l,m}\, Y_l^m(\theta, \varphi) \tag{11.2.55}$$

where the coefficients are given by

$$c_{l,m} = \int_0^{2\pi}\int_0^{\pi} (Y_l^m(\theta, \varphi))^* f(\theta, \varphi)\sin\theta\, d\theta d\varphi. \tag{11.2.56}$$

The closure relation for spherical harmonics is

$$\sum_{l=0}^{\infty}\sum_{m=-l}^{l} Y_l^m(\theta, \varphi)(Y_l^m(\theta', \varphi'))^* = \frac{\delta(\theta - \theta')\delta(\varphi - \varphi')}{\sin\theta}. \tag{11.2.57}$$

We conclude the formal discussion of spherical harmonics by stating a fortuitous theorem that separates a dependence on the angle between two directions into a dependence on the directions themselves. Suppose that we have two coordinate vectors r and r' with spherical coordinates (r, θ, φ) and (r', θ', φ'), respectively. Trigonometry determines the cosine of the angle α between the two vectors to be

$$\cos\alpha = \cos\theta\cos\theta' + \sin\theta\sin\theta'\cos(\varphi - \varphi'). \tag{11.2.58}$$

Remarkably, when $\cos\alpha$ becomes the argument of a Legendre polynomial, the dependence on θ and φ separates totally from that on θ' and φ'. In fact, what happens is

$$P_l(\cos\alpha) = \frac{4\pi}{2l + 1}\sum_{m=-l}^{l}(Y_l^m(\theta', \varphi'))^* Y_l^m(\theta, \varphi). \tag{11.2.59}$$

This identity is known as the **addition theorem.** In the special case of $\alpha = 0$, it produces the sum rule

$$\sum_{m=-l}^{l} |Y_l^m(\theta, \varphi)|^2 = \frac{2l + 1}{4\pi}. \tag{11.2.60}$$

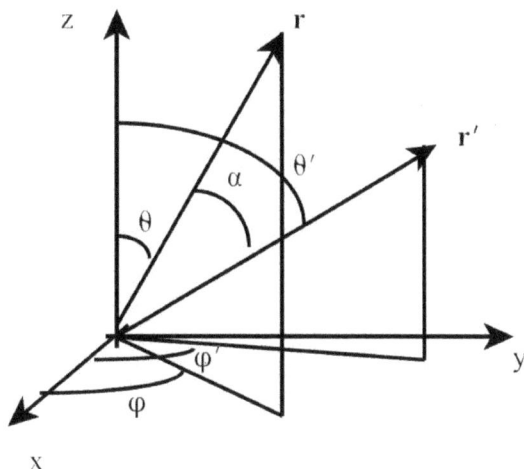

A far more important application, however, is in the expression of the inverse distance $\frac{1}{|r-r'|}$ between two points in terms of spherical harmonics. This arises in a wide range of potential problems, from electromagnetic and gravitational theory to quantum mechanics. A further application of some importance involves the rotation of coordinate axes.

11.2.7 Applications: Problems Without Azimuthal Symmetry

We begin by pursuing the observation about the inverse distance between two points. The electrostatic potential at r due to a point charge q at r' is

$$\psi(r) = \frac{1}{4\pi\epsilon_0} \frac{q}{|\mathbf{r}-\mathbf{r}'|}.$$

If instead of a point charge we have a distribution with charge density $\rho(r')$ confined to a region $r' < R$, the electrostatic potential at $\mathbf{r}, r > R$ is

$$\psi(\mathbf{r}) = \frac{1}{4\pi\epsilon_0} \int_V \frac{\rho(\mathbf{r}')}{|\mathbf{r}-\mathbf{r}'|} d^3 r'. \tag{11.2.61}$$

In either case, knowledge of the inverse distance $\frac{1}{|r-r'|}$ is critical.

The generating function for Legendre polynomials is

$$\frac{1}{\sqrt{1 - 2xt + t^2}} = \sum_{l=0}^{\infty} t^l P_l(x), \quad |t| < 1$$

and the inverse distance can be cast into exactly this form:

$$\frac{1}{|\mathbf{r}-\mathbf{r}'|} = \frac{1}{\sqrt{(r-r')^2}} = \frac{1}{\sqrt{r^2 - 2rr'\cos\alpha + r'^2}} = \frac{1}{r\sqrt{1 - 2\frac{r'}{r}\cos\alpha + \left(\frac{r'}{r}\right)^2}}.$$

Thus, identifying $\cos \alpha$ with x and $\frac{r'}{r}$ with t, we have

$$\frac{1}{|\boldsymbol{r} - \boldsymbol{r}'|} = \sum_{l=0}^{\infty} \frac{r'^{l}}{r^{l+1}} P_l(\cos \alpha), \quad r' < r. \tag{11.2.62}$$

Invoking the addition theorem, this becomes the fully separated expression

$$\frac{1}{|\boldsymbol{r} - \boldsymbol{r}'|} = \sum_{l=0}^{\infty} \sum_{m=-l}^{l} \frac{4\pi}{2l+1} r'^{l} \left(Y_l^m(\theta', \varphi')\right)^* \frac{1}{r^{l+1}} Y_l^m(\theta, \varphi) \tag{11.2.63}$$

which, when substituted into equation (11.2.61), produces a representation of the potential due to a distributed charge that has an immediate physical interpretation. The representation is

$$\psi(\boldsymbol{r}) = \frac{1}{\epsilon_0} \sum_{l=0}^{\infty} \sum_{m=-l}^{l} \frac{1}{2l+1} \frac{Q_l^m}{r^{l+1}} Y_l^m(\theta, \varphi) \tag{11.2.64}$$

where

$$Q_l^m = \int_V \left(Y_l^m(\theta', \varphi')\right)^* r'^{l} \rho(\boldsymbol{r}') \, d^3 \boldsymbol{r}'. \tag{11.2.65}$$

Evidently, the Q_l^m are spherical components of the multipole moments of the charge distribution. Therefore, the representation of $\psi(\boldsymbol{r})$ in terms of them is called a **multipole expansion**. Notice that each term has a distinctive angular distribution and an inverse dependence on r that falls off more rapidly with increasing l. The lowest order multipole moments are the monopole,

$$Q_0^0 = \frac{1}{\sqrt{4\pi}} \int_V \rho(\boldsymbol{r}') \, d^3 \boldsymbol{r}' = \frac{1}{\sqrt{4\pi}} q = \frac{1}{\sqrt{4\pi}} \times \text{(the total charge present)},$$

the dipole,

$$Q_1^{\pm 1} = \mp \sqrt{\frac{3}{8\pi}} (p_1 \mp i p_2), \quad Q_1^0 = \sqrt{\frac{3}{4\pi}} p_3 \quad \text{where} \quad \boldsymbol{p} = \int_V \boldsymbol{r}' \rho(\boldsymbol{r}') \, d^3 \boldsymbol{r}',$$

and the quadrupole,

$$Q_2^{\pm 2} = \frac{1}{12} \sqrt{\frac{15}{2\pi}} (Q_{11} \mp 2i Q_{12} - Q_{22}), \quad Q_2^{\pm 1} = \mp \frac{1}{3} \sqrt{\frac{15}{4\pi}} (Q_{13} \mp i Q_{23}),$$

$$Q_2^0 = \frac{1}{2} \sqrt{\frac{5}{4\pi}} Q_{33} \quad \text{where} \quad Q_{ij} = \int_V (3 x_i' x_j' - r'^2 \delta_{ij}) \rho(\boldsymbol{r}') \, d^3 \boldsymbol{r}'.$$

Thus, with respect to a Cartesian basis, our multipole expansion is

$$\psi(\boldsymbol{r}) = \frac{1}{4\pi \epsilon_0} \left\{ \frac{q}{r} + \frac{\boldsymbol{p} \cdot \boldsymbol{r}}{r^3} + \frac{1}{2} \sum_{i,j=1}^{3} Q_{ij} \frac{x_i x_j}{r^5} + \dots \right\}. \tag{11.2.66}$$

Evidently an expansion of a potential in terms of spherical harmonics is effectively the same as an expansion of the source charge distribution in terms of its multipole components. To reinforce our intuitive appreciation of this, we shall determine the potential due to specific multipole configurations of discrete charges.

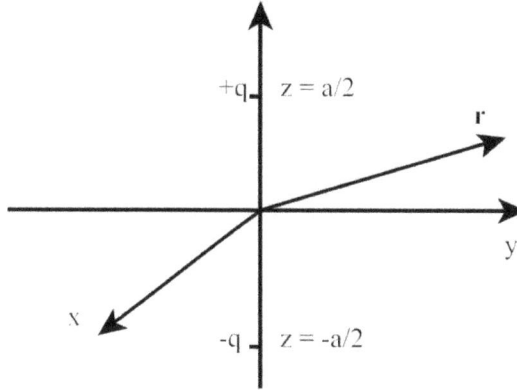

We start with a dipole constructed from a charge q located at $z = \frac{a}{2}$ and charge $-q$ at $z = -\frac{a}{2}$. The resulting potential at a point $\mathbf{r}$ is

$$\psi(\mathbf{r}) = \frac{q}{4\pi \epsilon_0} \left[\frac{1}{|r - \frac{a}{2}k|} - \frac{1}{|r + \frac{a}{2}k|} \right]$$

$$= \frac{q}{4\pi \epsilon_0} \frac{1}{r} \left[\left\{ 1 - 2\frac{a}{2r}\cos\theta + \left(\frac{a}{2r}\right)^2 \right\}^{-\frac{1}{2}} - \left\{ 1 + 2\frac{a}{2r}\cos\theta + \left(\frac{a}{2r}\right)^2 \right\}^{-\frac{1}{2}} \right]$$

or,

$$\psi(\mathbf{r}) = \frac{q}{4\pi \epsilon_0} \frac{1}{r} \left[\sum_{l=0}^{\infty} \left(\frac{a}{2r}\right)^l P_l(\cos\theta) - \sum_{l=0}^{\infty} \left(\frac{-a}{2r}\right)^l P_l(\cos\theta) \right]$$

$$= \frac{2q}{4\pi \epsilon_0} \frac{1}{r} \left[\left(\frac{a}{2r}\right) P_1(\cos\theta) + \left(\frac{a}{2r}\right)^3 P_3(\cos\theta) + \ldots \right].$$

Thus, at a distant point, $r \gg a$, the potential is

$$\psi(\mathbf{r}) \simeq \frac{qa}{4\pi \epsilon_0} \frac{P_1(\cos\theta)}{r^2} = \frac{1}{4\pi \epsilon_0} \frac{qar\cos\theta}{r^3} = \frac{1}{4\pi \epsilon_0} \frac{\mathbf{p} \cdot \mathbf{r}}{r^3}$$

where $\mathbf{p} = qa\mathbf{k}$ is the dipole moment that we learned to associate with such a charge distribution in introductory electricity and magnetism.

We shall now move the dipole off the z-axis so that it can be combined with a second dipole to form an electric quadrupole. To be as general as possible, we will use an arbitrary azimuthal orientation and locate charge q at $\mathbf{r}' = \left(\frac{a}{\sqrt{2}}, \frac{\pi}{4}, \varphi' \right)$ and $-q$

at $\boldsymbol{r}'' = \left(\frac{a}{\sqrt{2}}, \frac{3\pi}{4}, \varphi' \right)$. Again, we have

$$\psi(\boldsymbol{r}) = \frac{q}{4\pi\,\epsilon_0} \left[\frac{1}{|\boldsymbol{r} - \boldsymbol{r}'|} - \frac{1}{|\boldsymbol{r} - \boldsymbol{r}''|} \right]$$

but this time we have to use the addition theorem in the form

$$\frac{1}{|\boldsymbol{r} - \boldsymbol{r}'|} = \sum_{l=0}^{\infty} \sum_{m=-l}^{l} \frac{4\pi}{2l+1} \frac{r'^l}{r^{l+1}} \left(Y_l^m (\theta', \varphi') \right)^* Y_l^m (\theta, \varphi), \quad r > r'$$

to expand the two inverse distances. Then, since $\cos \frac{3\pi}{4} = -\cos \frac{\pi}{4}$, we have $P_l^m(\cos\frac{3\pi}{4}) = P_l^m(-\cos\frac{\pi}{4}) = (-1)^{l+m} P_l^m(\cos\frac{\pi}{4})$ and hence,

$$\psi(\boldsymbol{r}) = \frac{q}{4\pi\,\epsilon_0} \sum_{l=0}^{\infty} \sum_{m=-l}^{l} \frac{4\pi}{2l+1} \frac{a^l}{2^{l/2}\,r^{l+1}} [1 - (-1)^{l+m}] \left(Y_l^m (\tfrac{\pi}{4}, \varphi') \right)^* Y_l^m (\theta, \varphi).$$

The factor $\varepsilon_{l,m} = 1 - (-1)^{l+m}$ vanishes for even values of $l + m$. Therefore, we find

$$\psi(\boldsymbol{r}) = \frac{q}{\epsilon_0} \left[\frac{1}{3\sqrt{2}} \frac{a}{r^2} Y_1^0 (\tfrac{\pi}{4}, \varphi') Y_1^0 (\theta, \varphi) + \frac{1}{5} \frac{a^2}{r^3} Re \left[(Y_2^1(\tfrac{\pi}{4}, \varphi'))^* Y_2^1(\theta, \varphi) \right] + O \left(\frac{a^3}{r^4} \right) \right]$$

or, for $r \ll a$

$$\psi(\boldsymbol{r}) \simeq \frac{\sqrt{2}qa}{4\pi\,\epsilon_0} \frac{1}{r^2} P_1 \left(\cos \tfrac{\pi}{4} \right) P_1 \left(\cos\theta \right) = \frac{qa}{4\pi\,\epsilon_0} \frac{\cos\theta}{r^2}$$

As we would expect, this is the same as the result obtained for a dipole on the z-axis. A dependence on the azimuthal angles enters the picture only when the relative size of r and a justifies inclusion of the $\frac{a^2}{r^3}$ term. However, as we shall now see this becomes the leading term when we combine two dipoles to form a quadrupole.

The quadrupole is constructed by placing a second dipole a distance a from the first and with its charges oriented so that the dipoles are anti-parallel. So, if the first is at an azimuthal angle φ', the second is at $\varphi' + \pi$. Therefore, since

$$e^{im(\varphi' + \pi)} = e^{im\pi} e^{im\varphi'} = (-1)^m e^{im\varphi'},$$

the potential due to the two dipoles is

$$\psi(\boldsymbol{r}) = \frac{q}{4\pi\,\epsilon_0} \sum_{l=0}^{\infty} \sum_{m=-l}^{l} \frac{4\pi}{2l+1} \frac{a^l}{2^{l/2}\,r^{l+1}} [1 - (-1)^{l+m}][1 - (-1)^m] \left(Y_l^m (\tfrac{\pi}{4}, \varphi') \right)^* Y_l^m (\theta, \varphi).$$

The numerical factor $\chi_{l,m} = [1 - (-1)^{l+m}][1 - (-1)^m]$ results in all the lowest order terms vanishing with the first non-zero term corresponding to $l = 2$, $m = \pm 1$ and the next to $l = 4$, $m = \pm 3$ and ± 1. Thus, for $r \gg a$

$$\psi(\boldsymbol{r}) \simeq \frac{q}{4\pi\,\epsilon_0} \frac{4\pi}{5} \frac{a^2}{2\,r^3} 8Re \left[(Y_2^1(\tfrac{\pi}{4}, \varphi'))^* Y_2^1 (\theta, \varphi) \right] = \frac{3q}{4\pi\,\epsilon_0} \frac{a^2}{r^3} \sin\theta \cos\theta \cos(\varphi - \varphi').$$

An electric octupole can similarly be constructed from two quadrupoles and one finds that a further cancellation occurs leaving the $l = 3$ term as the leading one in the spherical harmonic expansion of its potential. Generalizing, we can assert that the potential due to an electric 2^l-pole falls off like r^{-l} and has an angular dependence determined by $Y_l^m(\theta, \varphi)$ for $r \gg a$.

We now turn our attention to solving problems involving Laplace's equation and non-homogeneous boundary conditions. Suppose that we have a spherical shell of radius R which is maintained at a potential $V_0 \cos 2\varphi$. Let us find the potential at any point inside the sphere.

This is an **interior problem** and so the solution must have the characteristic r-dependence that ensures boundedness at the origin. However, unlike the interior problem of Section 11.2.5, this one manifestly lacks azimuthal symmetry. Therefore, rather than use a simple Fourier- Legendre representation of the potential, we now must work with the spherical harmonic expansion

$$\psi(r, \theta, \varphi) = \sum_{l=0}^{\infty} \sum_{m=-l}^{l} A_{lm}\, r^l\, Y_l^m(\theta, \varphi).$$

(This problem is more amenable to use of the equivalent explicitly real expansion

$$\psi(r, \theta, \varphi) = \sum_{l=0}^{\infty} \sum_{m=0}^{l} r^l\, P_l^m(\cos\theta)(a_{lm} \cos m\varphi + b_{ml} \sin m\varphi).$$

However, the object of the exercise is to gain experience working with spherical harmonics.)

Using (11.2.56) and the boundary condition $\psi(R, \theta, \varphi) = V_0 \cos 2\varphi$, we have

$$A_{lm} = \frac{V_0}{R^l} \int_0^{2\pi}\int_0^{\pi} (Y_l^m(\theta, \varphi))^* \cos 2\varphi \sin\theta\, d\theta\, d\varphi$$

or,

$$A_{lm} = \frac{V_0}{R^l} \sqrt{\frac{2l+1}{4\pi}\frac{(l-|m|)!}{(l+|m|)!}} \int_0^{2\pi}\int_0^{\pi} P_l^{|m|}(\cos\theta)\, e^{-im\varphi} \cos 2\varphi \sin\theta\, d\theta\, d\varphi.$$

The integration over φ gives us

$$\int_0^{2\pi} e^{-im\varphi} \cos 2\varphi\, d\varphi = \frac{1}{2} \int_0^{2\pi} e^{-im\varphi}(e^{2i\varphi} + e^{-2i\varphi})\, d\varphi = \begin{cases} 0 & \text{if } m \neq \pm 2 \\ \pi & \text{if } m = \pm 2 \end{cases}$$

This means, of course, that in addition to the restriction on m, we must restrict l to $l \geq 2$. Proceeding with the θ integration, we have

$$\int_0^{2\pi} P_l^2(\cos\theta) \sin\theta\, d\theta = \int_{-1}^{1} P_l^2(x)\, dx = \int_{-1}^{1} (1-x^2)\frac{d^2 P_l}{dx^2}\, dx$$

where we have made the substitution $x = \cos\theta$. Integrating twice by parts, we find

$$\int_{-1}^{1} (1 - x^2)\frac{d^2 P_l}{dx^2}dx = 2x\, P_l(x)\Big|_{-1}^{1} - 2\int_{-1}^{1} P_l(x)dx = 2[1 + (-1)^l] - 2\int_{-1}^{1} P_l(x)\,P_0(x)dx.$$

The last integral is zero because $l \geq 2$ and the Legendre polynomials form an orthogonal set. Therefore,

$$\int_{0}^{\pi} P_l^2(\cos\theta)\sin\theta\, d\theta = \begin{cases} 0 & \text{if } l \text{ is odd} \\ 4 & \text{if } l \text{ is even} \end{cases}$$

Collecting all this information, we conclude that

$$\psi(r, \theta, \varphi) = \sum_{l=2,4,...}^{\infty} \frac{4\pi V_0}{R^l}\sqrt{\frac{(2l+1)}{4\pi}\frac{(l-2)!}{(l+2)!}}\, r^l[Y_l^2(\theta, \varphi) + Y_l^{-2}(\theta, \varphi)]$$

$$= \sum_{l=2,4,...}^{\infty} 2\,V_0(2l+1)\frac{(l-2)!}{(l+2)!}\left(\frac{r}{R}\right)^l P_l^2(\cos\theta)\cos 2\varphi.$$

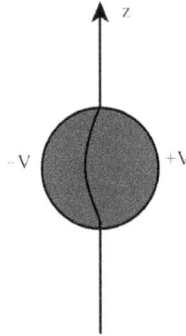

As a further application, consider two hemispherical shells of radius b separated by a thin ring of insulation and maintained at potentials V and $-V$ as shown in the diagram above. The boundary condition at the surface of the sphere is

$$\psi(b, \theta, \varphi) = \begin{cases} +V & \text{if } 0 \leq \varphi < \pi \\ -V & \text{if } \pi < \varphi \leq 2\pi \end{cases}$$

which is certainly not azimuthally symmetric. Therefore, if we wish to find the potential at any point inside the sphere, we will have to resort a second time to the expansion

$$\psi(r, \theta, \varphi) = \sum_{l=0}^{\infty}\sum_{m=-l}^{l} A_{lm}\, r^l\, Y_l^m(\theta, \varphi).$$

Using (11.2.56) plus our boundary condition at $r = b$, we determine the coefficients A_{lm} to be

$$A_{lm} = \frac{1}{b^l} \int\limits_{-1}^{1} \int\limits_{0}^{2\pi} (Y_l^m(\theta, \varphi))^* \, \psi(b, \theta, \varphi) d(\cos\theta) d\varphi$$

$$= \frac{V}{b^l} \sqrt{\frac{2l+1}{4\pi} \frac{(l-m)!}{(l+m)!}} \int\limits_{-1}^{1} P_l^m(\cos\theta) d(\cos\theta) \left[\int\limits_{0}^{\pi} e^{-im\varphi} d\varphi - \int\limits_{\pi}^{2\pi} e^{-im\varphi} d\varphi \right]$$

The factor in square brackets is

$$\int\limits_{0}^{\pi} e^{-im\varphi} d\varphi - \int\limits_{\pi}^{2\pi} e^{-im\varphi} d\varphi = \frac{2i}{m}[(-1)^m - 1].$$

Thus, m can only assume **odd** integer values. Recalling the parity $(-1)^{l+m}$ of associated Legendre functions, we realize that this means that l must also be restricted to **odd** values. To evaluate the integral over $\cos\theta$ we can either resort to integral tables or we can evaluate term by term starting from $l = 1, m = 1$ and try to deduce an expression for the general term by interpolation. The first option sounds simple until we discover that this is an uncommon integral and when listed, it is expressed in terms of hypergeometric series. The second option sounds tedious at best and at worst an opportunity to commit egregious arithmetical errors. And so, we search for a third option.

Comparing the diagram for this problem with that for the hemispherical shells in Section 11.2.5, we realize that this is the same problem but with either the sphere or the coordinate axes rotated through 90°. Therefore, we can write down the solution. It is

$$\psi(r, \theta, \varphi) = V \sum_{l=1,3,5,\dots}^{\infty} \frac{(2l+1)(-1)^{\frac{l-1}{2}}(l+1)!}{2^{l+1} \left(\frac{l+1}{2}\right)^2 l} \left(\frac{r}{b}\right)^l P_l(\cos\alpha),$$

but the angle α that appears here is not the polar angle but rather the angle between $\mathbf{r}$ and the y-axis. Nevertheless, this is a major breakthrough because we can use the addition theorem to relate $P_l(\cos\alpha)$ to spherical harmonics in the angular coordinates of $\mathbf{r}$ and of the y-axis. In fact, from (11.2.59) we have

$$P_l(\cos\alpha) = \frac{4\pi}{2l+1} \sum_{m=-l}^{l} \left(Y_l^m \left(\frac{\pi}{2}, \frac{\pi}{2}\right)\right)^* Y_l^m(\theta, \varphi).$$

Since $\cos\frac{\pi}{2} = 0$, this becomes

$$P_l(\cos \alpha) = \sum_{m=-l}^{l} \frac{(l - |m|)!}{(l + m)!} P_l^{|m|}(0) \, e^{-im\frac{\pi}{2}} P_l^{|m|}(\cos \theta) \, e^{im\varphi}$$

$$= 2 \sum_{m=0}^{l} \frac{(l - m)!}{(l + m)!} P_l^m(0) P_l^m(\cos \theta) \sin m\varphi.$$

The special value $P_l^m(0)$ can be worked out and one finds

$$P_l^m(0) = \frac{2^m \sqrt{\pi}}{\Gamma\left(\frac{l-m}{2} + 1\right) \Gamma\left(\frac{-l-m+1}{2}\right)}.$$

In our case, both l and m are odd. Therefore, the first gamma function in the denominator can be replaced by $\left(\dfrac{l-m}{2}\right)! = \dfrac{(l-m)!!}{2^{\frac{l-m}{2}}}$ and the second by $\dfrac{\sqrt{\pi}(-1)^{\frac{l+m}{2}} 2^{\frac{l+m}{2}}}{(l+m-1)!!}$.
Collecting all this information and substituting into the expression for the potential, we conclude that

$$\psi(r) = V \sum_{l=\text{odd}}^{\infty} \sum_{m=\text{odd}}^{l} \frac{(2l+1)(l+1)!}{2^l \left(\frac{l+1}{2}!\right)^2 l} (-1)^{\frac{m+1}{2}} \frac{(l-m-1)!!}{(l+m)!!} \left(\frac{r}{b}\right)^l P_l(\cos \theta) \sin m\varphi$$

where $(-1)!!$ and $(0)!!$ are both understood to be 1.

11.3 Bessel Functions: Problems Possessing Cylindrical Symmetry

11.3.1 Properties of Bessel and Neumann Functions

As we saw in Section 10.4, separation of variables applied to the Helmholtz and Laplace's equation when cylindrical coordinates are used results in a radial equation that can be transformed into Bessel's DE

$$\frac{d^2 R}{d\rho^2} + \frac{1}{\rho}\frac{dR}{d\rho} + \left(1 - \frac{m^2}{\rho^2}\right) R = 0, \quad m = 0, 1, 2, \ldots \tag{11.3.1}$$

by the simple expedient of replacing the radial variable r by $\rho = \alpha r$ where $\alpha^2 = k^2 - \lambda_2$, k^2 is the Helmholtz equation parameter and λ_2 is the separation constant associated with the z-dependence. The parameter m^2 is the separation constant associated with the θ-dependence and was determined by imposition of the homogeneous boundary condition that we have solutions that are single-valued functions of θ. The general solution of (11.3.1) is the linear combination

$$R(r) = c_1 J_m(\alpha r) + c_2 N_m(\alpha r) \tag{11.3.2}$$

where $J_m(x)$ and $N_m(x)$ are the Bessel and Neumann functions of order m, respectively. We solved for them explicitly in Chapter 9 and found the series representations (9.7.11) and (9.7.33). What we need to do now is relate those of their properties that are most germane to the solution of boundary value problems.

The Bessel functions are well behaved both at the origin and as $x \to \infty$. In fact, they "look" like lightly damped sine or cosine functions. Like sines and cosines, they are oscillatory functions with infinitely many zeros. However, the Bessel function zeros are not equally spaced. With the sole exception of $J_0(x)$ all of the $J_m(x)$ are zero at $x = 0$. Moreover, using the first term in the power series (9.7.11), we see that

$$J_m(x) \simeq \frac{1}{\Gamma(m+1)} \left(\frac{x}{2} \right)^m \text{ as } x \to 0. \tag{11.3.3}$$

The Neumann functions are not well behaved at $x = 0$. $N_0(z)$ has a logarithmic branch point there and $N_m(z)$, $m > 0$, has a pole of order m. Thus,

$$N_0(x) \simeq \frac{2}{\pi} \ln x \text{ as } x \to 0, \tag{11.3.4}$$

and

$$N_m(x) \simeq -\frac{(m-1)!}{\pi} \left(\frac{2}{x} \right)^m \text{ as } x \to 0. \tag{11.3.5}$$

The asymptotic or large x behaviour of the Bessel and Neumann functions can be deduced from integral representations and the method of steepest descents; (see Section 6.3). One finds for $x \gg m$

$$J_m(x) \sim \sqrt{\frac{2}{\pi x}} \cos \left(x - \frac{m\pi}{2} - \frac{\pi}{4} \right) \tag{11.3.6}$$

and

$$N_m(x) \sim \sqrt{\frac{2}{\pi x}} \sin \left(x - \frac{m\pi}{2} - \frac{\pi}{4} \right). \tag{11.3.7}$$

The complementarity of these two expressions reflects the judiciousness of the choice made for the definition of the Neumann functions.

Notice that the linear combinations

$$H_m^{(1)}(x) = J_m(x) + i N_m(x) \text{ and } H_m^{(2)}(x) = J_m(x) - i N_m(x) \tag{11.3.8}$$

have the asymptotic forms

$$H_m^{(1)}(x) \sim \sqrt{\frac{2}{\pi x}} \exp \left[i \left(x - \frac{m\pi}{2} - \frac{\pi}{4} \right) \right] \tag{11.3.9}$$

and

$$H_m^{(2)}(x) \sim \sqrt{\frac{2}{\pi x}} \exp \left[-i \left(x - \frac{m\pi}{2} - \frac{\pi}{4} \right) \right]. \tag{11.3.10}$$

These are called **Hankel functions** or Bessel functions of the **third kind**.

Hankel functions are named for the complex analyst Hermann Hankel (1839-1873). He worked with a who's who of nineteenth century German mathematicians including Kronecker, Möbius, Riemann and Weierstrass.

Recurrence relations for the Bessel functions can be derived directly from their power series representation (9.7.11). Dividing it by x^m and then differentiating, we find

$$\frac{d}{dx}\left(\frac{J_m(x)}{x^m}\right) = \sum_{k=1}^{\infty} \frac{(-1)^k}{(k-1)!(k+m)!}\frac{x^{2k-1}}{2^{2k+m-1}}. \tag{11.3.11}$$

This can be related to $J_{m+1}(x)$ by replacing k by $k+1$:

$$\frac{d}{dx}\left(\frac{J_m(x)}{x^m}\right) = \sum_{k=0}^{\infty} \frac{(-1)^{k+1}}{k!(k+m+1)!}\frac{x^{2k+1}}{2^{2k+m+1}} = -\frac{J_{m+1}(x)}{x^m}$$

which is valid for all $m \geq 0$. A particularly useful special case occurs for $m = 0$:

$$\frac{d}{dx}J_0(x) = -J_1(x). \tag{11.3.12}$$

Notice that repeated application of (11.3.11) , starting with the $m = 0$ case, allows us to relate each $J_m(x)$ back to $J_0(x)$. In fact, we can write down a Rodrigues-like formula,

$$J_m(x) = x^m\left(-\frac{1}{x}\frac{d}{dx}\right)^m J_0(x), \tag{11.3.13}$$

which means that the differential operator $-\frac{1}{x}\frac{d}{dx}$ is to be applied m times to $J_0(x)$ and the result is then multiplied by x^m .

Similarly, multiplying the power series for $J_m(x)$ by x^m and differentiating, we have

$$\frac{d}{dx}[x^m J_m(x)] = \sum_{k=0}^{\infty} \frac{(-1)^k}{k!(k+m-1)!}\frac{x^{2k+2m-1}}{2^{2k+m-1}} = x^m J_{m-1}(x) \tag{11.3.14}$$

which is valid for $m \geq 1$.

Adding $J_{m+1}(x) = -x^m\frac{d}{dx}\left(\frac{J_m(x)}{x^m}\right)$ to $J_{m-1}(x) = \frac{1}{x^m}\frac{d}{dx}[x^m J_m(x)]$, we establish the recurrence relation

$$J_{m+1}(x) + J_{m-1}(x) = \frac{2m}{x}J_m(x). \tag{11.3.15}$$

Subtracting them gives us

$$J_{m+1}(x) - J_{m-1}(x) = -2\frac{dJ_m}{dx}. \tag{11.3.16}$$

One can show that Bessel functions of integral order have a generating function of the form

$$\exp\left[\frac{z}{2}\left(t-\frac{1}{t}\right)\right] = \sum_{m=-\infty}^{\infty} J_m(z)\, t^m, \quad t \neq 0 \tag{11.3.17}$$

where we have switched to a complex independent variable for reasons that will become obvious when we look at applications. The proof is straightforward. From the Taylor series representation of the exponential, we have

$$\exp\left[\frac{z}{2}\left(t - \frac{1}{t}\right)\right] = \sum_{j=0}^{\infty} \frac{1}{j!}\left(\frac{z}{2}\right)^j t^j \cdot \sum_{k=0}^{\infty} \frac{(-1)^k}{k!}\left(\frac{z}{2}\right)^k \left(\frac{1}{t}\right)^k$$

$$= \sum_{j=0}^{\infty}\sum_{k=0}^{\infty} \frac{(-1)^k}{j!k!}\left(\frac{z}{2}\right)^{j+k} t^{j-k}.$$

All that remains is to replace the sum over j by one over $m = j - k$ since that transforms our power series into

$$\exp\left[\frac{z}{2}\left(t - \frac{1}{t}\right)\right] = \sum_{m=-\infty}^{\infty}\sum_{k=0}^{\infty} \frac{(-1)^k}{k!(k+m)!}\left(\frac{z}{2}\right)^{2k+m} t^m = \sum_{m=-\infty}^{\infty} J_m(z)\, t^m.$$

Notice that the generating function series is a Laurent rather than a Taylor series and that Laurent's theorem gives us an immediate contour integral representation of $J_m(z)$:

$$J_m(z) = \frac{1}{2\pi i} \int_C t^{-m-1}\, e^{\frac{z}{2}\left(t - \frac{1}{t}\right)}\, dt \tag{11.3.18}$$

where C is a closed contour about the origin. Changing the integration variable to $u = \frac{zt}{2}$ this becomes

$$J_m(z) = \frac{1}{2\pi i}\left(\frac{z}{2}\right)^m \int_C u^{-m-1} \exp\left(u - \frac{z^2}{4u}\right) du \tag{11.3.19}$$

which we recognize as the integral representation (9.9.33) of Chapter 9.

A Fourier integral representation is obtained by setting $t = e^{i\theta}$. The generating function becomes

$$e^{iz\sin\theta} = \sum_{m=-\infty}^{\infty} e^{im\theta} J_m(z) \tag{11.3.20}$$

with Fourier coefficients

$$J_m(z) = \frac{1}{2\pi} \int_{-\pi}^{\pi} e^{iz\sin\theta}\, e^{-im\theta}\, d\theta. \tag{11.3.21}$$

Because Bessel's equation is of the Sturm-Liouville form, the radial equation (11.3.1)

$$\frac{d^2 R}{d\rho^2} + \frac{1}{\rho}\frac{dR}{d\rho} + \left(1 - \frac{m^2}{\rho^2}\right) R = 0, \quad \rho = \alpha r = \sqrt{k^2 - \lambda_2}\, r \tag{11.3.22}$$

or

$$\frac{d}{dr}\left(r\frac{dR}{dr}\right) - \frac{m^2}{r}R(r) = -\alpha^2\, rR(r)$$

becomes an eigenvalue equation with eigenvalue α^2 as soon as we impose homogeneous boundary conditions on the general solution $R(r) = c_1 J_m(\alpha r) + c_2 N_m(\alpha r)$. Let us suppose that the range of r is $0 \leq r \leq a$. Typically, one boundary condition will be that $R(r)$ be bounded at the origin: $|R(0)| < \infty$. This immediately eliminates the Neumann functions from consideration and we conclude that $R(r) \propto J_m(\alpha r)$. The second condition will most likely be

$$\text{either}\ \ R(a) = 0\ \ \text{or}\ \ \left.\frac{dR}{dr}\right|_{r=a} = 0$$

which implies that

$$\text{either}\ \ J_m(\alpha a) = 0\ \ \text{or}\ \ J'_m(\alpha a) = 0.$$

Denoting the nth zero of $J_m(x)$ by $x_{m,n}$ and the nth zero of $J'_m(x)$ by $y_{m,n}$, we see that the eigenvalues are

$$\text{either}\ \ \alpha^2_{m,n} = \frac{x^2_{m,n}}{a^2}\ \ \text{or}\ \ \alpha^2_{m,n} = \frac{y^2_{m,n}}{a^2}$$

and the corresponding (unnormalized) eigenfunctions are

$$R_n(r) = J_m(\alpha_{m,n}\, r), \quad n = 1, 2, 3, \ldots. \tag{11.3.23}$$

These must comprise an orthogonal set with their orthogonality (with respect to weight function $w(r) = r$) expressed by

$$\int_0^a R_n(r)\, R_p(r) r\, dr = \int_0^a J_m(\alpha_{m,n}\, r) J_m(\alpha_{m,p}\, r) r\, dr = 0\ \ \text{for}\ \ n \neq p. \tag{11.3.24}$$

The zeros of $J_m(x)$ and $J'_m(x)$ are tabulated in standard references such as Abramowicz and Stegun. For future convenience, we provide below a limited table containing the first four zeros of the first five Bessel functions.

	$n = 1$	$n = 2$	$n = 3$	$n = 4$
$m = 0$	2.404	5.520	8.654	11.792
$m = 1$	3.832	7.016	10.173	13.323
$m = 2$	5.135	8.417	11.620	14.796
$m = 3$	6.379	9.760	13.017	16.224
$m = 4$	7.586	11.064	14.373	17.616

The determination of the norm of $J_m(\alpha_{m,n}\, r)$ requires evaluation of the integral

$$N_{m,n} = \int_0^a [\,J_m(\alpha_{m,n}\, r)\,]^2\, r\,dr = \frac{1}{\alpha_{m,n}^2} \int_0^{\alpha_{m,n}\, a} [\,J_m(x)\,]^2\, x\,dx = \frac{1}{\alpha_{m,n}^2}\, I_{m,n}$$

where we have set $x = \alpha_{m,n}\, r$. Integrating once by parts, we have

$$I_{m,n} = \frac{1}{2}[\,J_m(x)]^2 \Big|_{x=0}^{x=\alpha_{m,n}\, a} - \int_0^{\alpha_{m,n}\, a} J_m(x)\, J_m'(x)\, x^2\, dx.$$

But, from Bessel's equation we have

$$x^2\, J_m(x) = m^2\, J_m(x) - x\, J_m'(x) - x^2\, J_m''(x).$$

Therefore,

$$I_{m,n} = \frac{1}{2}[\,J_m(x)]^2 \Big|_{x=0}^{x=\alpha_{m,n}\, a} - \int_0^{\alpha_{m,n}\, a} J_m'(x)[m^2\, J_m(x) - x\, J_m'(x) - x^2\, J_m''(x)]dx,$$

or

$$I_{m,n} = \left\{ \frac{x^2}{2}[\,J_m(x)\,]^2 - \frac{m^2}{2}[\,J_m(x)\,]^2 + \frac{x^2}{2}[\,J_m'(x)\,]^2 \right\} \Bigg|_{x=0}^{x=\alpha_{m,n}\, a} .$$

Thus, if $J_m(\alpha_{m,n}\, a) = 0$,

$$N_{m,n} = \frac{a^2}{2}[\,J_m'(\alpha_{m,n}\, a)\,]^2 = \frac{a^2}{2}[\,J_{m+1}(\alpha_{m,n}\, a)\,]^2 \tag{11.3.25}$$

where we have made use of the recurrence relation

$$\frac{d\,J_m}{dx} = -J_{m+1}(x) + \frac{m}{x}\, J_m(x).$$

On the other hand, if $J_m'(\alpha_{m,n}\, a) = 0$,

$$N_{m,n} = \frac{a^2}{2}\left(1 - \frac{m^2}{\alpha_{m,n}^2\, a^2}\right)[\,J_m(\alpha_{m,n}\, a)\,]^2 . \tag{11.3.26}$$

Thus, combining the orthogonality and normalization results, we have either

$$\int_0^a J_m(\alpha_{m,n}\, r)\, J_m(\alpha_{m,p}\, r)r\,dr = \frac{a^2}{2}[\,J_{m+1}(\alpha_{m,n}\, a)\,]^2\, \delta_{n,p} \tag{11.3.27}$$

or

$$\int_0^a J_m(\alpha_{m,n}\, r)\, J_m(\alpha_{m,p}\, r)r\,dr = \frac{a^2}{2}\left(1 - \frac{m^2}{\alpha_{m,n}^2\, a^2}\right)[\,J_m(\alpha_{m,n}\, a)\,]^2\, \delta_{n,p} . \tag{11.3.28}$$

The further consequence of being solutions of a Sturm-Liouville problem is the $J_m(\alpha_{m,n}\, r), n = 1, 2, \ldots$, are a basis for the space of functions that are square-integrable with respect to the weight function $w(r) = r$ on the interval $0 \leq r \leq a$. Thus, any such function $f(r)$ can be represented by the (mean convergent) series

$$f(r) = \sum_{n=1}^{\infty} c_{m,n} J_{m,n}(\alpha_{m,n}\, r)$$

where

$$c_{m,n} = \frac{2}{a^2 [J_{m+1}(\alpha_{m,n}\, a)]^2} \int_0^a J_m(\alpha_{m,n}\, r) f(r) r dr, \qquad (11.3.29)$$

or

$$c_{m,n} = \frac{2}{\left(a^2 - \frac{m^2}{\alpha_{m,n}^2}\right) [J_m(\alpha_{m,n}\, a)]^2} \int_0^a J_m(\alpha_{m,n}\, r) f(r) r dr, \qquad (11.3.30)$$

depending on whether $\alpha_{m,n} = \frac{x_{m,n}}{a}$ or $\alpha_{m,n} = \frac{y_{m,n}}{a}$, $n = 1, 2, \ldots$.

11.3.2 Applications

Free Vibrations of a Circular Drum Head:

The transverse vibrations of a (two-dimensional) drum head are described by the wave equation

$$\nabla^2 \psi = \frac{1}{c^2} \frac{\partial^2 \psi}{\partial t^2} \quad \text{where} \quad c = \sqrt{\frac{T}{\mu}}, \quad T = \text{tension/length}, \quad \mu = \text{mass/area}.$$

As we saw in Sections 10.4 and 10.8, the solution of this equation can be expressed as

$$\psi(\mathbf{r}, t) = \sum_n [\, a_n \cos k_n\, ct + b_n \sin k_n\, ct\,] \, u_n(\mathbf{r})$$

with

$$\nabla^2 u_n + k_n^2 u_n = 0 \quad \text{and} \quad u_n(\mathbf{r}) = 0 \text{ for } \mathbf{r} \text{ on the edge of the drum head,}$$

where the latter condition follows from an assumption that the drum head is fixed along its edges. If it is a circular drum head, we should use cylindrical coordinates for $\mathbf{r}$. The solutions of the Helmholtz equation in these coordinates were obtained in Section 10.4. Since we have no z-dependence, the separation constant λ_2 in that discussion must be zero and so, $\alpha^2 = k^2$. Thus, the solutions of the separated angular and radial equations are

$$\Theta_m(\theta) = A_m \cos m\theta + B_m \sin m\theta \text{ and } R_m(r) = C_m J_m(kr) + D_m N_m(kr), \quad m = 0, 1, 2, \ldots$$

where we have imposed the boundary condition in θ but not those in r.

We shall take the radius of the drum head to be a and its centre to be located at $r = 0$. We then have as boundary conditions $|u(0, \theta)| < \infty$ and $u(a, \theta) = 0$. The first of these implies that $D_m = 0$ for all m. The second implies that $J_m(ka) = 0$ which has solutions $k_{m,n} = \frac{x_{m,n}}{a}$ where $x_{m,n}$ is the nth zero of $J_m(x)$. This means that we obtain as eigenfunctions of the Helmholtz equation the **normal modes**

$$u_{m,n}(r, \theta) = \begin{cases} J_m(k_{m,n}\, r) \cos m\theta \\ J_m(k_{m,n}\, r) \sin m\theta \end{cases} \quad m = 0, 1, 2, \ldots n = 1, 2, , \ldots.$$

The frequencies $w_{m,n} = k_{m,n}\, c = x_{m,n}\,\frac{c}{a}, m = 0, 1, 2, \ldots, n = 1, 2, \ldots,$ are the **natural frequencies** of the drum head. Each frequency has two normal modes, one with $\cos m\theta$ and the other with $\sin m\theta$, and so is twofold degenerate. The exceptions are the frequencies with $m = 0$ each of which has a single mode possessing radial symmetry. From the values of $x_{m,n}$ given above, we see that the lowest modes in order of frequency are

m, n	normal mode	frequency
(0, 1)	$J_0(k_{0,1}\, r)$	$2.404\frac{c}{a}$
(1, 1)	$J_1(k_{1,1}\, r) \cos \theta, J_1(k_{1,1}\, r) \sin \theta$	$3.832\frac{c}{a}$
(2, 1)	$J_2(k_{2,1}\, r) \cos 2\theta, J_2(k_{2,1}\, r) \sin 2\theta$	$5.135\frac{c}{a}$
(0, 2)	$J_0(k_{0,2}\, r)$	$5.520\frac{c}{a}$
(3, 1)	$J_3(k_{3,1}\, r) \cos 3\theta, J_3(k_{3,1}\, r) \sin 3\theta$	$6.379\frac{c}{a}$
(1, 2)	$J_1(k_{1,2}\, r) \cos \theta, J_1(k_{1,2}\, r) \sin \theta$	$7.016\frac{c}{a}$
(4, 1)	$J_4(k_{4,1}\, r) \cos 4\theta, J_4(k_{4,1}\, r) \sin 4\theta$	$7.586\frac{c}{a}$
(2, 2)	$J_2(k_{2,2}\, r) \cos 2\theta, J_2(k_{2,2}\, r) \sin 2\theta$	$8.417\frac{c}{a}$
(0, 3)	$J_0(k_{0,3}\, r)$	$8.654\frac{c}{a}$
(5, 1)	$J_5(k_{5,1}\, r) \cos 5\theta, J_5(k_{5,1}\, r) \sin 5\theta$	$8.779\frac{c}{a}$

To impose initial conditions on the transverse displacement and velocity of the drum head, we form the superposition of normal modes

$$\psi(r, \theta, t) = \sum_{n=1}^{\infty} \sum_{m=0}^{\infty} J_m(k_{m,n}\, r)[(a_{m,n} \cos m\theta + b_{m,n} \sin m\theta) \cos w_{m,n}\, t$$

$$+ (c_{m,n} \cos m\theta + d_{m,n} \sin m\theta) \sin w_{m,n}\, t].$$

At $t = 0$ this gives us

$$u_0(r, \theta) \equiv \psi(r, \theta, 0) = \sum_{n=1}^{\infty} \sum_{m=0}^{\infty} J_m(k_{m,n}\, r)[a_{m,n} \cos m\theta + b_{m,n} \sin m\theta]$$

and

$$v_0(r, \theta) \equiv \left.\frac{\partial \psi}{\partial t}\right|_{t=0} = \sum_{n=1}^{\infty} \sum_{m=0}^{\infty} J_m(k_{m,n}\, r)\, w_{m,n}[c_{m,n} \cos m\theta + d_{m,n} \sin m\theta].$$

These are Fourier-Bessel series in r as well as Fourier series in θ. Thus, invoking (11.3.29) for Fourier-Bessel coefficients and the Euler formulae for Fourier coefficients, we find

$$a_{m,n} = \frac{2}{\pi a^2 [J_{m+1}(x_{m,n})]^2} \begin{Bmatrix} 1 \\ \frac{1}{2} \end{Bmatrix} \int_0^a \int_0^{2\pi} u_0(r,\theta) \cos m\theta \, J_m(k_{m,n}\, r) r dr d\theta \text{ for } \begin{Bmatrix} m \neq 0 \\ m = 0 \end{Bmatrix}$$

$$b_{m,n} = \frac{2}{\pi a^2 [J_{m+1}(x_{m,n})]^2} \int_0^a \int_0^{2\pi} u_0(r,\theta) \sin m\theta \, J_m(k_{m,n}\, r) r dr d\theta$$

$$c_{m,n} = \frac{2}{\pi a^2 \omega_{m,n} [J_{m+1}(x_{m,n})]^2} \begin{Bmatrix} 1 \\ \frac{1}{2} \end{Bmatrix} \int_0^a \int_0^{2\pi} v_0(r,\theta) \cos m\theta \, J_m(k_{m,n}\, r) r dr d\theta \text{ for } \begin{Bmatrix} m \neq 0 \\ m = 0 \end{Bmatrix}$$

$$d_{m,n} = \frac{2}{\pi a^2 \omega_{m,n} [J_{m+1}(x_{m,n})]^2} \int_0^a \int_0^{2\pi} v_0(r,\theta) \sin m\theta \, J_m(k_{m,n}\, r) r dr d\theta.$$

Heat Conduction in a Cylinder of Finite Length:

For our next application, we shall consider a metal cylinder of radius R and length L whose surface is maintained at a constant temperature T_1. Initially the cylinder is at a uniform temperature T_0. We want to find out how the temperature changes as a function of time and position.

We locate the cylinder so that its central axis lies along the z-axis and its ends correspond to $z = 0$ and $z = L$. The temperature $\psi(\mathbf{r}, t)$ at any point within the cylinder and at any time t will be a solution of the heat conduction equation

$$\nabla^2 \psi = \frac{1}{D} \frac{\partial \psi}{\partial t} \text{ where } D = \frac{\kappa}{c\rho},$$

κ is the thermal conductivity, c is the specific heat and ρ is the density of the cylinder. Since $\psi(\mathbf{r}, t) = T_1 =$ a constant is a solution of this equation, we can set $\psi(\mathbf{r}, t) = T_1 + \psi_1(\mathbf{r}, t)$ where $\psi_1(\mathbf{r}, t)$ is a solution of the PDE that satisfies **homogeneous** boundary conditions at the surface of the cylinder. Separating variables, $\psi_1(\mathbf{r}, t)$ can be expressed as

$$\psi_1(\mathbf{r}, t) = \sum_\gamma u_\gamma(\mathbf{r}) e^{-D k_\gamma^2 t} \text{ where } \nabla^2 u_\gamma + k_\gamma^2 u_\gamma = 0$$

and $u_\gamma(\mathbf{r})$ is subject to homogeneous boundary conditions in all three coordinates. As ever, the requirement of single-valuedness implies that the angular dependence of the u_γ is given by a linear combination of $\cos m\theta$ and $\sin m\theta$. Then, since $|u_\gamma(0, \theta, z)| <$

∞ and $u_\gamma(R, \theta, z) = 0$, the radial dependence of the u_γ is determined by the Bessel functions $J_m(\alpha_{m,n} r)$, $\alpha_{m,n} = \frac{x_{m,n}}{R}$. Thus, the u_γ have the series representation

$$u_\gamma(r, \theta, z) = \sum_{n=1}^{\infty} \sum_{m=0}^{\infty} J_m(\alpha_{m,n} r)[a_{m,n} \cos m\theta + b_{m,n} \sin m\theta]$$

$$\times [c_{\gamma,m,n} e^{\sqrt{\alpha_{m,n}^2 - k_\gamma^2} z} + d_{\gamma,m,n} e^{-\sqrt{\alpha_{m,n}^2 - k_\gamma^2} z}].$$

Now we impose the requirement that $u_\gamma(r, \theta, 0) = u_\gamma(r, \theta, L) = 0$. It then follows that

$$c_{\gamma,m,n} = -d_{\gamma,m,n} \quad \text{and} \quad \sqrt{\alpha_{m,n}^2 - k_\gamma^2} = i\beta, \quad \beta L = p\pi, \quad p = 1, 2, \ldots$$

so that the final term in square brackets in u_γ becomes proportional to $\sin \frac{p\pi z}{L}$. Thus,

$$k_\gamma^2 = k_{m,n,p}^2 = \alpha_{m,n}^2 + \beta^2 = \frac{x_{m,n}^2}{R^2} + \frac{p^2 \pi^2}{L^2}$$

and the eigenfunctions (normal modes) corresponding to these eigenvalues are

$$u_{m,n,p}(r, \theta, z) = J_m(\alpha_{m,n} r) \sin \frac{p\pi z}{L}[a_{m,n,p} \cos m\theta + b_{m,n,p} \sin m\theta].$$

To complete the solution we now must impose the initial condition on

$$\psi(r, \theta, z, t) = T_1 +$$

$$\sum_{p=1}^{\infty} \sum_{n=1}^{\infty} \sum_{m=0}^{\infty} J_m(\alpha_{m,n} r) \sin \frac{p\pi z}{L}[a_{m,n,p} \cos m\theta + b_{m,n,p} \sin m\theta] e^{-D k_{m,n,p}^2 t}.$$

Setting $t = 0$, we have

$$T_0 - T_1 = \sum_{p=1}^{\infty} \sum_{n=1}^{\infty} \sum_{m=0}^{\infty} J_m(\alpha_{m,n} r) \sin \frac{p\pi z}{L}[a_{m,n,p} \cos m\theta + b_{m,n,p} \sin m\theta].$$

Since the left hand side is constant, m must be restricted to zero. Thus, we lose one summation and are left with the double series

$$T_0 - T_1 = \sum_{p=1}^{\infty} \sum_{n=1}^{\infty} a_{0,n,p} J_0(\alpha_{0,n} r) \sin \frac{p\pi z}{L}$$

which is a Fourier Bessel and Fourier sine series. Using the formulae for the coefficients of both such series, we find

$$a_{0,n,p} = \frac{4(T_0 - T_1)}{L R^2 [J_1(x_{0,n})]^2} \int_0^R \int_0^L J_0(\alpha_{0,n} r) \sin \frac{p\pi z}{L} dz r dr.$$

Since $x J_0(x) = \frac{d}{dx}[x J_1(x)]$, we have $\int_0^a J_0(x) x dx = a J_1(a)$. Therefore,

$$\int_0^R J_0(\alpha_{0,n} r) r dr = \frac{1}{\alpha_{0,n}^2} \int_0^{\alpha_{0,n} R} J_0(x) x dx = \frac{R}{\alpha_{0,n}} J_1(\alpha_{0,n} R) = \frac{R^2}{x_{0,n}} J_1(x_{0,n}).$$

Further, the integral $\int\limits_0^L \sin \frac{p\pi z}{L} dz = \frac{L}{p\pi}[1 - (-1)^p]$ and so we obtain

$$a_{0,n,p} = \frac{4(T_0 - T_1)}{p\pi x_{0,n} J_1(x_{0,n})}[1 - (-1)^p].$$

Our final solution for the temperature is thus

$$\psi(r, \theta, z) = T_1 + \frac{8(T_0 - T_1)}{\pi} \sum_{p=1,3,5,\dots}^{\infty} \sum_{n=1}^{\infty} \frac{1}{p\,x_{0,n} J_1(x_{0,n})} J_0(\alpha_{0,n}\,r) \sin \frac{p\pi z}{L} e^{-D\,k_{0,n,p}^2\,t}$$

where $k_{0,n}^2 = \alpha_{0,n}^2 + \frac{p^2\pi^2}{L^2} = \frac{x_{0,n}^2}{R^2} + \frac{p^2\pi^2}{L^2}$.

At the centre of the cylinder, $r = 0$ and $z = \frac{L}{2}$. Therefore, since $J_0(0) = 1$ and $\sin \frac{p\pi}{2} = (-1)^{\frac{p+1}{2}}$ for p = an odd integer, we find a temperature

$$T_c \equiv \psi\left(0, \theta, \frac{L}{2}\right) = T_1 + \frac{8(T_0 - T_1)}{\pi} \sum_{p=\text{odd}}^{\infty} \sum_{n=1}^{\infty} \frac{1}{p\,x_{0,n} J_1(x_{0,n})}(-1)^{\frac{p-1}{2}} e^{-D\,k_{0,n,p}^2\,t}.$$

For most metals, this series converges rapidly. For example, let us consider a steel cylinder of radius $R = 0.1$ m, length $L = 1$ m and $D = 0.126 \times 10^{-4} \frac{m^2}{\sec}$. The tables of Bessel functions provide

$$x_{0,1} \simeq 2.40, \quad x_{0,2} \simeq 5.52, \quad J_1(2.40) \simeq 0.52 \quad \text{and} \quad J_1(5.52) \simeq -0.34.$$

This means that successive values of $k_{0,n,p}^2$ are

$$k_{0,1,1}^2 = 576, k_{0,1,3}^2 = 665., \quad k_{0,1,5}^2 = 823., \quad k_{0,2,1}^2 = 3.06 \times 10^3, \quad k_{0,1,3}^2 = 3.14 \times 10^3$$

and, at time $t = 3$ mins, the corresponding exponents in successive terms of the series are

$$-1.31, \quad -1.51, \quad -1.87, \quad -6.93, \quad -7.11.$$

This means in turn that $\exp\left[-(k_{0,2,1}^2 - k_{0,1,1}^2)Dt\right] = e^{-5.61} \times 0.004$ and so the $n = 2$ terms can be ignored. Therefore, after 3 minutes, the central temperature will have decreased to

$$T_c = T_1 + \frac{8(T_0 - T_1)}{\pi} \frac{1}{(2.40)(0.52)}\left[e^{-1.30} - \frac{1}{3}e^{-1.51} + \frac{1}{5}e^{-1.87} - \frac{1}{7}e^{-2.40} + - \dots\right]$$

$$\simeq T_1 + \frac{8(T_0 - T_1)}{\pi} \frac{(0.223)}{(2.40)(0.52)}$$

$$\simeq T_1 + (0.455)(T_0 - T_1).$$

So, if the initial temperature is 500°C and the surface temperature is 20°C, the centre will have cooled to $T_c = 238$°C in just 3 minutes.

Particle in a Cylindrical Box:

The Schrödinger equation for a particle of mass m confined to a box but otherwise not interacting with a field or with other particles is

$$-\frac{\hbar^2}{2m}\nabla^2\psi = E\psi$$

with $\psi = 0$ at the walls of the box. Setting $E = \frac{\hbar^2 k^2}{2m}$ and cancelling out common factors, we convert the PDE to Helmholtz' equation. Therefore, if the box is cylindrical, with radius R and length L, we have precisely the same eigenvalue problem that we solved in the heat conduction problem. So, without further effort, we can assert that the energies available to the particle are $E_{m,n,p} = \frac{\hbar^2}{2m}\left[\frac{x_{m,n}^2}{R^2} + \frac{p^2\pi^2}{L^2}\right]$ corresponding to the (unnormalized) wave functions

$$\psi_{m,n,p}(r,\theta,z) = J_m\left(x_{m,n}\frac{r}{R}\right)\sin\frac{p\pi z}{L}\left\{\begin{array}{c}\cos m\theta \\ \sin m\theta\end{array}\right\}$$

Acoustic Radiation:

As a final application, we shall consider sound waves in a gas contained within a cylindrical box or wave guide. One way of describing them is in terms of condensations and rarefactions or density fluctuations of the gas relative to a uniform background. The fluctuations, $\psi(r,t) \equiv \frac{\rho(r,t)-\rho_0}{\rho_0}$, where ρ_0 is the uniform background density, can be shown to satisfy the three-dimensional wave equation. Moreover, since there can be no motion normal to the (rigid) walls of the container, we know that the component of the gradient of the density that is normal to each wall must vanish at that wall. Thus, our problem is to solve

$$\nabla^2\psi = \frac{1}{c^2}\frac{\partial^2\psi}{\partial t^2} \text{ subject to } \left.\boldsymbol{n}\cdot\nabla\psi(r,t)\right|_{\text{at the walls}} = 0$$

where c is the speed of sound in the gas and is determined by $c^2 = \frac{P_0\gamma}{\rho_0}$, P_0 is the background pressure and γ is the ratio of heat capacities $\frac{C_p}{C_v}$.

Separating the time dependence, we again have

$$\psi(r,t) = \sum_\gamma u_\gamma(r)[a_\gamma\cos k_\gamma ct + b_\gamma\sin k_\gamma ct]$$

where the normal modes $u_\gamma(r)$ are determined by the eigenvalue problem

$$\nabla^2 u_\gamma + k_\gamma^2 u_\gamma = 0 \text{ with } \left.\boldsymbol{n}\cdot\nabla u_\gamma(r)\right|_{\text{at the walls}} = 0.$$

Assuming a cylindrical container of radius R and finite length L and proceeding as in the preceding applications, our new boundary condition translates into the requirements

$$\left.\frac{d}{dr}J_m(ar)\right|_{r=R} = 0 \text{ and } \left.\frac{d}{dz}\left[c_\gamma e^{\sqrt{a^2-k_\gamma^2}} + d_\gamma e^{\sqrt{a^2-k_\gamma^2}}\right]\right|_{z=0,L} = 0$$

rather than those requiring the undifferentiated functions to vanish. Therefore, in this case the eigenfunctions (normal modes) are

$$u_{mnp}(r, \theta, z) = J_m\left(y_{mn}\frac{r}{R}\right)\cos\frac{p\pi z}{L}\begin{Bmatrix}\cos m\theta \\ \sin m\theta\end{Bmatrix}$$

with eigenvalues $k^2_{mnp} = \frac{y^2_{mn}}{R^2} + \frac{p^2\pi^2}{L^2}$ where y_{mn} is the nth zero of $J'_m(x)$. Initial values $\psi(r, 0)$ and $\left.\frac{\partial\psi}{\partial t}\right|_{t=0}$ can now be fit to superpositions of these modes.

Of more interest is the propagation of acoustic waves along a very long cylindrical wave guide of radius R. Suppose that we generate the waves with a harmonic time dependence and frequency ω : $\psi(r, \theta, z, t) = u(r, \theta, z)e^{-i\omega t}$. Substituting this into the wave equation, we obtain the Helmholtz equation $\nabla^2 u + k^2 u = 0$ again **but** with k^2 already determined via $k^2 = \frac{\omega^2}{c^2}$. Therefore, the solutions are

$$u_{mn}(r, \theta, z) = J_m(\alpha_{mn}r)[a_{mn}\cos m\theta + b_{mn}\sin m\theta][c_{mn}e^{\sqrt{\alpha^2_{mn} - k^2}z} + d_{mn}e^{-\sqrt{\alpha^2_{mn} - k^2}z}].$$

In order that these represent waves propagating down the wave guide in the positive z-direction, we require $d_{mn} = 0$ and $\sqrt{\alpha^2_{mn} - \frac{\omega^2}{c^2}} = i\kappa_{mn}$, κ_{mn} real, so that

$$\psi_{mn}(r, \theta, z, t) = v_{mn}(r, \theta)e^{i(\kappa_{mn}z - \omega t)} \quad \text{where} \quad v_{mn}(r, \theta) = J_m\left(\frac{y_{mn}}{R}r\right)\begin{Bmatrix}\cos m\theta \\ \sin m\theta\end{Bmatrix}.$$

The modes that are **allowed** to propagate down the guide are those for which κ_{mn} is indeed real. Since $\kappa^2_{mn} = \frac{\omega^2}{c^2} - \frac{y^2_{mn}}{R^2}$, we see that κ^2_{mn} becomes negative for frequencies below the **cut-off frequency** $\omega_{mn}(\text{min}) = \frac{c}{R}y_{mn}$ and the (m, n) mode is not propagated. In fact, if one attempts to propagate the (m, n) mode at a frequency $\omega < \omega_{mn}(\text{min})$, one will have $\kappa^2_{mn} = \frac{\omega^2 - \omega^2_{mn}(\text{min})}{c^2} = -\beta^2_{mn} < 0$ resulting in a wave number $\kappa_{mn} = \pm i\beta_{mn}$ that is pure imaginary and a wave

$$\psi_{mn}(r, \theta, z, t) = v_{mn}(r, \theta)e^{-\beta_{mn}z - i\omega t}$$

that is exponentially damped or **attenuated**. Notice that the $m = 0, n = 1$ mode is **always** propagated. This is because $y_{01} = 0$ $(J'_0(x) = -J_1(x))$ and since $J_0(0) = 1$,

$$\psi_{01}(r, \theta, z, t) \propto e^{i(kz - \omega t)}$$

which is a **plane wave** propagating in the $+$ z-direction with wave number $\kappa_{01} = k = \frac{\omega}{c}$.

11.3.3 Modified Bessel Functions

The differential equation

$$\frac{d^2 y}{d x^2} + \frac{1}{x}\frac{dy}{dx} - \left(1 + \frac{m^2}{x^2}\right)y = 0 \tag{11.3.31}$$

is the same as Bessel's DE but with x replaced by ix. Thus, its solutions are

$$y(x) = \left\{ \begin{array}{c} J_m(ix) \\ N_m(ix) \end{array} \right\}.$$

However, in physical applications it is convenient to have the solutions expressed in a form that is explicitly **real** for **real** values of x. Therefore, we define the **modified Bessel functions**

$$I_m(x) = e^{-im\frac{\pi}{2}} J_m(ix) = (-i)^m J_m(ix) = \sum_{k=0}^{\infty} \frac{1}{k!(k+m)!} \left(\frac{x}{2}\right)^{2k}, \tag{11.3.32}$$

and

$$K_m(x) = \frac{\pi}{2} i^{m+1} H_m^{(1)}(ix). \tag{11.3.33}$$

The choice of $K_m(x)$ as the second linearly independent solution is made to ensure that the two functions exhibit complementary asymptotic behaviour. Specifically, for $x \gg 1$,

$$I_m(x) \sim \sqrt{\frac{1}{2\pi x}} e^x \quad \text{and} \quad K_m(x) \sim \sqrt{\frac{\pi}{2x}} e^{-x}. \tag{11.3.34}$$

For small values of x, $x \ll 1$, the modified Bessel functions have the limiting forms

$$I_m(x) \simeq \left\{ \begin{array}{ll} \frac{1}{m!} \left(\frac{x}{2}\right)^m & \text{for } m > 0 \\ 1 & \text{for } m = 0 \end{array} \right., \quad \text{and} \tag{11.3.35}$$

$$K_m(x) \simeq \left\{ \begin{array}{ll} \frac{(m-1)!}{2} \left(\frac{x}{2}\right)^{-m} & \text{for } m > 0 \\ -(\ln \frac{x}{2} + \gamma) & \text{for } m = 0 \end{array} \right.. \tag{11.3.36}$$

Note that the $I_m(x)$ is well behaved at the origin but diverges at infinity while the reverse is true for $K_m(x)$. Like the hyperbolic functions, neither $I_m(x)$ nor $K_m(x)$ has multiple zeros.

Not surprisingly, the recurrence relations satisfied by the modified Bessel functions are similar to those satisfied by $J_m(x)$. The most important ones are

$$I_{m+1}(x) + I_{m-1}(x) = 2 I_m'(x), \tag{11.3.37}$$

$$\frac{d}{dx}[x^m I_m(x)] = x^m I_{m-1}(x) \quad \text{and} \quad \frac{d}{dx}\left[\frac{I_m(x)}{x^m}\right] = \frac{I_{m+1}(x)}{x^m} \tag{11.3.38}$$

$$I_{m-1}(x) - I_{m+1}(x) = \frac{2m}{x} I_m(x), \tag{11.3.39}$$

for $I_m(x)$ and

$$K_{m+1}(x) + K_{m-1}(x) = -2\,K_m'(x), \tag{11.3.40}$$

$$\frac{d}{dx}[x^m\,K_m(x)] = -x^m\,K_{m-1}(x) \ \text{ and } \ \frac{d}{dx}\left[\frac{K_m(x)}{x^m}\right] = -\frac{K_{m+1}(x)}{x^m} \tag{11.3.41}$$

$$K_{m-1}(x) - K_{m+1}(x) = -\frac{2m}{x}\,K_m(x), \tag{11.3.42}$$

for $K_m(x)$.

Because the modified Bessel functions do not have multiple zeros, they cannot satisfy homogeneous boundary conditions of the type found in Sturm-Liouville problems and so do not comprise complete orthogonal sets. Thus, when they figure in the solution of a boundary value problem, they are always coupled with functions that do form complete orthogonal sets. As we shall see in the next sub-section, potential problems with cylindrical symmetry provide a graphic illustration of this point.

11.3.4 Electrostatic Potential in and around Cylinders

Consider a cylinder of radius R and height L. One surface, either the top or the curved lateral wall of the cylinder, is maintained at a non-zero and perhaps variable electrostatic potential V. We want to find the potential $\psi(r, \theta, z)$ at any point inside the cylinder. Since there are no charges present, the potential must satisfy Laplace's equation, $\nabla^2\psi = 0$. Using cylindrical coordinates and then separating variables in Section 8.4, we found that the solutions of this equation could be expressed as one of two possible superpositions: either

$$\psi(r, \theta, z) = \sum_{\alpha, m} \left\{ \begin{array}{c} J_m(\alpha r) \\ N_m(\alpha r) \end{array} \right\} \left\{ \begin{array}{c} \cosh\alpha z \\ \sinh\alpha z \end{array} \right\} \left\{ \begin{array}{c} \cos m\theta \\ \sin m\theta \end{array} \right\}, \alpha^2 > 0$$

or

$$\psi(r, \theta, z) = \sum_{\alpha, m} \left\{ \begin{array}{c} I_m(|\alpha|r) \\ K_m(|\alpha|r) \end{array} \right\} \left\{ \begin{array}{c} \cos|\alpha|z \\ \sin|\alpha|z \end{array} \right\} \left\{ \begin{array}{c} \cos m\theta \\ \sin m\theta \end{array} \right\}, \alpha^2 < 0$$

where, as usual, each set of braces is understood to be a linear combination of the functions they contain. As we noted then, both superpositions have the r and z dependence coupled in such a way that one or the other but not both is oscillatory in behaviour. Therefore, the choice between the two options is made for us by the boundary conditions in the problem: to satisfy homogeneous conditions at both boundaries associated with a particular variable, we require a function with multiple zeros and thus an oscillatory dependence on that variable.

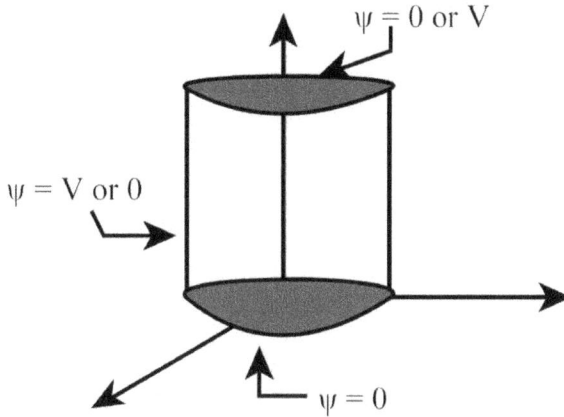

So, let us start with the boundary conditions

$$\psi(r, \theta, 0) = \psi(r, \theta, L) = 0 \quad \text{and} \quad \psi(R, \theta, z) = V(\theta, z).$$

To satisfy the homogeneous boundary conditions at the z boundaries, we must choose the **second** superposition, discard the $\cos |\alpha| z$ possibility and set $\alpha^2 = -\frac{n^2 \pi^2}{L^2}$, $n = 1, 2, \ldots$. Next, the (implied) boundary condition $|\psi(0, \theta, z)| < \infty$ requires that we discard the $K_m(|\alpha|r)$ possibility and so we arrive rather quickly at the following expression for the potential inside the "can":

$$\psi(r, \theta, z) = \sum_{m=0}^{\infty} \sum_{n=1}^{\infty} I_m\left(\frac{n\pi r}{L}\right) \sin\frac{n\pi z}{L}[a_{mn} \cos m\theta + b_{mn} \sin m\theta].$$

This double Fourier series can now be made to fit the remaining boundary condition $\psi(R, \theta, z) = V(\theta, z)$ and thus determine the coefficients a_{mn}, b_{mn}:

$$a_{mn} = \frac{2}{\pi L \, I_m\left(\frac{n\pi R}{L}\right)} \left\{ \begin{array}{c} 1 \\ \frac{1}{2} \end{array} \right\} \int_0^L \int_0^{2\pi} V(\theta, z) \sin\frac{n\pi z}{L} \cos m\theta \, d\theta \, dz \quad \text{for} \quad \left\{ \begin{array}{c} m \neq 0 \\ m = 0 \end{array} \right\},$$

$$b_{mn} = \frac{2}{\pi L \, I_m\left(\frac{n\pi R}{L}\right)} \int_0^L \int_0^{2\pi} V(\theta, z) \sin\frac{n\pi z}{L} \sin m\theta \, d\theta \, dz.$$

Suppose, for example, that we halve the cylinder vertically and insert the usual very thin strip of perfect insulator to permit a potential $V(\theta, z) = \left\{ \begin{array}{ll} V_0 & \text{for } 0 < \theta < \pi \\ -V_0 & \text{for } \pi < \theta < 2\pi \end{array} \right.$ where V_0 is a constant. Since it is an odd function of θ, $a_{mn} = 0$ for all m and

$$b_{mn} = \frac{4 V_0}{\pi L \, I_m\left(\frac{n\pi R}{L}\right)} \int_0^L \int_0^{\pi} \sin\frac{n\pi z}{L} \sin m\theta \, d\theta \, dz.$$

Since $\int\limits_0^L \sin\frac{n\pi z}{L}\,dz = \frac{L}{n\pi}[1-(-1)^n]$ and $\int\limits_0^\pi \sin m\theta\,d\theta = \frac{1}{m}[1-(-1)^m]$, we have

$$b_{m,n} = \begin{cases} \dfrac{1}{mn}\dfrac{16V_0}{\pi^2 I_m\left(\frac{n\pi R}{L}\right)} & \text{for } m \text{ and } n \text{ odd} \\ 0 & \text{for } m \text{ and } n \text{ even} \end{cases}.$$

Thus,

$$\psi(r,\theta,z) = \frac{16\,V_0}{\pi^2}\sum_{m=1,3,\ldots}^\infty \sum_{n=1,3,\ldots}^\infty \frac{1}{mn}\frac{I_m\left(\frac{n\pi r}{L}\right)}{I_m\left(\frac{n\pi R}{L}\right)}\sin m\theta \sin\frac{n\pi z}{L}.$$

If, on the other hand, we had started with the boundary conditions $\psi(R,\theta,z) = 0$, $\psi(r,\theta,0) = 0$ and $\psi(r,\theta,L) = V(r,\theta)$, we would have been obliged to go with the first superposition, discarded the $N_m(\alpha r)$ possibility (to ensure boundedness at the origin) and the $\cosh \alpha z$ possibility (to meet the condition at $z = 0$), and set $\alpha = \alpha_{mn} = \frac{x_{mn}}{R}$ where x_{mn} is the nth zero of $J_m(x)$. Thus, our expression for the potential inside the "can" would become

$$\psi(r,\theta,z) = \sum_{m=0}^\infty \sum_{n=1}^\infty J_m(\alpha_{mn}\,r)\sinh\alpha_{mn}\,z[a_{mn}\cos m\theta + b_{mn}\sin m\theta]$$

where

$$a_{mn} = \frac{2\operatorname{cosech}\alpha_{mn}L}{\pi R^2[J_{m+1}(\alpha_{mn}R)]^2}\begin{Bmatrix} 1 \\ \frac{1}{2} \end{Bmatrix}\int\limits_0^{2\pi}\int\limits_0^R V(r,\theta)J_m(\alpha_{mn}\,r)\cos m\theta\,r\,dr\,d\theta \ \text{ for }\ \begin{Bmatrix} m \neq 0 \\ m = 0 \end{Bmatrix}$$

and

$$b_{mn} = \frac{2\operatorname{cosech}\alpha_{mn}L}{\pi R^2[J_{m+1}(\alpha_{mn}R)]^2}\int\limits_0^{2\pi}\int\limits_0^R V(r,\theta)J_m(\alpha_{mn}\,r)\sin m\theta\,r\,dr\,d\theta.$$

As an example, let us take $V(r,\theta) = V_0 =$ a constant. We would then have $b_{mn} = 0$ for all m and $a_{mn} = 0$ for all $m \neq 0$. Thus,

$$\psi(r,\theta,z) = \sum_{n=1}^\infty a_{0,n} J_0(\alpha_{0,n}\,r)\sinh\alpha_{0,n}\,z$$

with

$$a_{0n} = \frac{2\,V_0\operatorname{cosech}\alpha_{0n}L}{R^2[J_1(\alpha_{0n}R)]^2}\int\limits_0^R J_0(\alpha_{0n}\,r)r\,dr.$$

As we learned in sub-section 11.3.2, $\int\limits_0^R J_0(\alpha_{0n}\,r)r\,dr = \frac{R}{\alpha_{0n}}J_1(\alpha_{0n}R)$. Therefore, our final expression for the potential under these boundary conditions is

$$\psi(r,\theta,z) = 2\,V_0\sum_{n=1}^\infty \frac{J_0\left(x_{0n}\frac{r}{R}\right)}{x_{0n}J_1(x_{0n})}\frac{\sinh\left(x_{0n}\frac{z}{R}\right)}{\sinh\left(x_{0n}\frac{L}{R}\right)}.$$

As a final example, we shall allow our cylinder to extend infinitely far in both the positive and negative z directions. The boundary conditions at the z boundaries must become $|\psi(r, \theta, z)| < \infty$ which immediately implies that $\alpha = ik, -\infty < k < \infty$. The appropriate superposition to represent ψ must then be

$$\psi(r, \theta, z) = \sum_{m=0}^{\infty} \left\{ \int_{-\infty}^{\infty} [A_m(k) I_m(|k|r) + B_m(k) K_m(|k|r)] e^{ikz} dk \right\} \left\{ \begin{array}{c} \cos m\theta \\ \sin m\theta \end{array} \right\}$$

where the summation over α has become a continuous sum or **integral** over k.

Suppose that we are interested in finding the potential **outside** the cylinder, given that it is maintained at a value $V(\theta, z)$ on the curved surface. The boundary conditions in r must then be $\psi(R, \theta, z) = V(\theta, z)$ plus the requirement that the potential be bounded as $r \to \infty$, $|\psi(\infty, \theta, z)| < \infty$. But we know from the asymptotic behaviour of the modified Bessel functions given in (11.3.34) that this latter requirement eliminates the I_m' s from consideration. Thus,

$$\psi(r, \theta, z) = \sum_{m=0}^{\infty} \int_{-\infty}^{\infty} K_m(|k|r) e^{ikz} [a_m(k) \cos m\theta + b_m(k) \sin m\theta] dk$$

where

$$a_m(k) = \frac{1}{2\pi^2 K_m(|k|R)} \left\{ \begin{array}{c} 1 \\ \frac{1}{2} \end{array} \right\} \int_0^{2\pi} \int_{-\infty}^{\infty} V(\theta, z) e^{-ikz} \cos m\theta d\theta dz \quad \text{for} \quad \left\{ \begin{array}{c} m \neq 0 \\ m = 0 \end{array} \right\}$$

and,

$$b_m(k) = \frac{1}{2\pi^2 K_m(|k|R)} \int_0^{2\pi} \int_{-\infty}^{\infty} V(\theta, z) e^{-ikz} \sin m\theta d\theta dz.$$

Taking the relatively simple functional form $V(\theta, z) = V_0 e^{-c|z|}$, V_0 and c constants, results in a restriction to $m = 0$ and yields

$$a_0(k) = \frac{V_0}{2\pi K_m(|k|R)} \int_{-\infty}^{\infty} e^{-c|z|} e^{-ikz} dz = \frac{V_0}{\pi K_m(|k|R)} \frac{c}{c^2 + k^2}.$$

Thus, in this case, our solution for the potential outside the cylinder becomes

$$\psi(r, \theta, z) = \frac{V_0}{\pi} \int_{-\infty}^{\infty} \frac{K_m(|k|r)}{K_m(|k|R)} \frac{c}{c^2 + k^2} e^{ikz} dk.$$

11.3.5 Fourier-Bessel Transforms

As we have seen repeatedly, including the preceding example, when homogeneous boundary conditions are imposed at the limits of an infinite or semi-infinite range the eigenvalue spectrum that results is continuous rather than discrete and the superposition of eigensolutions requires an integration or, more precisely, an integral transformation. A further example is provided by Laplace's equation in cylindrical coordinates with the homogeneous boundary conditions that ψ be bounded both at $r = 0$ and as $r \to \infty$. Bessel's DE becomes a Sturm-Liouville eigenvalue equation again but this time the eigenfunction solutions are the set $J_m(\alpha r)$, $0 < \alpha < \infty$. To arrive at this conclusion, we simply have to note that neither of the modified Bessel functions nor the Neumann function can be bounded at both limits while the Bessel function of the first kind obviously can.

To conform with more conventional notation we shall set $\alpha = k$, $0 < k < \infty$. We then have, from the properties of Sturm-Liouville eigenfunctions, orthogonality and completeness with respect to the weight function $\rho(r) = r$. As it happens, the Bessel functions are appropriately normalized already for this range of r and so the statement of orthogonality is

$$\int_0^\infty J_m(kr) J_m(k'r) r\, dr = \frac{1}{k}\delta(k - k') \tag{11.3.43}$$

while the closure relation is

$$\int_0^\infty J_m(kr) J_m(kr') k\, dk = \frac{1}{r}\delta(r - r'). \tag{11.3.44}$$

The proof of the orthogonality statement follows from an application of equation (10.5.25) and the generalized Green's identity (10.5.18) to these eigenfunctions of the Bessel differential operator. The result is

$$r\left[J_m(kr)\frac{\partial J_m(k'r)}{\partial r} - J_m(k'r)\frac{\partial J_m(kr)}{\partial r} \right]\Bigg|_0^\infty = [k^2 - (k')^2]\int_0^\infty J_m(kr) J_m(k'r) r\, dr.$$

If we now use the recurrence relation (11.3.16) and the asymptotic form (11.3.6), we can express the left hand side of this equation as

$$\lim_{r\to\infty} \frac{1}{\pi}\left[\frac{k+k'}{\sqrt{kk'}}\sin(k-k')r - \frac{k-k'}{\sqrt{kk'}}(-1)^m\cos(k+k')r \right].$$

Thus, dividing through by $[k^2 - (k')^2]$, we have

$$\int_0^\infty J_m(kr) J_m(k'r) r\, dr = \lim_{r\to\infty} \frac{1}{\pi}\left[\frac{1}{\sqrt{kk'}}\frac{\sin(k-k')r}{k-k'} + \frac{(-1)^m}{\sqrt{kk'}}\frac{\cos(k+k')r}{k+k'} \right].$$

The first term on the right hand side, in that limit, is a representation of $\frac{1}{\sqrt{kk'}}\delta(k-k') = \frac{1}{k}\delta(k - k')$ while the second has a limit of zero. Thus, we get the result in equation (11.3.43).

The statement of completeness is that any function $f(r)$ that is square-integrable with respect to the weight function $\rho(r) = r$ can be represented by the (mean) convergent integral

$$f(r) = \int_0^\infty F(k) J_m(kr) k\, dk \qquad (11.3.45)$$

where

$$F(k) = \int_0^\infty f(r) J_m(kr) r\, dr. \qquad (11.3.46)$$

The functions $f(r)$ and $F(k)$ are **Fourier-Bessel Transforms** of each other.

An application for these transforms is the solution of Laplace's equation in the space between two infinite planes located at $z = 0$ and $z = L$ with ψ specified on the planes themselves and with the assumption that $|\psi| \rightarrow 0$ as x and y or $r \rightarrow \infty$. In a problem like this we have the option of using either Cartesian or cylindrical coordinates. If we choose the former, ψ will be represented by a double Fourier integral transform involving the x and y variables. If the latter, we will have a Fourier-Bessel transform for the r-dependence and a Fourier series for the θ-dependence. We shall illustrate with a specific example.

The electrostatic potential on a plane at $z = 0$ is given by the function

$$V(r, \theta) = \begin{cases} 0 & \text{for } r < 1 \\ \dfrac{V_0}{\sqrt{r^2 - 1}} & \text{for } r > 1 \end{cases}.$$

We seek the potential everywhere above the plane given that it goes to zero uniformly as $z \rightarrow \infty$ and as $r \rightarrow \infty$. Evidently, cylindrical coordinates are appropriate and so we represent the solution with the superposition

$$\psi(r, \theta, z) = \sum_{m=0}^\infty \left\{ \begin{array}{c} \cos m\theta \\ \sin m\theta \end{array} \right\} \int_0^\infty J_m(kr) \left\{ \begin{array}{c} e^{kz} \\ e^{-kz} \end{array} \right\} k\, dk. \qquad (11.3.47)$$

The boundary conditions in this problem provide considerable simplification: because $|\psi| \rightarrow 0$ as $z \rightarrow \infty$ we discard the e^{kz} possibility and because ψ is independent of θ when $z = 0$ we restrict m to be zero. Thus,

$$\psi(r, \theta, z) = \int_0^\infty A_0(k) e^{-kz} J_0(kr) k\, dk. \qquad (11.3.48)$$

Therefore, since $\psi(r, \theta, 0) = V(r, \theta) = \int_0^\infty A_0(k) J_0(kr) k\, dk$, we have

$$A_0(k) = \int_1^\infty \frac{V_0}{\sqrt{r^2 - 1}} J_0(kr) r\, dr = \frac{V_0}{k} \cos k.$$

where we have used integral 6.554#3 from Gradshteyn and Ryzhik. Substituting this back into (11.3.48), we obtain the solution

$$\psi(r, \theta, z) = V_0 \int_0^\infty \cos k \, e^{-kz} J_0(kr) dk.$$

Using Gradshteyn and Ryzhik one more time, formula 6.611#1 in this case, we can evaluate this explicitly and find

$$\psi(r, \theta, z) = V_0 \, \text{Re} \left[\frac{1}{\sqrt{(z^2 + r^2 - 1) + 2iz}} \right], \quad z \geq 0, \quad r > 1.$$

11.4 Spherical Bessel Functions: Spherical Waves

11.4.1 Properties of Spherical Bessel Functions

In Section 10.4 we discovered that separation of variables applied to the Helmholtz equation $\nabla^2 u(r) + k^2 u(r) = 0$ when spherical coordinates are used results in a radial equation

$$\frac{d^2 R}{dr^2} + \frac{2}{r} \frac{dR}{dr} + \left[k^2 - \frac{l(l+1)}{r^2} \right] R = 0 \tag{11.4.1}$$

with general solution

$$R(r) = \left\{ \begin{array}{c} j_l(kr) \\ n_l(kr) \end{array} \right\}. \tag{11.4.2}$$

The functions $j_l(kr)$ and $n_l(kr)$ are called **spherical** Bessel and Neumann functions and are defined by

$$j_l(x) = \sqrt{\frac{\pi}{2x}} J_{l+\frac{1}{2}}(x) \quad \text{and} \quad n_l(x) = \sqrt{\frac{\pi}{2x}} N_{l+\frac{1}{2}}(x) = (-1)^{l+1} \sqrt{\frac{\pi}{2x}} J_{-l-\frac{1}{2}}(x). \tag{11.4.3}$$

From the power series representation of $J_m(x)$ we have

$$J_{l+\frac{1}{2}}(x) = \sum_{k=0}^\infty \frac{(-1)^k}{k! \Gamma(k+l+\frac{3}{2})} \left(\frac{x}{2} \right)^{2k+l+\frac{1}{2}}.$$

Thus, using

$$\Gamma(\frac{1}{2} + n + 1) = \frac{(2n+1)(2n-1)\ldots 1}{2^{n+1}} \Gamma(\frac{1}{2}) = \frac{(2n+1)!}{2^{2n+1} n!} \sqrt{\pi},$$

we find

$$j_l(x) = 2^l \sum_{k=0}^\infty \frac{(-1)^k (k+l)!}{k!(2k+2l+1)!} x^{2k+l}. \tag{11.4.4}$$

Similarly, starting from the power series representation of $J_{-l-\frac{1}{2}}(x)$, one finds

$$n_l(x) = \frac{(-1)^l}{2^l} \sum_{k=0}^{\infty} \frac{(-1)^k \Gamma(k - l + 1)}{k! \Gamma(2k - 2l + 1)} x^{2k-l-1} . \tag{11.4.5}$$

These two power series are instantly recognizable in the special case of $l = 0$:

$$j_0(x) = \sum_{k=0}^{\infty} \frac{(-1)^k}{(2k + 1)!} x^{2k} = \frac{\sin x}{x}, \tag{11.4.6}$$

and

$$n_0(x) = (-1) \sum_{k=0}^{\infty} \frac{(-1)^k}{(2k)!} x^{2k-1} = -\frac{\cos x}{x} \tag{11.4.7}$$

where we have used $\Gamma(2n + 1) = (2n)!$ and $\Gamma(n + 1) = n!$. These two identities offer a valuable aid to one's intuitive appreciation of $j_0(x)$ and $n_0(x)$.

From the recurrence relations satisfied by $J_{l+\frac{1}{2}}(x)$ and $J_{-l-\frac{1}{2}}(x)$, one can derive corresponding relations for $j_l(x)$ and $n_l(x)$. One finds that they **both** satisfy

$$j_{l-1}(x) + j_{l+1}(x) = \frac{2l + 1}{x} j_l(x), \tag{11.4.8}$$

and

$$l j_{l-1}(x) - (l + 1) j_{l+1}(x) = (2l + 1) \frac{d j_l(x)}{dx} . \tag{11.4.9}$$

Multiplying the first of these by l and then subtracting the second from it, we find

$$j_{l+1}(x) = -x^l \frac{d}{dx} \left(\frac{j_l(x)}{x^l} \right) . \tag{11.4.10}$$

We can generate successive $j_l's$ and $n_l's$ by applying this relation repeatedly to $j_0(x)$ and $n_0(x)$, respectively. Thus, we write formally

$$j_l(x) = x^l \left(-\frac{1}{x} \frac{d}{dx} \right)^l j_0(x) = (-1)^l x^l \left(\frac{1}{x} \frac{d}{dx} \right)^l \left(\frac{\sin x}{x} \right), \quad l = 1, 2, \ldots \tag{11.4.11}$$

and

$$n_l(x) = x^l \left(-\frac{1}{x} \frac{d}{dx} \right)^l n_0(x) = (-1)^{l+1} x^l \left(\frac{1}{x} \frac{d}{dx} \right)^l \left(\frac{\cos x}{x} \right), l = 1, 2, \ldots . \tag{11.4.12}$$

Evidently all spherical Bessel and Neumann functions can be expressed in terms of sines and cosines. For example,

$$j_1(x) = \frac{\sin x}{x^2} - \frac{\cos x}{x}, \quad n_1(x) = -\frac{\cos x}{x^2} - \frac{\sin x}{x}$$

$$j_2(x) = \left(\frac{3}{x^3} - \frac{1}{x}\right) \sin x - \frac{3}{x^2} \cos x, \quad n_2(x) = -\left(\frac{3}{x^3} - \frac{1}{x}\right) \cos x - \frac{3}{x^2} \sin x.$$

The spherical Hankel functions are defined by analogy with ordinary Hankel functions:

$$h_l^{(1)}(x) = j_l(x) + i\, n_l(x) = (-x)^l \left(\frac{1}{x}\frac{d}{dx}\right)^l \left(\frac{e^{ix}}{ix}\right), \tag{11.4.13}$$

$$h_l^{(2)}(x) = j_l(x) - i\, n_l(x) = (-x)^l \left(\frac{1}{x}\frac{d}{dx}\right)^l \left(\frac{e^{-ix}}{-ix}\right). \tag{11.4.14}$$

The small x behaviour of these functions is easily obtained from their power series representations:

$$j_l(x) \simeq \frac{2^l\, l!}{(2l+1)!}\, x^l = \frac{x^l}{(2l+1)!!} = \frac{x^l}{(2l+1)\ldots 5 \cdot 3 \cdot 1}, \quad \text{for } x \ll 1, \tag{11.4.15}$$

$$n_l(x) \simeq -\frac{(2l)!}{2^l\, l!}\frac{1}{x^{l+1}}, \quad \text{for } x \ll 1. \tag{11.4.16}$$

To obtain the asymptotic behaviour, we could start from the behaviour reported for the Bessel and Neumann functions in Section 11.3.1. A more direct approach however is to use equations (11.4.13) and (11.4.14) for the Hankel functions. For very large x, $x \gg 1$ or l, the largest contributions in these formulas results from applying all derivatives to e^{ix} rather than the inverse powers of x. Thus, we find

$$h_l^{(1)}(x) \sim (-1)^l (i)^l \frac{e^{ix}}{ix} = \frac{1}{x}\, e^{i[x-(l+1)\frac{\pi}{2}]} \quad \text{for } x \gg 1, l, \tag{11.4.17}$$

$$h_l^{(2)}(x) \sim \frac{1}{x}\, e^{-i[x-(l+1)\frac{\pi}{2}]} \quad \text{for } x \gg 1, l. \tag{11.4.18}$$

Combining these to construct $j_l(x)$ and $n_l(x)$ we obtain

$$j_l(x) \sim \frac{1}{x}\cos\left(x-(l+1)\frac{\pi}{2}\right) = \frac{1}{x}\sin\left(x-l\frac{\pi}{2}\right) \quad \text{for } x \gg 1, l, \tag{11.4.19}$$

$$n_l(x) \sim \frac{1}{x}\sin\left(x-(l+1)\frac{\pi}{2}\right) = -\frac{1}{x}\cos\left(x-l\frac{\pi}{2}\right) \quad \text{for } x \gg 1, l. \tag{11.4.20}$$

The spherical Bessel and Neumann functions $j_l(kr)$ and $n_l(kr)$ are solutions of

$$\frac{d}{dr}\left(r^2\frac{dR_l}{dr}\right) - l(l+1)\, R_l(r) = -k^2\, r^2\, R_l(r) \tag{11.4.21}$$

which is of the Sturm-Liouville form. If we impose homogeneous boundary conditions such as

$$|R_l(0)| < \infty \quad \text{and} \quad R_l(a) = 0, \tag{11.4.22}$$

it becomes an eigenvalue equation with eigenvalue $\lambda = k^2$ and weight function $\rho(r) = r^2$. The first boundary condition requires us to discard the possibility of $n_l(kr)$ figuring in the solution. The second condition requires $j_l(ka) = 0$ or

$$ka = k_{l,n}\, a = z_{l,n} \equiv \{\text{the nth zero of } j_l(x)\} = x_{l+\frac{1}{2},n} \equiv \{\text{the nth zero of } J_{l+\frac{1}{2}}(x)\}.$$

Thus, the eigenfunctions that correspond to these boundary conditions are $j_l(k_{l,n}\, r)$.

Eigenfunctions associated with different eigenvalues are orthogonal with respect to the weight function $\rho(r) = r^2$:

$$\int_0^a j_l(k_{l,n}\, r)\, j_l(k_{l,m}\, r)\, r^2\, dr = 0 \ \text{ for } \ n \neq m. \tag{11.4.23}$$

The normalization of the eigenfunctions follows from that for ordinary Bessel functions:

$$\int_0^a [j_l(k_{l,n}\, r)]^2\, r^2\, dr = \frac{\pi}{2\, k_{l,n}} \int_0^a [J_{l+\frac{1}{2}}(k_{l,n}\, r)]^2\, r\, dr = \frac{\pi}{2\, k_{l,n}} \frac{a^2}{2} [J'_{l+\frac{1}{2}}(k_{l,n}\, a)]^2$$

$$= \frac{a^3}{2} [j'_l(k_{l,n}\, a)]^2. \tag{11.4.24}$$

If we now use the recurrence relation $j'_l(x) = \frac{l}{x}\, j_l(x) - j_{l+1}(x)$, we can give this result the alternative expression

$$\int_0^a [j_l(k_{l,n}\, r)]^2\, r^2\, dr = \frac{a^3}{2} [j_{l+1}(k_{l,n}\, a)]^2. \tag{11.4.25}$$

The eigenfunctions form a complete as well as an orthogonal set. Thus, any function that is square integrable with respect to r^2 on the interval $0 \leq r \leq a$ can be represented by the (mean) convergent series

$$f(r) = \sum_{n=1}^{\infty} c_n\, j_l(k_{l,n}\, r) \tag{11.4.26}$$

where

$$c_n = \frac{2}{a^3 [j_{l+1}(k_{l,n}\, a)]^2} \int_0^a j_l(k_{l,n}\, r) f(r)\, r^2\, dr. \tag{11.4.27}$$

As with ordinary Bessel functions, when $a \to \infty$ and the homogeneous boundary conditions become $|R_l(0)| < \infty$ and $\lim_{r \to \infty} |R_l(r)| < \infty$, the eigenfunction solutions of (11.4.21) are $j_l(kr)$ where k is now a continuous variable with range $0 \leq k < \infty$. The orthogonality/normalization statement becomes

$$\int_0^{\infty} j_l(kr)\, j_l(k'r)\, r^2\, dr = \frac{\pi}{2\, k^2}\, \delta(k - k'), \tag{11.4.28}$$

the closure relation is

$$\int_0^\infty j_l(kr)\, j_l(kr')\, k^2 \, dk = \frac{\pi}{2\, r^2}\, \delta(r - r'),\tag{11.4.29}$$

and any function $f(r)$ that is square integrable with respect to r^2 on $0 \le r < \infty$ can be represented by the Fourier Bessel transform

$$f(r) = \sqrt{\frac{2}{\pi}} \int_0^\infty j_l(kr) F(k)\, k^2 \, dk\tag{11.4.30}$$

where

$$F(k) = \sqrt{\frac{2}{\pi}} \int_0^\infty j_l(kr) f(r)\, r^2 \, dr.\tag{11.4.31}$$

11.4.2 Applications: Spherical Waves

Particle in a Spherical Box

The independent particle "shell model" of the atomic nucleus postulates that each nucleon in a medium to large nucleus experiences an effective central potential due to the sum of all of the pair-wise interactions it has with the other nucleons. In the simplest version, the central potential is taken to be $V(r) = - V_0$ for $0 \le r < R$ and, to insure that the nucleons cannot escape, infinite at $r = R$ where R is the radius of the nucleus in question. Under these circumstances, the time-independent Schrödinger equation for each nucleon is just the Helmholtz equation

$$\nabla^2 \psi + k^2\, \psi = 0 \text{ where } k^2 = \frac{2m}{\hbar^2} \sqrt{E - V_0}$$

and its solutions are subject to the boundary conditions $\lim_{r \to 0} |\psi(r, \theta, \varphi)| < \infty$ and $\psi(R, \theta, \varphi) = 0$. From the foregoing analysis, we see immediately that the (unnormalized) energy eigenfunctions are $\psi_{lmn}(r, \theta, \varphi) = j_l(k_{l,n}\, r)\, Y_l^m(\theta, \varphi)$ and the corresponding energy levels are $E_{ln} = \frac{\hbar^2}{2m}\, k_{l,n}^2 - V_0$ where $k_{l,n} = \frac{z_{l,n}}{R}$. Notice that the levels are $(2l+1)$-fold degenerate. Allowing for the spin of the nucleons, they are in fact $2(2l+1)$-fold degenerate. It is this degeneracy that produces the so-called "magic numbers" of nucleons associated with increased nuclear stability and was the basis for developing and elaborating on this simple model.

Acoustic Radiation

The energy eigenfunctions of the preceding application will, with one modification, also describe the normal modes of sound waves in a gas that is contained in a spherical cavity. The one modification is due to a change in boundary condition at the wall of the container: we now have $\boldsymbol{n} \cdot \nabla \psi(r,\theta,\varphi,t)\big|_{r=R} = \frac{\partial \psi}{\partial r}\big|_{r=R} = 0$ and so $k_{ln} = \frac{w_{ln}}{R}$ where w_{ln} is the nth zero of $j_l'(x)$. Thus, the waves are described by the superposition

$$\psi(r,\theta,\varphi,t) = \sum_{n=1}^{\infty} \sum_{l=0}^{\infty} \sum_{m=-l}^{l} j_l(k_{ln}\, r)\, Y_l^m(\theta,\varphi)[a_{lmn} \cos w_{ln}\, t + b_{lmn} \sin w_{ln}\, t]$$

where $w_{ln} = c\, k_{ln}$. The coefficients are determined by the initial conditions $\psi(r,\theta,\varphi,0) = u_0(r,\theta,\varphi)$ and $\frac{\partial \psi}{\partial t}\big|_{t=0} = v_0(r,\theta,\varphi)$. Specifically,

$$a_{lmn} = \frac{2}{R^3[j_{l+1}(k_{ln}\,R)]^2} \int_0^R \int_0^\pi \int_0^{2\pi} u_0(r,\theta,\varphi)\, j_l(k_{ln}\, r)(Y_l^m(\theta,\varphi))^*\, r^2\, dr \sin\theta\, d\theta\, d\varphi,$$

$$b_{lmn} = \frac{2}{w_{ln}\, R^3[j_{l+1}(k_{ln}\,R)]^2} \int_0^R \int_0^\pi \int_0^{2\pi} v_0(r,\theta,\varphi)\, j_l(k_{ln}\, r)(Y_l^m(\theta,\varphi))^*\, r^2\, dr \sin\theta\, d\theta\, d\varphi.$$

To describe travelling rather than standing waves we use spherical Hankel functions in place of the spherical Bessel functions. Thus, an lth partial wave in a superposition will be either

$$\psi_{lm}(r,t) = h_l^{(1)}(kr)\, Y_l^m(\theta,\varphi)\, e^{-i\omega t} \quad \text{or} \quad \psi_{lm}(r,t) = h_l^{(2)}(kr)\, Y_l^m(\theta,\varphi)\, e^{-i\omega t}$$

depending on whether we want the wave to look asymptotically like an **outgoing** or an **incoming** wave. To understand this point, we need look no further than equations (11.4.17) and (11.4.18) which give the large argument behaviour of $h_l^{(1)}(x)$ and $h_l^{(2)}(x)$. With the choice of $e^{-i\omega t}$ to describe the time dependence of waves generated with frequency ω, the $h_l^{(1)}(kr)$ combination will have the limit

$$\psi_{lm}(r,t) \sim (-i)^{l+1} \frac{e^{i(kr-\omega t)}}{kr} Y_l^m(\theta,\varphi) \quad \text{as} \quad r \to \infty$$

which represents an **outgoing** spherical wave. The $h_l^{(2)}(kr)$ combination on the other hand has the limit

$$\psi_{lm}(r,t) \sim i^{l+1} \frac{e^{-i(kr+\omega t)}}{kr} Y_l^m(\theta,\varphi) \quad \text{as} \quad r \to \infty$$

which corresponds to an **incoming** spherical wave. (If we had chosen an $e^{+i\omega t}$ time dependence, the incoming and outgoing roles would be reversed.)

Suppose that we have a monochromatic (fixed ω) source that generates at $r = a$ sound waves of the form

$$\psi(r,\theta,\varphi,t) = F(\theta,\varphi)\, e^{-i\omega t}$$

where $F(\theta, \varphi)$ is a known function determined by the nature of the source. For $r > a$, the radiation is a superposition of **outgoing** waves with the same frequency,

$$\psi(r, \theta, \varphi, t) = e^{-i\omega t} \sum_{l=0}^{\infty} \sum_{m=-l}^{l} a_{lm} h_l^{(1)}(kr) Y_l^m(\theta, \varphi), \quad k = \frac{\omega}{c}.$$

The coefficients a_{lm} can be determined from the boundary condition at $r = a$:

$$F(\theta, \varphi) = \sum_{l=0}^{\infty} \sum_{m=-l}^{l} a_{lm} h_l^{(1)}(ka) Y_l^m(\theta, \varphi),$$

and so

$$a_{lm} = \frac{1}{h_l^{(1)}(ka)} \int_0^{2\pi} \int_0^{\pi} (Y_l^m(\theta, \varphi))^* F(\theta, \varphi) \sin\theta\, d\theta\, d\varphi.$$

As an example let us assume that the waves are produced by a "split-sphere" antenna for which

$$F(\theta, \varphi) \equiv F(\theta) = \begin{cases} f_0 & 0 < \theta < \frac{\pi}{2} \\ -f_0 & \frac{\pi}{2} < \theta < \pi \end{cases}.$$

The waves will share the azimuthal symmetry of the antenna and so we need only retain the $m = 0$ terms in the superpositions. Thus,

$$\psi(r, \theta, \varphi, t) = \psi(r, \theta, t) = e^{-i\omega t} \sum_{l=0}^{\infty} a_l h_l^{(1)}(kr) P_l(\cos\theta)$$

with

$$F(\theta) = \sum_{l=0}^{\infty} a_l h_l^{(1)}(ka) P_l(\cos\theta)$$

and

$$a_l = \frac{2l+1}{2} \frac{1}{h_l^{(1)}(ka)} \left[f_0 \int_0^{\frac{\pi}{2}} P_l(\cos\theta) \sin\theta\, d\theta - f_0 \int_{\frac{\pi}{2}}^{\pi} P_l(\cos\theta) \sin\theta\, d\theta \right].$$

We have evaluated the term in square brackets in an earlier example. It yields zero for even values of l and

$$a_l = (-1)^{\frac{l-1}{2}} \frac{f_0}{h_l^{(1)}(ka)} \frac{(2l+1)(l+1)(l-1)!}{2^{l+1} \left[\left(\frac{l+1}{2} \right)! \right]^2}$$

for odd values of l. Therefore, our final solution is

$$\psi(r, \theta, \varphi, t) = f_0\, e^{-i\omega t} \sum_{l=1,3,\ldots}^{\infty} (-1)^{\frac{l-1}{2}} \frac{(2l+1)(l+1)(l-1)!}{2^{l+1} \left[\left(\frac{l+1}{2} \right)! \right]^2} \frac{h_l^{(1)}(kr)}{h_l^{(1)}(ka)} P_l(\cos\theta).$$

When we are far from the source, we can replace $h_l^{(1)}(kr)$ by its asymptotic form. The wave then becomes

$$\psi(r, \theta, \varphi, t) \sim f(\theta, \varphi) \frac{e^{i(kr-\omega t)}}{r}$$

where

$$f(\theta, \varphi) = \frac{1}{k} \sum_{l=0}^{\infty} \sum_{m=-l}^{l} a_{lm}(-i)^{l+1} Y_l^m(\theta, \varphi).$$

For the special case of waves produced by a split-sphere antenna, we have

$$f(\theta, \varphi) = -\frac{f_0}{k} \sum_{l=1,3,5,\ldots}^{\infty} \frac{(2l+1)(l+1)(l-1)!}{2^{l+1}\left[\left(\frac{l+1}{2}\right)!\right]^2} \frac{1}{h_l^{(1)}(ka)} P_l(\cos\theta).$$

The relevance of the function $f(\theta, \varphi)$ becomes clear when we realize that the energy density at a point r associated with a monochromatic wave $\psi(r, t)$ is proportional to $|\psi(r, t)|^2$. Thus, the energy flux in the (θ, φ) direction is proportional to $c|\psi(r, \theta, \varphi, t)|^2$ which means when we are far from the source

$$\text{the energy flux} \propto c \frac{|f(\theta, \varphi)|^2}{r^2}.$$

To obtain the rate of energy flow through solid angle $d\Omega = \sin\theta\, d\theta\, d\varphi$ in the direction (θ, φ) at $\mathbf{r}$, we multiply the flux by the subtending area $r^2\, d\Omega$. The two factors of r^2 cancel and we find that the wave energy per unit time (or power) per unit solid angle is

$$\frac{dP}{d\Omega} \propto c|f(\theta, \varphi)|^2.$$

As we will see, the physical significance which this relation attaches to $f(\theta, \varphi)$ becomes particularly important in the analysis of scattering of waves.

Expansion of a Plane Wave in Spherical Waves

For a plane wave $e^{i(\mathbf{k}\cdot\mathbf{r}-\omega t)}$ propagating in the z direction, $\mathbf{k}$ will be parallel to the z-axis and so the spatial part of the wave will be $u(\mathbf{r}) = e^{i\mathbf{k}\cdot\mathbf{r}} = e^{ikr\cos\theta}$. The spatial part must also be a solution of the Helmholtz equation, $\nabla^2 u + k^2 u = 0$, and so must have a representation in spherical coordinates of the form

$$u(\mathbf{r}) = \sum_{l=0}^{\infty} \sum_{m=-l}^{l} \left\{ \begin{array}{c} j_l(kr) \\ n_l(kr) \end{array} \right\} Y_l^m(\theta, \varphi).$$

However, because it has no φ dependence and because it is finite at the origin, this simplifies to

$$u(\mathbf{r}) = e^{ikr\cos\theta} = \sum_{l=0}^{\infty} c_l\, j_l(kr)\, P_l(\cos\theta)$$

which is a Fourier-Legendre series with coefficients given by

$$c_l j_l(kr) = \frac{2l+1}{2} \int_0^{\pi} e^{ikr \cos \theta} P_l(\cos \theta) \sin \theta d\theta.$$

To simplify the evaluation of the coefficients, we introduce the variables $x = \cos \theta$ and $y = kr$. We then have

$$e^{ixy} = \sum_{l=0}^{\infty} c_l j_l(y) P_l(x) \quad \text{with} \quad c_l j_l(y) = \frac{2l+1}{2} \int_{-1}^{1} e^{ixy} P_l(x) dx.$$

We shall go through two rather different evaluations of the coefficients since both offer insights into the tricks of manipulating special functions.

The first evaluation makes use of recurrence relations. Since

$$P_l(x) = \frac{1}{2l+1} [P'_{l+1}(x) - P'_{l+1}(x)],$$

we find

$$c_l j_l(y) = \frac{1}{2} \int_{-1}^{1} e^{ixy} [P'_{l+1}(x) - P'_{l-1}(x)] dx$$

$$= \frac{iy}{2} \int_{-1}^{1} e^{iyx} [P_{l-1}(x) - P_{l+1}(x)] dx$$

$$= iy \left[\frac{1}{2l-1} c_{l-1} j_{l-1}(y) - \frac{1}{2l+3} c_{l+1} j_{l+1}(y) \right].$$

However, we already know a unique relationship that links these three spherical Bessel functions, the recurrence relation

$$j_l(y) = \frac{y}{2l+1} [j_{l+1}(y) + j_{l-1}(y)].$$

Therefore, consistency requires that

$$c_l = i \left(\frac{2l+1}{2l-1} \right) c_{l-1} = -i \left(\frac{2l+1}{2l+3} \right) c_{l+1}.$$

Applied l times, this relation gives us $c_l = i^l (2l + 1) c_0$ and

$$e^{ixy} = c_0 \sum_{l=0}^{\infty} i^l (2l + 1) j_l(y) P_l(x).$$

The final step sets $y = 0$. Then, since $j_0(0) = 1$ and $j_l(0) = 0$ for $l > 0$, we deduce that $c_0 = 1$ and the spherical wave expansion of a plane wave is

$$e^{ikz} = e^{ikr \cos \theta} = \sum_{l=0}^{\infty} i^l (2l + 1) j_l(kr) P_l(\cos \theta).$$

This, like the expansion of the monochromatic wave in the preceding example, is often referred to as a **partial wave expansion**.

The second approach to evaluating the expansion coefficients makes use of Rodrigues' formulas. Starting with the formula for $P_l(x)$, we have

$$c_l j_l(y) = \frac{2l+1}{2} \frac{1}{2^l \, l!} \int_{-1}^{1} e^{ixy} \frac{d^l}{dx^l} (x^2 - 1)^l \, dx$$

$$= \frac{2l+1}{2^{l+1} \, l!} (-iy)^l \int_{-1}^{1} e^{ixy} (x^2 - 1)^l \, dx \equiv \frac{2l+1}{2^{l+1} \, l!} (-iy)^l \, I_l$$

where we have performed l integrations by parts. The integral I_l can be subjected to further integrations by parts. Substituting $\frac{d}{dx} \frac{e^{ixy}}{iy}$ for e^{ixy}, we have

$$I_l = \int_{-1}^{1} \left(\frac{d}{dx} \frac{e^{ixy}}{iy} \right) (x^2 - 1)^l \, dx$$

$$= (-2l) \int_{-1}^{1} \frac{e^{ixy}}{iy} x (x^2 - 1)^{l-1} \, dx$$

$$= (2l) \frac{1}{y} \frac{d}{dy} \int_{-1}^{1} e^{ixy} (x^2 - 1)^l \, dx.$$

Performing a second integration by parts, this becomes

$$I_l = (2l)2(l-1) \frac{1}{y} \frac{d}{dy} \frac{1}{y} \frac{d}{dy} \int_{-1}^{1} e^{ixy} (x^2 - 1)^l \, dx$$

from which we deduce that l integrations will yield

$$I_l = 2^l \, l! \left(\frac{1}{y} \frac{d}{dy} \right)^l \int_{-1}^{1} e^{ixy} \, dx = 2^{l+1} \, l! \left(\frac{1}{y} \frac{d}{dy} \right)^l \left(\frac{\sin y}{y} \right).$$

But we know that

$$j_l(y) = (-y)^l \left(\frac{1}{y} \frac{d}{dy} \right)^l \left(\frac{\sin y}{y} \right).$$

Therefore, returning to our Fourier-Legendre coefficients we have

$$c_l j_l(y) = \frac{2l+1}{2^{l+1} \, l!} (-iy)^l \, 2^{l+1} \, l! \frac{j_l(y)}{(-y)^l} = (2l+1) \, i^l j_l(y),$$

or

$$c_l = (2l+1) \, i^l.$$

Thus, we again find that the partial wave expansion of a plane wave is

$$e^{ikz} = e^{ikr \cos \theta} = \sum_{l=0}^{\infty} (2l + 1) \, i^l \, j_l(kr) \, P_l(\cos \theta).$$

Notice that we have derived an interesting integral relationship linking spherical Bessel functions and Legendre polynomials

$$j_l(y) = \frac{1}{2 \, i^l} \int_{-1}^{1} e^{ixy} \, P_l(x) dx.$$

Since this looks like an integral transform, one expects that there must be an inverse relation as well. To find it, one requires the orthogonality and normalization condition

$$\int_{-\infty}^{\infty} j_l(x) \, j_m(x) dx = \frac{\pi}{2l + 1} \, \delta_{l,m} \, .$$

Then, treating the expansion of e^{ixy} as a Fourier-Bessel series, we find

$$P_l(x) = \frac{1}{\pi \, i^l} \int_{-\infty}^{\infty} e^{ixy} \, j_l(y) dy.$$

Scattering of Waves by a Sphere

We shall now make use of the partial wave expansions of the two preceding examples to analyse what happens when a plane wave impinges on a spherical object. We will centre the object at the origin and take its radius to be R. The plane wave is assumed to be propagating in the z direction with unit amplitude and so it is described by

$$\psi_{inc}(\mathbf{r}, t) = e^{i(\mathbf{k} \cdot \mathbf{r} - \omega t)} = e^{i(kr \cos \theta - \omega t)} \, .$$

Collision with the obstacle will result downstream in an outgoing spherical scattered wave superimposed on the incident plane wave. Since the incident wave is monochromatic, we can assume quite reasonably that the scattered wave will be monochromatic too and with the same frequency. Thus, the downstream wave will be of the form

$$\psi(\mathbf{r}, t) = e^{i(kr \cos \theta - \omega t)} + \psi_{sc}(\mathbf{r}, t)$$

where ψ_{sc} is the superposition

$$\psi_{sc}(\mathbf{r}, t) = e^{-i\omega t} \sum_{l=0}^{\infty} \sum_{m=-l}^{l} a_{lm} \, h_l^{(1)}(kr) \, Y_l^m (\theta, \varphi).$$

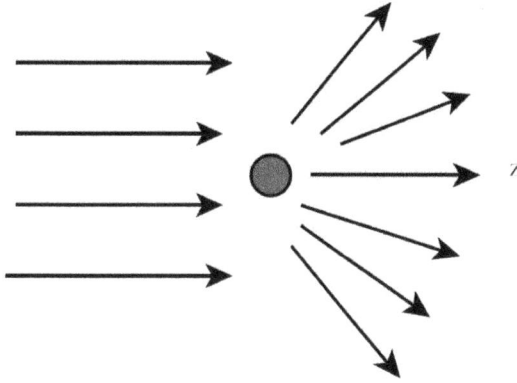

Far from the obstacle, $r \to \infty$ and the scattered wave assumes the asymptotic form

$$\psi_{sc}(\mathbf{r}, t) \sim f(\theta, \varphi) \frac{e^{i(kr-\omega t)}}{r} \quad \text{with } f(\theta, \varphi) = \frac{1}{k} \sum_{l=0}^{\infty} \sum_{m=-l}^{l} (-i)^{l+1} a_{lm} Y_l^m(\theta, \varphi).$$

In this context, the function $f(\theta, \varphi)$ is called the scattering amplitude and because of its connection to the power per unit solid angle associated with the scattered wave, it can be determined from what we call the **differential cross-section** for scattering into solid angle $d\Omega$ in direction (θ, φ):

$$d\sigma \equiv \frac{\text{power of scattered wave in solid angle } d\Omega}{\text{power of incident wave in unit area}} = \frac{c |f(\theta, \varphi)|^2}{c |e^{i(kz-\omega t)}|^2},$$

or

$$\frac{d\sigma}{d\Omega} = |f(\theta, \varphi)|^2.$$

The problem as we have set it up thus far has azimuthal symmetry about the z-axis and so, unless the boundary condition at $r = a$ breaks this symmetry, the solution should be independent of φ. This will certainly be the case if the boundary condition is homogeneous as happens for sound waves scattering from a rigid sphere and for "hard sphere scattering" in quantum mechanics which have the boundary conditions $\left. \frac{\partial \psi}{\partial r} \right|_{r=R} = 0$ and $\psi(R, \theta, \varphi) = 0$, respectively. Therefore, for such problems, we can write

$$\psi_{sc}(\mathbf{r}, t) = e^{-i\omega t} \sum_{l=0}^{\infty} a_l h_l^{(1)}(kr) P_l(\cos\theta)$$

and

$$\psi(\mathbf{r}, t) = e^{i(kr\cos\theta - \omega t)} + \psi_{sc}(\mathbf{r}, t) = e^{-i\omega t} \sum_{l=0}^{\infty} [(2l+1) i^l j_l(kr) + a_l h_l^{(1)}(kr)] P_l(\cos\theta)$$

where, in the last equation, we have used the partial wave expansion of the plane wave derived in the preceding application. Imposing the boundary condition appropriate to sound waves on this superposition, we find

$$\sum_{l=0}^{\infty} [(2l + 1)i^l j_l'(kR) + a_l h_l^{(1)\prime}(kR)]\, P_l(\cos\theta) = 0$$

and hence,

$$a_l = -(2l + 1)\, i^l\, \frac{j_l'(kR)}{h_l^{(1)\prime}(kR)}.$$

This completely determines the scattered wave and, using its asymptotic form for $r \gg a$, the scattering amplitude as well. Specifically, we have

$$f(\theta, \varphi) \equiv f(\theta) = \frac{1}{k} \sum_{l=0}^{\infty} a_l(-i)^{l+1} P_l(\cos\theta) = \frac{i}{k} \sum_{l=0}^{\infty}(2l + 1)\frac{j_l'(kR)}{h_l'(kR)} P_l(\cos\theta).$$

This can be simplified by introducing a parameter $\delta_l(k)$ that is defined by the equation

$$\tan \delta_l(k) \equiv \frac{j_l'(kR)}{n_l'(kR)}.$$

The reason for this somewhat unusual looking choice becomes apparent from two considerations. First, the factor

$$i\frac{j_l'(kR)}{h_l'(kR)} = i\frac{j_l'(kR)}{j_l'(kR) + i\, n_l'(kR)} = \frac{1}{\frac{n_l'(kR)}{j_l'(kR)} - i} = \frac{1}{\cotan \delta_l(k) - i} = e^{i\,\delta_l(k)} \sin \delta_l(k)$$

and so the scattering amplitude becomes

$$f(\theta) = \frac{1}{k} \sum_{l=0}^{\infty}(2l + 1)\, e^{i\,\delta_l(k)} \sin \delta_l(k)\, P_l(\cos\theta).$$

Second, when we express the asymptotic form of the wave in terms of δ_l we find

$$\psi(\mathbf{r}, t) = e^{i(kr\cos\theta - \omega t)} + \psi_{sc}(\mathbf{r}, t)$$

$$\sim e^{-i\omega t} \sum_{l=0}^{\infty}(2l + 1)\left[i^l\, \frac{\cos\left(kr - (l+1)\frac{\pi}{2}\right)}{kr} + \frac{e^{i\,\delta_l(k)} \sin \delta_l(k)}{k}\frac{e^{ikr}}{r}\right] P_l(\cos\theta)$$

$$= \frac{e^{-i\omega t}}{2ik} \sum_{l=0}^{\infty}(2l + 1)\left[\frac{e^{ikr}}{r} + (-1)^{l+1}\frac{e^{-ikr}}{r} + (e^{2i\,\delta_l(k)} - 1)\frac{e^{ikr}}{r}\right] P_l(\cos\theta)$$

$$= \frac{e^{-i\omega t}}{2ik} \sum_{l=0}^{\infty}(2l + 1)\left[e^{2i\,\delta_l(k)}\frac{e^{ikr}}{r} + (-1)^{l+1}\frac{e^{-ikr}}{r}\right] P_l(\cos\theta)$$

which tells us that the effect of the sphere has been to **shift the phase of the outgoing wave** by an amount $2\,\delta_l(k)$. Thus, $\delta_l(k)$ has an intuitively satisfying interpretation and one that is the source of its name: $\delta_l(k)$ is called the lth partial wave **phase shift**.

Notice that the **total** scattering cross section which is obtained by integrating $\frac{d\sigma}{d\Omega}$ over all solid angles has a particularly simple expression in terms of phase shifts. The integration gives

$$\sigma = (2\pi) \int_0^\pi |f(\theta)|^2 \sin\theta \, d\theta = \frac{4\pi}{k^2} \sum_{l=0}^\infty \sin^2 \delta_l(k) = \sum_{l=0}^\infty \sigma_l \quad \text{where} \quad \sigma_l = \frac{4\pi}{k^2} \sin^2 \delta_l(k).$$

These measures of the scattering can be easily estimated only in the "large wavelength limit", $kR \ll 1$ or $\lambda \gg R$. From the small argument behaviour of the spherical Bessel and Neumann functions we have

$$\tan \delta_l(k) \simeq \frac{l}{l+1} \frac{(2^l \, l!)^2}{(2l)!(2l+1)!} (kR)^{2l+1} \quad \text{for} \quad l \geq 1,$$

and

$$\tan \delta_0(k) = \frac{j_0'(kR)}{n_0'(kR)} = \frac{j_1(kR)}{n_1(kR)} \simeq -\frac{(kR)^3}{3}.$$

Therefore, $\delta_l(k)$ is itself $\ll 1$, and so

$$e^{i\delta_l(k)} \sin \delta_l(k) \simeq \delta_l(k) \simeq \tan \delta_l(k).$$

Moreover, $\delta_l(k)$ decreases rapidly with increasing $l > 1$ so we need to retain only the first few partial waves. Thus,

$$f(\theta) \simeq \frac{1}{k} \left[-\frac{(kR)^3}{3} P_0(\cos\theta) + 3\frac{(kR)^3}{6} P_1(\cos\theta) + \dots \right] \simeq -\frac{k^2 R^3}{3} [1 - \frac{3}{2}\cos\theta]$$

and

$$\sigma \simeq \frac{5\pi}{9} k^4 R^6.$$

This formalism for the analysis of scattering problems is very general. The only thing that changes from problem to problem is the boundary condition at the obstacle and the consequent definition of the phase shift. For example, if instead of $\left.\frac{\partial\psi}{\partial r}\right|_{r=R} = 0$ we had used the boundary condition for quantum mechanical hard sphere scattering, $\psi(R, \theta, \varphi) = 0$, we would replace our definition of $\delta_l(k)$ above with the identity

$$\tan \delta_l(k) \equiv \frac{j_l(kR)}{n_l(kR)}$$

but everything else would be unchanged.

11.5 The Classical Orthogonal Polynomials

11.5.1 The Polynomials and Their Properties

We have completed a study of the special functions that arise from the solution of the Laplace and Helmholtz equations in spherical and cylindrical coordinates. However, this is not an inclusive set: there are further special functions that arise, for example, in quantum mechanics. One such is the set of Hermite polynomials which we encountered in Chapter 9 as the solutions of a DE that we attributed to a one-dimensional harmonic oscillator. Others include the Laquerre polynomials (which feature in the theory of the hydrogen atom and of the three-dimensional harmonic oscillator), the Tchebichef polynomials and the Gegenbauer polynomials. Together with the Legendre polynomials, they are examples of what have come to be known as the **classical orthogonal polynomials**.

A common feature of these functions is that each class of polynomials is orthogonal on an interval $a \le x \le b$ with respect to a weight function $\rho(x)$. This could be used to define the polynomials since they could be generated from a Schmidt orthogonalization process applied to the monomials $1, x, x^2, x^3, \ldots$. However, there is a more elegant approach based on a second common feature: each class can be computed from a Rodrigues' formula.

We define a **generalized Rodrigues' formula** by means of the identity

$$C_n(x) = \frac{1}{K_n} \frac{1}{\rho(x)} \frac{d^n}{dx^n}[\rho(x)\, s^n(x)] \tag{11.5.1}$$

where
- K_n is a standardizing constant that is chosen by convention
- $\rho(x)$ is a real, positive (weight) function that is integrable on the interval $a \le x \le b$
- $s(x)$ is a polynomial of degree ≤ 2 with real roots and satisfies the boundary conditions

$$\rho(a)s(a) = \rho(b)s(b) = 0. \tag{11.5.2}$$

In addition, we require that $C_1(x)$ be a **first**-degree polynomial.

We know of course that Legendre polynomials satisfy these constraints with

$$a = -1, \quad b = 1, \quad \rho(x) = 1, \quad K_n = 2^n n!, \quad s(x) = x^2 - 1 \text{ and } C_1 = x.$$

The question is, do all $C_n(x)$ that satisfy these requirements comprise a set of polynomials of degree n that are orthogonal with respect to weight function $\rho(x)$ on the interval $a \le x \le b$? As we will now prove, the answer is yes.

We start by noting that setting $n = 1$ in (11.5.1) gives us

$$C_1(x)\rho(x) K_1 = \frac{d}{dx}[\rho(x)s(x)] \text{ or } s(x)\frac{d\rho}{dx} = \rho(x)\left(C_1(x) - \frac{ds}{dx}\right). \tag{11.5.3}$$

Next, we calculate $\frac{d}{dx}[\rho(x)\, s^n(x)\, p_k]$ where $p_k(x)$ is an arbitrary polynomial of degree k:

$$\frac{d}{dx}[\rho(x)\, s^n(x)\, p_k(x)] = \frac{d\rho}{dx} s^n p_k + \rho n s^{n-1}\frac{ds}{dx} p_k + \rho s^n \frac{dp_k}{dx}$$

$$= \rho\left(K_1 C_1 - \frac{ds}{dx}\right)s^{n-1} p_k + n\rho s^{n-1}\frac{ds}{dx} p_k + \rho s^n(x)\frac{dp_k}{dx}$$

$$= \rho(x)\, s^{n-1}(x) \left[\left(K_1\, C_1(x) + (n-1)\frac{ds}{dx} \right) p_k(x) + s(x)\frac{d\, p_k}{dx} \right]$$

where we have used (11.5.3) to obtain the second line of the calculation. Since $s(x)$ is of degree ≤ 2, $\frac{ds}{dx}$ is of degree ≤ 1, and $\frac{d\, p_k}{dx}$ is of degree $k-1$, the term in square brackets on the right hand side is a polynomial of degree $\leq k+1$. Denoting the latter by $q_{k+1}(x)$, we write

$$\frac{d}{dx}[\rho(x)\, s^n(x)\, p_k(x)] = \rho(x)\, s^{n-1}(x)\, q_{k-1}(x)$$

from which we see that differentiating m times will give us

$$\frac{d^m}{d\, x^m}[\rho(x)\, s^n(x)\, p_k(x)] = \rho(x)\, s^{n-m}(x)\, q_{k+m}(x) \tag{11.5.4}$$

where $q_{n+m}(x)$ is a polynomial of degree $\leq n+m$.

Notice that insertion of $p_0(x) = 1$ into (11.5.4) gives us

$$\frac{d^m}{d\, x^m}[\rho(x)\, s^n(x)] = \rho(x)\, s^{n-m}(x)\, q_m(x)$$

and, because of the boundary conditions satisfied by $\rho(x)s(x)$, this means that all such derivatives with $m < n$ vanish at $x = a$ and $x = b$.

Moreover, if $m = n$, we have the derivative that appears in the Rodrigues' formula which allows us to assert that

$$C_n(x) = q_n(x) \equiv \text{ a polynomial of degree } \leq n$$
$$= q_{n-1}(x) + a_n\, x^n \text{ where } a_n \text{ is a coefficient to be determined.} \tag{11.5.5}$$

The next step involves the evaluation of the integral $\int_a^b p_m(x)\, C_n(x)\rho(x)dx$ where $p_m(x)$ is a polynomial of degree $m < n$. Invoking Rodrigues' formula for $C_n(x)$ and integrating by parts n times, we find

$$\int_a^b p_m(x)\, C_n(x)\rho(x)dx = \frac{1}{K_n} \int_a^b p_m(x)\frac{d^n}{d\, x^n}[\rho(x)\, s^n(x)]dx$$

$$= \frac{(-1)^n}{K_n} \int_a^b [\rho(x)\, s^n(x)]\frac{d^n}{d\, x^n}\, p_m(x) = 0 \tag{11.5.6}$$

where we have used the vanishing of derivatives of $\rho(x)\, s^n(x)$ at $x = a$ and $x = b$ to dispose of all integrated terms.

Now we make use of (11.5.5). Using it to replace one factor of $C_n(x)$ in $\int_a^b C_n^2(x)\rho(x)dx$ we find

$$N_n \equiv \int_a^b C_n^2(x)\rho(x)dx = \int_a^b q_{n-1}(x)\, C_n(x)dx + a_n \int_a^b x^n\, C_n(x)\rho(x)dx = a_n \int_a^b x^n\, C_n(x)\rho(x)dx$$

by the preceding result. Since the left hand side cannot be zero, it follows that a_n cannot be zero either and hence, that $C_n(x)$ **is a polynomial of degree** n. Moreover, using the Rodrigues' formula, we can evaluate its normalization integral:

$$N_n = a_n \int_a^b x^n\, C_n(x)\rho(x)dx = \frac{a_n}{K_n} \int_a^b x^n \frac{d^n}{dx^n}[s^n(x)\rho(x)]dx = \frac{(-1)^n\, n!\, a_n}{K_n} \int_a^b s^n(x)\rho(x)dx$$

(11.5.7)

where we have performed n integrations by parts to obtain the final line.

Finally, we substitute $C_m(x)$ for $p_m(x)$ in (11.5.6) to obtain

$$\int_a^b C_m(x)\, C_n(x)\rho(x)dx = 0 \quad \text{for} \quad m < n.$$

(11.5.8)

Thus, not only does the generalized Rodrigues' formula define a sequence of polynomials, the sequence forms an **orthogonal set of polynomials with respect to a weight** $\rho(x)$ **on the interval** $a \le x \le b$.

At first sight, it might appear that this is a prescription for generating any number of distinct sets of orthogonal polynomials. In fact, there are only three classes of polynomial corresponding to the three possibilities for the degree of $s(x)$.

We can always define the polynomial $C_1(x)$ to be

$$C_1(x) = -\frac{x}{K_1}$$

since any other first-degree polynomial can be obtained from it by a linear transformation of the variable x. Substituting this into the Rodrigues' formula gives us a relationship between $\rho(x)$ and $s(x)$:

$$\frac{1}{\rho(x)}\frac{d\rho}{dx} = -\frac{1}{s(x)}\left(\frac{ds}{dx} + x\right).$$

(11.5.9)

Suppose that $s(x)$ is of zeroth-degree. Without loss of generality, we can set $s(x) = 1$. Equation (11.5.9) becomes

$$\frac{1}{\rho(x)}\frac{d\rho}{dx} = -x$$

with solution $\rho(x) = \text{const}\, e^{-\frac{x^2}{2}}$. The product $s(x)\rho(x)$ vanishes only at $x = \pm\infty$. Therefore, transforming x by $\frac{x}{\sqrt{2}} \to x$ and setting the multiplicative constant equal to one, we conclude that a consistent set of parameters is

$$s(x) = 1, \quad \rho(x) = e^{-x^2}, \quad a = -\infty \text{ and } b = \infty.$$

(11.5.10)

With appropriate standardization, $K_n = (-1)^n$, the set $\{C_n(x)\}$ defined by these parameters consists of the **Hermite polynomials** $H_n(x)$.

For the case of a first-degree $s(x)$, we set $s(x) = (x - \alpha)$. Equation (11.5.9) now yields the DE

$$\frac{1}{\rho(x)}\frac{d\rho}{dx} = -\frac{x+1}{x-\alpha}$$

with solution (having set the multiplicative constant of integration equal to one)

$$\rho(x) = (x - \alpha)^{-\alpha-1} e^{-x}.$$

If $-\alpha - 1 \equiv \nu > 1$, then $s(x)\rho(x)$ vanishes at $x = \alpha$ and $x = \infty$ and $\rho(x)$ is integrable on $\alpha \leq x < \infty$. Thus, transforming x by means of the substitution $x \to x + \alpha$ and ignoring multiplicative constants, we conclude that the appropriate parameters in this case are

$$s(x) = x, \quad \rho(x) = x^{\nu} e^{-x}, \quad \nu > -1, \quad a = 0, \quad b = \infty \tag{11.5.11}$$

which correspond to $C_1(x) = \frac{1}{K_1}(\nu + 1 - x)$ and $\frac{1}{\rho(x)}\frac{d\rho}{dx} = \frac{\nu-x}{x}$. Used in the generalized Rodrigues' formula with $K_n = n!$, these generate the **Laguerre polynomials** $L_n^{\nu}(x)$.

Finally, if $s(x)$ is of second degree, we can assign it the quadratic form

$$s(x) = (x - \alpha)(\beta - x), \quad \beta > \alpha.$$

Equation (11.5.9) then reads

$$\frac{1}{\rho(x)}\frac{d\rho}{dx} = \frac{(x - \alpha) - (\beta - x) - x}{(x - \alpha)(\beta - x)}$$

which has the solution (ignoring multiplicative constants)

$$\rho(x) = (x - \alpha)^{-\frac{\beta}{\beta-\alpha}}(\beta - x)^{-\frac{\alpha}{\beta-\alpha}}.$$

Evidently $s(x)\rho(x)$ vanishes at $x = \alpha$ and $x = \beta$ and $\rho(x)$ is integrable on $\alpha \leq x \leq \beta$ provided that the exponents in $\rho(x)$ are both > 1. This will be the case if we choose $\alpha = -1$ and $\beta = 1$. Therefore, we can take our parameters to be

$$s(x) = (1 - x^2), \rho(x) = (x + 1)^{\mu}(1 - x)^{\nu}, \mu, \nu > -1, a = -1 \text{ and } b = 1 \tag{11.5.12}$$

which result in the redefinition

$$C_1(x) = \frac{1}{K_1}[(\mu - \nu) - (\mu + \nu + 2)x] \text{ and } \frac{1}{\rho(x)}\frac{d\rho}{dx} = \frac{\mu(1 - x) - \nu(1 + x)}{1 - x^2}.$$

These produce the **Jacobi polynomials** $P_n^{(\mu,\nu)}(x)$ which have a number of special cases to which other names have been attached:

- the **Gegenbauer polynomials** $C_n^{\mu+\frac{1}{2}}(x)$ which correspond to $\mu = \nu$;
- the **Legendre polynomials** $P_n(x)$ which correspond to $\mu = \nu = 0$;
- the **Tchebichef polynomials of the first kind** $T_n(x)$ which correspond to $\mu = \nu = \frac{1}{2}$;
- the **Tchebichef polynomials of the second kind** $U_n(x)$ which correspond to $\mu = \nu = -\frac{1}{2}$.

We state without proof that any three consecutive orthogonal polynomials satisfy a recurrence relation of the form

$$C_{n+1}(x) = (A_n x + B_n) C_n(x) - D_n C_{n-1}(x) \tag{11.5.13}$$

where A_n, B_n and D_n are constants that depend only on n and the class of polynomials under consideration. Indeed, one can show that

$$A_n = \frac{a_{n+1}}{a_n}, \quad B_n = \frac{a_{n+1}}{a_n} \left(\frac{a'_{n+1}}{a_{n+1}} - \frac{a'_n}{a_n} \right), \quad D_n = \frac{N_n}{N_{n-1}} \frac{a_{n+1} a_{n-1}}{a_n^2} \tag{11.5.14}$$

where, as before, N_n is the normalization integral, a_n is the coefficient of x^n, and a'_n is the coefficient of x^{n-1}, for the polynomial $C_n(x)$.

In addition, one can show that $C_n(x)$ satisfies the Sturm-Liouville DE

$$\frac{d}{dx} \left(\rho(x) s(x) \frac{d C_n}{dx} \right) = -\lambda_n \rho(x) C_n(x) \tag{11.5.15}$$

with

$$\lambda_n = -n \left[K_1 \frac{d C_1}{dx} + \frac{1}{2}(n-1) \frac{d^2 s}{dx^2} \right]. \tag{11.5.16}$$

The equation has regular singular points at the roots of $\rho(x)s(x)$ which means at $x = a$ and $x = b$. Thus, since the polynomials $C_n(x)$ are finite there, they must be eigenfunction solutions of the more general problem

$$\frac{d}{dx} \left(\rho(x) s(x) \frac{du}{dx} \right) = -\lambda \rho(x) u(x) \quad \text{with} \quad |u(b)| < \infty \quad \text{and} \quad |u(a)| < \infty. \tag{11.5.17}$$

We list below the salient features of Hermite and Laguerre polynomials which, along with the Legendre polynomials, are the ones most frequently encountered.

Hermite Polynomials:
Rodrigues' formula

$$H_n(x) = (-1)^n e^{x^2} \frac{d^n}{dx^n} (e^{-x^2}) \tag{11.5.18}$$

Orthogonality and Normalization

$$\int_{-\infty}^{\infty} H_m(x) H_n(x) e^{-x^2} dx = N_n \delta_{m,n} = \sqrt{\pi} \, 2^n \, n! \, \delta_{m,n}. \tag{11.5.19}$$

Differential equation

$$\frac{d^2}{dx^2} H_n(x) - 2x \frac{d}{dx} H_n(x) + 2n H_n(x) = 0 \tag{11.5.20}$$

Recurrence relation

$$H_{n+1}(x) = 2x\, H_n(x) - 2n\, H_{n-1}(x) \tag{11.5.21}$$

Series

$$H_n(x) = \sum_{k=0}^{[n/2]} (-1)^k \frac{n!}{(n-2k)!k!} (2x)^{n-2k} \tag{11.5.22}$$

Laguerre Polynomials:
Rodrigue's formula

$$L_n^v(x) = \frac{1}{n!}\, x^{-v}\, e^x\, \frac{d^n}{dx^n}(e^{-x} x^{v+n}) \tag{11.5.23}$$

Orthogonality and Normalization

$$\int_0^\infty L_m^v(x) L_n^v(x)\, x^v\, e^{-x}\, dx = N_n\, \delta_{m,n} = \frac{\Gamma(n+v+1)}{n!}\, \delta_{m,n} \tag{11.5.24}$$

Differential equation

$$x\frac{d^2}{dx^2} L_n^v(x) + (v+1-x)\frac{d}{dx} L_n^v(x) + n L_n^v(x) = 0 \tag{11.5.25}$$

Recurrence relation

$$(n+1) L_{n+1}^v(x) = (2n+v+1-x) L_n^v(x) - (n+v) L_{n-1}^v(x) \tag{11.5.26}$$

Series

$$L_n^v(x) = \sum_{k=0}^{n} (-1)^k \frac{(n+v)!}{(n-k)!(v+k)!k!} x^k \quad v > -1 \tag{11.5.27}$$

11.5.2 Applications

The Quantum Mechanical Simple Harmonic Oscillator
As mentioned earlier, the Hermite polynomials arise in the quantum mechanical description of simple harmonic oscillators. The potential energy of the oscillator is $V(z) = \frac{1}{2}K z^2$ corresponding to a restoring force $\mathbf{F} = -\nabla V = -\frac{d}{dz} V \mathbf{k} = -Kz\mathbf{k}$. Therefore, if we take the mass of the oscillating particle to be m, its time-independent Schrödinger equation is

$$-\frac{\hbar^2}{2m}\frac{d^2\psi}{dz^2} + \frac{1}{2}K z^2\, \psi(z) = E\psi(z) \tag{11.5.28}$$

where E is the total energy. In addition, the particle's wave function must be bounded and normalizable and so we require $\lim\limits_{z \to \pm\infty} |\psi(z)| = 0$.

Introducing the parameters

$$x = \alpha z \quad \text{with} \quad \alpha^4 = \frac{mK}{\hbar^2} = \frac{m^2 \omega^2}{\hbar^2} \quad \lambda = \frac{2E}{\hbar}\left(\frac{m}{K}\right)^{\frac{1}{2}} = \frac{2E}{\hbar\omega}, \tag{11.5.29}$$

where $\omega = \left(\frac{K}{m}\right)^{\frac{1}{2}}$ is the angular frequency of the corresponding classical oscillator, this equation becomes

$$\frac{d^2 u}{dx^2} + (\lambda - x^2)u(x) = 0 \quad \text{with} \quad u(x) = \psi\left(\frac{x}{\alpha}\right) = \psi(z). \tag{11.5.30}$$

This is the same equation, (9.8.4), that we solved in Section 9.8.2 and found that λ must be set equal to $2n + 1$ yielding a set of solutions $u_n(x) \propto e^{-\frac{x^2}{2}} H_n(x)$. Normalizing so that the wave functions give unit probability of finding the particle in the interval $-\infty < x < \infty$, we find

$$u_n(x) = \pi^{-\frac{1}{4}}(2^n\, n!)^{-\frac{1}{2}}\, e^{-\frac{x^2}{2}}\, H_n(x)$$

or,

$$\psi_n(z) = \pi^{-\frac{1}{4}}(2^n\, n!)^{-\frac{1}{2}}\, e^{-\frac{\alpha^2 z^2}{2}}\, H_n(\alpha z). \tag{11.5.31}$$

The corresponding energies are

$$E_n = \left(n + \frac{1}{2}\right)\hbar\omega \quad n = 0, 1, 2, \dots . \tag{11.5.32}$$

Notice that the minimum or zero point energy is **not** zero but rather $E_{\min} = E_0 = \frac{1}{2}\hbar\omega$. This is a quantum phenomenon connected with the uncertainty principle.

The three-dimensional harmonic oscillator has played an important role in understanding sub-atomic spectroscopy. In Section 11.4.2 we noted that the nuclear shell model assumes that the individual two-nucleon interactions in a medium to large nucleus sum to produce an effective central potential in which each nucleon moves independently. The next question is what functional form can be assigned to the central potential that will not only be simple to work with but will be reasonably realistic in the sense that its predictions agree at least qualitatively with experiment. The guess of Section 11.4.2, a "hard" or infinite spherical well, is very simple to work with but has only limited success. Far and away more successful is the almost as simple guess that the potential is that of a three-dimensional oscillator: $V(r) = -V_0 + \frac{1}{2}Kr^2$.

Another application of the oscillator potential is in understanding the bonding of quarks in sub-nuclear matter. Quarks interact with each other by means of gluon exchange which produces a force that increases with increasing separation and thus prevents the production of free quarks. At the level of non-relativistic quantum mechanics, an oscillator potential is a very convenient way of representing this effect and hence, of approximating the quark-quark interaction.

The Nuclear Shell Model

We can absorb the constant part of the potential by redefining the zero of energy. Therefore, the (time-independent) Schrödinger equation that must be solved for a nucleon of mass m in a potential $V(r) = \frac{1}{2}Kr^2$ is

$$-\frac{\hbar^2}{2m}\nabla^2\psi + \frac{1}{2}Kr^2\,\psi(\mathbf{r}) = E\,\psi(\mathbf{r}). \qquad (11.5.33)$$

If we separate variables using Cartesian coordinates and the fact that $r^2 = x^2 + y^2 + z^2$, this becomes three one-dimensional harmonic oscillator equations all with the same "spring constant" K and with energies E_x, E_y, and E_z that sum to E : $E = E_x + E_y + E_z$. Thus, the eigenfunction solutions of (11.5.33) that satisfy $\lim_{r\to\infty}\psi(r) = 0$ may be expressed as

$$\psi_N(\mathbf{r}) = \psi_{n_1}(x)\,\psi_{n_2}(y)\,\psi_{n_3}(z)$$
$$= \pi^{-\frac{3}{4}}\left(2^N\,n_1!\,n_2!\,n_3!\right)^{-\frac{1}{2}}e^{-\frac{\alpha^2 r^2}{2}}\,H_{n_1}(\alpha x)\,H_{n_2}(\alpha y)\,H_{n_3}(\alpha z) \qquad (11.5.34)$$

and correspond to the energies

$$E_N = E_{n_1} + E_{n_2} + E_{n_3} = \left(n_1 + n_2 + n_3 + \frac{3}{2}\right)\hbar\omega = \left(N + \frac{3}{2}\right)\hbar\omega \qquad (11.5.35)$$

where

$$N = n_1 + n_2 + n_3, \quad n_1, n_2, n_3 = 0, 1, 2, \ldots \qquad (11.5.36)$$

$$\omega = \sqrt{\frac{K}{m}} \quad \text{and} \quad \alpha^2 = \frac{m\omega}{\hbar}.$$

Notice that the degeneracy of the eigenvalues increases fairly rapidly with increasing N. The ground state energy $E_0 = \frac{3}{2}\hbar\omega$ is non-degenerate since $N = 0$ implies uniquely $n_1 = n_2 = n_3 = 0$. However, the next energy $E_1 = \frac{5}{2}\hbar\omega$ is three-fold degenerate because it is associated with the three distinct states associated with the triads $n_1 = 1, n_2 = n_3 = 0$, $n_2 = 1, n_1 = n_3 = 0$ and $n_3 = 1, n_1 = n_2 = 0$. The $N = 2$ energy is 6-fold degenerate and since in general there are $N + 1$ choices for n_1, $\frac{N}{2} + 1$ choices for n_2 and 1 choice for n_3, the degeneracy of the N th level is

$$d_N = \frac{(N + 1)(N + 2)}{2}. \qquad (11.5.37)$$

The existence of degeneracy always indicates that the eigenvalue problem possesses symmetry of some sort. In the present case, it is spherical symmetry and whenever we have encountered spherical symmetry before we have used it as a justification for using spherical rather than Cartesian coordinates. Doing so here, we set $\psi(\mathbf{r}) = R(r)\,Y_l^m(\theta, \varphi)$ and substitute into (11.5.33) to find

$$-\frac{\hbar^2}{2m}\frac{1}{r}\frac{d^2}{dr^2}[rR(r)] + \left[\frac{l(l+1)\hbar^2}{2m r^2} + \frac{1}{2}Kr^2\right]R(r) = ER(r). \qquad (11.5.38)$$

In quantum mechanics, unlike the description of classical waves, the l has a physical association: it is a measure of the nucleon's orbital angular momentum. Evidently, it is a conserved quantity when there is spherical symmetry.

Making the replacement $u(r) = rR(r)$ equation (11.5.38) can be cast in the somewhat simpler form:

$$\frac{d^2 u}{d r^2} - \left[\alpha^4 r^2 + \frac{l(l+1)}{r^2} \right] u(r) = -\frac{2mE}{\hbar^2} u(r) \quad \text{where} \quad \alpha^2 = \frac{m\omega}{\hbar}.$$

Solution of this DE requires a number of tricks from Section 9.8 . Our boundary conditions are boundedness at zero and infinity. Therefore, the first step is to examine the leading behaviour there and factor it out. As $r \to 0$, the DE assumes the limiting form

$$\frac{d^2 u}{d r^2} - \frac{l(l+1)}{r^2} u \simeq 0$$

with solutions r^{l+1} and r^{-l} . At the other extreme, as $r \to \infty$, the DE becomes

$$\frac{d^2 u}{d r^2} - \alpha^4 r^2 u \simeq 0$$

with approximate solutions $e^{\pm \frac{\alpha^2 r^2}{2}}$. Therefore, to assure boundedness, we set

$$u(r) = r^{l+1} e^{-\frac{\alpha^2 r^2}{2}} v(r). \tag{11.5.39}$$

Substituting this into the DE for $u(r)$ gives us

$$\frac{d^2 v}{d r^2} + 2 \left[\frac{l+1}{r} - \alpha^2 r \right] \frac{dv}{dr} + \left[\frac{2mE}{\hbar^2} - \alpha^2 (2l+3) \right] v = 0 \tag{11.5.40}$$

which we expect to admit polynomial solutions. However, it is not readily apparent what kind of polynomials these may be. To save the effort of attempting a Frobenius solution, we make one further simplification by introducing the new independent variable $x = \alpha^2 r^2$. Setting $y(x) = v(r)$, we obtain

$$x \frac{d^2 y}{d x^2} + \left(l + \frac{3}{2} - x \right) \frac{dy}{dx} + \frac{1}{4} \left(\frac{2E}{\hbar\omega} - 2l - 3 \right) y = 0 \tag{11.5.41}$$

which we can now compare with the DE's satisfied by classical orthogonal polynomials. A case by case comparison is quickly rewarded: this is precisely the form found in (11.5.25), the DE for Laguerre polynomials. Thus, $y(x) = L_n^\nu(x)$ with $n = \frac{1}{4} \left(\frac{2E}{\hbar\omega} - 2l - 3 \right)$ and $\nu = l + \frac{1}{2}$. In other words, the quantized energies are

$$E_N = E_{nl} = \left(2n + l + \frac{3}{2} \right) \hbar\omega = \left(N + \frac{3}{2} \right) \hbar\omega, \quad N = 0, 1, 2, \ldots \tag{11.5.42}$$

in agreement with the conclusion (11.5.35) obtained with the use of Cartesian coordinates.

The wave functions that correspond to these energies are

$$\psi_{nl}(r) = N_{nl} \, r^l \, e^{-\frac{\alpha^2 r^2}{2}} \, L_n^{l+\frac{1}{2}}(\alpha^2 r^2) \, Y_l^m(\theta, \varphi)$$

where N_{nl} is a constant determined by normalizing to unity the probability of finding the nucleon somewhere in space. In fact, using (11.5.24) we find

$$1 = \int_0^\infty \int_0^\pi \int_0^{2\pi} |\psi_{nl}(r)|^2 \, r^2 \, dr \sin\theta \, d\theta \, d\varphi = \frac{|N_{nl}|^2}{2} \frac{1}{\alpha^{2l+3}} \frac{\Gamma(n+l+3/2)}{n!}$$

and so,

$$\psi_{nl}(r) = \left(\frac{2\,\alpha^3\,n!}{\Gamma(n+l+3/2)} \right)^{\frac{1}{2}} (\alpha r)^l \, e^{-\frac{\alpha^2 r^2}{2}} \, L_n^{l+\frac{1}{2}}(\alpha^2 r^2) \, Y_l^m(\theta, \varphi). \tag{11.5.43}$$

Let us now check the degeneracy question. For $N = 0$ there is only one possibility: $n = 0$ and $l = 0$. For $N = 1$ there is again one possibility, $n = 0$ and $l = 1$, but for nonzero l there is automatically a degeneracy of $2l+1$ corresponding to the allowed values of m. Thus, we recover the same three-fold degeneracy that we found with Cartesian coordinates. Similarly, for $N = 2$ we can have $n = 0$ and $l = 2$ **plus** $n = 1$ and $l = 0$ for a six-fold degeneracy and, in general, since l can assume the values $N, N-2, N-4, \ldots$ down to 1 or 0,

$$d_N = \frac{(N+1)(N+2)}{2}.$$

Allowing for a spin-degeneracy of two, this predicts a shell structure for nuclei in which there are successive energy levels or shells that can be filled successively with up to 2, 6 , 12, 20, 30, 42, 56 , ... like nucleons, neutrons or protons. (The **exclusion principle** prohibits more than one identical nucleon in each state). By analogy with atomic physics, one would expect nuclei to be particularly stable when they have exactly the number of neutrons or protons or both to fill closed shells. Therefore the model predicts "magic numbers" of stability equal to 2, 8, 20, 40, 70, 112, ... nucleons of either type. The first three of these agree rather well with experiment as the relative stability of He_2^4 and O_8^{16} attests. However, to match the higher magic numbers that are observed in nature it is necessary to add what is called a spin-orbit coupling interaction to the central harmonic oscillator potential. This has the effect of depressing the energies of states with total angular momentum $j = l + \frac{1}{2}$ relative to those with $j = l - \frac{1}{2}$ with the result that some states get re-assigned to either higher or lower shells. The development of this model resulted in the award of a Nobel prize to Mayer and Jensen in 1966.

The Hydrogenic Atom

Another important application of Laguerre polynomials is in the solution of the Schrödinger equation for the hydrogenic atom. The equation is

$$-\frac{\hbar^2}{2m} \nabla^2 \psi - \frac{Z e^2}{r} \psi = E\psi \tag{11.5.44}$$

where $Z = 1$ for hydrogen, $Z = 2$ for singly ionized helium, and so on. Because the Coulomb potential is central, we have spherical symmetry again and can separate variables by setting

$$\psi(\mathbf{r}) = R(r)\, Y_l^m\,(\theta,\,\varphi) = \frac{u(r)}{r}\, Y_l^m\,(\theta,\,\varphi). \tag{11.5.45}$$

Once again, l is a (conserved) orbital angular momentum. Substituting (11.5.45) into (11.5.44) we obtain the radial equation

$$-\frac{\hbar^2}{2m}\frac{1}{r^2}\frac{d}{dr}\left(r^2\frac{dR}{dr}\right) - \frac{Z e^2}{r}R(r) + \frac{\hbar^2}{2m}\frac{l(l+1)}{r^2}R(r) = ER(r), \tag{11.5.46}$$

or

$$-\frac{\hbar^2}{2m}\frac{d^2 u}{dr^2} - \frac{Z e^2}{r}u(r) + \frac{\hbar^2}{2m}\frac{l(l+1)}{r^2}u(r) = Eu(r). \tag{11.5.47}$$

To render this into a recognizable form, we introduce some parameters similar to those used with the harmonic oscillator. We define

$$\beta^2 = -\frac{8mE}{\hbar^2} \quad \text{and} \quad \lambda = \frac{2mZ e^2}{\beta\hbar^2} \tag{11.5.48}$$

where, recognizing that bound states correspond to negative values of E, we have put in a minus sign to ensure that β is real. Then we introduce the new variables

$$x = \beta r \quad \text{and} \quad y(x) = u(r) = u(x/\beta)$$

in terms of which (11.5.47) becomes

$$\frac{d^2 y}{dx^2} - \left(\frac{l(l+1)}{x^2} - \frac{\lambda}{x} + \frac{1}{4}\right)y(x) = 0. \tag{11.5.49}$$

The next step is to deduce the large and small x behaviour of $y(x)$ by solving the asymptotic forms of the DE (11.5.49). For large x, we have

$$\frac{d^2 y}{dx^2} - \frac{1}{4}y(x) \simeq 0$$

with solutions $y(x) \sim e^{\pm\frac{x}{2}}$. Similarly, for small x the DE becomes

$$\frac{d^2 y}{dx^2} - \frac{l(l+1)}{x^2}y(x) \simeq 0$$

with solutions x^{l+1} and x^{-l}. Since we require solutions that are bounded at both zero and infinity, we set

$$y(x) = x^{l+1}\, e^{-\frac{x}{2}}\, v(x) \tag{11.5.50}$$

and substitute into (11.5.49) to obtain the DE

$$x\frac{d^2 v}{dx^2} + (2l + 2 - x)\frac{dv}{dx} + (\lambda - l - 1)v(x) = 0. \tag{11.5.51}$$

Once again this is the Laguerre DE (11.5.25) with polynomial solutions $L_N^\nu(x)$ provided that $2l + 1 = \nu > -1$ and $\lambda - l - 1 = N$ is an integer or zero. The first condition is obviously satisfied and the second implies that $\lambda = N + l + 1 = n$, $n = 1, 2, \ldots$. This gives us the quantization rule

$$\lambda = \frac{2mZ e^2}{\beta \hbar^2} = \left(\frac{m}{-2E}\right)^{\frac{1}{2}} \frac{Z e^2}{\hbar} = n,$$

or

$$E_n = -\frac{1}{2} m \frac{Z^2 e^4}{\hbar^2} \frac{1}{n^2}, \quad n = 1, 2, 3, \ldots \tag{11.5.52}$$

or, expressed in terms of the "fine structure constant" $\alpha = \frac{e^2}{\hbar c}$,

$$E_n = -\frac{1}{2} m c^2 (Z \alpha^2) \frac{1}{n^2}, n = 1, 2, 3, \ldots.$$

This agrees with the (Nobel prize winning) result that Neils Bohr obtained from an application of the somewhat heuristic Bohr-Sommerfeld quantization rules .

From (11.5.48), we have

$$\beta \equiv \beta_n = 2 \frac{mZ e^2}{\hbar^2} \frac{1}{n} = \frac{2Z}{n a_0} \quad \text{where} \quad a_0 = \frac{\hbar^2}{m e^2} \equiv \text{ the Bohr radius.} \tag{11.5.53}$$

Then, reconstructing the solution of the Schrödinger equation and using the Laguerre polynomial normalization given in (11.5.24) , we find that the normalized hydrogenic wave function is

$$\psi_{nlm}(\mathbf{r}) = \left[\left(\frac{2Z}{n a_0}\right)^3 \frac{(n - l - 1)!}{2n(n + l)!}\right]^{\frac{1}{2}} e^{-\frac{\beta_n r}{2}} (\beta_n r)^l \, L_{n-l-1}^{2l+1} (\beta_n r) \, Y_l^m (\theta, \varphi). \tag{11.5.54}$$

Since E_n depends only on n, there is degeneracy with respect to both m (due to the spherical symmetry of the problem) and l (due to a higher, "accidental" symmetry of the wave equation). For each value of n, l can vary from 0 to $n - 1$ and for each of these l values m can vary from $-l$ to l. Thus, the degree of degeneracy is

$$d_n = \sum_{l=0}^{n-1} (2l + 1) = 2 \frac{n(n - 1)}{2} + n = n^2 . \tag{11.5.55}$$

Allowing for spin degeneracy, this means that the nth level (shell) of the hydrogenic atom contains $2 n^2$ states.

12 Non-Homogeneous Boundary Value Problems: Green's Functions

12.1 Ordinary Differential Equations

12.1.1 Definition of a Green's Function

Suppose that we wish to solve the **non-homogeneous** DE

$$\mathfrak{L}u(x) \equiv \frac{d}{dx}\left(p(x)\frac{du}{dx}\right) - q(x)u(x) = f(x) \tag{12.1.1}$$

in the interval $a \leq x \leq b$ with $f(x)$ a known function and with **non-homogeneous** boundary conditions

$$\alpha_1 u(a) + \alpha_2 \frac{du}{dx}\Big|_{x=a} = \alpha_3$$

$$\beta_1 u(b) + \beta_2 \frac{du}{dx}\Big|_{x=b} = \beta_3 \tag{12.1.2}$$

for given values of the constants $\alpha_1, \alpha_2, \alpha_3, \beta_1, \beta_2, and\,\beta_3$.

We recognize that $\mathfrak{L}$ is the Sturm Liouville differential operator and so we know that it satisfies the generalized green's identity. This means that if $u(x)$ and $v(x)$ are any twice- differentiable functions defined on $a \leq x \leq b$,

$$\int_a^b [v(x)\mathfrak{L}u(x) - u(x)\mathfrak{L}v(x)]dx = \left[p(x)\left(v(x)\frac{du}{dx} - u(x)\frac{dv}{dx}\right)\right]_{x=a}^{x=b}. \tag{12.1.3}$$

Let us apply this to functions of our choosing. First we choose $u(x)$ to be the solution of our non-homogeneous boundary value problem. Then, we choose $v(x)$ to be the function $G(x; x')$ where

$$\mathfrak{L}G(x; x') = \delta(x - x'). \tag{12.1.4}$$

With these choices, the identity (12.1.3) gives us

$$\int_a^b \left[G(x; x')f(x) - u(x)\delta(x - x')\right]dx = \left[p(x)\left(G(x; x')\frac{du}{dx} - u(x)\frac{dG}{dx}\right)\right]_{x=a}^{x=b}.$$

The second term on the left hand side is just $-u(x')$. Therefore, we are going to interchange x and $x', x \leftrightarrow x'$, and then rearrange all the terms to yield an expression for $u(x)$:

$$u(x) = \int_a^b G(x'; x)f(x')dx' - \left[p(x')\left(G(x'; x)\frac{d}{dx'}u(x') - u(x')\frac{d}{dx'}G(x'; x)\right)\right]_{x'=a}^{x'=b}. \tag{12.1.5}$$

This is a very auspicious result for it tells us that if we can determine $G(x; x')$, we can simply write down the solution of **any** other non-homogeneous problem involving $\mathcal{L}$. The determination of $G(x; x')$ begins with the specification of boundary conditions to complement our knowledge of its differential equation. The "surface term" in (12.1.5) tells us that we must impose the **homogeneous** counterparts to the boundary conditions imposed on $u(x)$ since otherwise we will be unable to make a full evaluation of that term. For example, if the boundary conditions on $u(x)$ are $u(a) = \alpha$ and $u(b) = \beta$, we have to require $G(a; x') = 0$ and $G(b; x') = 0$ to eliminate the unknown quantities $\frac{du}{dx}\big|_{x=a}$ and $\frac{du}{dx}\big|_{x=b}$. Thus, the **Green's function** $G(x; x')$ is the unique solution of

$$\mathcal{L}G(x; x') = \delta(x - x') \tag{12.1.6}$$

subject to

$$\alpha_1 G(a; x') + \alpha_2 \frac{dG}{dx}\bigg|_{x=a} = 0$$

$$\beta_1 G(b; x') + \beta_2 \frac{dG}{dx}\bigg|_{x=b} = 0. \tag{12.1.7}$$

12.1.2 Direct Construction of the Sturm Liouville Green's Function

The Green's function DE (12.1.7) is as close to a homogeneous equation as a non-homogeneous one can be. In fact, $G(x; x')$ satisfies $\mathcal{L}u(x) = 0$ everywhere except the point $x = x'$. Therefore, we should feel reasonably optimistic about our capacity to solve it. We start by noting some basic attributes of the solution that flow from the properties of the DE and the boundary conditions.

Applying the generalized Green's identity (12.1.3) to $u(x) = G(x; x')$ and $v(x) = G(x; x'')$ we find

$$G(x'; x'') - G(x''; x') = \left[p(x) \left(G(x; x'') \frac{d}{dx} G(x; x') - G(x; x') \frac{d}{dx} G(x; x'') \right) \right]_{x=a}^{x=b} = 0$$

because G satisfies homogeneous boundary conditions. Thus, we conclude that $G(x; x')$ is symmetric under $x \leftrightarrow x'$:

$$G(x; x') = G(x'; x). \tag{12.1.8}$$

As we will see again in Section 12.3.4, the boundary conditions are replaced by an **initial** condition

$$G\left(t; t'\right) = 0 \text{ for } t < t'$$

when the independent variable is time t, $-\infty < t < \infty$. This condition gives expression to the **causality principle**: G is the response of a system to an instantaneous disturbance at $t = t'$ and that response cannot precede its cause. Used in an analogous application of the Green's identity, it changes the **symmetry property** (12.1.8) from $G\left(x; x'\right) = G\left(x'; x\right)$ to $G\left(t; t'\right) = G\left(-t'; -t\right)$.

Next, we note that integrating the DE from $x' - \varepsilon$ to $x' + \varepsilon$ gives us

$$p(x)\frac{d}{dx}G(x;x')\Big|_{x=x'-\varepsilon}^{x=x'+\varepsilon} - \int_{x'-\varepsilon}^{x'+\varepsilon} q(x)G(x;x')dx = 1.$$

Equation (12.1.8) tells us that $G(x;x')$ is symmetric about $x = x'$,

$$G(x' + \varepsilon;x') = G(x';x' + \varepsilon) = G(x' - \varepsilon;x'),$$

and therefore that it is continuous there. So is $q(x)$. This means that when we take the limit as $\varepsilon \to 0$, we will have

$$\lim_{\varepsilon \to 0} \int_{x'-\varepsilon}^{x'+\varepsilon} q(x)G(x;x')dx = 0.$$

Thus, while $G(x;x')$ is continuous at $x = x'$, its first derivative must have a discontinuity there:

$$\lim_{\varepsilon \to 0} \frac{d}{dx}G(x;x')\Big|_{x=x'-\varepsilon}^{x=x'+\varepsilon} = \frac{1}{p(x')}. \tag{12.1.9}$$

We shall now use (12.1.8) and (12.1.9) to construct $G(x;x')$ from non-trivial solutions of the **homogeneous** DE $\mathfrak{L}u(x) = 0$.

Let $u_<(x)$ be such a solution that satisfies one additional constraint, the **homogeneous** boundary condition at $x = a$,

$$\alpha_1 u_<(a) + \alpha_2 \frac{d\,u_<}{dx}\Big|_{x=a} = 0.$$

Since we also have

$$\alpha_1 G(a;x') + \alpha_2 \frac{dG}{dx}\Big|_{x=a} = 0$$

and since these two algebraic equations have to admit a non-trivial solution for at least one of α_1 and α_2, the determinant of their coefficients must vanish. In other words, we require

$$u_<(a) \frac{dg(x;x')}{dx}\Big|_{x=a} - G(a;x') \frac{du_<(x)}{dx}\Big|_{x=a} = 0.$$

But this is just the Wronskian of two solutions of the **same** DE and since it vanishes at a particular point $(x = a)$, it must be identically zero on $a \le x < x'$ and the two solutions must be **linearly dependent** there. Therefore, we can write

$$G(x;x') = c_< u_<(x) \quad \text{for } a \le x < x' \tag{12.1.10}$$

where $c_<$ is some constant.

Similarly, if $u_>(x)$ is a non-trivial solution of $\mathcal{L}u(x) = 0$ that satisfies the **homogeneous** boundary condition

$$\beta_1\, u_>(b) + \beta_2\, \frac{d\,u_>}{dx}\bigg|_{x=b} = 0,$$

then

$$G(x; x') = c_> \, u_>(x) \quad \text{for } x' < x \le b \tag{12.1.11}$$

where $c_>$ is some other constant.

Now we shall impose the continuity of $G(x; x')$ and the discontinuity of $\frac{d}{dx} G(x; x')$ at $x = x'$:

$$c_< \, u_<(x') = c_> \, u_>(x')$$

and

$$c_< \frac{d\,u_<}{dx}\bigg|_{x=x'} - c_> \frac{d\,u_>}{dx}\bigg|_{x=x'} = -\frac{1}{p(x')}.$$

Solving for $c_<$ and $c_>$, we find

$$c_< = \frac{u_>(x')}{p(x')W(x')} \quad \text{and} \quad c_> = \frac{u_<(x')}{p(x')W(x')}$$

where $W(x) = u_<(x)\frac{d\,u_>}{dx} - u_>(x)\frac{d\,u_<}{dx}$ is the Wronskian of $u_<(x)$ and $u_>(x)$. Thus, our final expression for the Green's function is

$$G(x; x') = \begin{cases} \dfrac{u_<(x)\,u_>(x')}{p(x')W(x')} & \text{for } a \le x < x' \\[2mm] \dfrac{u_<(x')\,u_>(x)}{p(x')W(x')} & \text{for } x' < x \le b \end{cases} \tag{12.1.12}$$

This is called the **direct construction** formula for determining $G(x; x')$ and it results in a **closed form** expression that contains no summation or integration to perform. The function $p(x')W(x')$ in the denominator is in fact a constant as can be seen from equation (9.2.6)

$$W(x) = W(x_0)\exp\left\{-\int_{x_0}^{x} a(\xi)d\xi\right\}$$

for the Wronskian of the DE

$$\frac{d^2 u}{d\,x^2} + a(x)\frac{du}{dx} + b(x)u(x) = 0.$$

Converting

$$\frac{d}{dx}\left(p(x)\frac{du}{dx}\right) - q(x)u(x) = 0$$

into this other canonical form, we see that

$$a(\xi) = \frac{1}{p(\xi)}\frac{dp}{d\xi}$$

and so,

$$W(x) = W(x_0)\frac{p(x_0)}{p(x)} \text{ or } p(x)W(x) = a \text{ constant.} \qquad (12.1.13)$$

Equation (12.1.12) provides a simple prescription for determining Sturm Liouville Green's functions and thence for determining the solutions of non-homogeneous Sturm Liouville problems. The latter, obtained by inserting (12.1.12) into (12.1.5), have a form reminiscent of what we found in Chapter 9 from application of the variation of constants approach to solving non-homogeneous ODE's. However, the present approach represents a major advance on variation of constants because it
– incorporates boundary conditions, and
– can be generalized to apply to PDE's in any number of dimensions.

There is a possible complication that can arise: the Wronskian that appears in the denominator of (12.1.12) could be zero leaving our Green's function undefined. However, the lack of mathematical definition does not prevent interpretation. If the Wronskian of $u_<(x)$ and $u_>(x)$ is zero, the two functions have to be proportional to each other and so there must exist a single solution of the homogeneous DE that satisfies **both** boundary conditions. Since the function $q(x)$ can be written $q(x) = r(x) - \lambda\rho(x)$ for some fixed λ and $\rho(x) \geq 0$, the homogeneous DE can be converted into the eigenvalue equation

$$\frac{d}{dx}\left(p(x)\frac{du}{dx}\right) - r(x)u(x) = -\lambda\rho(x)u(x)$$

and we see that the vanishing of the Wronskian implies that λ is an eigenvalue of $\mathfrak{L} \equiv \frac{d}{dx}\left(p(x)\frac{d}{dx}\right) - r(x)$. As we will see in the following example, this in turn implies a physical interpretation in terms of resonant behaviour.

12.1.3 Application: The Bowed Stretched String

Suppose that we have a stretched string that is subjected to a transverse bowing force. If the force per unit length at position x and time t is $F(x, t)$, the transverse displacement of the string is a solution of the non-homogeneous wave equation

$$\frac{\partial^2 \psi}{\partial x^2} - \frac{1}{c^2}\frac{\partial^2 \psi}{\partial t^2} = -\frac{F(x, t)}{T}$$

where T is the tension in the string. Suppose further that the bowing force is harmonic so that $F(x, t) = -Tf(x)e^{-i\omega t}$. The forced response of the string will then be $\psi(x, t) = u(x)e^{-i\omega t}$ where

$$\frac{d^2 u}{dx^2} + k^2 u(x) = f(x), \quad k = \frac{\omega}{c}.$$

If, as we usually do, require the ends of the string to be fixed, this non-homogeneous ODE will be accompanied by the boundary conditions $u(0) = u(L) = 0$.

The Green's function that we need for this problem is the solution of

$$\frac{d^2}{dx^2}G(x;x') + k^2\,G(x;x') = \delta(x-x')$$

subject to $G(0;x') = G(L;x') = 0$. To apply the direct construction formula we require a solution $u_<(x)$ of

$$\frac{d^2u}{dx^2} + k^2\,u(x) = 0$$

that satisfies $u_<(0) = 0$. The general solution of this equation is $u(x) = \left\{\begin{matrix} \cos kx \\ \sin kx \end{matrix}\right\}$
and so a solution that meets our boundary condition is

$$u_<(x) = \sin kx, \quad 0 \le x.$$

A solution $u_>(x)$ that clearly meets the condition at the other boundary, namely $u_>(L) = 0$, is

$$u_>(x) = \sin k(L-x), \quad x \le L.$$

Calculating their Wronskian, we have

$$W(x) = u_<(x)\frac{du_>}{dx} - \frac{du_>}{dx}u_>(x) = -\sin kx\,k\cos k(L-x) - k\cos kx\sin k(L-x)$$

$$= -k\sin k(x+L-x) = -k\sin kL.$$

Therefore, since $p(x) \equiv 1$ in this case, the desired Green's function is

$$G(x;x') = \left\{ \begin{array}{ll} -\dfrac{\sin kx\sin k(L-x')}{k\sin kL} & \text{for } 0 \le x < x' \\[2mm] -\dfrac{\sin kx'\sin k(L-x)}{k\sin kL} & \text{for } x' < x \le L \end{array} \right. \tag{12.1.14}$$

Thus, the solution to our bowed string problem is $\psi(x,t) = u(x)\,e^{-i\omega t}$ where

$$u(x) = \int_0^L G(x;x')f(x')dx'$$

$$= -\frac{\sin k(L-x)}{k\sin kL}\int_0^x \sin kx'f(x')dx' - \frac{\sin kx}{k\sin kL}\int_x^L \sin k(L-x')f(x')dx'.$$

We can use the same Green's function to solve a different type of non-homogeneous string problem. Suppose that there is no bowing force, $f(x) \equiv 0$, but one end of the string is vibrated with frequency ω and amplitude A. This means that we will have boundary conditions $u(0) = 0$ and $u(L) = A$. Equation (12.1.5) still applies as does the Green's function (12.1.14). In fact, inserting the latter along with the new boundary conditions on $u(x)$, (12.1.5) gives us the solution $\psi(x,t) = u(x)\,e^{-i\omega t}$ with

$$u(x) = D\frac{d}{dx'}G(x;x')\Big|_{x'=L} = A\frac{1}{k\sin kL}\sin kx\,k\cos k(L-L) = A\frac{\sin kx}{\sin kL}.$$

Notice that the Wronskian of $u_<(x)$ and $u_>(x)$ provides this particular Green's function with simple poles located at the zeros of $\sin kL$ which means at the values $k = \frac{\omega}{c} = \frac{m\pi}{L}$, $m = 1, 2, \ldots$. As we know and as the last sub-section suggested should happen, these values correspond to the normal modes of vibration of the string. Thus, when we stimulate the string at values of ω approaching one of its natural frequencies, the Green's function and hence the transverse displacement of the string increases without limit. In other words, we produce a **resonant** response or **resonance.**

12.1.4 Eigenfunction Expansions : The Bilinear Formula

The origin of resonant behaviour and even of the Green's function method itself becomes much more transparent when one attempts solution by means of an eigenfunction expansion.

We shall use the modification to the non-homogeneous Sturm-Liouville equation that was introduced at the end of Section 12.1.2. Specifically, we shall write it in the form

$$\mathcal{L}u(x) + \lambda\rho(x)u(x) = f(x) \quad \text{where} \quad \mathcal{L} \equiv \frac{d}{dx}\left(p(x)\frac{d}{dx}\right) - r(x) \qquad (12.1.15)$$

so that the correspondence with (12.1.1) is brought about by the replacement of $q(x)$ in that equation by $r(x) - \lambda\rho(x)$. Here λ is a constant and $\rho(x)$ is the (positive-definite) weight function that is defined by the eigenvalue equation

$$\mathcal{L}u_m(x) = -\lambda_m \rho(x) u_m(x) \qquad (12.1.16)$$

and the (homogeneous) boundary conditions

$$\alpha_1 u_m(a) + \alpha_2 \left.\frac{d\,u_m}{dx}\right|_{x=a} = 0 \quad \text{and} \quad \beta_1 u_m(b) + \beta_2 \left.\frac{d\,u_m}{dx}\right|_{x=b} = 0. \qquad (12.1.17)$$

We know that the eigenfunctions $\{u_m(x)\}$ form a complete orthogonal set for the space of functions that are square integrable with respect to the weight function $\rho(x)$ over the interval $a \le x \le b$. Therefore, assuming the eigenfunctions to be normalized to 1 over $a \le x \le b$, we expand $u(x)$ and $\frac{f(x)}{\rho(x)}$ in terms of them and write

$$u(x) = \sum_{m=1}^{\infty} a_m u_m(x) \quad \text{where} \quad a_m = \int_a^b u_m^*(x')u(x')\rho(x')dx', \qquad (12.1.18)$$

and

$$f(x) = \rho(x)\sum_{m=1}^{\infty} b_m u_m(x) \quad \text{where} \quad b_m = \int_a^b u_m^*(x')f(x')dx'. \qquad (12.1.19)$$

Notice that we are allowing for the possibility that the eigenfunctions are complex. Substituting these expansions into the DE in (12.1.15), we obtain the algebraic equation

$$\rho(x)\sum_{m=1}^{\infty} a_m(\lambda - \lambda_m)\, u_m(x) = \rho(x)\sum_{m=1}^{\infty} b_m\, u_m(x).$$

Because of the orthogonality of the $\{u_m(x)\}$, this implies

$$a_m = \frac{b_m}{\lambda - \lambda_m} \quad \text{for each } m = 1, 2, 3, \ldots.$$

Substituting back into the series in (12.1.18) and using the definition of b_m in (12.1.19), this yields a solution to the non-homogeneous DE, namely

$$u(x) = \sum_{m=1}^{\infty} \frac{u_m(x)}{\lambda - \lambda_m} \int_a^b u_m^*(x')f(x')dx'.$$

Both the integral and the series should be uniformly convergent and so we interchange their order to obtain

$$u(x) = \int_a^b G(x;x')f(x')dx' \quad \text{where} \quad G(x;x') = \sum_{m=1}^{\infty} \frac{u_m(x)\, u_m^*(x')}{\lambda - \lambda_m}. \qquad (12.1.20)$$

This is called the **bilinear formula** for the Green's function.

Note that if $f(x') = \delta(x' - x_0)$, (12.1.20) yields the solution $u(x) = G(x;x_0)$. This verifies that $G(x;x')$ as defined by the bilinear formula is a solution of the Green's function DE

$$\mathcal{L}G(x;x') + \lambda\rho(x)G(x;x') = \delta(x - x').$$

Moreover, since each eigenfunction $u_m(x)$ satisfies the homogeneous boundary conditions (12.1.17), so does $G(x;x')$.

We remarked at the beginning of this sub-section that the origin of resonant behaviour becomes particularly transparent when the bilinear formula is used to construct a Green's function. Indeed, it corresponds to $\lambda = \lambda_m$ for some m and when that happens $G(x;x')$ clearly becomes undefined and there is **no solution** to the original non-homogeneous problem unless, by chance,

$$\int_a^b u_m^*(x')f(x')dx' = 0$$

for that particular value of m.

The normalized eigenfunctions of the bowed string differential operator, $\mathcal{L} \equiv \frac{d^2}{dx^2}$, corresponding to the homogeneous boundary conditions $u(0) = u(L) = 0$, are

$$u_m(x) = \sqrt{\frac{2}{L}} \sin\frac{m\pi x}{L} \quad \text{with} \quad \lambda_m = \frac{m^2 \pi^2}{L^2} \quad m = 1, 2, \ldots.$$

Therefore, substituting into (12.1.20) and using $\lambda = k^2$, we find the bilinear form

$$G(x; x') = \frac{2}{L} \sum_{m=1}^{\infty} \frac{\sin \frac{m\pi x}{L} \sin \frac{m\pi x'}{L}}{k^2 - \frac{m^2 \pi^2}{L^2}} \tag{12.1.21}$$

which exhibits explicitly the poles at $k = \frac{m\pi}{L}$. Evidently, this series is the Fourier sine series expansion of (12.1.14)

$$G(x; x') = \begin{cases} -\dfrac{\sin kx \sin k(L - x')}{k \sin kL} & \text{for } 0 \le x < x' \\[2ex] -\dfrac{\sin kx' \sin k(L - x)}{k \sin kL} & \text{for } x' < x \le L \end{cases}$$

12.1.5 Application: the Infinite Stretched String

In many problems that are amenable to use of the bilinear formula, the eigenvalue spectrum is continuous. To illustrate what happens in such a situation, we shall consider a one-dimensional analogue of acoustic and electromagnetic radiation.

Suppose that we have an infinitely long stretched string that is subjected to a transverse harmonic force per unit length $F(x, t) = -Tf(x) e^{-i\omega_0 t}$ where T is, as usual, the tension in the string. Here, "infinitely long" means long enough that the ends of the string have a negligible effect on the behaviour of points in any neighbourhood of the middle. The PDE satisfied by the transverse displacement is the same as for the finite string,

$$\frac{\partial^2 \psi}{\partial x^2} - \frac{1}{c^2} \frac{\partial^2 \psi}{\partial t^2} = -\frac{1}{T} F(x, t),$$

but the boundary conditions are, of course, quite different. We shall assume that the displacement is everywhere bounded: $|\psi(x, t)| < \infty$ for all $-\infty < x < \infty$.

As with the finite string, we shall seek solutions of the form $\psi(x, t) = u(x) e^{-i\omega_0 t}$ which reduces the problem to one of solving the non-homogeneous ODE

$$\frac{d^2 u}{dx^2} + k_0^2 u(x) = f(x) \quad \text{where} \quad k_0 = \frac{\omega_0}{c}$$

subject to $|u(x)| < \infty$ for all $-\infty < x < \infty$. Therefore, the Green's function for this problem must be the solution of

$$\frac{d^2 G}{dx^2} + k_0^2 G(x; x') = \delta(x - x') \tag{12.1.22}$$

that satisfies the same (homogeneous) boundary conditions. To find it by means of the bilinear formula, we must first solve the eigenvalue problem

$$\frac{d^2}{dx^2} u_\lambda(x) = -\lambda u_\lambda(x) \quad \text{with} \quad |u_\lambda(x)| < \infty \quad \text{for all} \quad -\infty < x < \infty. \tag{12.1.23}$$

The boundedness condition can only be met if $\lambda < 0$. Therefore, we set $\lambda = -k^2$ and obtain the (normalized) eigenfunctions $u_k(x) = \frac{1}{\sqrt{2\pi}} e^{-ikx}$. This means that the bilinear formula for our Green's function is

$$G(x; x') = \frac{1}{2\pi} \int_{-\infty}^{\infty} \frac{e^{ikx'} e^{-ikx}}{k_0^2 - k^2} dk. \tag{12.1.24}$$

Evidently, this is an inverse Fourier transform and so must be the solution of (12.1.22) that we would have obtained had we used Fourier transforms. To confirm this, we set $\mathcal{F}\{G(x; x')\} \equiv g(k; x')$, and transform (12.1.22) to obtain

$$-k^2 g(k; x') + k_0^2 g(k; x') = \frac{1}{\sqrt{2\pi}} \int_{-\infty}^{\infty} e^{ikx} \delta(x - x') dx = \frac{1}{\sqrt{2\pi}} e^{ikx'},$$

or

$$g(k; x') = \frac{1}{\sqrt{2\pi}} \frac{e^{ikx'}}{k_0^2 - k^2}.$$

Thus, since

$$G(x; x') = \frac{1}{\sqrt{2\pi}} \int_{-\infty}^{\infty} e^{-ikx} g(k; x') dk,$$

we do indeed recover the bilinear formula (12.1.24).

We have some experience in evaluating Fourier integrals and so (12.1.24) should lead to a closed form expression for $G(x; x')$ that can be compared to the expression obtainable from the direct construction technique. There is a slight complication however: the integrand has (simple) poles on the real axis at $k = \pm k_0$. This means that we require an additional piece of information that instructs us how to deform the contour of integration to avoid them. What that information may be becomes apparent as soon as we investigate the residues at the poles. The residues at $k = \pm k_0$ are

$$\mp \frac{1}{4\pi} \frac{e^{\pm ik(x'-x)}}{k_0}$$

respectively. The first of these would make a contribution to $G(x; x') e^{-i\omega_0 t}$ that contains the factor $e^{-i[k_0(x-x')+\omega_0 t]}$ while the contribution from the second would contain $e^{i[k_0(x-x')-\omega_0 t]}$. These are waves travelling to the left and to the right, respectively, from the source point $x = x'$. But the role of the Green's function is to give us the the response at a point x due to a disturbance at a point x'. Therefore, if x is to the left of $x', x < x'$, then $G(x; x')$ should **not** include the wave travelling to the right. Consequently, the contour should **not** enclose the pole $k = -k_0$. On the other hand, if x is to the right of $x', x > x'$, the wave travelling to the left must be excluded and so now the contour should **not** enclose $k = k_0$. Thus, the **physical** identity of the Green's function provides a key piece of information. We shall now complement it with the requirements of Jordan's Lemma to come up with a unique prescription for the contour of integration. If $x < x'$, we are obliged to close the contour in the upper half plane and so we avoid the poles by means of small semi-circle closing above $k = -k_0$ and a second semi-circle closing below $k = k_0$. If $x > x'$, the contour is closed in the lower half-plane and so the poles are avoided by exactly the same means.

The evaluation of the Fourier integral is now straightforward: one finds

$$G(x; x') = 2\pi i \left[-\frac{1}{2\pi} \frac{e^{ik(x'-x)}}{k + k_0} \right]_{k=k_0} = -\frac{i}{2 k_0} e^{i k_0 (x'-x)} \quad \text{for } x < x',$$

and

$$G(x; x') = 2\pi i \left[\frac{1}{2\pi} \frac{e^{ik(x'-x)}}{k - k_0} \right]_{k=-k_0} = -\frac{i}{2 k_0} e^{-i k_0 (x'-x)} \quad \text{for } x > x',$$

or

$$G(x; x') = -\frac{i}{2 k_0} e^{i k_0 |x-x'|} \quad \text{for all } -\infty < x < \infty. \tag{12.1.25}$$

Therefore, the solution to the infinite string problem is $\psi(x, t) = u(x) e^{-i \omega_0 t}$ where

$$u(x) = \int_{-\infty}^{\infty} G(x; x') f(x') dx' = -\frac{i}{2 k_0} \left\{ \int_{-\infty}^{x} e^{i k_0 (x-x')} f(x') dx' + \int_{x}^{\infty} e^{i k_0 (x'-x)} f(x') dx' \right\}. \tag{12.1.26}$$

Notice that if we had chosen the time dependence to be $e^{i \omega_0 t}$, we would have had the reverse correspondence between the waves travelling to the left and right. In that case, the Green's function to be used is the complex conjugate of the one in (12.1.25) If the applied force is real, $F(x, t) = -Tf(x) \begin{Bmatrix} \cos \omega_0 t \\ \sin \omega_0 t \end{Bmatrix}, f^*(x) = f(x)$, we can express the transverse displacement in an explicitly real form by setting it equal to the real or imaginary parts, respectively of $u(x) e^{-i \omega_0 t}$ (the real part yielding an even function of t and the imaginary part an odd function of t to match the parity of $\cos \omega_0 t$ and $\sin \omega_0 t$). In the first case, this yields

$$\psi(x, t) = \int_{-\infty}^{\infty} \text{Re} \left\{ e^{-i \omega_0 t} G(x; x') \right\} f(x') dx' = \frac{1}{2 k_0} \int_{-\infty}^{\infty} \sin(k_0 |x - x'| - \omega_0 t) f(x') dx' \tag{12.1.27}$$

and in the second,

$$\psi(x, t) = \int_{-\infty}^{\infty} \text{Im} \left\{ e^{-i \omega_0 t} G(x; x') \right\} f(x') dx' = -\frac{1}{2 k_0} \int_{-\infty}^{\infty} \cos(|x - x'| - \omega_0 t) f(x') dx'. \tag{12.1.28}$$

We shall complete our analysis of the infinite stretched string by constructing the Green's function directly. The relevant homogeneous DE is

$$\frac{d^2 u}{dx^2} + k_0^2\, u(x) = 0.$$

We require a solution $u_<(x)$ which meets the boundary condition that as $x \to -\infty$ $u_<(x)\, e^{-i\,\omega_0\, t}$ is a wave **travelling to the left.** Since the general solution is $\left\{ \begin{array}{c} e^{i\,k_0\, x} \\ e^{-i\,k_0\, x} \end{array} \right\}$, this means that $u_<(x) = e^{-i\,k_0\, x}$. Next, we seek a solution $u_>(x)$ such that as $x \to +\infty$ $u_>(x)\, e^{-i\,\omega_0\, t}$ is a wave **travelling to the right.** The obvious choice is $u_>(x) = e^{i\,k_0\, x}$.

The Wronskian of $u_<(x)$ and $u_>(x)$ is

$$W(x) = u_<(x)\frac{d\,u_>}{dx} - \frac{d\,u_>}{dx}\, u_<(x) = i2\, k_0 \,.$$

Thus, since $p(x) \equiv 1$, equation (12.1.12) yields

$$G(x; x') = -\frac{i}{2\, k_0} \left\{ \begin{array}{ll} e^{i\,k_0(x-x')} & \text{for } -\infty < x < x' \\ e^{i\,k_0(x'-x)} & \text{for } x' < x < \infty \end{array} \right.$$

or,

$$G(x; x') = -\frac{i}{2\, k_0}\, e^{i\,k_0\,|x-x'|} \quad \text{for all } -\infty < x < \infty.$$

in full agreement with (12.1.25).

12.2 Partial Differential Equations

12.2.1 Green's Theorem and Its Consequences

In more than one dimension a non-homogeneous boundary value problem generally involves the solution of a PDE

$$\mathcal{L}\psi(\mathbf{r}) = f(\mathbf{r}) \quad \text{with} \quad \mathcal{L} \equiv \nabla\cdot[p(\mathbf{r})\nabla] + s(\mathbf{r}) \tag{12.2.1}$$

inside a volume V that is bounded by a surface S on which either $\psi(\mathbf{r})$ or $\mathbf{n}\cdot\nabla\psi$ is specified. The partial differential operator $\mathcal{L}$ in (12.2.1) is self-adjoint and satisfies **Green's Theorem** which states that if $u(\mathbf{r})$ and $v(\mathbf{r})$ are any two twice-differentiable functions,

$$\int_V [u(\mathbf{r})\mathcal{L}v(\mathbf{r}) - v(\mathbf{r})\mathcal{L}u(\mathbf{r})]dV = \int_S p(\mathbf{r})[u(\mathbf{r})\nabla v(\mathbf{r}) - v(\mathbf{r})\nabla u(\mathbf{r})] \cdot d\mathbf{S}. \tag{12.2.2}$$

The proof of the theorem follows from a consideration of the integrals

$$\int_S [u(r)p(r)\nabla v(r)] \cdot dS = \int_V \nabla \cdot [u(r)p(r)\nabla v(r)] dV$$

$$= \int_V (\nabla u(r)) \cdot (p(r)\nabla v(r)) dV + \int_V u(r)\nabla \cdot (p(r)\nabla)v(r) dV$$

and

$$\int_S [v(r)p(r)\nabla u(r)] \cdot dS = \int_V \nabla \cdot [v(r)p(r)\nabla u(r)] dV$$

$$= \int_V (\nabla v(r)) \cdot (p(r)\nabla u(r)) dV + \int_V v(r)\nabla \cdot (p(r)\nabla)u(r) dV$$

where we have used the divergence theorem in the first line of both equations. Subtracting the second from the first of these equations, there is a cancellation that gives us

$$\int_S p(r)[u(r)\nabla v(r) - v(r)\nabla u(r)] \cdot dS = \int_V [u(r)\nabla \cdot (p(r)\nabla)v(r) - v(r)\nabla \cdot (p(r)\nabla)u(r)] dV.$$

Adding $u(r)s(r)v(r) - v(r)s(r)u(r)$ to the integrand on the left hand side completes the derivation of (12.2.2).

Let us introduce a Green's function $G(r;r')$ by defining it to be a solution of

$$\mathcal{L}G(r;r') = \delta(r - r') \tag{12.2.3}$$

in V subject to suitable boundary conditions on S. Applying Green's Theorem with $u(r)$ replaced by $\psi(r)$ (the solution of (12.2.1)) and $v(r)$ by $G(r;r')$ and using their respective PDE's, we find

$$\int_V [\psi(r)\mathcal{L}G(r;r') - G(r;r')\mathcal{L}\psi(r)] dV = \int_V [\psi(r)\delta(r - r') - G(r;r')f(r)] dV$$

$$= \psi(r') - \int_V G(r;r')f(r) dV = \int_S p(r)[\psi(r)\nabla G(r;r') - G(r;r')\nabla \psi(r)] \cdot dS.$$

Interchanging r and r' and rearranging terms, the last two lines of this equation become

$$\psi(r) = \int_V G(r';r)f(r') dV' + \int_S p(r')[\psi(r')\nabla' G(r';r) - G(r';r)\nabla' \psi(r')] \cdot dS' \tag{12.2.4}$$

which is the multi-dimensional analogue of equation (12.1.5).

We now choose boundary conditions for $G(r;r')$ that will eliminate unknown quantities from the surface integral on the right hand side of (10.2.4). Normally, there are only two cases to consider.

Case 1 (Dirichlet Boundary Conditions): $\psi(r)$ is given on S.

The obvious choice for the Green's function under this circumstance is the homogeneous condition

$$G(r;r') = 0 \text{ for } r \text{ on } S. \tag{12.2.5}$$

Equation (12.2.4) then becomes

$$\psi(r) = \int_V G(r';r)f(r')dV' + \int_S p(r')\psi(r')\nabla'G(r';r) \cdot dS'. \tag{12.2.6}$$

Case 2 (Neumann Boundary Conditions): $n \cdot \nabla \psi(r)$ is given on S.

The choice here is not quite so obvious. If we apply the divergence theorem to

$$\int_V \mathcal{L}G(r;r')dV = \int_V \delta(r-r')dV = 1,$$

we find

$$\int_S p(r)\nabla G(r;r') \cdot dS + \int_V s(r)G(r;r')dV = 1.$$

This means that if $s(r) \equiv 0$, we cannot require $n \cdot \nabla G(r;r') = 0$ for r on S since that would produce a contradiction. Therefore, we are obliged to recognize two sub-cases: if $s(r) \neq 0$, we make the obvious choice and impose the homogeneous condition

$$n \cdot \nabla G(r;r') = 0 \text{ for } r \text{ on } S \tag{12.2.7}$$

and if $s(r) \equiv 0$, we require the next best thing,

$$n \cdot \nabla G(r;r') = \frac{1}{A_p}, \text{ where } A_p = \int_S p(r)dS, \text{ for } r \text{ on } S. \tag{12.2.8}$$

In the first instance, (12.2.4) becomes

$$\psi(r) = \int_V G(r';r)f(r')dV' - \int_S p(r')G(r';r)\nabla'\psi(r') \cdot dS' \tag{12.2.9}$$

and in the second,

$$\psi(\boldsymbol{r}) = \langle \psi \rangle_S + \int\limits_V G(\boldsymbol{r}';\boldsymbol{r})f(\boldsymbol{r}')dV' - \int\limits_S G(\boldsymbol{r}';\boldsymbol{r})\boldsymbol{\nabla}'\psi(\boldsymbol{r}') \cdot d\boldsymbol{S}' \tag{12.2.10}$$

where $\langle \psi \rangle_S = \frac{1}{A_p} \int\limits_S p(\boldsymbol{r}')\psi(\boldsymbol{r}')dS'$ is the weighted average of $\psi(\boldsymbol{r})$ over the whole surface $\boldsymbol{S}$.

Poisson's equation is an important example of a PDE for which $s(\boldsymbol{r}) \equiv 0$ and whose Green's function must therefore meet the non-homogeneous Neumann boundary condition (12.2.8). In that case, $\langle \psi \rangle_S = \frac{1}{A} \int\limits_S \psi(\boldsymbol{r}')dS'$ where A is the area of the surface S. This means that $\psi(\boldsymbol{r})$ is determined only to within an additive constant by the boundary condition. On the other hand, we know from electromagnetic theory that the definition of zero potential is arbitrary and exercising that arbitrariness, we can set $\langle \psi \rangle_S$ to zero.

Now that we know how to solve for $\psi(\boldsymbol{r})$ in terms of $G(\boldsymbol{r};\boldsymbol{r}')$, it is time to turn our attention to the construction of Green's functions in more than one dimension. We will do so by considering specific PDE's and boundary conditions. But first, we shall deduce a property common to all Green's functions. What is involved is another application of Green's Theorem. Setting $u(\boldsymbol{r}) = G(\boldsymbol{r};\boldsymbol{r}')$ and $v(\boldsymbol{r}) = G(\boldsymbol{r};\boldsymbol{r}'')$ in equation (12.2.2), we note that the surface term vanishes and leaves us with the result $G(\boldsymbol{r}';\boldsymbol{r}'') - G(\boldsymbol{r}'';\boldsymbol{r}') = 0$ or,

$$G(\boldsymbol{r}';\boldsymbol{r}'') = G(\boldsymbol{r}'';\boldsymbol{r}'). \tag{12.2.11}$$

In other words, $G(\boldsymbol{r};\boldsymbol{r}')$ is **symmetric** under $\boldsymbol{r} \leftrightarrow \boldsymbol{r}'$.

12.2.2 Poisson's Equation in Two Dimensions and With Rectangular Symmetry

Suppose that we wish to find the **static** deflection $u(x, y)$ of a rectangular membrane due to an external force. Using $f(x, y)$ to denote the external force per unit area divided by the tension, this will require that we solve

$$\frac{\partial^2 u}{\partial x^2} + \frac{\partial^2 u}{\partial y^2} = f(x, y) \tag{12.2.12}$$

subject to (homogeneous) Dirichlet conditions at the fixed edges which we shall locate at $x = 0, x = a, y = 0$ and $y = b$.

The Green's function for this problem is the solution of

$$\frac{\partial^2 G}{\partial x^2} + \frac{\partial^2 G}{\partial y^2} = \delta(x - x')\delta(y - y') \tag{12.2.13}$$

that satisfies

$$G(0, y; x', y') = G(a, y; x', y') = G(x, 0; x', y') = G(x, b; x', y') = 0. \tag{12.2.14}$$

In general, it is either not possible or not useful to find a closed form expression for a multi-dimensional Green's function when the boundaries are **finite** closed surfaces. However, as in the one-dimensional case, there are two construction methods: the eigenfunction expansion and the direct construction approach. The first of these yields an expression, the bilinear formula, with as many summations or integrations as there are dimensions (or separable differential operators in $\mathfrak{L}$) . The second eliminates one of these summations by making use of our ability to directly construct closed form expressions for one dimensional Green's functions.

Eigenfunction Expansion Method:

Let $u_\lambda(x, y)$ denote the normalized eigenfunctions of ∇^2 that satisfy homogeneous Dirichlet conditions at $x = 0, x = a, y = 0$ and $y = b$.

In other words, let

$$\nabla^2 u_\lambda(x, y) = -\lambda u_\lambda(x, y) \text{ with } \| u_\lambda(x, y) \| = 1$$
$$\text{and } u_\lambda(0, y) = u_\lambda(a, y) = u_\lambda(x, 0) = u_\lambda(x, b) = 0. \tag{12.2.15}$$

Since their closure relation must be

$$\sum_\lambda u_\lambda(x, y) u_\lambda^*(x', y') = \delta(x - x')\delta(y - y'), \tag{12.2.16}$$

it is clear that the bilinear formula of Section 12.1.3 applies here too and yields

$$G(x, y; x', y') = \sum_\lambda \frac{u_\lambda(x, y) u_\lambda^*(x', y')}{-\lambda}. \tag{12.2.17}$$

We know already (from Section 10.8) that the eigenvalues for the rectangular membrane are $\lambda_{m,n} = \frac{m^2 \pi^2}{a^2} + \frac{n^2 \pi^2}{b^2}, m, n = 1, 2, 3, \ldots$ corresponding to the (normalized) eigenfunctions $u_{m,n}(x, y) = \frac{2}{\sqrt{ab}} \sin \frac{m\pi x}{a} \sin \frac{n\pi y}{b}$. Therefore, the bilinear formula representation of our green's function is

$$G(x, y; x', y') = -\frac{4}{ab} \sum_{m=1}^{\infty} \sum_{n=1}^{\infty} \frac{\sin \frac{m\pi x}{a} \sin \frac{n\pi y}{b} \sin \frac{m\pi x'}{a} \sin \frac{n\pi y'}{b}}{\frac{m^2 \pi^2}{a^2} + \frac{n^2 \pi^2}{b^2}}. \tag{12.2.18}$$

Direct Construction Method:

We start with a partial eigenfunction expansion. Choosing to do so in the y variable, we write

$$G(x, y; x', y') = \sqrt{\frac{2}{b}} \sum_{n=1}^{\infty} G_n(x; x', y') \sin \frac{n\pi y}{b} \tag{12.2.19}$$

and invoke closure for the normalized eigenfunctions $\sqrt{\frac{2}{b}}\sin\frac{n\pi y}{b}$,

$$\delta(y-y') = \frac{2}{b}\sum_{n=1}^{\infty}\sin\frac{n\pi y}{b}\sin\frac{n\pi y'}{b}.$$

Substituting these two expansions into the Green's function PDE and equating on a term by term basis, we conclude that $G_n(x;x',y')$ factors according to

$$G_n(x;x',y') = \sqrt{\frac{2}{b}}\sin\frac{n\pi y'}{b}g_n(x;x') \text{ where } \frac{d^2 g_n}{dx^2} - \frac{n^2\pi^2}{b^2}g_n(x;x') = \delta(x-x').$$
(12.2.20)

This is a one-dimensional Green's function DE which we can solve by the direct construction method.

Remembering that we have homogeneous boundary conditions, $g_n(0;x') = g_n(a;x') = 0$, we proceed by seeking a solution $u_<(x)$ of the homogeneous DE

$$\frac{d^2 u}{dx^2} - \frac{n^2\pi^2}{b^2}u(x) = 0$$

that satisfies $u_<(0) = 0$. Since the general solution is $u(x) = \left\{\begin{array}{c}\cosh\frac{n\pi x}{b}\\ \sinh\frac{n\pi x}{b}\end{array}\right\}$, the simplest choice is $u_<(x) = \sinh\frac{n\pi x}{b}$.

Next, we need a solution $u_>(x)$ of the homogeneous DE that satisfies $u_>(a) = 0$. Again, the simplest choice is pretty obvious: $u_>(x) = \sinh\frac{n\pi}{b}(x-a)$.

The Wronskian of $u_<(x)$ and $u_>(x)$ is

$$W(x) = u_<(x)u_>'(x) - u_>(x)u_<'(x) = \frac{n\pi}{b}\sinh\frac{n\pi a}{b}.$$

Therefore, using (12.1.12) we have

$$g_n(x;x') = \left\{\begin{array}{l} -\dfrac{b}{n\pi}\dfrac{\sinh\frac{n\pi x}{b}\sinh\frac{n\pi}{b}(a-x')}{\sinh\frac{n\pi a}{b}},\quad 0\le x < x'\\[3mm] -\dfrac{b}{n\pi}\dfrac{\sinh\frac{n\pi}{b}(a-x)\sinh\frac{n\pi x'}{b}}{\sinh\frac{n\pi a}{b}},\quad x' < x \le a\end{array}\right.$$
(12.2.21)

Substituting back into the expansion (12.2.19) for $G(x,y;x',y')$ we conclude that

$$G(x,y;x',y') = \left\{\begin{array}{l} -\sum_{n=1}^{\infty}\dfrac{2}{n\pi}\dfrac{\sinh\frac{n\pi x}{b}\sinh\frac{n\pi}{b}(a-x')}{\sinh\frac{n\pi a}{b}}\sin\frac{n\pi y}{b}\sin\frac{n\pi y'}{b}\\[3mm] -\sum_{n=1}^{\infty}\dfrac{2}{n\pi}\dfrac{\sinh\frac{n\pi}{b}(a-x)\sinh\frac{n\pi x'}{b}}{\sinh\frac{n\pi a}{b}}\sin\frac{n\pi y}{b}\sin\frac{n\pi y'}{b}\end{array}\right.$$
(12.2.22)

for $\left\{\begin{array}{l}0\le x < x'\\ x' < x \le a\end{array}\right.$, respectively. This is equivalent to the **double** Fourier representation (12.2.18) found by application of the bilinear formula but with the sum over m actually performed.

Had our starting point been substitution of

$$\delta(x - x') = \frac{2}{a} \sum_{m=1}^{\infty} \sin \frac{m\pi x}{a} \sin \frac{m\pi x'}{a},$$

and

$$G(x, y; x', y') = \sqrt{\frac{2}{a}} \sum_{m=1}^{\infty} G_m(y; x', y') \sin \frac{m\pi x}{a}$$

into the PDE (12.2.13), we would have found $G_m(y; x', y') = g_m(y; y') \sin \frac{m\pi x'}{a}$ where

$$g_m(y; y') = \begin{cases} -\dfrac{a}{m\pi} \dfrac{\sinh \frac{m\pi y}{a} \sinh \frac{m\pi}{a}(b - y')}{\sinh \frac{m\pi b}{a}}, & 0 \le y < y' \\[3mm] -\dfrac{a}{m\pi} \dfrac{\sinh \frac{m\pi}{a}(b - y) \sinh \frac{m\pi y'}{a}}{\sinh \frac{m\pi b}{a}}, & y' < y \le b \end{cases} \qquad (12.2.23)$$

This, of course, yields an expression for $G(x, y; x', y')$ that is equivalent to summing over n in the double Fourier series (12.2.18). It could have been obtained directly from (12.2.22) by invoking the problem's symmetry under $x \leftrightarrow y$, $(x' \leftrightarrow y')$, and $a \leftrightarrow b$.

12.2.3 Potential Problems in Three Dimensions and the Method of Images

Coulomb's Law is an implicit expression of the solution of the Green's function PDE

$$\nabla^2 G(r; r') = \delta(r - r') \qquad (12.2.24)$$

plus the (Dirichlet) boundary condition $\lim_{r \to \infty} G(r; r') = 0$. It tells us that the potential due to unit charge located at the point $r = r'$ is

$$\psi(r) = -\frac{1}{4\pi \varepsilon_0} \frac{1}{|r - r'|}$$

and, since the charge density associated with the charge is $\rho(r) = \delta(r - r')$, this means that

$$\nabla^2 \psi(r) = -\frac{\rho(r)}{\varepsilon_0} = -\frac{1}{\varepsilon_0} \delta(r - r').$$

Comparing this with (12.2.24), we deduce that

$$G(r; r') = -\frac{1}{4\pi} \frac{1}{|r - r'|}. \qquad (12.2.25)$$

This is confirmed by direct integration of (12.2.24). Integrating over a spherical volume centred at $r = r'$ and applying the divergence theorem, we find

$$\int_S \nabla G(r; r') \cdot dS = 1$$

where S denotes the boundary surface of the sphere. But the normal gradient on a spherical surface is just the partial derivative with respect to the radial coordinate. Therefore, we can rewrite this last equation as

$$\int_S \frac{\partial G}{\partial r} dS = \int_0^{2\pi} \int_0^\pi \frac{\partial G}{\partial r} r^2 \sin\theta d\theta d\varphi = 1$$

where $r = |\boldsymbol{r} - \boldsymbol{r}'|$. With the origin of coordinates at $\boldsymbol{r}=\boldsymbol{r}'$, the delta function in the PDE and the boundary condition to be imposed on its solution both depend only on the radial variable and so the same must be true of the solution itself. Thus, the integration can be performed to give us

$$\frac{dG}{dr} = \frac{1}{4\pi r^2} = \frac{1}{4\pi} \frac{1}{|\boldsymbol{r} - \boldsymbol{r}'|^2}.$$

Integrating once more and using the boundary condition to dispose of the integration constant, we obtain, as expected,

$$G(\boldsymbol{r};\boldsymbol{r}') = -\frac{1}{4\pi} \frac{1}{|\boldsymbol{r} - \boldsymbol{r}'|}.$$

This result can also be obtained by use of Fourier transforms as will be demonstrated in Section 12.3.3.

If the Dirichlet condition is imposed on a finite surface, we can still solve for the Green's function in closed form by using a trick that is called, in electrostatic theory, the **method of images**. Any Green's function can be set equal to a superposition of solutions of the non-homogeneous and homogeneous PDE's. In the present case, this means that we can set

$$G(\boldsymbol{r};\boldsymbol{r}') = G_s (\boldsymbol{r};\boldsymbol{r}') + G_o (\boldsymbol{r};\boldsymbol{r}') \quad \text{where} \quad G_s(\boldsymbol{r};\boldsymbol{r}') = -\frac{1}{4\pi} \frac{1}{|\boldsymbol{r} - \boldsymbol{r}'|} \tag{12.2.26}$$

and

$$\nabla^2 G_o(\boldsymbol{r};\boldsymbol{r}') = 0 \quad \text{with} \quad [G_s(\boldsymbol{r};\boldsymbol{r}') + G_o (\boldsymbol{r};\boldsymbol{r}')]|_{r \text{ on surface}} = 0. \tag{12.2.27}$$

For specificity, let us take the boundary surface to be a sphere of radius R. If $\boldsymbol{r}$ is constrained to vary **inside** the spherical volume, $r < R$, and $\boldsymbol{r}''$ is a point **outside**, $r'' > R$, the delta function $\delta(\boldsymbol{r} - \boldsymbol{r}'')$ is zero and $\frac{1}{|\boldsymbol{r}-\boldsymbol{r}''|}$ is a solution of the **homogeneous** PDE there. This means we can set

$$G_o(\boldsymbol{r};\boldsymbol{r}') = -\frac{1}{4\pi} \frac{k}{|\boldsymbol{r} - \boldsymbol{r}''|}, \quad r \le R \text{ and } r'' > R,$$

where k is a constant and $\boldsymbol{r}''$ is chosen to lie along the same radius vector as $\boldsymbol{r}'$. The values of k and r'' are to be determined by imposing the boundary condition (12.2.27).

Since $\boldsymbol{r}'' = \frac{r''}{r'}\boldsymbol{r}'$, our Green's function is

$$G(\boldsymbol{r}; \boldsymbol{r}') = G_s(\boldsymbol{r}; \boldsymbol{r}') + G_0(\boldsymbol{r}; \boldsymbol{r}') = -\frac{1}{4\pi} \left[\frac{1}{|\boldsymbol{r} - \boldsymbol{r}'|} + \frac{k}{|\boldsymbol{r} - \frac{r''}{r'}\boldsymbol{r}'|} \right].$$

This vanishes at $r = R$ if

$$r'' = \frac{R^2}{r'} \quad \text{and} \quad k = -\frac{r''}{R}$$

which gives us the unique solution

$$G(r; r') = -\frac{1}{4\pi} \left[\frac{1}{|r - r'|} + \frac{R}{r'} \frac{1}{\left| r - \frac{rR^2}{r'^2} r' \right|} \right] , r \text{ and } r' \le R. \tag{12.2.28}$$

Visual inspection of this result reveals why its construction is called the method of images. We know that $G_s(r; r')$ has the physical significance of the potential due to an isolated point charge located at $r = r'$. Similarly, the Green's function we are trying to find is the potential due to that same charge when it is enclosed within a conducting sphere of radius R. What we have found is that the effect of the sphere is the same as that of adding an "image" point charge located at the **inverse** of $r = r'$ with respect to the spherical surface.

This approach works well for any simple boundary surface, a plane and a cylinder of infinite length being two other examples. However, finding image points is a challenge with more complicated surfaces and is generally not worth the effort. Even (12.2.28) is difficult to work with in the context of finding the potential due to a continuous charge distribution via (12.2.6). A more fruitful approach is to proceed as we did in Section 12.2.2 and expand the Green's function in series (or integrals).

12.2.4 Expansion of the Dirichlet Green's Function for Poisson's Equation When There Is Spherical Symmetry

Suppose that we have Dirichlet conditions imposed on a surface consisting of two concentric spheres of radii $r = a$ and $r = b$, $b > a$. This means that we wish to solve

$$\nabla^2 G(r; r') = \delta(r - r') \quad \text{subject to} \quad G(r; r')\Big|_{r=a} = G(r; r')\Big|_{r=b} = 0. \tag{12.2.29}$$

In spherical coordinates, the delta function in the PDE can be expanded according to

$$\delta(r - r') = \frac{1}{r^2} \delta(r - r')\delta(\cos\theta - \cos\theta')\delta(\varphi - \varphi')$$

$$= \frac{1}{r^2} \delta(r - r') \sum_{l=0}^{\infty} \sum_{m=-l}^{l} (Y_l^m(\theta', \varphi'))^* \, Y_l^m(\theta, \varphi) \tag{12.2.30}$$

where we have used the closure relation for spherical harmonics. Similarly, we can write

$$G(r; r') = \sum_{l=0}^{\infty} \sum_{m=-l}^{l} G_{lm}(r; r') \, Y_l^m(\theta, \varphi) \tag{12.2.31}$$

and substitute both expansions into the PDE of (12.2.29) to obtain

$$\sum_{l=0}^{\infty} \sum_{m=-l}^{l} \left\{ \frac{1}{r} \frac{d^2}{dr^2} (r\, G_{lm}(r; r')) - \frac{l(l+1)}{r^2} G_{lm}(r;r') \right\} Y_l^m(\theta, \varphi)$$

$$= \frac{1}{r^2} \delta(r - r') \sum_{l=0}^{\infty} \sum_{m=-l}^{l} (Y_l^m(\theta', \varphi'))^* \, Y_l^m(\theta, \varphi). \qquad (12.2.32)$$

Invoking the orthogonality of the spherical harmonics to set up equations on a term by term basis, we deduce that $G_{lm}(r; r')$ must factor according to

$$G_{lm}(r; r') = g_l(r; r') \, Y_l^m(\theta', \varphi') \qquad (12.2.33)$$

where

$$r^2 \frac{d^2}{dr^2} g_l(r; r') + 2r \frac{d}{dr} g_l(r; r') - l(l+1) g_l(r; r') = \delta(r - r'). \qquad (12.2.34)$$

The homogeneous counterpart of (12.2.34) is Euler's equation

$$r^2 \frac{d^2 u}{dr^2} + 2r \frac{du}{dr} - l(l+1)u(r) = 0$$

which has the general solution $\left\{ \begin{matrix} r^l \\ r^{-l-1} \end{matrix} \right\}$. Therefore, a solution $u_<(r)$ that satisfies the boundary condition $u_<(a) = 0$ is

$$u_<(r) = \left(r^l - \frac{a^{2l+1}}{r^{l+1}} \right), \quad a \le r$$

while one that satisfies $u_>(b) = 0$ is

$$u_>(r) = \left(\frac{1}{r^{l+1}} - \frac{r^l}{b^{2l+1}} \right), \quad r \le b.$$

The Wronskian of $u_<(r)$ and $u_>(r)$ is

$$W(r) = u_<(r) u_>'(r) - u_>(r) u_<'(r) = \frac{2l+1}{r^2} \left[\left(\frac{a}{b} \right)^{2l+1} - 1 \right]$$

and the Sturm-Liouville function $p(r)$ is r^2. Thus, from the direct construction formula (12.1.12) for one-dimensional Green's functions, we have

$$g_l(r;r') = \frac{(-1)}{(2l+1)\left[1 - \left(\frac{a}{b}\right)^{2l+1}\right]} \begin{cases} \left(r^l - \dfrac{a^{2l+1}}{r^{l+1}} \right) \left(\dfrac{1}{r'^{l+1}} - \dfrac{r'^l}{b^{2l+1}} \right), & a \le r < r' \\[4mm] \left(\dfrac{1}{r^{l+1}} - \dfrac{r^l}{b^{2l+1}} \right) \left(r'^l - \dfrac{a^{2l+1}}{r'^{l+1}} \right), & r' < r \le b \end{cases}$$

or,

$$g_l(r;r') = \frac{(-1)}{(2l+1)\left[1 - \left(\frac{a}{b}\right)^{2l+1}\right]} \left(r_<^l - \frac{a^{2l+1}}{r_<^{l+1}} \right) \left(\frac{1}{r_>^{l+1}} - \frac{r_>^l}{b^{2l+1}} \right) \qquad (12.2.35)$$

where $r_< \equiv$ the smaller of r and r' and $r_> \equiv$ the larger of r and r'. Substituting this into (12.2.33) and the latter into (12.2.31) gives us a final expression for the Poisson equation Green's function for a spherical shell bounded by $r = a$ and $r = b$:

$$G(r;r') = -\sum_{l=0}^{\infty}\sum_{m=-l}^{l} \frac{Y_l^m(\theta,\varphi)(Y_l^m(\theta',\varphi'))^*}{(2l+1)\left[1-\left(\frac{a}{b}\right)^{2l+1}\right]} \left(r_<^l - \frac{a^{2l+1}}{r_<^{l+1}}\right)\left(\frac{1}{r_>^{l+1}}\right). \qquad (12.2.36)$$

There are three **special cases:**
1. if $a = 0$ and $b \to \infty$, we have

$$G(r;r') = \sum_{l=0}^{\infty}\sum_{m=-l}^{l} \frac{(-1)}{2l+1} \frac{r_<^l}{r_>^{l+1}} Y_l^m(\theta,\varphi)(Y_l^m(\theta',\varphi'))^* = -\frac{1}{4\pi}\frac{1}{|r-r'|} \qquad (12.2.37)$$

where the last equality was derived earlier as an application of the addition theorem of spherical harmonics;
2. if a remains finite and $b \to \infty$, we have an **exterior** problem and the Green's function is

$$G(r;r') = \sum_{l=0}^{\infty}\sum_{m=-l}^{l} \frac{(-1)}{2l+1} \frac{1}{r_>^{l+1}} \left(r_<^l - \frac{a^{2l+1}}{r_<^{l+1}}\right) Y_l^m(\theta,\varphi)(Y_l^m(\theta',\varphi'))^*; \qquad (12.2.38)$$

3. if $a = 0$ and b remains finite, we have an **interior** problem and the Green's function is

$$G(r;r') = \sum_{l=0}^{\infty}\sum_{m=-l}^{l} \frac{(-1)}{2l+1} r_<^l \left(\frac{1}{r_>^{l+1}}\right) Y_l^m(\theta,\varphi)(Y_l^m(\theta',\varphi'))^*. \qquad (12.2.39)$$

12.2.5 Applications

Solution of Laplace's Equation
We know already that the potential inside a sphere of radius b with no charges present but subject to $\psi(b,\theta,\varphi) = V(\theta,\varphi)$ is

$$\psi(r,\theta,\varphi) = \sum_{l=0}^{\infty}\sum_{m=-l}^{l} c_{lm}\, r^l\, Y_l^m(\theta,\varphi)$$

with

$$c_{lm} = \frac{1}{b^l} \int_0^{2\pi}\int_{-1}^{1} (Y_l^m(\theta',\varphi'))^* V(\theta',\varphi')\, d(\cos\theta')\, d\varphi'.$$

What we wish to verify now is that the Green's function we have just derived yields exactly the same solution. To do so, we require the normal gradient

$$n \cdot \nabla G(r;r')\Big|_{r=b} = \frac{\partial}{\partial r} G(r;r')\Big|_{r=b}.$$

From (12.2.39) we have

$$\frac{\partial}{\partial r} G(\mathbf{r}; \mathbf{r}')\bigg|_{r=b} = \frac{\partial}{\partial r_>} G(\mathbf{r}; \mathbf{r}')\bigg|_{r_>=b} = \frac{1}{b^2} \sum_{l=0}^{\infty} \sum_{m=-l}^{l} \left(\frac{r}{b}\right)^l (Y_l^m(\theta', \varphi'))^* \, Y_l^m(\theta, \varphi).$$

Thus, since $\rho(\mathbf{r}) \equiv 0$ and $dS' = b^2 \, d(\cos\theta')d\varphi'$, we obtain from equation (12.2.9)

$$\psi(\mathbf{r}) = -\frac{1}{\varepsilon_0} \int_V G(\mathbf{r}; \mathbf{r}')\rho(\mathbf{r}')dV' + \int_S \psi(\mathbf{r}')\mathbf{n} \cdot \nabla' G(\mathbf{r}; \mathbf{r}')dS'$$

$$= \sum_{l=0}^{\infty} \sum_{m=-l}^{l} \left[\int_0^{2\pi} \int_{-1}^{1} V(\theta', \varphi')(Y_l^m(\theta', \varphi'))^* \, d(\cos\theta')d\varphi' \right] \left(\frac{r}{b}\right)^l Y_l^m(\theta, \varphi)$$

which is identical to our earlier result as required.

Note that if $V(\theta, \varphi)$ is independent of φ (that is, if we have azimuthal symmetry), only the $m = 0$ terms are retained in $G(\mathbf{r}; \mathbf{r}')$. Thus, since

$$Y_l^m(\theta, \varphi)(Y_l^m(\theta', \varphi'))^* \rightarrow \frac{2l+1}{4\pi} P_l(\cos\theta) \, P_l(\cos\theta') \quad \text{for } m = 0,$$

the electrostatic potential becomes

$$\psi(r, \theta) = \sum_{l=0}^{\infty} \left[\frac{2l+1}{2} \int_{-1}^{1} V(\theta') P_l(\cos\theta')d(\cos\theta') \right] \left(\frac{r}{b}\right)^l P_l(\cos\theta).$$

Solution of Poisson's Equation

Consider a hollow, grounded sphere of radius b with a concentric ring of charge, of radius $c < b$ and total charge Q, inside it. Taking the ring to lie in the xy-plane, we can assert that the charge density inside the sphere will be independent of φ and have a delta function dependence on r and θ. Thus, we can write

$$\rho(\mathbf{r}) = A\delta(r - c)\delta(\cos\theta) \quad \text{where A is a constant of proportionality.}$$

The constant A can be determined from the normalizing condition

$$Q = \int_0^b \int_0^\pi \int_0^{2\pi} \rho(\mathbf{r}) \, r^2 \, dr \sin\theta d\theta d\varphi.$$

Thus, $A = \frac{Q}{2\pi c^2}$. This means that the electrostatic potential inside the sphere is the solution of

$$\nabla^2 \psi(\mathbf{r}) = -\frac{Q}{2\pi c^2 \, \varepsilon_0} \delta(r - c)\delta(\cos\theta)$$

that satisfies $\psi(b, \theta, \varphi) = 0$.

Using the interior problem Green's function (12.2.39) modified for azimuthal symmetry, we have

$$\psi(r) = \int_0^{2\pi} \int_{-1}^{1} \int_0^{b} \left[-\frac{Q}{2\pi c^2 \varepsilon_0} \delta(r' - c)\delta(\cos\theta') \right] G(r;r')r'^2 \, dr' d(\cos\theta')d\varphi'$$

and so,

$$\psi(r) = \sum_{l=0}^{\infty} \frac{P_l(\cos\theta)}{4\pi} \int_0^{2\pi} \int_{-1}^{1} \int_0^{b} \frac{Q}{2\pi c^2 \varepsilon_0} \delta(r' - c)\delta(\cos\theta') \times r_<^l \left(\frac{1}{r_>^{l+1}}\right) r'^2 \, dr' d(\cos\theta')d\varphi'$$

or,

$$\psi(r) = \frac{Q}{4\pi \varepsilon_0} \sum_{l=0}^{\infty} P_l(0) r_<^l \left(\frac{1}{r_>^{l+1}}\right) P_l(\cos\theta)$$

where $r_<(r_>)$ is now the smaller (larger) of r and c. Making use of

$$P_{2l+1}(0) = 0 \quad \text{and} \quad P_{2l}(0) = \frac{(-1)^l(2l)!}{2^{2l}(l!)^2},$$

this becomes

$$\psi(r) = \frac{Q}{4\pi \varepsilon_0} \sum_{l=0}^{\infty} \frac{(-1)^l(2l)!}{2^{2l}(l!)^2} r_<^{2l} \left(\frac{1}{r_>^{2l+1}}\right) P_{2l}(\cos\theta).$$

Notice that for $b \to \infty$ and $r > c$, this reduces to

$$\psi(r) = \frac{Q}{4\pi \varepsilon_0} \sum_{l=0}^{\infty} \frac{(-1)^l(2l)!}{2^{2l}(l!)^2} \frac{c^{2l}}{r^{2l+1}} P_{2l}(\cos\theta)$$

which for a point on the z-axis converges to the well-known consequence of Coulomb's law,

$$\psi(z, 0) = \frac{Q}{4\pi \varepsilon_0} \frac{1}{\sqrt{z^2 + c^2}}.$$

12.3 The Non-Homogeneous Wave and Diffusion Equations

12.3.1 The Non-Homogeneous Helmholtz Equation

The non-homogeneous wave equation is

$$\nabla^2 \psi(r, t) - \frac{1}{c^2} \frac{\partial^2 \psi(r, t)}{\partial t^2} = \sigma(r, t). \tag{12.3.1}$$

As in the bowed string problem, if the source is monochromatic and harmonic, that is if $\sigma(r, t) = \sigma(r) e^{-i\omega t}$, we can assume the same time dependence for $\psi(r, t)$,

$\psi(\boldsymbol{r}, t) = \psi(\boldsymbol{r})\, e^{-i\omega t}$. This means that the problem is effectively **time-independent** and so no initial conditions are needed. We proceed by substitution into the original PDE which yields

$$(\nabla^2 + k^2)\psi(\boldsymbol{r}) = \sigma(\boldsymbol{r}), \quad k = \omega/c. \tag{12.3.2}$$

From equation (12.2.4) we know that the solution of the non-homogeneous Helmholtz equation is

$$\psi(\boldsymbol{r}) = \int_V G(\boldsymbol{r}';\boldsymbol{r})\sigma(\boldsymbol{r}')dV' + \int_S [\psi(\boldsymbol{r}')\boldsymbol{n}\cdot\nabla' G(\boldsymbol{r}';\boldsymbol{r}) - G(\boldsymbol{r}';\boldsymbol{r})\boldsymbol{n}\cdot\nabla'\psi(\boldsymbol{r}')]dS' \tag{12.3.3}$$

where

$$(\nabla^2 + k^2)G(\boldsymbol{r};\boldsymbol{r}') = \delta(\boldsymbol{r} - \boldsymbol{r}') \tag{12.3.4}$$

and $G(\boldsymbol{r};\boldsymbol{r}')$ is subject to the homogeneous counterparts of whatever boundary conditions are imposed on $\psi(\boldsymbol{r})$. One expression for $G(\boldsymbol{r};\boldsymbol{r}')$ is provided by the bilinear formula,

$$G(\boldsymbol{r};\boldsymbol{r}') = \sum_n \frac{u_n(\boldsymbol{r})(u_n(\boldsymbol{r}'))^*}{k^2 - k_n^2} \tag{12.3.5}$$

where the functions $u_n(\boldsymbol{r})$ are the normalized **normal modes** defined by $(\nabla^2 + k^2)u_n(\boldsymbol{r}) = 0$ with boundary conditions $u_n(\boldsymbol{r}) = 0$ or $\boldsymbol{n}\cdot\nabla u_n(\boldsymbol{r}) = 0$ for $\boldsymbol{r}$ on S. Another is provided by the same kind of partial expansion and **direct construction** technique that we used for Poisson's equation.

12.3.2 The Forced Drumhead

If the external force per unit area applied normal to a drumhead is $F(\boldsymbol{r}, t)$, its transverse displacement will obey the equation

$$\nabla^2 \psi(\boldsymbol{r}, t) - \frac{1}{c^2}\frac{\partial^2}{\partial t^2}\psi(\boldsymbol{r}, t) = -\frac{1}{c^2 \mu}F(\boldsymbol{r}, t) \equiv f(\boldsymbol{r}, t) \tag{12.3.6}$$

where, as before, μ is the mass per unit area. We assume that $f(\boldsymbol{r}, t) = f(\boldsymbol{r})\, e^{-i\omega t}$ and set the forced displacement of the drumhead equal to $\psi(\boldsymbol{r}, t) = \psi(\boldsymbol{r})\, e^{-i\omega t}$. Substituting into (10.3.5), we get the two dimensional non-homogeneous Helmholtz equation

$$(\nabla^2 + k^2)\psi(\boldsymbol{r}) = f(\boldsymbol{r}), \quad k = \frac{\omega}{c}.$$

This is to be solved subject to the (Dirichlet) boundary condition $\psi(\boldsymbol{r}) = 0$ for $\boldsymbol{r}$ on the perimeter of the drumhead. Thus, if the drumhead is circular with radius a, we have the condition $\psi(a, \theta) = 0$.

With both $\psi(r)$ and $G(r;r')$ satisfying homogeneous Dirichlet conditions, the solution is provided by the integral

$$\psi(r) = \int\limits_0^{2\pi} \int\limits_0^a G(r;r')f(r')r'\,dr'\,d\theta'.$$

Thus, all we need to do is construct the Green's function.

Trying the direct construction method first, we expand $G(r;r')$ in a Fourier series in θ:

$$G(r;r') = \sum_{m=-\infty}^{\infty} G_m(r;r',\theta')\,e^{im\theta}. \tag{12.3.7}$$

Next, we use closure to expand the delta function in θ:

$$\delta(r - r') = \frac{1}{r}\delta(r - r')\delta(\theta - \theta') = \frac{1}{r}\delta(r - r')\frac{1}{2\pi}\sum_{m=-\infty}^{\infty} e^{im(\theta-\theta')}. \tag{12.3.8}$$

Substituting these expansions into (12.3.4) expressed in terms of two dimensional polar coordinates and using the orthogonality of the Fourier functions, we find that the coefficients $G_m(r;r',\theta')$ factor according to

$$G_m(r;r',\theta') = \frac{1}{2\pi}\,e^{-im\theta'}\,g_m(r;r') \tag{12.3.9}$$

where

$$\frac{d^2}{dr^2}g_m(r;r') + \frac{1}{r}\frac{d}{dr}g_m(r;r') + \left(k^2 - \frac{m^2}{r^2}\right)g_m(r;r') = \frac{1}{r}\delta(r - r'). \tag{12.3.10}$$

The homogeneous version of this is Bessel's equation with general solution $u(r) = \begin{Bmatrix} J_{|m|}(kr) \\ N_{|m|}(kr) \end{Bmatrix}$. A solution that satisfies the boundary condition $|u_<(0)| < \infty$ is $u_<(r) = J_{|m|}(kr), 0 \le r$, while one that satisfies $u_>(a) = 0$ is $u_>(r) = N_{|m|}(ka)J_{|m|}(kr) - J_{|m|}(ka)N_{|m|}(kr), r \le a$. Their Wronskian is

$$W(r) = -kJ_{|m|}(ka)[J_{|m|}(kr)N'_{|m|}(kr) - N_{|m|}(kr)J'_{|m|}(kr)].$$

We can show from their small x behaviour that the Wronskian of the Bessel and Neumann functions is

$$J_m(x)N'_m(x) - J'_m(x)N_m(x) = \frac{2}{\pi x}.$$

Therefore, $W(r) = -\frac{2}{\pi r}J_{|m|}(ka)$ which with $p(r) = r$ and equation (12.1.12) gives us

$$g_m(r;r') = -\frac{\pi}{2}\frac{1}{J_{|m|}(ka)}\begin{cases} J_{|m|}(kr)[N_{|m|}(ka)J_{|m|}(kr') - J_{|m|}(ka)N_{|m|}(kr')], & 0 \le r < r' \\ [N_{|m|}(ka)J_{|m|}(kr) - J_{|m|}(ka)N_{|m|}(kr)]J_{|m|}(kr'), & r' < r \le a \end{cases}$$

Substituting back into (12.3.7), this yields the Green's function

$$G(r;r') = \frac{1}{4} \frac{J_0(k\,r_<)}{J_0(ka)}[J_0(ka)N_0(k\,r_>) - N_0(ka)J_0(k\,r_>)]$$

$$+ \frac{1}{2} \sum_{m=1}^{\infty} \frac{J_m(k\,r_<)}{J_m(ka)}[J_m(ka)N_m(k\,r_>) - N_m(ka)J_m(k\,r_>)]\cos m(\theta - \theta')$$

$$(12.3.11)$$

where, as usual, $r_<(r_>)$ is the smaller (larger) of r and r'.

To find the equivalent bilinear formula expression for the Green's function is quite straight forward since we already know what are the circular drumhead eigenfunctions or normal modes. Specifically, from Section 11.3.2 we have

$$u_{mn}(r, \theta) = \begin{cases} J_m(k_{mn}\,r)\cos m\theta \\ J_m(k_{mn}\,r)\sin m\theta \end{cases} \quad m = 0, 1, 2, \ldots, \quad n = 1, 2, \ldots.$$

To normalize these functions over the area of the drumhead, we multiply them by the normalization constants for the Bessel and Fourier functions which means multiplying them by

$$N_{0n} = \frac{1}{\sqrt{\pi}\,a\,J_0'(k_{0n}\,a)} \quad \text{for } m = 0 \quad \text{and} \quad N_{mn} = \sqrt{\frac{2}{\pi}}\frac{1}{a\,J_m'(k_{mn}\,a)} \quad \text{for } m = 1, 2, \ldots.$$

Substituting into (12.3.5), we obtain the bilinear formula

$$G(r;r') = \frac{1}{\pi a^2} \sum_{n=1}^{\infty} \frac{1}{[J_0'(k_{0n}\,a)]^2} \frac{J_0(k_{0n}\,r)J_{0n}(k_{0n}\,r')}{k^2 - k_{0n}^2}$$

$$+ \frac{2}{\pi a^2} \sum_{m=1}^{\infty}\sum_{n=1}^{\infty} \frac{1}{[J_m'(k_{mn}\,a)]^2} \frac{J_m(k_{mn}\,r)J_m(k_{mn}\,r')\cos m(\theta - \theta')}{k^2 - k_{mn}^2} \quad (12.3.12)$$

where we have used the identity $\cos m\theta \cos m\theta' + \sin m\theta \sin m\theta' = \cos m(\theta - \theta')$.

12.3.3 The Non-Homogeneous Helmholtz Equation With Boundaries at Infinity

When there are no finite boundaries (the source is isolated), the boundary condition that accompanies the Green's function PDE

$$(\nabla^2 + k^2)G(r;r') = \delta(r - r')$$

is simply that the solution be bounded and, in particular, that $\lim_{|r-r'|\to\infty} |G(r;r')| < \infty$. This can be solved by using Fourier transforms in much the same way that we did in Section 12.1.5.

We start by putting a subscript on the wave number in the Green's function PDE to distinguish it from the transform variable:

$$(\nabla^2 + k_0^2)G(\boldsymbol{r};\boldsymbol{r}') = \delta(\boldsymbol{r} - \boldsymbol{r}') \tag{12.3.13}$$

and so the time dependence associated with $G(\boldsymbol{r};\boldsymbol{r}')$ is now $e^{-i\omega_0 t}$, $\omega_0 = c\,k_0$. Taking the three-dimensional transform of (12.3.13), we obtain

$$g(\boldsymbol{k};\boldsymbol{r}') = \frac{1}{(2\pi)^{3/2}} \frac{e^{i\boldsymbol{k}\cdot\boldsymbol{r}'}}{k_0^2 - k^2}$$

where $k^2 = \boldsymbol{k}\cdot\boldsymbol{k}$ and

$$g(\boldsymbol{k};\boldsymbol{r}') \equiv \mathcal{F}\{G(\boldsymbol{r};\boldsymbol{r}')\} = \frac{1}{(2\pi)^{3/2}} \int_0^{2\pi}\int_0^{\pi}\int_0^{\infty} e^{i\boldsymbol{k}\cdot\boldsymbol{r}}\, G(\boldsymbol{r};\boldsymbol{r}')\, r^2\, dr \sin\theta d\theta d\varphi.$$

Thus,

$$G(\boldsymbol{r};\boldsymbol{r}') = \frac{1}{(2\pi)^3} \int_0^{2\pi}\int_0^{\pi}\int_0^{\infty} \frac{e^{i\boldsymbol{k}\cdot(\boldsymbol{r}'-\boldsymbol{r})}}{k_0^2 - k^2}\, k^2\, dk \sin\theta_k\, d\theta_k\, d\varphi_k \tag{12.3.14}$$

which we recognize as the bilinear formula for the Green's function.

To evaluate this integral, we take the k_3 axis in the direction of $\boldsymbol{r} - \boldsymbol{r}'$ so that

$$\boldsymbol{k}\cdot(\boldsymbol{r} - \boldsymbol{r}') = kR\cos\theta_k, \quad R = |\boldsymbol{r} - \boldsymbol{r}'|.$$

Then,

$$G(\boldsymbol{r};\boldsymbol{r}') = -\frac{1}{(2\pi)^3} \int_0^{2\pi}\int_0^{\pi}\int_0^{\infty} \frac{e^{-ikR\cos\theta_k}}{k^2 - k_0^2}\, k^2\, dk \sin\theta_k\, d\theta_k\, d\varphi_k$$

$$= -\frac{1}{(2\pi)^2}\frac{2}{R} \int_0^{\infty} \frac{\sin kR}{k^2 - k_0^2}\, k\, dk$$

$$= -\frac{1}{(2\pi)^2}\frac{1}{iR} \int_0^{\infty} \frac{e^{ikR} - e^{-ikR}}{k^2 - k_0^2}\, k\, dk$$

$$= -\frac{1}{(2\pi)^2}\frac{1}{iR} \int_{-\infty}^{\infty} \frac{e^{ikR}}{k^2 - k_0^2}\, k\, dk.$$

The final integral can be evaluated by contour integration. When combined with time dependence $e^{-i\omega_0 t}$, inclusion of the pole at $k = -k_0$ within the contour results in an incoming wave. On the other hand, the residue of the pole at $k = k_0$ yields an outgoing wave. Therefore, we use the same contours that were used in Section 12.1.5 and obtain the solutions

$$G_{out}(\boldsymbol{r};\boldsymbol{r}') = -\frac{1}{4\pi}\frac{e^{ik_0\,|\boldsymbol{r}-\boldsymbol{r}'|}}{|\boldsymbol{r}-\boldsymbol{r}'|},$$ (12.3.15)

and

$$G_{in}(\boldsymbol{r};\boldsymbol{r}') = -\frac{1}{4\pi}\frac{e^{-ik_0\,|\boldsymbol{r}-\boldsymbol{r}'|}}{|\boldsymbol{r}-\boldsymbol{r}'|}.$$ (12.3.16)

These are used to generate purely outgoing or incoming wave solutions, respectively. Thus, for example, if the source $\sigma(\boldsymbol{r}, t) = \sigma(\boldsymbol{r})e^{-i\omega_0 t}$ is an isolated loudspeaker or acoustic antenna, the waves that it emits will be described by

$$\psi(\boldsymbol{r}, t) = -\frac{e^{-i\omega_0 t}}{4\pi}\int_{\text{all space}}\frac{e^{ik_0\,|\boldsymbol{r}-\boldsymbol{r}'|}}{|\boldsymbol{r}-\boldsymbol{r}'|}\sigma(\boldsymbol{r})dV'.$$

Far from the source, $r \gg r'$, we have $|\boldsymbol{r}-\boldsymbol{r}'| \sim r - \boldsymbol{n}\cdot\boldsymbol{r}'$ where $\boldsymbol{n}=\frac{\boldsymbol{r}}{r}$ and so $\frac{e^{ik_0\,|\boldsymbol{r}-\boldsymbol{r}'|}}{|\boldsymbol{r}-\boldsymbol{r}'|}\sim\frac{e^{ik_0 r}}{r}e^{-i\boldsymbol{k}'\cdot\boldsymbol{r}'}$ where $\boldsymbol{k}'=k_0\,\boldsymbol{n}$. In that case,

$$\psi(\boldsymbol{r}) \sim f(\boldsymbol{k}')\frac{e^{ikr}}{r}\text{ where }f(\boldsymbol{k}') = \int_{\text{all space}}e^{-i\boldsymbol{k}'\cdot\boldsymbol{r}'}\sigma(\boldsymbol{r}')dV'.$$ (12.3.17)

This is analogous to the formalism we introduced in Section 11.4.2 in our discussion of spherical waves.

As one would expect, (12.3.15) and (12.3.16) both become the Green's function (12.2.25) for Poisson's equation in the limit as $k_0 \to 0$.

12.3.4 General Time Dependence

If the source term $\sigma(\boldsymbol{r}, t)$ has a **general time dependence,** we cannot separate the space and time dependence so easily and we do have to worry about initial conditions. Let us start by re-stating the problem. It consists of solving

$$\nabla^2\psi(\boldsymbol{r}, t) - \frac{1}{c^2}\frac{\partial^2\psi(\boldsymbol{r}, t)}{\partial t^2} = \sigma(\boldsymbol{r}, t)$$ (12.3.18)

where the source term $\sigma(\boldsymbol{r}, t)$ has a general time dependence and the solution $\psi(\boldsymbol{r}, t)$ is subject to

$$\psi(\boldsymbol{r}, t)\text{ or }\boldsymbol{n}\cdot\nabla\psi(\boldsymbol{r}, t)\text{ given for }\boldsymbol{r}\text{ on }S$$ (12.3.19)

plus initial conditions

$$\psi(\boldsymbol{r}, t)\text{ and }\frac{\partial\psi(\boldsymbol{r}, t)}{\partial t}\text{ given at an initial time }t = \tau\text{ throughout }V.$$ (12.3.20)

To tackle this we need the Green's function $G(r, t; r', t')$ which is the solution of

$$\nabla^2 G(r, t; r', t') - \frac{1}{c^2} \frac{\partial^2}{\partial t^2} G(r, t; r', t') = \delta(r - r')\delta(t - t') \tag{12.3.21}$$

that satisfies the boundary condition $G = 0$ or $n \cdot \nabla G = 0$ for r on S **plus** the **initial condition** $G(r, t; r', t') = 0$ for $t < t'$. The latter condition follows from the principle of causality: G is a response to a stimulus at $t = t'$ and so should be zero **prior** to that time. With the usual Green's theorem manipulations, one can show that $G(r, t; r', t') = G(r', -t'; r, -t)$ and that the solution to our original problem is

$$\psi(r, t) = \int_\tau^t \int_V G(r, t; r', t')\sigma(r', t')dV'dt' - \int_\tau^t \int_S [G(r, t; r', t')n \cdot \nabla'\psi(r', t')$$
$$- \psi(r', t')n \cdot \nabla' G(r, t; r', t')dS'dt']$$
$$- \frac{1}{c^2} \int_V [G(r, t; r', T)\frac{\partial}{\partial t'}\psi(r', t')\Big|_{t'=\tau} - \psi(r', T)\frac{\partial}{\partial t'} G(r, t; r', t')\Big|_{t'=\tau}]dV'$$

$$\tag{12.3.22}$$

An analogous approach applies to the diffusion equation. To solve

$$\nabla^2 \psi(r, t) - \frac{1}{\kappa} \frac{\partial \psi(r, t)}{\partial t} = \sigma(r, t) \tag{12.3.23}$$

with $\psi(r, t)$ or $n \cdot \nabla\psi(r, t)$ given for r on S **plus** the initial condition $\psi(r, t)$ given at $t = \tau$, we first determine the Green's function $G(r, t; r', t')$ which is the solution of

$$\nabla^2 G(r, t; r', t') - \frac{1}{\kappa} \frac{\partial}{\partial t} G(r, t; r', t') = \delta(r - r')\delta(t - t') \tag{12.3.24}$$

that satisfies the boundary condition

$$G(r, t; r', t') = 0 \quad \text{or} \quad n \cdot \nabla G = 0 \quad \text{for} \quad r \quad \text{on} \quad S$$

plus the initial condition $G(r, t; r', t') = 0$ for $t < t'$.
One can again show that $G(r, t; r', t') = G(r', -t'; r, -t)$ and

$$\psi(r, t) = \int_\tau^t \int_V G(r, t; r', t')\sigma(r', t')dV'dt'$$
$$- \int_\tau^t \int_S [G(r, t; r', t')n \cdot \nabla'\psi(r', t') - \psi(r', t')n \cdot \nabla' G(r, t; r', t')]dS'dt'$$
$$- \frac{1}{\kappa} \int_V G(r, t; r', \tau)\psi(r', \tau)dV'. \tag{12.3.25}$$

12.3.5 The Wave and Diffusion Equation Green's Functions for Boundaries at Infinity

The Green's function associated with the wave equation PDE and Dirichlet boundary conditions at infinity is a solution of

$$\left(\nabla^2 - \frac{1}{c^2}\frac{\partial^2}{\partial t^2}\right) G(\mathbf{r}, t; \mathbf{r}', t') = \delta(\mathbf{r} - \mathbf{r}')\delta(t - t') \tag{12.3.26}$$

that is everywhere bounded as a function of both $\mathbf{r}$ and t. The solution by means of Fourier transforms proceeds exactly as in Section 12.3.3 and yields

$$G(\mathbf{r}, t; \mathbf{r}', t') = -\frac{c^2}{(2\pi)^4}\int_{-\infty}^{\infty}\int_{\text{all space}} e^{-i\mathbf{k}\cdot(\mathbf{r}-\mathbf{r}')}\frac{e^{-i\omega(t-t')}}{c^2 k^2 - \omega^2} k^2\, dk \sin\theta_k\, d\theta_k\, d\varphi_k\, d\omega. \tag{12.3.27}$$

As we saw in Chapter 3, the integral

$$\Delta \equiv \int_{-\infty}^{\infty} \frac{e^{-i\omega(t-t')}}{c^2 k^2 - \omega^2} d\omega \tag{12.3.28}$$

has four different values depending on how one avoids the poles at $\omega = \pm ck$. However, only one of these four satisfies our (causal) initial condition $G(\mathbf{r}, t; \mathbf{r}', t') = 0$ for $t < t'$. This is the so-called **retarded** solution

$$\Delta_{ret} = \begin{cases} \dfrac{2\pi \sin ck(t - t')}{ck} & \text{for } t > t' \\ 0 & \text{for } t < t' \end{cases} \tag{12.3.29}$$

which arises from deforming the contour of integration to pass **above** both poles. The name is a reflection of the correspondence to a signal emitted at a time t' that is earlier or retarded compared to the time of arrival t.

Inserting (12.3.29) into (12.3.27) yields the (retarded) Green's function

$$G(\mathbf{r}, t; \mathbf{r}', t') = \begin{cases} -\dfrac{c}{2\pi}\int_{\text{all space}} e^{-i\mathbf{k}\cdot(\mathbf{r}-\mathbf{r}')}\dfrac{\sin ck(t - t')}{k} k^2\, dkd(\cos\theta_k)d\varphi_k \\ 0 \end{cases} \tag{12.3.30}$$

for $t > t'$ and $t < t'$, respectively. Choosing the k_3 axis to be in the direction of $\mathbf{r} - \mathbf{r}'$, the angular integration is easy to perform and gives us

$$\int_{\text{all space}} e^{-i\mathbf{k}\cdot(\mathbf{r}-\mathbf{r}')}\frac{\sin ck(t - t')}{k} k^2\, dkd(\cos\theta_k)d\varphi_k = \frac{4\pi}{R}\int_0^{\infty} \sin kR \sin ckT dk$$

where $R = |\mathbf{r} - \mathbf{r}'|$ and $T = t - t'$. Converting the sines on the right hand side into exponentials, we obtain

$$\int_0^\infty \sin kR \sin ckT dk = -\frac{1}{4} \int_0^\infty [e^{ik(R+cT)} + e^{-ik(R+cT)} - e^{ik(R-cT)} - e^{-ik(R-cT)}]dk$$

$$= -\frac{1}{4} \int_{-\infty}^\infty [e^{ik(R+cT)} - e^{ik(R-cT)}]dk$$

$$= \frac{1}{4}[\delta(R - cT) - \delta(R + cT)].$$

Since the second delta function does not contribute for $T > 0$ and since

$$\delta(R - cT) = \frac{1}{c}\delta\left(\frac{R}{c} - T\right),$$

substitution back into (12.3.30) gives us

$$G(\mathbf{r}, t; \mathbf{r}', t') = \begin{cases} -\frac{1}{4\pi}\delta\left(\frac{|\mathbf{r} - \mathbf{r}'|}{c} - (t - t')\right) & \text{for } t > t' \\ 0 & \text{for } t < t' \end{cases} \tag{12.3.31}$$

as our final result for the (retarded) Green's function.

Given the unusual appearance of this Green's function, it is helpful to keep in mind its physical interpretation. It is a spherical wave produced by an instantaneous disturbance at the single point $\mathbf{r}=\mathbf{r}'$ and time $t = t'$. Prior to $t = t'$, at times $t < t'$, there is no wave. After $t = t'$, at times $t > t'$, the wave spreads out (propagates) to the location $|\mathbf{r} - \mathbf{r}'| = c(t - t')$ and as it spreads, its amplitude decreases as $\frac{1}{|\mathbf{r}-\mathbf{r}'|}$. A close analogue is the wave produced by dropping a small but massive pebble into a quiet pool of water.

Equation (12.3.22) is the prescription for superposing all of these elementary waves, part originating from the continuous source described by the function $\sigma(\mathbf{r}, t)$ and part from the boundary and initial conditions. Maximal simplification is obtained for the case of boundaries at infinity, an initial time $\tau \to -\infty$, and the requirement that $\psi(\mathbf{r}, t) \to 0$ as $r \to \infty$ and $t \to -\infty$. What results is a single integral of the form

$$\psi(\mathbf{r}, t) = -\frac{1}{4\pi} \int_{\text{all space}} \frac{\sigma\left(\mathbf{r}' - \frac{|t-t'|}{c}\right)}{|\mathbf{r} - \mathbf{r}'|} dV' \tag{12.3.32}$$

which is referred to in electrodynamics as the "retarded potential" solution of the wave equation.

Let us now turn our attention to the diffusion equation. Using Fourier transforms again to solve

$$\nabla^2 G(\mathbf{r}, t; \mathbf{r}', t') - \frac{1}{\kappa}\frac{\partial}{\partial t} G(\mathbf{r}, t; \mathbf{r}', t') = \delta(\mathbf{r} - \mathbf{r}')\delta(t - t'),$$

we find

$$G(r, t; r', t') = -\frac{i\kappa}{(2\pi)^4} \int\limits_{-\infty}^{\infty} \int\limits_{\text{all space}} \frac{e^{-i\mathbf{k}\cdot(\mathbf{r}-\mathbf{r}')-i\omega(t-t')}}{\omega + i\kappa k^2} k^2 \, dk d(\cos\theta_k) d\varphi_k \, d\omega.$$

The integral over ω can be evaluated by a simple application of the residue calculus. The integrand has a single simple pole located at $\omega = -i\kappa k^2$ in the lower half of the complex ω plane. If $t < t'$, the contour will have to be closed in the upper half plane excluding the pole and resulting in a null result. But, for $t > t'$, we must close the contour in the lower half plane and thus pick up a non-zero contribution from its residue. Thus,

$$G(r, t; r', t') = \begin{cases} -\frac{\kappa}{(2\pi)^3} \int\limits_{\text{all space}} e^{-i\mathbf{k}\cdot(\mathbf{r}-\mathbf{r}')-\kappa k^2(t-t')} k^2 \, dk d(\cos\theta_k) d\varphi_k \\ 0 \end{cases} \tag{12.3.33}$$

for $t > t'$ and $t < t'$, respectively. This result inherently meets the initial condition that we wanted to impose on this Green's function.

The integral is a three-dimensional version of one we evaluated in Chapter 3. If we set $\mathbf{R} = \mathbf{r} - \mathbf{r}'$ and $T = t - t'$ and complete the square in the exponent, it becomes

$$\int\limits_{\text{all space}} e^{-i\mathbf{k}\cdot(\mathbf{r}-\mathbf{r}')-\kappa k^2(t-t')} k^2 \, dk d(\cos\theta_k) d\varphi_k = e^{-\frac{R^2}{4\kappa T}} \int\limits_{\text{all space}} e^{-\kappa T \left(k - i\frac{R}{2\kappa T}\right)^2} \, dk_1 \, dk_2 \, dk_3$$

$$= e^{-\frac{R^2}{4\kappa T}} \left(\int\limits_{-\infty}^{\infty} e^{-\kappa T \left(k_1 - i\frac{X}{2\kappa T}\right)^2} \, dk_1 \right)^3$$

where we have switched from spherical to rectangular coordinates and set $x - x' = X$. The integral in the second line was evaluated in Chapter 3 and we found that it equals $\sqrt{\frac{\pi}{\kappa T}}$. Therefore, substituting all this information back into (12.3.33), we obtain as our final expression

$$G(r, t; r', t') = \begin{cases} -\dfrac{\kappa}{[4\pi\kappa(t-t')]^{3/2}} \exp\left[-|\mathbf{r}-\mathbf{r}'|^2/4\kappa(t-t')\right] & \text{for } t > t' \\ 0 & \text{for } t < t' \end{cases} \tag{12.3.34}$$

This is a sharply peaked (Gaussian) function of $|\mathbf{r} - \mathbf{r}'|$ for small values of $t - t'$ and broad for large values of $t - t'$. In fact, in the limit as $t \to t'$, the Green's function $G(r, t; r', t') \to -\kappa\delta(\mathbf{r} - \mathbf{r}')$. Thus, it describes how a delta function impulse at $t = t'$ diffuses through the medium at later times.

We can now use (12.3.25) to write down the solution to a diffusion problem involving a source $\sigma(r, t)$, boundary condition $\psi \to 0$ (and $G \to 0$) as $|\mathbf{r} - \mathbf{r}'| \to \infty$, and initial condition $\psi(r, t)$ given at $t = \tau$: it is

$$\psi(\mathbf{r}, t) = \frac{1}{\sqrt{4\pi\kappa}} \int_\tau^t \int_{\text{all space}} \frac{1}{(t - t')^{3/2}} \exp\left[-|\mathbf{r} - \mathbf{r}'|^2 / 4\kappa(t - t')\right]\sigma(\mathbf{r}', t')dt'dV'$$

$$+ \frac{1}{[4\pi\kappa(t - \tau)]^{3/2}} \int_{\text{all space}} \exp\left[-|\mathbf{r} - \mathbf{r}'|^2 / 4\kappa(t - \tau)\right]\psi(\mathbf{r}', \tau)dV'. \quad (12.3.35)$$

In particular, if the source $\sigma(\mathbf{r}, t) \equiv 0$ and $\psi(\mathbf{r}, 0) = \delta(x - a)$, the solution is

$$\psi(\mathbf{r}, t) = \psi(x, t) = \frac{1}{\sqrt{4\pi\kappa t}} e^{-\frac{(x-a)^2}{4\kappa t}}$$

which describes the diffusion in both x-directions of a substance initially confined in the yz-plane at $x = a$.

13 Integral Equations

13.1 Introduction

The Green's function method can transform a differential equation into an integral equation if the source term actually contains the solution being sought. For example, in the quantum theory of scattering the Schrödinger equation can be cast in the form

$$(\nabla^2 + k^2)\psi(\mathbf{r}) = U(\mathbf{r})\psi(\mathbf{r}) \tag{13.1.1}$$

where $U(\mathbf{r}) = \frac{2m}{\hbar^2}V(\mathbf{r})$ and $V(\mathbf{r})$ is the potential describing the interaction responsible for the scattering. If we describe the incident particle by means of a plane wave $e^{i\mathbf{k}_0 \cdot \mathbf{r}}$ where $|\mathbf{k}_0| = k$, the solution of (13.1.1) that we seek must satisfy the boundary condition

$$\psi(\mathbf{r}) \rightarrow e^{i\mathbf{k}_0 \cdot \mathbf{r}} + f(\theta, \varphi)\frac{e^{ikr}}{r} \text{ as } r \rightarrow \infty. \tag{13.1.2}$$

As we learned in an earlier discussion of scattering, the second term of this expression is the outgoing spherical scattered wave produced by the interaction. The Green's function appropriate to such a problem is the outgoing Green's function for the Helmhotz equation given by (12.3.15):

$$G(\mathbf{r};\mathbf{r}') = -\frac{1}{4\pi}\frac{e^{ik|\mathbf{r}-\mathbf{r}'|}}{|\mathbf{r} - \mathbf{r}'|}.$$

Using it with a source function $\sigma(\mathbf{r}) = U(\mathbf{r})\psi(\mathbf{r})$ and imposing the boundary condition (13.1.2), we see that the desired solution of the PDE can be expressed as

$$\psi(\mathbf{r}) = e^{i\mathbf{k}_0 \cdot \mathbf{r}} - \int_{\text{all space}} \frac{e^{ik|\mathbf{r}-\mathbf{r}'|}}{|\mathbf{r} - \mathbf{r}'|}U(\mathbf{r}')\psi(\mathbf{r}')dV'. \tag{13.1.3}$$

Notice that this **integral equation** incorporates **both** the PDE **and** the boundary condition that we wish to impose. This is a standard feature of integral equations and one that often renders them a more convenient or more powerful tool than differential equations. Add to this the fact that some problems, notably those involving transport phenomena, admit solution only by means of integral equations and we conclude that they warrant some attention. Since any multi-dimensional integral equation can be reduced to an equivalent set of one-dimensional equations by making use of eigenfunction expansions, we shall focus that attention on the one-dimensional case.

13.2 Types of Integral Equations

Integral equations are classified in terms of their integration limits and in terms of whether the unknown function appears inside and outside the integral or only outside. We call equations with fixed integration limits **Fredholm equations.** Those with

one limit that is variable are called **Volterra equations**. If the unknown function appears **only** under the integral sign, the equation is said to be an integral equation of the **first kind;** otherwise it is an equation of the **second kind.**
Thus,
1.

$$f(x) = \int_a^b K(x, y)\psi(y)dy \tag{13.2.1}$$

is a **Fredholm equation of the first kind;**
2.

$$\psi(x) = f(x) + \lambda \int_a^b K(x, y)\psi(y)dy, \quad \lambda = \text{ a constant} \tag{13.2.2}$$

is a **Fredholm equation of the second kind;**
3.

$$f(x) = \int_a^x K(x, y)\psi(y)dy \tag{13.2.3}$$

is a **Volterra equation of the first kind;** and
4.

$$\psi(x) = f(x) + \int_a^x K(x, y)\psi(y)dy \tag{13.2.4}$$

is a **Volterra equation of the second kind.**
We say that (13.2.2) and (13.2.4) are **homogeneous** if $f(x) \equiv 0$. In all four cases $f(x)$ and $K(x, y)$ are presumed to be known and $\psi(y)$ is the function whose identity is sought. As with integral representations, $K(x, y)$ is called the **kernel.**

Evidently, equation (13.1.3) is a (three-dimensional) Fredholm equation of the second kind. Volterra equations arise when the solution **and** its first derivative are specified at a single point as is the case with initial conditions. For example, if we start with the second-order non-homogeneous DE

$$\frac{d^2\psi}{dx^2} + a(x)\frac{d\psi}{dx} + b(x)\psi(x) = g(x) \tag{13.2.5}$$

complimented by (initial) conditions $\psi(x_0) = u_0$ and $\psi'(x_0) = v_0$, we can use indefinite integration to write

$$\frac{d\psi}{dx} = -\int_{x_0}^x a(x)\frac{d\psi}{dx}dx - \int_{x_0}^x b(x)\psi(x)dx + \int_{x_0}^x g(x)dx + v_0 \,.$$

Integrating the first integral on the right by parts gives us

$$\frac{d\psi}{dx} = -a(x)\psi(x) - \int_{x_0}^{x} [b(x) - a'(x)]\psi(x)dx + \int_{x_0}^{x} g(x)dx + a(x_0)u_0 + v_0$$

which incorporates both of our initial conditions. Integrating the complete equation a second time, we obtain

$$\psi(x) = -\int_{x_0}^{x} a(x)\psi(x)dx - \int_{x_0}^{x}\int_{x_0}^{x} [b(y) - a'(y)]\psi(y)dydx$$

$$+ \int_{x_0}^{x}\int_{x_0}^{x} g(y)dydx + [a(x_0)u_0 + v_0](x - x_0) + u_0. \tag{13.2.6}$$

This can be simplified by using the identity (the two sides have the same first derivative and the same value at x_0)

$$\int_{x_0}^{x}\int_{x_0}^{x} f(y)dydx = \int_{x_0}^{x} (x - y)f(y)dy.$$

Applying it to (13.2.6) we find

$$\psi(x) = -\int_{x_0}^{x} \{a(y) + (x - y)[b(y) - a'(y)]\}\psi(y)dy$$

$$+ \int_{x_0}^{x} (x - y)g(y)dy + [a(x_0)u_0 + v_0](x - x_0) + v_0. \tag{13.2.7}$$

Evidently, this is a Volterra integral equation of the second kind and comparing it with (13.2.4) we can identify

$$K(x, y) \equiv (y - x)[b(y) - a'(y)] - a(y)$$

as the kernel and

$$f(x) \equiv \int_{x_0}^{x} (x - y)g(y)dy + [a(x_0)u_0 + v_0](x - x_0) + u_0$$

as the non-homogeneous term.

13.3 Convolution Integral Equations

If the kernel and integration limits of a Fredholm equation are those of a known integral transform then it can be solved simply by invoking the corresponding inverse transform. Beyond this almost trivial application, an integral transform can also be useful in the solution of an integral equation when the kernel has the symmetry properties needed to exploit the transform's convolution theorem. For example, suppose that we have a Fredholm equation of the form

$$\psi(x) = f(x) + \int_{-\infty}^{\infty} K(x - y)\psi(y)dy \tag{13.3.1}$$

where the kernel depends only on the **difference** between the variables x and y. The integral is a convolution integral of the type that arises for Fourier transforms. Therefore, when we transform the equation we find

$$\Psi(k) = F(k) + 2\pi\kappa(k)\Psi(k) \tag{13.3.2}$$

where

$$\Psi(k) \equiv \mathcal{F}\{\psi(x)\} = \frac{1}{\sqrt{2\pi}} \int_{-\infty}^{\infty} f(x)\, e^{ikx}\, dx,$$

$$F(k) \equiv \mathcal{F}\{f(x)\} = \frac{1}{\sqrt{2\pi}} \int_{-\infty}^{\infty} f(x)\, e^{ikx}\, dx$$

and

$$\kappa(k) \equiv \mathcal{F}\{K(x)\} = \frac{1}{\sqrt{2\pi}} \int_{-\infty}^{\infty} K(x)\, e^{ikx}\, dx.$$

Solving (13.3.2), we have

$$\Psi(k) = \frac{F(k)}{1 - 2\pi\kappa(k)} \tag{13.3.3}$$

and so,

$$\psi(x) = \frac{1}{\sqrt{2\pi}} \int_{-\infty}^{\infty} \frac{F(k)}{1 - 2\pi\kappa(k)}\, e^{-ikx}\, dk. \tag{13.3.4}$$

Laplace transforms are equally useful in the solution of Volterra equations of the form

$$\psi(x) = f(x) + \int_{0}^{x} K(x - y)\psi(y)dy. \tag{13.3.5}$$

For example, Abel's integral equation is

$$f(x) = \int_{0}^{x} \frac{\psi(y)}{(x - y)^{\alpha}}dy, \quad 0 < \alpha < 1 \tag{13.3.6}$$

where $f(x)$ is known but $\psi(x)$ is not. The left hand side of this equation is a convolution integral of the type that arises with Laplace transforms and taking the transform of both sides of the equation, we obtain

$$\mathcal{L}\{f(x)\} = \mathcal{L}\{x^{-\alpha}\}\mathcal{L}\{\psi(x)\}.$$

Since $\mathcal{L}\{x^{-\alpha}\} = \frac{\Gamma(1-\alpha)}{s^{1-\alpha}}$, this means that

$$\mathcal{L}\{\psi(x)\} = \frac{s^{1-\alpha}\,\mathcal{L}\{f(x)\}}{\Gamma(1-\alpha)}. \tag{13.3.7}$$

The inverse $\mathcal{L}^{-1}\{s^{1-\alpha}\}$ does not exist but $\mathcal{L}^{-1}\{s^{-\alpha}\} = \frac{x^{\alpha-1}}{\Gamma(\alpha)}$. Therefore, we divide through by s to obtain

$$\frac{1}{s}\mathcal{L}\{\psi(x)\} = \frac{\mathcal{L}\{x^{\alpha-1}\}\mathcal{L}\{f(x)\}}{\Gamma(\alpha)\Gamma(1-\alpha)} = \frac{\sin\pi\alpha}{\pi}\mathcal{L}\{x^{\alpha-1}\}\mathcal{L}\{f(x)\}$$

where we have invoked the gamma function identity (5.2.10). This can be inverted by means of a further application of the convolution theorem which yields

$$\int_0^x \psi(y)dy = \frac{\sin\pi\alpha}{\pi}\int_0^x \frac{f(y)}{(x-y)^{1-\alpha}}dy,$$

or

$$\psi(x) = \frac{\sin\pi\alpha}{\pi}\frac{d}{dx}\int_0^x \frac{f(y)}{(x-y)^{1-\alpha}}dy. \tag{13.3.8}$$

13.4 Integral Equations With Separable Kernels

Fredholm equations of the second kind can be solved in closed form if their kernels are separable, that is if they have the form

$$K(x, y) = u(x)v(y). \tag{13.4.1}$$

To demonstrate this we substitute (13.4.1) into (13.2.2) to obtain

$$\psi(x) = f(x) + \lambda\int_a^b K(x,y)\psi(y)dy = f(x) + \lambda\int_a^b u(x)v(y)\psi(y)dy = f(x) + \lambda Cu(x) \tag{13.4.2}$$

where

$$C = \int_a^b v(y)\psi(y)dy. \tag{13.4.3}$$

Next, we substitute the final line of (13.4.2) into (13.4.3) to create a linear equation for C:

$$C = \int_a^b v(y)[f(y) + \lambda Cu(y)]dy.$$

Solving for C, we find

$$C = \frac{\int_a^b v(y)f(y)dy}{1 - \lambda \int_a^b u(y)v(y)dy} \tag{13.4.4}$$

which, when substituted into (13.4.2), yields the solution

$$\psi(x) = f(x) + \lambda u(x)\left[\frac{\int_a^b v(y)f(y)dy}{1 - \lambda \int_a^b u(y)v(y)dy}\right]. \tag{13.4.5}$$

The homogeneous version of (13.2.2) is the **eigenvalue equation**

$$\psi(x) = \lambda \int_a^b K(x, y)\psi(y)dy. \tag{13.4.6}$$

This too is readily solved when the kernel is separable. Substituting (13.4.1) into it gives us

$$\psi(x) = \lambda u(x) \int_a^b v(y)\psi(y)dy = \lambda Cu(x) = (\text{constant}) \times u(x). \tag{13.4.7}$$

In this case, substitution of (13.4.7) into the definition of C in (13.4.3) does not determine C (because it cancels out of the resulting equation) but it does determine the eigenvalue λ:

$$\lambda = \left[\int_a^b u(y)v(y)dy\right]^{-1}. \tag{13.4.8}$$

Thus, the eigenvalue is uniquely determined and the corresponding eigenfunction is determined to within a multiplicative constant that can be fixed through a normalization condition placed on $\psi(x)$ over the interval $a \le x \le b$.

Notice that if λ in (13.4.5) is equal to the eigenvalue defined by (13.4.8), C is infinite and the non-homogeneous equation has no solution unless

$$\int_a^b v(y)f(y)dy = 0$$

in which case the solution is

$$\psi(x) = f(x) + (\text{constant}) \times u(x).$$

We have seen the same phenomenon arise with the solution of non-homogeneous **differential** equations.

Kernels that are sums of separable pieces, such as

$$K(x, y) = \sum_{j=1}^{N} u_j(x) \, u_j(y), \tag{13.4.9}$$

also give rise to integral equations that can be solved explicitly. To confirm this, we substitute (13.4.9) into (13.2.2) to obtain

$$\psi(x) = f(x) + \lambda \int_a^b K(x, y)\psi(y)dy = f(x) + \lambda \sum_{j=1}^{N} u_j(x) \int_a^b v_j(y)\psi(y)dy$$

$$= f(x) + \lambda \sum_{j=1}^{N} c_j \, u_j(x) \tag{13.4.10}$$

where the constants c_j are given by

$$c_j = \int_a^b v_j(y)\psi(y)dy. \tag{13.4.11}$$

Substitution of the third line of (13.4.10) in place of $\psi(y)$ under the integral in (13.4.11) generates a set of linear algebraic equations to determine the c_j :

$$c_i = \int_a^b v_i(y)f(y)dy + \lambda \sum_{j=1}^{N} \left[\int_a^b v_i(y) \, u_j(y)dy \right] c_j . \tag{13.4.12}$$

This can be re-written in matrix notation by introducing the $n \times n$ matrix M with elements

$$M_{ij} = \int_a^b v_i(y) \, u_j(y)dy$$

and the vectors C and F with elements C_i and

$$F_i = \int_a^b v_i(y)f(y)dy,$$

respectively. Specifically, (13.4.12) becomes

$$\sum_{j=1}^{N} \delta_{ij} \, c_j = F_i + \lambda \sum_{j=1}^{N} M_{ij} \, c_j,$$

or

$$\sum_{j=1}^{N}(\delta_{ij} - \lambda M_{ij})c_j = F_i,$$

or

$$(\mathbb{I} - \lambda M)C = F. \tag{13.4.13}$$

Thus, the values of the constants c_i are given by

$$C = (\mathbb{I} - \lambda M)^{-1}F \tag{13.4.14}$$

and these, in turn, completely determine the solution of the non-homogeneous integral equation via equation (13.4.10). In the case of the (homogeneous) eigenvalue problem (13.4.6), equation (13.4.10) still obtains but with $f(x) \equiv 0$. Therefore, we set $F \equiv 0$ in (13.4.13) and so obtain the matrix eigenvalue problem

$$(\mathbb{I} - \lambda M)C = 0. \tag{13.4.15}$$

This has a non-trivial solution if and only if

$$\det(\mathbb{I} - \lambda M) = 0 \tag{13.4.16}$$

which determines at least one and at most N eigenvalues λ_n. Substituting these back into (13.4.15), one can find the corresponding eigenvectors $C^{(n)}$ whose components $c_i^{(n)}$ determine the eigenfunction solutions of the integral equation via

$$\psi_n(x) = (\text{constant}) \times \sum_{j=1}^{N} c_j^{(n)} u_j(x). \tag{13.4.17}$$

By way of illustration, consider the simple example presented by the equation

$$\psi(x) = \lambda \int_{-1}^{1}(y + x)\psi(y)dy$$

from which we can read off $u_1(x) = 1, v_1(y) = y, u_2(x) = x, v_2(y) = 1$. Thus,

$$M_{11} = M_{22} = \int_{-1}^{1} ydy = 0, \quad M_{12} = \int_{-1}^{1} y^2\, dy = \frac{2}{3}, \quad M_{21} = \int_{-1}^{1} dy = 2$$

and equation (13.4.16) reads

$$\begin{vmatrix} 1 & -\frac{2\lambda}{3} \\ -2\lambda & 1 \end{vmatrix} = 0.$$

Expanding, we obtain

$$1 - \frac{4\lambda^2}{3} = 0$$

and hence, $\lambda = \pm\frac{\sqrt{3}}{2}$. As we shall see in Section 13.6, the reality of the eigenvalues is a consequence of the symmetry of the kernel under $x \leftrightarrow y$. If the kernel were to be replaced by the skew-symmetric function $K(x, y) = (y - x)$, the eigenvalues would be the pure imaginary $\lambda = \pm i\frac{\sqrt{3}}{2}$. Substitution of $\lambda = \pm\frac{\sqrt{3}}{2}$ into (13.4.15) gives us

$$c_2^{(1)} = \sqrt{3}\,c_1^{(1)} \quad \text{and} \quad c_2^{(2)} = -\sqrt{3}\,c_1^{(2)} \,.$$

Thus, the normalized eigenfunctions are

$$\psi_1(x) = \pm\frac{1}{2}(1 + \sqrt{3}x) \quad \text{and} \quad \psi_2(x) = \pm\frac{1}{2}(1 - \sqrt{3}x)$$

corresponding to eigenvalues $\lambda_1 = \frac{\sqrt{3}}{2}$ and $\lambda_2 = -\frac{\sqrt{3}}{2}$. Notice that the sign (or phase) of the eigenfunctions remains arbitrary.

Explicit solutions are not so readily obtained when the kernel is not separable. Therefore, one approach to dealing with a non-separable kernel is to approximate it with one that is separable. An obvious choice is to expand $K(x, y)$ in terms of an appropriate complete set $\{u_j(x)\}$,

$$K(x, y) = \sum_{j=1}^{\infty} u_j(x)\,v_j(y),$$

and then retain only the first N terms.

13.5 Solution by Iteration: Neumann Series

A second approach to solving equations with non-separable kernels is to use a method of successive approximations. The method applies equally well to Fredholm and Volterra equations but, for the sake of specificity, we shall demonstrate it using the Fredholm equation

$$\psi(x) = f(x) + \lambda \int_a^b K(x, y)\psi(y)dy. \tag{13.5.1}$$

We start by approximating the unknown function $\psi(x)$ by

$$\psi_0(x) = f(x)$$

which, when substituted **under the integral** in (13.5.1), gives us the improved approximation

$$\psi_1(x) = f(x) + \lambda \int_a^b K(x, y)f(y)dy. \tag{13.5.2}$$

A still better approximation $\psi_2(x)$ is obtained by replacing the $\psi(x)$ **under the integral** in (13.5.1) by $\psi_1(x)$:

$$\psi_2(x) = f(x) + \lambda \int_a^b K(x, y_1) f(y_1) dy_1 + \lambda^2 \int_a^b \int_a^b K(x, y_1) K(y_1, y_2) dy_1 dy_2 . \quad (13.5.3)$$

Repeating this process of substituting the new $\psi_n(x)$ under the integral in (13.5.1), we develop the sequence

$$\psi_m(x) = f(x) + \sum_{j=1}^m \lambda^j u_j(x) \quad (13.5.4)$$

where

$$u_j(x) = \int_a^b \int_a^b \cdots \int_a^b K(x, y_1) K(y_1, y_2) \cdots K(y_{j-1}, y_j) f(y_j) dy_j \cdots dy_1 . \quad (13.5.5)$$

Thus, the solution to our integral equation should be the limit

$$\psi(x) = \lim_{m\to\infty} \psi_m(x) = f(x) + \lim_{m\to\infty} \sum_{j=1}^\infty \lambda^j u_j(x) = f(x) + \sum_{j=1}^\infty \lambda^j u_j(x)$$

provided that the infinite series converges. This is called the **Neumann series** for $\psi(x)$ and its convergence occurs only for sufficiently small λ, $\lambda < \lambda_{min}$ where λ_{min} is the magnitude of the eigenvalue of smallest magnitude of the corresponding homogeneous equation (i.e. when $f(x) \equiv 0$).

As an application of the Neumann method of solution by iteration, let us consider the integral equation

$$\psi(x) = x + \frac{1}{2} \int_{-1}^1 (y - x)\psi(y) dy. \quad (13.5.6)$$

In this case, $\lambda = \frac{1}{2}$ and, as was noted in the example at the end of the preceding Section, $\lambda_{min} = \frac{\sqrt{3}}{2}$. Therefore, we expect the Neumann series to converge. From (13.5.5), we find that the series is

$$\psi_0(x) = x$$

$$\psi_1(x) = x + \frac{1}{3}$$

$$\psi_2(x) = x + \frac{1}{3} - \frac{x}{3}$$

$$\psi_3(x) = x + \frac{1}{3} - \frac{x}{3} - \frac{1}{3^2}$$

and, by induction,

$$\psi_{2m}(x) = x + \sum_{j=1}^{m}(-1)^{j-1}\frac{1}{3^j} - x\sum_{j=1}^{m}(-1)^{j-1}\frac{1}{3^j} \qquad (13.5.7)$$

with a similar expression for $\psi_{2m+1}(x)$. The series in (13.5.7) does indeed converge in the limit as $m \to \infty$ and results in the solution

$$\psi(x) = \frac{3}{4}x + \frac{1}{4}.$$

A more interesting application arises in connection with the scattering problem that we used to introduce this Chapter. The integral equation to be solved in that context is

$$\psi(\mathbf{r}) = e^{i\mathbf{k_0}\cdot\mathbf{r}} - \int_{all\ space} \frac{e^{ik|\mathbf{r}-\mathbf{r'}|}}{|\mathbf{r}-\mathbf{r'}|}U(\mathbf{r'})\psi(\mathbf{r'})dV'.$$

The Neumann series solution of this equation is

$$\psi(\mathbf{r}) = e^{i\mathbf{k_0}\cdot\mathbf{r}} + \int_{all\ space} \frac{e^{ik|\mathbf{r}-\mathbf{r'}|}}{|\mathbf{r}-\mathbf{r'}|}U(\mathbf{r'})e^{i\mathbf{k_0}\cdot\mathbf{r'}}dV'$$

$$+ \int_{all\ space}\int \frac{e^{ik|\mathbf{r}-\mathbf{r'}|}}{|\mathbf{r}-\mathbf{r'}|}U(\mathbf{r'})\frac{e^{ik|\mathbf{r'}-\mathbf{r''}|}}{|\mathbf{r'}-\mathbf{r''}|}U(\mathbf{r''})e^{i\mathbf{k_0}\cdot\mathbf{r''}}dV'dV'' + \dots. \qquad (13.5.8)$$

In quantum mechanics, this series in powers of the potential $U(\mathbf{r})$ is called a perturbation series and its convergence depends on the strength of the interaction.

13.6 Hilbert Schmidt Theory

We shall now focus on the homogeneous Fredholm equation of the second kind

$$\psi(x) = \lambda \int_a^b K(x, y)\psi(y)dy \qquad (13.6.1)$$

which, as we noted in Section 13.4, is also an eigenvalue equation. We shall assume that the kernel is Hermitian,

$$K(x, y) = K^*(y, x). \qquad (13.6.2)$$

Our first objective is to prove that the eigenvalues, λ_m, are real and that the corresponding eigenfunctions, $\psi_m(x)$, are orthogonal. Let $\psi_m(x)$ and $\psi_n(x)$ be two different eigenfunctions and λ_m and λ_n be the corresponding eigenvalues. Then, according to (13.6.1), we have

$$\psi_m(x) = \lambda_m \int_a^b K(x, y)\,\psi_m(y)dy, \quad \text{and}$$

$$\psi_n(x) = \lambda_n \int_a^b K(x, y)\, \psi_n(y)\,dy. \tag{13.6.3}$$

If we complex conjugate the first of these and multiply it by $\lambda_n\, \psi_n(x)$ while multiplying the second by $\lambda_m^*\, \psi_m^*(x)$ and then integrate with respect to x, the two equations become

$$\lambda_n \int_a^b \psi_m^*(x)\, \psi_n(x)\,dx = \lambda_m^*\, \lambda_n \int_a^b \int_a^b K^*(x, y)\, \psi_m^*(y)\, \psi_n(x)\,dy\,dx, \quad \text{and} \tag{13.6.4}$$

$$\lambda_m^* \int_a^b \psi_m^*(x)\, \psi_n(x)\,dx = \lambda_m^*\, \lambda_n \int_a^b \int_a^b K(x, y)\, \psi_n(y)\, \psi_m^*(x)\,dy\,dx. \tag{13.6.5}$$

Invoking the symmetry property (13.6.2), we can rewrite (13.6.5) as

$$\lambda_m^* \int_a^b \psi_m^*(x)\, \psi_n(x)\,dx = \lambda_m^*\, \lambda_n \int_a^b \int_a^b K^*(y, x)\, \psi_m^*(x)\, \psi_n(y)\,dy\,dx. \tag{13.6.6}$$

Subtracting (13.6.6) from (13.6.4), we have

$$(\lambda_n - \lambda_m^*) \int_a^b \psi_m^*(x)\, \psi_n(x)\,dx = 0. \tag{13.6.7}$$

When $m = n$, this becomes

$$(\lambda_m - \lambda_m^*) \int_a^b |\psi_m(x)|^2\, dx = 0.$$

The integral cannot vanish and so we conclude that

$$\lambda_m^* = \lambda_m; \tag{13.6.8}$$

the **eigenvalues are real**. Notice that if the kernel were anti-Hermitian, $K(x, y) = -K^*(x, y)$, or even real skew-symmetric, we would have found $\lambda_m^* = -\lambda_m$ which implies eigenvalues that are pure imaginary.

For $\lambda_m \neq \lambda_n$ when $m \neq n$, the integral has to vanish which means that the **eigenfunctions are mutually orthogonal**:

$$\int_a^b \psi_m^*(x)\, \psi_n(x)\,dx = 0 \quad \text{for} \ \ m \neq n. \tag{13.6.9}$$

For $\lambda_m = \lambda_n$ when $m \neq n$, the eigenvalue is degenerate. Equation (13.6.7) is then indeterminate but the eigenfunctions can be orthogonalized by the Schmidt orthogonalization method.

Thus the outcome of a Hilbert-Schmidt eigenvalue problem involving an integral equation with a Hermitian kernel is just like that of a Sturm-Liouville problem with a self-adjoint differential operator. This similarity is of course no accident. The solution of the latter can always be expressed as

$$\psi(x) = \lambda \int_a^b G(x;y)\rho(y)\psi(y)dy \tag{13.6.10}$$

where $G(x;y)$ is the Green's function associated with $\mathfrak{L} \equiv \frac{d}{dx}\left(p(x)\frac{d}{dx}\right) + q(x)$ together with appropriate homogeneous boundary conditions at $x = a$ and b and $\rho(x)$ is the weight function in the Sturm-Liouville equation

$$\mathfrak{L}\psi(x) = -\lambda\rho(x)\psi(x).$$

As we know, this Green's function satisfies the requirement $G(x;y) = G^*(y;x)$ and so we obtain a properly symmetrized Hilbert-Schmidt equation by introducing the function $\varphi(x) = \sqrt{\rho(x)}\psi(x)$ since it converts (13.6.10) to read

$$\varphi(x) = \lambda \int_a^b G(x;y)\sqrt{\rho(x)\rho(y)}\varphi(y)dy. \tag{13.6.11}$$

The kernel is $K(x,y) = \sqrt{\rho(x)\rho(y)}G(x;y)$.

The eigenfunctions of a Hilbert-Schmidt integral equation form a complete set in the sense that any function $g(x)$ that can be expressed by the integral

$$g(x) = \int_a^b K(x,y)h(y)dy \tag{13.6.12}$$

in which $h(y)$ is a piecewise continuous function, can be represented by the eigenfunction expansion

$$g(x) = \sum_{m=1}^{\infty} c_m \psi_m(x).$$

Suppose that we extend this result to apply to the kernel itself and set

$$K(x,y) = \sum_{m=1}^{\infty} c_m(x)\psi_m(y).$$

The coefficients $c_m(x)$ are

$$c_m(x) = \int_a^b K(x,y)\psi_m^*(y)dy = \frac{1}{\lambda_m}\psi_m^*(x)$$

which yields the bilinear formula

$$K(x, y) = \sum_{m=1}^{\infty} \frac{\psi_m^*(x)\,\psi_m(y)}{\lambda_m}. \tag{13.6.13}$$

We shall use this completeness to solve the non-homogeneous Fredholm equation

$$\psi(x) = f(x) + \lambda \int_a^b K(x, y)\psi(y)dy. \tag{13.6.14}$$

The difference $\psi(x) - f(x)$ has an integral representation of the form (13.6.12) and so we can write

$$\psi(x) - f(x) = \sum_{m=1}^{\infty} c_m \psi_m(x)$$

where

$$c_m = \int_a^b \psi_m^*(x)(\psi(x) - f(x))dx$$

$$= \lambda \int_a^b \int_a^b \psi_m^*(x)K(x, y)\psi(y)dydx$$

$$= \lambda \int_a^b \int_a^b \psi_m^*(x)K(x, y)dx\psi(y)dy$$

$$= \frac{\lambda}{\lambda_m} \int_a^b \psi_m^*(y)\psi(y)dy.$$

Adding and subtracting $f(y)$ to $\psi(y)$ under the integral, this becomes

$$c_m = \frac{\lambda}{\lambda_m} \int_a^b \psi_m^*(y)\{[\psi(y) - f(y)] + f(y)\}dy = \frac{\lambda}{\lambda_m} c_m + \frac{\lambda}{\lambda_m} \int_a^b \psi_m^*(y)f(y)dy.$$

Thus,

$$c_m = \frac{\lambda}{\lambda_m - \lambda} \int_a^b \psi_m^*(y)f(y)dy$$

and the solution of the integral equation (13.6.14) is

$$\psi(x) = f(x) + \sum_{m=1}^{\infty} \psi_m(x) \frac{\lambda}{\lambda_m - \lambda} \int_a^b \psi_m^*(y)f(y)\,dy. \tag{13.6.15}$$

This does not come as a surprise given our observation above regarding the similarity of the Hilbert-Schmidt and Sturm-Liouville theories. Note once again that if $\lambda = \lambda_m$ for some m the non-homogeneous problem has no solution unless $f(x)$ is orthogonal to all of the (degenerate) eigenfunctions $\psi_m(x)$ that correspond to λ_m.

Example: To illustrate the use of Hilbert-Schmidt theory, we shall consider the non-homogeneous Fredholm equation

$$\psi(x) = f(x) + \lambda \int\limits_{-1}^{1} (y + x)\, \psi(y)\, dy.$$

The eigenvalues and eigenfunctions for this kernel and interval of integration were found in Section 13.4. The normalized eigenfunctions are

$$\psi_1(x) = \pm\frac{1}{2}\left(1 + \sqrt{3}x\right) \quad \text{and} \quad \psi_2(x) = \pm\frac{1}{2}(1 - \sqrt{3}x)$$

corresponding to eigenvalues $\lambda_1 = \frac{\sqrt{3}}{2}$ and $\lambda_2 = -\frac{\sqrt{3}}{2}$. Thus, substituting into (13.6.15) we have

$$\psi(x) = f(x) + \sum_{m=1}^{2} \psi_m(x)\, \frac{\lambda}{\lambda_m - \lambda} \int\limits_{-1}^{1} \psi_m(y) f(y)\, dy$$

$$= f(x) + \frac{1}{4}\frac{1 + \sqrt{3}x}{\frac{\sqrt{3}}{2} - \lambda}\lambda \int\limits_{-1}^{1} \left(1 + \sqrt{3}y\right) f(y)\, dy$$

$$+ \frac{1}{4}\frac{1 - \sqrt{3}x}{-\frac{\sqrt{3}}{2} - \lambda}\lambda \int\limits_{-1}^{1} \left(1 - \sqrt{3}y\right) f(y)\, dy.$$

At this point, we need a specific functional form for $f(x)$. We choose $f(x) = x^2$.

Clearly, this is not orthogonal to either eigenfunction and so there will be no solution for $\lambda = \lambda_1$ or $\lambda = \lambda_2$. Substituting into the two integrals, we find

$$\int\limits_{-1}^{1} \left(1 \pm \sqrt{3}y\right) y^2\, dy = \frac{2}{3}$$

and so

$$\psi(x) = x^2 + \frac{2\lambda}{3}\frac{3x + 2\lambda}{3 - 4\lambda^2}, \qquad \lambda \neq \pm\frac{\sqrt{3}}{2}.$$

Explicit substitution into the original Fredholm equation comfirms that this is indeed the solution. For the specific value of $\lambda = \frac{1}{2}$ it becomes

$$\psi(x) = x^2 + \frac{x}{2} + \frac{1}{6}.$$

Notice that if we add $\pm\frac{x}{\sqrt{3}}$ to x^2 we obtain an $f(x)$ that is orthogonal to one of the eigenfunctions. In particular, $f(x) = x^2 - \frac{x}{\sqrt{3}}$ is orthogonal to $\psi_1(x)$ while $x^2 + \frac{x}{\sqrt{3}}$ is orthogonal to $\psi_2(x)$. Choosing the former, we find

$$\int_{-1}^{1} (1 - \sqrt{3}y)\left(y^2 - \frac{y}{\sqrt{3}}\right) dy = \frac{4}{3}$$

and so

$$\psi(x) = x^2 - \frac{x}{\sqrt{3}} - \frac{2\lambda}{3}\frac{1 - \sqrt{3}x}{\sqrt{3} + 2\lambda}, \quad \lambda \neq -\frac{\sqrt{3}}{2}.$$

In particular, if $\lambda = \lambda_1 = \frac{\sqrt{3}}{2}$,

$$\psi(x) = f(x) \mp \frac{1}{3}\psi_2(x) = x^2 - \frac{\sqrt{3}x}{6} - \frac{1}{6}.$$

Again, explicit substitution into the non-homogeneous Fredholm equation confirms that this is the solution.

Bibliography

Abramowitz, M. and Stegun, I.A., 1964. Handbook of Mathematical Functions. National Bureau of Standards Applied Mathematics Series, Washington, D. C.

Arfken, G.B., Weber, H.J. and Harris, E.J., 2013. Mathematical Methods for Physicists, 7th edition: A Comprehensive Guide. Academic Press, Waltham MA.

Baker, G.A., Jr. and Graves-Morris, P., 1996. Padé Approximants. Cambridge University Press, London.

Ball, W.W. Rousse, 1960. A Short Account of the History of Mathematics. Dover, New York.

Burkill, J.C., 1956. The Theory of Ordinary Differential Equations. Oliver and Boyd, Edinburgh and London.

Butkov, E., 1968. Mathematical Physics. Addison-Wesley, Reading, Mass.

Churchill, R.V., 1960. Complex Variables and Applications. McGraw-Hill, New York.

Churchill, R.V., 1963. Fourier Series and Boundary Value Problems. McGraw-Hill, New York.

Copson, E.T., 1935. An Introduction to the Theory of Functions of a Complex Variable. Oxford University Press, London.

Copson, E.T., 1965. Asymptotic Expansions, Cambridge University Press, London.

Courant, R. and Hilbert, D., 1989. Methods of Mathematical Physics, Vols. 1 and 2. John Wiley & Sons, Inc., New York.

Dennery, P. and Krzywicki, A., 1967. Mathematics for Physicists. Harper and Row, New York.

Fayyazuddin and Riazuddin, 2013. Quantum Mechanics, 2nd edition. World Scientific Publishing Co., Singapore.

Gasiorowicz, S., 1996. Quantum Physics. John Wiley & Sons, New York

Gottfried, K. and Yan, T.-M., 2003. Quantum Mechanics: Fundamentals, 2nd edition. Springer-Verlag, New York.

Gradshsteyn, I.S. and Ryzhik, I.M., 1980. Table of Integrals, Series and Products, corrected and enlarged edition. Academic Press, Orlando

Jackson, J.D.,1975. Classical Electrodynamics, 2nd ed. John Wiley & Sons, New York.

Jefferys, H. and Jefferys, B., 1956. Methods of Mathematical Physics. Cambridge University Press, London.

Kirkwood, J.R., 2013. Mathematical Physics With Partial Differential Equations. Academic Press, Waltham, MA.

Kreysig, E., 2006. Advanced Engineering Mathematics, 9th ed. John Wiley & Sons, New York.

Lea, Susan M., 2003. Mathematics for Physicists. Brooks-Cole Pub. Co.

Morse, P.M. and Feshbach, H., 1935. Methods of Theoretical Physics. McGraw-Hill, New York.

Phillips, E.G., 1961. Functions of a Complex Variable With Applications, 8th ed. Oliver and Boyd, Edinburgh and London.

Roman, P., 1965. Advanced Quantum Theory: An Outline of the Fundamental Ideas. Addison-Wesley Publishing Co., Boston.

Saleh, B.E.A. and Teich, M.C., 2006. Fundamentals of Photonics, 2nd edition. John Wiley & Sons, New York.

Sneddon, I.N., 1957. Elements of Partial Differential Eqations. McGraw-Hill, New York.

Sneddon,I.N., 1961. Special Functions of Mathematical Physics and Chemistry. Oliver and Boyd, Edinburgh and London.

Titchmarsh, E.C., 1939. The Theory of Functions. Oxford University Press, London.

Titchmarsh, E.C., 1948. Introduction to the Theory of Fourier Integrals, 2nd ed. Clarendon Press, Oxford.

Whittaker, E.T. and Watson, G.N., 1952. A Course of Modern Analysis. Cambridge University Press, London.

Watson, G.N., 1958. A Treatise on the Theory of Bessel Functions. Cambridge University Press, London.

Wong, W.W., 2013. Introduction to Mathematical Physics, Methods & Concepts, 2nd edition. Oxford University Press, Oxford.

Wyld, H.W., 1976. Mathematical Methods for Physics. W.A. Benjamin, Inc., Reading, Mass.

www.ingramcontent.com/pod-product-compliance
Lightning Source LLC
Chambersburg PA
CBHW040138200326

41458CB00025B/6303